GENETICS
a molecular approach

Second edition

T.A. Brown

Department of Biochemistry and Applied Molecular Biology, UMIST, Manchester, UK

CHAPMAN & HALL
University and Professional Division
London · Glasgow · New York · Tokyo · Melbourne · Madras

Published by Chapman & Hall, 2-6 Boundary Row, London SE1 8HN, UK

Chapman & Hall, 2-6 Boundary Row, London SE1 8HN, UK

Blackie Academic & Professional, Wester Cleddens Road, Bishopbriggs, Glasgow G64 2NZ, UK

Chapman & Hall Inc., One Penn PLaza, 41st Floor, NY 10119, USA

Chapman & Hall Japan, Thomson Publishing Japan, Hirakawacho Nemoto Building, 6F, 1-7-11 Hirakawa-cho, Chiyoda-ku, Tokyo 102, Japan

Chapman & Hall Australia, Thomas Nelson Australia, 102 Dodds Street, South Melbourne, Victoria 3205, Australia

Chapman & Hall India, R. Seshadri, 32 Second Main Road, CIT East, Madras 600 035, India

---

First edition 1989
Reprinted 1990
Second edition 1992
Reprinted 1993 (twice)

Typeset in 10.5/12pt Palatino by Best-set Typesetter Ltd, Hong Kong

Printed in Singapore by Fong & Sons Printers Pte Ltd

ISBN 0 412 44730 4

A catalogue record for this book is available from the British Library
Library of Congress Cataloging-in-Publication Data available

To my wife, Keri.

If not for you
my sky would fall.

Contents

Acknowledgements xv
Preface xvii
Preface to the First Edition xix
The organization of the book xxi

PART ONE GENES AND GENE EXPRESSION

1 The origins of genetics and molecular biology **3**
1.1 Mendel and the experimental approach to genetics 4
1.2 The birth of genetics – Mendel rediscovered 5
1.3 Genes, chromosomes and fruit flies 6
1.4 What is life? – the advent of molecular biology 8
Reading 9
Problems 10

2 Genes are made of DNA **12**
2.1 Chromosomes are made of protein and DNA 12
2.2 Experimental proof that DNA is the genetic material 13
2.2.1 The transforming principle 14
2.2.2 Bacteriophage genes are made of DNA 20
2.3 Acceptance of DNA as the genetic material 23
Reading 23
Problems 24

3 The structure of DNA **26**
3.1 DNA is a polymer 26
3.1.1 Nucleotides – the monomers in DNA 26
3.1.2 Polynucleotides 29
3.1.3 RNA is also a polynucleotide 31
3.2 The double helix 32
3.2.1 Complementary base pairing is the fundamental fact of molecular genetics **39**
3.2.2 The double helix exists in several different forms 40
Reading 41
Problems 41

4 Genes and biological information **44**
4.1 Genes are segments of DNA 44
4.2 The organization of genes on DNA molecules 46

4.2.1 Genes may occur in clusters 46
4.2.2 Some genes have lost their function 49
4.2.3 Some genes are discontinuous 50
4.3 Gene expression 52
4.3.1 The Central Dogma 52
4.3.2 Transcription – the first stage of gene expression 52
4.3.3 Translation – the second stage of gene expression 53
4.3.4 Protein synthesis is the key to expression of biological information 53
Reading 54
Problems 55

5 Transcription **57**
5.1 Eukaryotes and prokaryotes 57
5.2 Nucleotide sequences 59
5.3 RNA synthesis 60
5.3.1 RNA polymerases 62
5.4 Transcription in *E. coli* 63
5.4.1 Initiation 63
5.4.2 Elongation 67
5.4.3 Termination 68
5.5 Transcription in eukaryotes 71
5.5.1 Initiation of transcription by RNA polymerase II 71
5.5.2 Termination of transcription by RNA polymerase II 73
Reading 74
Problems 74

6 Types of RNA molecule: rRNA and tRNA **76**
6.1 Ribosomal RNA 77
6.1.1 The structure of ribosomes 77
6.1.2 Synthesis of rRNAs 83
6.2 Transfer RNA 86
6.2.1 Structure of tRNAs 86
6.2.2 Processing and modification of tRNA transcripts 89
Reading 92
Problems 93

7 Types of RNA molecule: mRNA **95**
7.1 Most mRNA molecules are unstable 95
7.2 Modification and processing of mRNA 96
7.3 Chemical modifications at the ends of eukaryotic mRNAs 98
7.3.1 All eukaryotic mRNAs are capped 98
7.3.2 Most eukaryotic mRNAs are polyadenylated 100
7.4 Intron splicing 101
7.4.1 Splicing of GT–AG introns 101
7.4.2 The self-splicing intron 107
7.4.3 Other types of intron 108

7.5 RNA editing 109
Reading 110
Problems 111

8 The genetic code 113
8.1 Polypeptides are polymers 114
8.1.1 Amino acids 114
8.1.2 Different levels of protein structure 116
8.1.3 The amino acid sequence is the key to protein structure and function 118
8.2 The genetic code 119
8.2.1 The code from first principles 120
8.2.2 Elucidation of the code 123
8.2.3 Features of the code 126
Reading 129
Problems 130

9 Translation 132
9.1 The role of tRNA in translation 132
9.1.1 Aminoacylation of tRNA 132
9.1.2 Codon recognition 135
9.2 The mechanics of protein synthesis in *E. coli* 138
9.2.1 Initiation of translation 138
9.2.2 Elongation of the polypeptide chain 142
9.2.3 Chain termination 145
9.3 Translation in eukaryotes 146
Reading 148
Problems 149

10 Control of gene expression 151
10.1 Why control gene expression? 151
10.1.1 Gene regulation in *E. coli* 151
10.1.2 Gene regulation in multicellular organisms 153
10.2 Possible strategies for controlling gene expression 154
10.3 Control of gene expression in bacteria 155
10.3.1 Regulation of lactose utilization 156
10.4 Upstream sites and DNA-binding proteins 164
10.4.1 Upstream sites for eukaryotic genes 165
10.4.2 DNA-binding proteins 167
10.5 Gene regulation during development 168
10.5.1 Developmental mutants in *Drosophila* 169
10.5.2 The homeobox 170
Reading 171
Problems 172

11 Replication of DNA molecules 174
11.1 The overall pattern of DNA replication 174

11.1.1 The Meselson–Stahl experiment 176
11.2 The mechanism of DNA replication in *E. coli* 179
11.2.1 DNA polymerase 179
11.2.2 Events at the replication fork 181
11.2.3 The topological problem 185
11.3 DNA replication in eukaryotes 186
11.4 Replication of molecules 187
Reading 188
Problems 189

12 Alterations in the genetic material **191**
12.1 Mutations 193
12.1.1 Different types of mutation 193
12.1.2 Reversing the effect of a mutation 198
12.1.3 Mutagens 200
12.2 DNA repair 205
12.3 Recombination 207
12.3.1 Generalized recombination 208
12.3.2 The role of recombination in genetics 210
Reading 212
Problems 213

PART TWO GENOMES

13 Viruses – the simplest forms of life **219**
13.1 Bacteriophages 220
13.1.1 Bacteriophage genomes can be DNA or RNA 221
13.1.2 Bacteriophage life cycles 222
13.1.3 Organization and expression of bacteriophage genes 227
13.2 Viruses of eukaryotes 233
13.2.1 Structures and infection cycles of eukaryotic viruses 233
13.2.2 Replication strategies for eukaryotic viruses 236
13.2.3 Retroviruses and cancer 240
Reading 242
Problems 243

14 Prokaryotic genomes **245**
14.1 The bacterial nucleoid 245
14.2 Prokaryotic genes 248
14.2.1 Eubacterial genes 248
14.2.2 Archaeal genes 250
14.3 Plasmids 251
14.3.1 Different types of plasmid 252
14.3.2 Plasmids and bacterial sex 253
14.3.3 Copy number and incompatibility 257
14.4 Bacterial transposons 258
Reading 262
Problems 263

15 Eukaryotic genomes **265**
15.1 The nuclear genome 265
15.1.1 The organization of nuclear genomes 267
15.2 Packaging DNA into chromosomes 268
15.2.1 Proteins, DNA and chromatin 268
15.3 Chromosome morphology 271
15.3.1 Centromeres 273
15.3.2 Telomeres 274
15.4 The cell cycle 277
15.4.1 Events during the cell cycle 278
15.4.2 Control of the cell cycle 281
15.5 Extrachromosomal genes 281
15.5.1 Organellar genetic systems 282
15.5.2 Organellar genomes 283
Reading 287
Problems 288

16 The human genome **290**
16.1 The structure of the human genome 290
16.1.1 Genes and gene families 290
16.1.2 Extragenic DNA 294
16.1.3 The Human Genome Project 296
16.2 The human genome as a tool to trace mankind's origins 299
16.2.1 Polymorphisms in the human genome 299
16.2.2 The ancient DNA time machine 304
Reading 305
Problems 306

PART THREE STUDYING GENES

17 What Mendel discovered **309**
17.1 The scientific examination of inheritance 309
17.1.1 The monohybrid crosses 311
17.1.2 Crosses involving two pairs of characteristics 316
17.2 How molecular genetics relates to Mendel 318
17.2.1 Each gene exists as a pair of alleles 318
17.2.2 Dominance and recessiveness 320
17.2.3 Mendel's Laws – molecular explanations 325
Reading 325
Problems 326

18 Using Mendelian genetics to study eukaryotic genes **328**
18.1 Linkage 328
18.1.1 Thomas Hunt Morgan and the development of gene mapping 330
18.2 Gene mapping with microbial eukaryotes 337
18.2.1 Gene mapping with *S. cerevisiae* 337

18.3 Taking Mendelian genetics beyond gene mapping 347
18.3.1 A single gene may give rise to a complex phenotype 348
18.3.2 Complex phenotypes that result from interactions between genes 349
Reading 352
Problems 353

19 Genetic analysis of bacteria **356**
19.1 Basic features of gene mapping in bacteria 356
19.2 Mapping by conjugation 359
19.2.1 The interrupted mating experiment 359
19.3 Mapping by transduction 363
19.3.1 The discovery of transduction 363
19.3.2 Mapping genes by transduction 366
19.4 Mapping by transformation 368
19.5 Relative merits of the three methods 372
Reading 372
Problems 373

20 Cloning genes **375**
20.1 What is gene cloning? 375
20.2 Constructing recombinant DNA molecules 378
20.2.1 Enzymes for cutting DNA: restriction endonucleases 378
20.2.2 Ligases join DNA molecules together 381
20.3 Cloning vectors and the way they work 382
20.3.1 Cloning with pBR322 383
20.3.2 Other types of cloning vector for *E. coli* 385
20.4 Cloning vectors for eukaryotes 389
20.4.1 Cloning vectors for *Saccharomyces cerevisiae* 390
20.4.2 Vectors for plants and animals 392
Reading 393
Problems 394

21 Studying cloned genes **395**
21.1 Identifying a gene in a genomic library 395
21.1.1 Hybridization probing 395
21.1.2 Chromosome walking 400
21.2 DNA sequencing: working out the structure of a gene 403
21.2.1 The Sanger–Coulson method: chain-terminating nucleotides 403
21.2.2 The power of DNA sequencing 405
21.3 Studying gene expression 406
21.3.1 Transcript analysis 406
21.3.2 Studying gene regulation 408
21.4 Determining the function of a protein coded by a cloned gene 413
21.5 Cloned genes in biotechnology 414
21.5.1 Bacteria are not ideal hosts for recombinant protein 414

21.5.2 Synthesis of recombinant proteins in eukaryotes 415
21.6 The future for molecular genetics 416
Reading 417
Problems 418

Answers to selected problems 419
Glossary 425
Index 457

Acknowledgements

I would like to say a general thank you to the many people who were kind enough to comment on the First Edition when it was published and more recently to give suggestions for the Second Edition. Several clear messages came through and have been incorporated into the new version. I am particularly grateful to David Baltimore, Thomas Cech, Erwin Chargaff, Robert Gallo, Robert Holley, Arthur Kornberg, Joshua Lederberg, Severo Ochoa, Frederick Sanger and Howard Temin for providing short quotes to append to their biographies. I suspect that this book has tried the patience of my friends and family more than the previous ones, so I must thank them, especially my wife Keri, for seeing it through with me. I would also like to thank the anonymous student who wrote 'I did not like genetics until I read your book, now I think it's great'. That makes it all worthwhile.

Preface

The underlying philosophy of the First Edition was that the teaching of genetics should begin with DNA rather than Mendel. Nothing has happened during the intervening 3 years to change my mind about the molecular approach: if anything I am more convinced than ever that an initial understanding of the gene as a piece of DNA provides the student with the confidence needed to deal successfully with the challenges and subtleties of the more 'classical' aspects of genetics. The Second Edition therefore retains the molecular approach, although with two important differences. The first is that my own confidence has been boosted to the extent that I have now taken the narrative slightly further, in an attempt to provide a more thorough introduction for degree programmes in which genetics will form a large part of the subsequent coursework. To this end the existing sections on gene analysis have been expanded and additional topics such as population genetics and evolution brought in at appropriate places. These changes make the book more complete in its coverage and should not detract from its popularity as a concise introductory text for the genetics component of general biology courses.

The second difference is that I have given eukaryotes rather more emphasis, especially in Part One. There has always been a temptation to base an introductory series of molecular biology lectures solely on *E. coli*, on the grounds that eukaryotes are more 'complicated' and that many of the key features of eukaryotic molecular biology are still poorly understood. The fact is that many of today's students are attracted to genetics through their awareness of cancer, AIDS and other 'eukaryotic' topics. This does not mean that we should abandon prokaryotes altogether and just teach medical genetics, but it does argue for eukaryotes, man in particular, being brought in at an early stage. The important thing is to recognize the value of prokaryotes in providing a less complicated introduction to a topic, such as transcription or DNA replication, but to draw out the unifying themes between prokaryotic and eukaryotic molecular biology so that the student can see how the study of *E. coli* has led to research in today's 'hot' topics. I have attempted to do this by giving rather more detail on eukaryotic events throughout the book, and by writing an entirely new chapter on the molecular genetics of man.

T. A. Brown

Preface to the First Edition

There are so many genetics texts available in the bookshops that the author of an entirely new one has a duty to explain why his own contribution should be necessary. In my case the decision to write a genetics text was prompted by the strong feeling that in reality the choice in the bookshops is rather limited. Most of the current books provide comprehensive and often exquisitely detailed accounts, usually starting with Mendel and ending with population genetics, and in between covering everything from trisomy to transposons. This approach is probably valuable for at least one group of undergraduates – those reading for degrees in genetics itself – but is somewhat inappropriate for the large number of students for whom genetics is a part of a broader degree course in, for example, biochemistry, microbiology, zoology or biotechnology. It seems to me that the standard texts fail these students in two ways. First, the advanced treatment of genetics means that many topics are covered in greater depth than required by the student, so that the text is only a supplement and not a backup to lecture material. The second and perhaps more important failing is that few of the standard texts make concessions to the fact that all students are novices at some stage. How many potential geneticists have been frightened away from the subject at first encounter because of the daunting intellectual challenge posed by several of the genetics texts available today?

The difficulty in attempting to write an introductory textbook, in any subject, lies in presenting the material in an understandable fashion without falling into the trap of over-simplification. To be of value the book should ensure that the basic facts and concepts are grasped by the reader, and yet should provide a sufficient depth of knowledge to stimulate the student's interest and to engender the desire to progress on to more advanced aspects of the subject. With an introductory text in genetics these objectives are perhaps relatively easy to attain, as even the most fundamental facts are fascinating and, in my experience at least, most undergraduates arrive already primed with a curiosity about genes. I hope that this book will help to turn that curiosity into a lifelong pursuit.

T. A. Brown

The organization of the book

Should genetics begin with Mendel or with DNA? My experience in teaching (and of once being taught) an introductory course in genetics has led me to the latter alternative. To begin the narrative with the concept of the gene as a piece of DNA provides a firm foundation that all students can grasp and on which their subsequent understanding can be built. To begin with the more abstract Mendelian concept of the gene risks losing the less keen intellects at the very outset. Of course the Mendelian concept of the gene is important, but the point is that the student is better equipped to understand Mendel's experiments, and to appreciate the relevance of classical genetic analysis in modern biology, after having already gained some confidence by dealing with the 'easier' molecular aspects of genetics.

The first of the three parts into which this book is divided is therefore called 'Genes and Gene Expression'. The first chapter provides an historical overview of the development of ideas about the gene up to 1944. So we do in fact start with Mendel after all. The historical byline continues through Chapters 2 and 3 with the chemical nature of the gene and the structure of DNA. Chapter 4 summarizes some of the features of genes as molecules and introduces gene expression, the details of which are described in Chapters 5 to 9. We then move on to the control of gene expression in prokaryotes and eukaryotes, and complete the first part of the book with chapters on DNA replication and mutation, repair and recombination.

Having started with the gene it seems logical to concentrate next on 'Genomes'. Part Two of this book runs through the features of genome organization and replication in viruses (Chapter 13), prokaryotes (Chapter 14) and eukaryotes (Chapter 15), with Chapter 16 devoted to a more detailed look at specific aspects of the human genome. The chapter on viruses centres on bacteriophages, as the best-studied systems, but I have attempted to balance this with a section on viruses of eukaryotes, in particular those of relevance to cancer. Chapter 14 allows bacterial sex to be introduced and in Chapter 15 mitosis and meiosis are outlined: the student is warned that these topics will assume extra importance when genetic analysis is encountered.

Genetic analysis is in fact encountered in Part Three, 'Studying Genes'. In Chapter 17 Mendel's experiments and his interpretations are described in some detail. A molecular explanation of Mendel's conclusions is provided in order to make the important link between the two concepts of the gene. In Chapter 18 the historical byline returns as we progress from Mendel to Morgan and tackle gene mapping in eukaryotes. Here I felt it valuable to explain how gene mapping is carried out with yeast, as many students will work with microbial eukaryotes in practical classes and later in research. Also in Chapter 18 I have tried to provide a succinct account of how Mendelian genetics can be used to understand the functions of genes and the way in which genes interact to produce a phenotype. Gene mapping returns in Chapter 19, concentrating now on prokaryotes, and then in the last two chapters I attempt to capture the excitement of gene cloning and the recombinant DNA revolution.

Supplementary material

I have tried to enhance the value of the book as a teaching aid by supplementing the text with boxes, problems, reading lists and a glossary.

Boxes are used to present snippets of information of relevance to the text. Some of the boxes provide important information, on key techniques for example, others are more esoteric and designed to stimulate the brighter students. Boxes are also used in places to provide a historical context and in addition to carry short biographies of the scientists who have made the most influential contributions to genetics. The boxes therefore serve several purposes but their main function is to illustrate that genetics is a living science, that real people are responsible for our present knowledge, and that this knowledge is advancing all the time.

Each chapter has a list of problems. The first problem of each set asks for short definitions of the key terms encountered in the chapter. All the terms are defined in the Glossary and the student should be encouraged to use these questions to check that the basic terminology has been grasped. The next few problems require short answers of 100 words or so; adequate answers are easily obtained from the text but the questions may prove useful in short closed-book tests. The problems marked with an asterisk cannot be answered from the text and require input from the teacher: these are designed for discussion in tutorials. I make no apologies for including in this category some problems that I do not know the answers to. Finally, where appropriate, there are traditional 'data-handling' questions, the answers to which are given at the back of the book.

Each chapter also has a reading list. Providing useful reading for an introductory course can be a problem, with *Scientific American* being the only journal that is consistently reliable at this level. Many of the articles I have listed are quite challenging but all should be within reach of the student who has understood the basic information provided in the book.

T. A. Brown

Part One
Genes and Gene Expression

1 The origins of genetics and molecular biology

Mendel • The birth of genetics • Genes, chromosomes and fruit flies • What is life?

Genetics is the name given to the study of **heredity**, the process by which characteristics are passed from parents to offspring so that all organisms, human beings included, resemble their ancestors. The central concept of genetics is that heredity is controlled by a vast number of factors, called **genes**, which are discrete physical particles present in all living organisms.

The first geneticists were mainly interested in how genes are transmitted from parents to their offspring during reproduction and how different genes act together to control variable traits such as height and eye colour. A change in emphasis occurred during the 1930s when it was recognized that if genes are physical entities then, like other cell components, they must be made of molecules and it should therefore be possible to study them directly by biophysical and biochemical methods. This led to a new branch of genetics, called **molecular biology**, which had as one of its initial aims the identification of the chemical nature of the gene. This new approach led to new concepts, and soon biologists ceased to regard individual genes simply as units of inheritance and instead began to look on them as units of **biological information**, with the entire complement of genes in an organism containing the total amount of information needed to construct a living, functioning example of that organism. The aim of geneticists and molecular biologists over the past 40 years has been to understand the way in which biological information is stored in genes and how that information is made available to the living cell.

Genetics and molecular biology are very closely related subjects, and although there are distinctions between them it is more constructive to treat them as one. For this reason the term **molecular genetics** is now often used to describe that branch of biology concerned with the study of all aspects of the gene. This book is about molecular genetics and as such covers some of the most fascinating discoveries and intriguing problems of modern science. However, before describing the discoveries and examining the problems we must first turn our attention to the way in which genetics and molecular biology developed during the late nineteenth and early twentieth centuries. This will set the scene for the subsequent chapters by providing the historical and intellectual backdrop against which our more recent endeavours to understand the gene have been set.

1.1 Mendel and the experimental approach to genetics

Genetics as we know it originated with Gregor Mendel, in particular with a paper he published in 1866 in the *Verhandlungen des naturforschenden Vereines in Brunn* – the Proceedings of the Society of Natural Sciences in Brno. Mendel was typical of many nineteenth-century scientists in that he had an insatiable curiosity about the natural and physical world and was keenly interested in the diversity of living things. If Mendel's family had been able to afford for their son an education fitting his intellectual talents then Gregor could very well have risen to become a noted Professor in one of the great European universities. Unfortunately his parents were poor and although Mendel spent four terms at the University

Gregor Mendel b. 22 July 1822, Heinzendorf, Silesia (now Hyncice, Czechoslovakia); d. 6 January 1884, Brunn, Austria (now Brno, Czechoslovakia)

The outline of Mendel's life is well known. He was born in a small village, the only son of peasant farmers, and displayed a keen intelligence at an early age. His village schoolmaster arranged a place for him at the Gymnasium in Troppau (now Opava) and one of his three sisters used part of her dowry to allow him to continue his studies at the Olmutz Philosophical Institute. He entered the Church in 1843, joining the Augustinian monastery at Brunn, and received probably the most influential part of his education at Vienna University between October 1851 and August 1853. He began his breeding experiments with pea plants in 1856, publishing his results in 1865, and continued his work until 1871 when his duties as abbot effectively curtailed his scientific career. Unfortunately his later work was with hawkweed (*Hieracium*), which proved much less amenable to study than the pea.

Despite our knowledge of Mendel's life, we know little about his ideas on heredity beyond what is written in his publications. This is partly because Mendel was not recognized as a scientific hero during his lifetime, but mainly because his private papers were destroyed after his death. Exactly how far-sighted Mendel was is still a subject of debate. A full account of Mendel's life and his impact is provided in *Origins of Mendelism* by R. Olby (see Reading).

of Vienna his education was piecemeal and what he obtained was achieved by perseverance rather than privilege.

It is well known that Mendel joined a monastery near Brno, which was at that time in Austria but is now in Czechoslovakia, and eventually in 1868 became abbot. This choice of career was a fortunate one, because it seems that his duties in the years 1856–1864 were not so onerous as to prevent him carrying out a lengthy series of experiments concerning the inheritance in garden peas of traits such as height, flower colour and seed shape. These experiments were presented to the Brno Society in papers read by Mendel on 8 February and 8 March 1865 and published the following year in the Society Proceedings.

The details of Mendel's experiments and his findings need not concern us at the moment; we will return to them in Chapter 17. But what should be appreciated at the outset is that Mendel's contribution was not only the basic Laws of Heredity but also the demonstration that heredity could be studied experimentally. Before Mendel, and indeed for the remaining 35 years of the nineteenth century, concepts about heredity were based almost entirely on simple observations of the living world. The idea that heredity could be studied systematically by straightforward experiments was totally alien to the mainstream of biological thought at that time. Mendel's experiments, which were carefully planned and meticulously executed, demonstrated that heredity works not by some mysterious process but in a predictable and logically consistent manner. His techniques and results provided a clear indication of how the problem of heredity could be unravelled by experimental means. Unfortunately Mendel remained until his death the only person who understood and appreciated the true significance of his work.

1.2 The birth of genetics – Mendel rediscovered

The suggestion that Mendel's paper was 'lost' because it was published in an obscure journal by an unknown monk in a remote monastery is not really consistent with the facts. The Proceedings of the Brno Society were fairly well circulated and copies of the 1866 issue containing Mendel's paper were sent to at least 55 European libraries and learned societies, including the Royal Society and Linnean Society in London. Mendel sent reprints of his paper to many of the leading botanists of the day, including Professor Kerner of the University of Innsbruck (who evidently did not read it as the reprint was found after his death with the pages still uncut) and Professor Nageli of Munich, with whom Mendel corresponded regularly between 1866 and 1873. Mendel's work was noticed to the extent that he or his paper was mentioned in 16 publications in the late nineteenth century, including the ninth edition of the *Encyclopaedia Britannica*.

The fact is that it was not until 1900 that biological thought had progressed to the point where other biologists could understand the importance of Mendel's work. Mendel was literally 35 years ahead of his time. Eventually in 1900 three botanists, Hugo De Vries, Carl Correns and Erich von Tschermak, each conceived the idea of experiments similar to Mendel's to test their own theories of heredity. Each performed their experiments and then, when studying the literature before publishing their results, each independently discovered Mendel's paper. After a lengthy gestation the science of genetics was finally born.

1.3 Genes, chromosomes and fruit flies

Mendel's experiments, and those of biologists such as De Vries, Correns and Tschermak, removed the mystique from heredity and showed that the process follows predictable rules. These rules can be rationalized as the passage of physical factors, each controlling a separate heritable trait, from the parents to offspring during reproduction. These factors went under a variety of names until 1909 when W. Johannsen proposed the term 'gene' which subsequently entered common usage. By this time it was understood that the genes are carried by the **chromosomes** of higher

Thomas Hunt Morgan b. 25 September 1866, Hopemont, Kentucky; d. 4 December 1945, Pasadena, California

Thomas Hunt Morgan was one of the first great American biologists. He was educated at Kentucky University and at Johns Hopkins University, which had been founded just 10 years earlier but already possessed an international reputation for academic excellence. Morgan's first interest was in embryology and animal development, and he pursued his research as a teacher at Bryn Mawr College, Pennsylvania, and on study leaves at the Marine Biological Stations at Woods Hole, Massachusetts and at Naples. In 1903 he moved to Columbia University in New York City and, inspired by the rediscovery of Mendel's work, began a number of experiments on heredity, several with the fruit fly *Drosophila melanogaster*. Much of this work came to nothing (he claimed to do three kinds of experiment: foolish ones, damned foolish ones, and ones that were worse) but it led to the discovery in May 1910 of a white-eyed fly, a striking contrast to the normal red-eyed form. His first Mendelian-type breeding experiments were with white-eyed and red-eyed flies, and he followed these with a large number of further crosses, using other variants that he obtained by treating flies with agents such as X-rays. He and his research group developed the methods for mapping gene positions on chromosomes and he remained for 30 years the leading intellectual light in genetics. He was awarded a Nobel Prize in 1933.

organisms, an idea that was prompted by the observation that the transmission of chromosomes during cell division and reproduction exactly parallels the behaviour of genes during these events. The chromosome theory was stated in its most convincing form in 1903 by W. S. Sutton, who at that time was a graduate student at Columbia University in New York and who subsequently became a surgeon before his early death in 1916.

Once the experimental basis of heredity had been established and the chromosome theory accepted, the way was open for a rapid advance in the understanding of genetics. That this advance occurred as rapidly as it did was mainly the result of the intuition and imagination of Thomas Hunt Morgan and the members of his research group, notably Calvin Bridges, Arthur Sturtevant and Hermann Muller. Morgan and his colleagues

Drosophila characteristics

Morgan's group managed to identify a remarkable number of different inherited characteristics of the fruit fly, most of which showed themselves as simple morphological features such as eye colour, eye shape, wing venation, and so on. The list of eye colours is particularly impressive:

white	purple	scarlet
ruby	cinnabar	pink
vermilion	light red	cardinal
garnet	brown	claret
carnation	sepia	

Clearly a fine perception of the different shades of red was needed in order to work with Morgan!

How are new discoveries in genetics to be made?

According to Thomas Hunt Morgan:

> By industry, trusting to luck for new openings, by the intelligent use of working hypotheses (by this I mean a readiness to reject any such hypotheses unless critical evidence can be found for their support), by a search for favourable material, which is often more important than plodding along the well-trodden path, hoping that something a little different may be found. And lastly by not holding genetic congresses too often.

At the Sixth International Congress of Genetics, 1932. Taken from G. E. Allen (1978) *Thomas Hunt Morgan*, Princeton University Press, New Jersey.

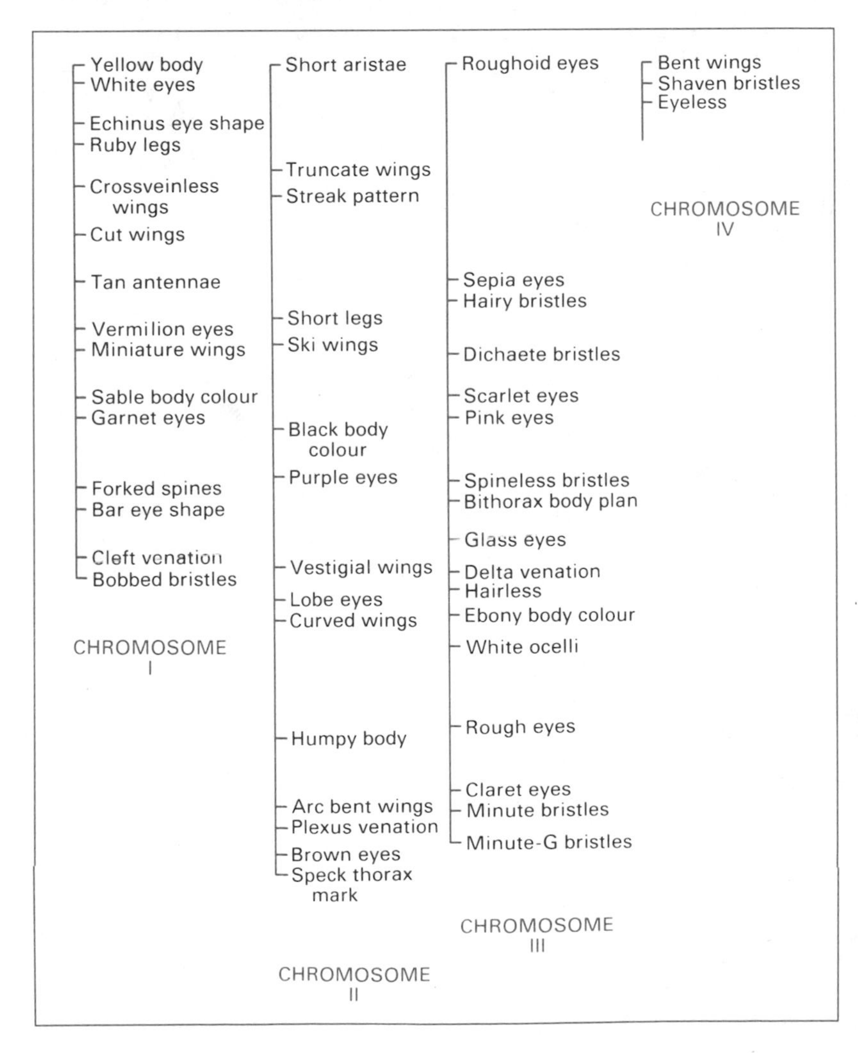

Figure 1.1 The positions of 50 different genes on the four chromosomes of the fruit fly, as published by Morgan's group in 1922. We now know the positions of over 1000 genes on these chromosomes.

achieved something that many biologists dream about: they discovered an organism that was ideally suited for the particular research programme that they wished to carry out. This organism was *Drosophila melanogaster*, the fruit fly. *Drosophila* possesses several features that make it very suitable for genetic analysis but most important from Morgan's point of view was the fact that a large number of stable variant forms of the fly could be obtained. The differences between these variants involve features such as wing shape and body colour, some traits having several different varieties. These varieties enabled Morgan to investigate the way in which different combinations of genes work together in controlling the inheritance of an individual characteristic. Morgan's group developed many of the techniques that have now become standard methods in genetic analysis, including those that map the relative positions of different genes on a chromosome (Fig. 1.1). Between 1911 and 1929 the 'fly-room' at Columbia University provided the data that remain the foundation to our knowledge of the gene as a unit of inheritance.

Two physicists' views of biology

Bohr's views on basic biological processes are summed up by:

> The wonderful features which are constantly revealed in physiological investigations and differ so strikingly from what is known of inorganic matter, have led many biologists to doubt that a real understanding of the nature of life is possible on a purely physical basis. . . . I think that we all agree with Newton that the real basis of science is the conviction that Nature under the same conditions will always exhibit the same regularities. Therefore, if we were able to push the analysis of the mechanism of living organisms as far as that of atomic phenomena, we should scarcely expect to find any features differing from the properties of inorganic matter.

N. Bohr (1933) Light and life. *Nature*, **131**, 421–3.

And from Schrödinger:

> Living matter, while not eluding the laws of physics as established to date, is likely to involve other laws of physics hitherto unknown which, however, once they have been revealed, will form as integral a part of this science as the former.

E. Schrödinger (1944) *What is Life?* Cambridge University Press.

1.4 What is life? – the advent of molecular biology

The rediscovery of Mendel's work and the remarkable advances made by Morgan and his colleagues attracted the attention of many biologists not actively engaged in genetic research, and during the first few decades of the twentieth century our understanding of the gene was advanced not only by geneticists but also by cytologists, physiologists, biochemists and biophysicists. But the contribution that in the long run has proved most influential was made not by biologists at all but by a group of physicists whose previous work had been in the area of quantum mechanics, far removed from Mendel and fruit flies. The involvement of physicists in biology was heralded by a lecture called 'Light and Life' presented by Niels Bohr to an international congress in Copenhagen in August 1932, and subsequently reached its culmination with the publication of the book *What is Life?* by Erwin Schrödinger in 1944. Bohr and Schrödinger both attempted to interpret the central question in biology – the nature of life itself – in physical terms, but both were frustrated by the apparent limitations of trying to explain life by orthodox physical principles.

Paradoxically, neither Bohr's lecture nor Schrödinger's book stand up to present-day scrutiny: both contain errors and inconsistencies that today seem almost naive. Their contributions are important not so much for the ideas put forward but more

Max Delbrück b. 6 September 1906, Berlin; d. 9 March 1981, New York

Delbrück first studied quantum chemistry and then theoretical nuclear physics, obtaining his PhD from the University of Göttingen in 1930. During spells at Copenhagen, Zurich and Berlin he became interested in the gene, especially after hearing Bohr's lecture 'Light and Life' in 1932. In 1935 he published with N. W. Timofeyev-Ressovsky and K. G. Zimmer a paper on the nature of the gene that became one of the landmarks of early molecular genetics. He moved to the California Institute of Technology in 1937 on a Rockefeller Fellowship in biology and stayed in the USA, becoming a citizen in 1945. During the early 1940s he set up the 'Phage Group', a loose association of the most important molecular biologists of the time, including those responsible between 1940 and 1953 for several of the major breakthroughs in understanding the molecular nature of the gene. The later years of his research career were spent at Cold Spring Harbor, the famous and unique research institute on the north shore of Long Island. Here he continued to influence the progress of molecular biology even though his central role as intellectual leader was taken over by Crick and others. Delbrück shared the 1969 Nobel Prize with Salvador Luria and Alfred Hershey, two other founder members of the Phage Group.

for the influence that they had on other people. Several young physicists listened to Bohr's lecture, or read the text published the following year in the scientific journal *Nature*, and were fascinated by the challenge provided by biology. Among these was Max Delbrück, a German who subsequently was forced by the political climate in Europe to emigrate to the USA. Delbrück learnt from Bohr about **bacteriophages** (**viruses** that attack **bacteria**) and decided that their infection cycle could be used as an experimental system with which to tackle the question of what genes are and how they work. He brought into existence in 1940 the 'Phage Group', an informal association of physicists, biologists and chemists, all working in different laboratories but all with a common interest in the gene. The formation of this group stimulated the development of molecular biology as a discipline in its own right and led within 20 years to an understanding of the chemical nature of the gene. But we must now curtail our historical narrative as we have reached the point at which a detailed description of the gene must begin.

Reading

R. Olby (1985) *Origins of Mendelism*, 2nd edn, University of Chicago Press, London – not only provides a comprehensive and authoritative account of Mendel's life and his work, but also describes the views and theories on heredity held by other nineteenth-century biologists.

Events in genetics and molecular biology up to 1944

1865	Mendel presents the results of his breeding experiments with peas
1900	Mendel's work is rediscovered by De Vries, Correns and Tschermak
1903	Sutton proposes that genes reside on chromosomes
1905	Bateson introduces the term 'genetics' to describe the study of heredity
1909	Johannsen proposes that Mendel's factors are called 'genes'
1910	Morgan starts his breeding experiments with the fruit fly
1911	Sturtevant maps the positions of five genes on a *Drosophila* chromosome
1915	The fruit fly gene map is extended to include 85 different genes on four chromosomes
1927	Müller discovers that X-rays induce alterations ('mutations') in genes
1932	Bohr delivers his 'Light and Life' lecture in Copenhagen
1935	Delbrück, Timofeyev-Ressovsky and Zimmer propose that the gene is a physical unit
1940	Delbrück founds the 'Phage Group'
1941	Beadle and Tatum propose that each gene controls the activity of a single enzyme
1944	Publication of *What is Life?* by Schrödinger

J. A. Peters (1959) *Classic Papers in Genetics*, Prentice Hall, New Jersey – the most accessible source of Mendel's original paper.

I. Shine and S. Wrobel (1976) *Thomas Hunt Morgan: Pioneer of Genetics*, University Press of Kentucky, Lexington – a very readable scientific biography.

R. Olby (1974) *The Path to the Double Helix*, Macmillan, London; F. H. Portugal and J. S. Cohen (1977) *A Century of DNA*, MIT Press, Cambridge; H. F. Judson (1979) *The Eighth Day of Creation*, Jonathan Cape, London – excellent books that chart the historical development of genetics and molecular biology during the twentieth century.

S. S. Cohen (1986) Finally, the beginnings of molecular biology. *Trends in Biochemical Sciences*, **11**, 92–3; R. Olby (1986) Biochemical origins of molecular biology: a discussion. *Trends in Biochemical Sciences*, **11**, 303–5 – provide more personal views on the origins of molecular biology.

J. A. Witkowski (1986) Schrödinger's 'What is Life?': entropy, order and hereditary code-scripts. *Trends in Biochemical Sciences*, **11**, 266–8; M. F. Perutz (1987) Physics and the riddle of life. *Nature*, **326**, 555–8 – reassessments of Schrödinger's book.

Problems

1. Define the following terms:

   genetics
   heredity
   gene
   molecular biology
   biological information
   molecular genetics
   chromosome theory

2. Describe the roles played by the following scientists in the development of genetics and molecular biology:

   Mendel
   De Vries, Correns and Tschermak
   Johannsen
   Sutton
   Morgan
   Bohr and Schrödinger
   Delbrück

3. Distinguish between the terms 'genetics', 'molecular biology' and 'molecular genetics'.

4. Describe the historical connection between Mendel's experiments and the work of Morgan and his group.

5. Describe the part played by physicists such as Bohr and Schrödinger in prompting a new approach to the study of genes.

6. *If Gregor Mendel lived in the 1990s and needed a research grant to carry out his experiments do you think he would have been able to obtain one?

7. *Is it possible today for the results of experiments as important as Mendel's to go unappreciated?

8. *What features would be desirable for an organism that is to be used for extensive studies of heredity?

2 Genes are made of DNA

Chromosomes are protein plus DNA • Proof that genes are made of DNA – the transforming principle – bacteriophage genes

The first question that we must address is the fundamental one that was uppermost in the minds of geneticists and molecular biologists during the first 50 years of this century. In its simplest form this question is 'What are genes made of?' To be more technical, the information we seek is the chemical nature of the **genetic material**.

2.1 Chromosomes are made of protein and DNA

Once it had been established that genes reside on the chromosomes it was realized that the chemical nature of the gene might be discovered by determining exactly what types of biochemicals chromosomes are made of. A range of experimental techniques, most importantly **cytochemistry**, were brought to bear on this problem and by 1920 it had become clear that chromosomes contain two biological compounds, **protein** and the **nucleic acid** called deoxyribonucleic acid, **DNA**. One or other (or perhaps a combination of both – nucleoprotein) must therefore be the genetic material.

In deciding which was the more likely candidate, biologists considered the properties of genes and how these properties might be provided for by proteins or by DNA. The most fundamental property required of the genetic material is that it must be able to exist in an almost infinite variety of forms: each cell contains a large number of different genes (several thousand in the simplest bacteria, tens of thousands in higher organisms), each controlling a different heritable trait and each presumably having a structure slightly different from any of the other genes in the cell. To satisfy this requirement for variability, the genetic material must be able to take on an equally large number of different chemical structures.

These speculations, although quite sound, unfortunately led biologists to conclude that protein and not DNA must be the

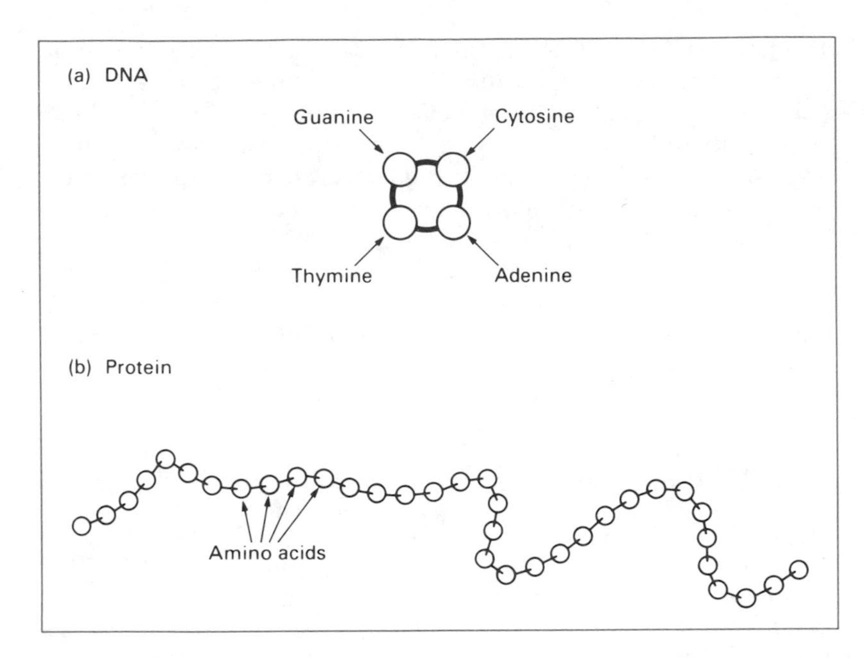

Figure 2.1 The 1930s views on DNA and protein structure. (a) The incorrect tetranucleotide model for DNA, proposed in 1935; all DNA molecules were thought to be exactly the same. (b) The correct polypeptide structure for proteins; polypeptides were known to exist in a variety of different forms, all with different amino acid sequences.

genetic material, an error that arose because at that time the structure of DNA was misunderstood. DNA was believed to be a relatively small, invariant molecule with a molecular weight of 1227. Each molecule of DNA was thought to be exactly the same as every other molecule of DNA (Fig. 2.1(a)), indicating that DNA did not have the variability required of the genetic material. In contrast, proteins were known to be **macromolecules** made up of long **polymers** of **amino acids**, with 20 different amino acids and apparently no restrictions on the order in which these could be linked together (Fig. 2.1(b)). It was known that there could be an almost infinite number of different types of protein, distinct from one another by virtue of their different amino acid sequences: proteins therefore possessed the potential variability required by the genetic material. Not surprisingly, biologists during the first half of this century concluded that genes were made of protein and looked on the DNA component of chromosomes as perhaps a structural material, needed to hold the protein genes together.

2.2 Experimental proof that DNA is the genetic material

Although the hypothesis that genes are made of protein was very widely held in the first half of the century, its supporters were aware that they had no solid experimental evidence to back their view. The necessity for an experimental identification of the genetic material became pressing during the late 1930s, when it gradually became apparent that DNA, rather than being a simple molecule unsuitable as the genetic material, was in fact a long polymer and, like protein, could exist in an almost infinite number of variable forms. If both protein and DNA satisfy the fundamental requirement of the genetic material, and both are present in chromosomes, then which are genes made of? Two critical experiments, radically different in their design, eventually led to the conclusion that the genetic material is DNA and not protein. The first of these experiments was the identification of the chemical nature of the **transforming principle**, a substance that can change the bacterium *Streptococcus pneumoniae* from one form to another.

2.2.1 The transforming principle

The fact that apparently identical bacteria can exist in a variety of forms was discovered in the nineteenth century and led to confusion as to whether the concept of species could be applied to bacteria in the same way as to higher organisms. During the 1880s the great microbiologist Robert Koch eventually persuaded his contemporaries that bacterial species do exist, but the variety of forms that can be found within a single species remained a perplexing issue for many years.

Variation is particularly important in bacteria that cause diseases as often a single species may include both virulent and avirulent forms. The study of such forms in *S. pneumoniae*, the causative agent of one type of pneumonia, led to the discovery of **bacterial transformation** by a London medical officer, Frederick Griffith, in 1928.

(a) The discovery of bacterial transformation. For some time three virulent **serotypes** of *S. pneumoniae*, called Types I, II and III, had been known. Bacteria of each serotype are distinguished by the nature of their capsule, a slimy coating that surrounds each cell (Fig. 2.2). This capsule is made of a polysaccharide secreted by the bacterium, with each serotype producing a capsule with a different polysaccharide composition. The capsule gives colonies of *S. pneumoniae* a smooth, shiny appearance; hence the designation 'S' (for 'smooth') form.

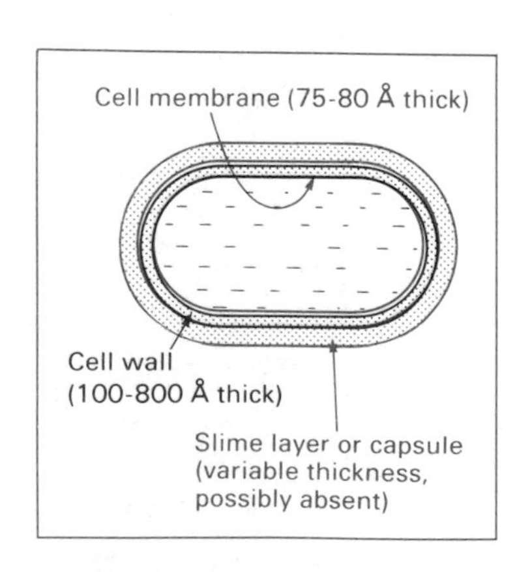

Figure 2.2 A typical bacterial cell showing the surface layers, including the capsule.

Each serotype can also exist in a harmless, avirulent form, distinguished from the virulent type by the absence of the capsule and a 'rough' appearance to the colonies. The avirulent types are therefore called 'R' forms. It was known that a particular serotype can change forms (e.g. Type IS can change to Type IR) either naturally or as a result of experimental treatment. This interested Griffith because his results showed that patients recovering from pneumonia often carried in their saliva the avirulent R form of the

Serotypes of *S. pneumoniae*

Different serotypes are distinguished by immunological tests. If Type I *S. pneumoniae* bacteria are injected into the bloodstream of a mouse, the animal will synthesize protective proteins, called antibodies, which will bind to the bacteria and aid attempts to destroy them. Antibodies are highly specific and will bind only to one type of bacterium: for example, an antibody for Type I *S. pneumoniae* will not cross-react with Type II or Type III cells. An antibody against Type II bacteria is similarly specific for that serotype and will react with no others. Purified antibodies can therefore be used in diagnostic tests for different *S. pneumoniae* serotypes.

Cells of two different serotypes carry different surface molecules but may in all other respects be exactly the same; they are still members of the same species. Each serotype of *S. pneumoniae* is in fact characterized by a different chemical composition to the polysaccharide capsule that surrounds the cell. Some examples:

Serotype	*Capsule components*
II	rhamnose, glucose, glucuronic acid
III	glucose, glucuronic acid
VI	galactose, glucose, rhamnose
XIV	galactose, glucose, *N*-acetyl-glucosamine
XVIII	rhamnose, glucose

With *S. pneumoniae*, the capsule is also responsible for eliciting the disease response in an infected animal. The rough forms of each serotype, which lack the capsule, are therefore avirulent.

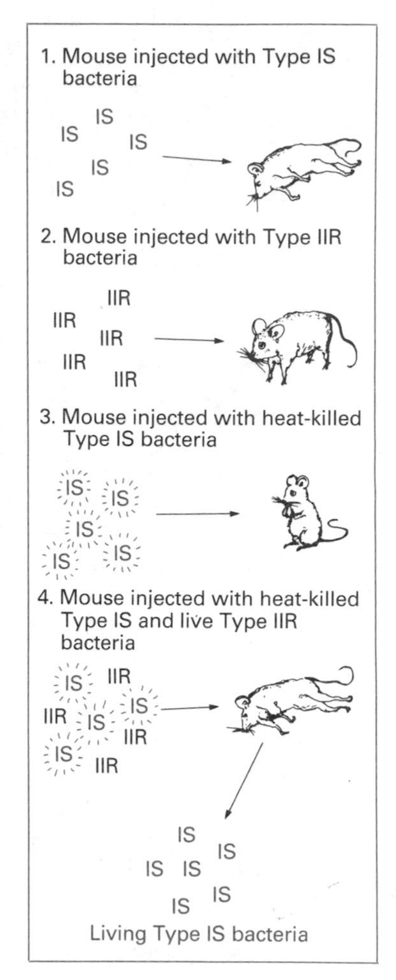

Figure 2.3 Griffith's experiments which demonstrated transformation of avirulent Type IIR bacteria into virulent Type IS forms.

serotype responsible for their infection, possibly, he suggested, because of attenuation of the bacterium during recovery from the illness. However, in 1928 Griffith published the results of a startling series of experiments that demonstrated that a more important type of transformation, from one serotype to another (e.g. Type I to Type II), can also occur, an event that would require a genetic rather than physiological change.

Griffith carried out four particularly important experiments (Fig. 2.3). These were as follows:

1. A mouse was injected with a sample of virulent, smooth bacteria of the Type I serotype (Type IS); as expected the mouse contracted pneumonia.
2. A second mouse was injected with avirulent, rough bacteria of the Type II serotype (Type IIR); again the expected result was obtained and the mouse remained healthy.
3. Next a sample of Type IS bacteria were killed by heating at 60°C for 3 h and then injected into the mouse; the result was that the animal remained healthy. Again this was the expected result because, although Type IS bacteria are virulent, only live bacteria can cause the disease.
4. Finally, a sample of heat-killed Type IS bacteria, incapable of causing the disease, were mixed with live avirulent Type IIR

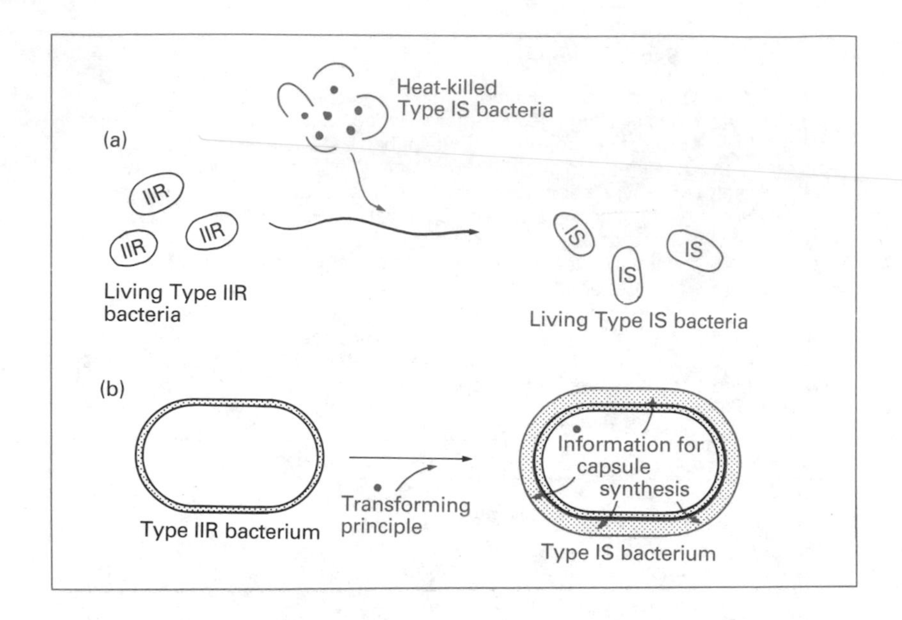

Figure 2.4 The concept of the transforming principle.

bacteria and the mixture injected into a mouse; this time a wholly unexpected result was obtained. Rather than remaining healthy, the mouse contracted pneumonia. In addition, from this animal large numbers of living Type IS bacteria were eventually isolated, even though the only living bacteria in the original inoculum were avirulent Type IIR forms.

(b) The transforming principle is genetic material. The last of Griffith's four experiments showed that live Type IIR bacteria and heat-killed Type IS bacteria, neither of which are virulent on their own, can cause pneumonia when mixed together. Importantly, the live virulent bacteria obtained at the end of this experiment were Type IS. The explanation cannot be that the Type IIR bacteria in the original inoculum regained their ability to synthesize their capsular polysaccharides. If this had happened then the resulting bacteria would have been Type IIS. Instead the Type IIR bacteria were transformed into Type IS: *somehow they had obtained capsules of a completely different type.*

The conclusion has to be that a component of the heat-killed Type IS bacteria transformed the Type IIR cells into the virulent IS serotype. These new, living type IS bacteria then caused pneumonia after injection into the mouse (Fig. 2.4(a)). The component responsible for the transformation (the 'transforming principle') could not be the capsular polysaccharide itself because the heat-killed Type IS cells would not yield enough material to coat all the living cells subsequently isolated from the mouse. Instead the transforming principle, when introduced into Type IIR bacteria, confers on these cells a new, permanent, heritable characteristic –

Oswald Avery b. 21 October 1877, Nova Scotia; d. 2 February 1955, Nashville

Avery was born in Canada but his father, a clergyman, moved his family to New York in the early 1880s. Avery studied medicine at Columbia University, New York City, obtaining his MD in 1904, and joined the Rockefeller Institute Hospital in 1913, remaining there until 1948, when he retired to Nashville, Tennessee. Avery was a bachelor and a dedicated bacteriologist who continued to spend long hours in the laboratory even towards the end of his career, the time when many scientists are happy to let their students do the hard work. He made several contributions to the study of pneumonia-causing bacteria, including identification in 1932 of the polysaccharides specific to the capsules of different serotypes. His meticulous attack on the transforming principle lasted several years before culminating in 1944 with the clear identification that the active agent is DNA. The Nobel Committee has been criticized for not recognizing Avery's achievement before his death but from all accounts a Nobel Prize would not have added to Avery's own personal satisfaction concerning his work. Rene Dubos has written a biography of Avery: *The Professor, the Institute, and DNA*, Rockefeller University Press, New York, 1976.

the ability to synthesize the capsular polysaccharides characteristic of the Type IS serotype (Fig. 2.4(b)). The transforming principle must therefore be genetic material. Identify the transforming principle and the chemical nature of the genetic material would be known.

(c) The transforming principle is DNA. Griffith did not himself attempt to identify the transforming principle. Instead, the work was carried out by Oswald Avery and his colleagues Colin MacLeod and Maclyn McCarty, at the Rockefeller Institute in New York. Their strategy was to prepare a filtrate from heat-killed S cells, containing the transforming principle, and then to digest the individual components of this filtrate with specific degradative enzymes (Fig. 2.5). For instance, a **protease** could be used to degrade all the protein in the filtrate, and a **ribonuclease** to degrade the ribonucleic acid (**RNA**), the second type of nucleic acid present in living cells (although, unlike DNA, not a major constituent of chromosomes). After enzymatic digestion of one or more components the filtrate was tested for retention of its transforming ability. Surprisingly (as genes were still thought to be made of protein) treatment with protease had no effect on the ability of the filtrate to transform R cells into the S form. Similarly, ribonuclease treatment had no effect on the transforming activity. However, digestion of the DNA with **deoxyribonuclease** totally destroyed the transforming ability so that the filtrate was no longer able to convert one cell type into the other.

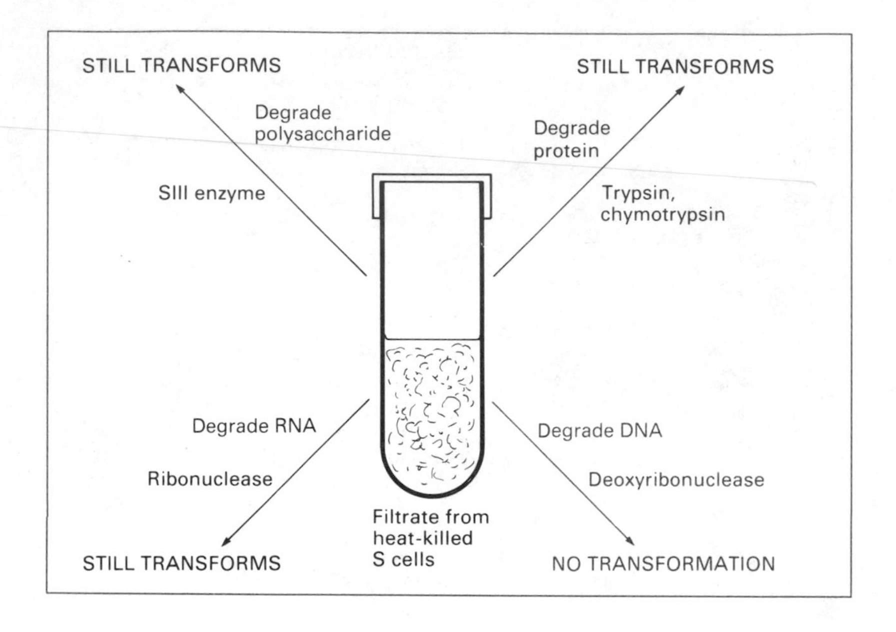

Figure 2.5 The transforming principle is DNA.

Gradually it became clear to Avery and his colleagues that the transforming principle must be DNA, but they needed more evidence. McCarty therefore set out to purify the DNA from heat-killed S cells, and by the use of techniques such as **electrophoresis**, **ultracentrifugation** and **ultraviolet spectroscopy** confirmed that the transforming principle and DNA were one and the same thing.

The paper by Avery, MacLeod and McCarty was published in February 1944, at the height of World War II and just 4 months before D-Day. On reading the paper today we have little doubt of the reliability of the results or the validity of the conclusions. The paper did not, however, convince the scientific community that genes are made of DNA. Many biologists, especially in Europe, did not see the paper until several years after it was published, and others who were aware of the work still remained sceptical. Doubts were raised about the efficiency of the enzymes used to digest the protein component of the S cell filtrate. It was suggested that 2% residual protein in the 'pure' preparation of transforming principle would be undetectable and might be responsible for the transformation of R cells to S cells – if genes were made of protein. These nagging doubts, although unfounded, prevented the results of Avery's group from being unanimously accepted by the scientific community. Perhaps, like with Mendel, scientists at that time were just not ready for this totally new idea. A second, independent, experimental identification of the genetic material was required.

Electrophoresis, ultracentrifugation and ultraviolet spectroscopy

These three techniques, which were in their infancy when used by Avery, MacLeod and McCarty to analyse the transforming principle, are now important methods regularly employed in research projects concerning DNA.

Electrophoresis is defined as the movement of charged molecules in an electric field. DNA molecules, like proteins and many other biological compounds, carry an electric charge, negative in the case of DNA. Consequently, when DNA molecules are placed in an electrical field they will migrate towards the positive pole. In an aqueous solution the rate of migration of a molecule depends on two factors: its shape and its electrical charge, meaning that most DNA molecules will migrate at the same speed in an aqueous electrophoresis system.

Nowadays, electrophoresis of DNA is usually carried out in a gel made of agarose, polyacrylamide or a mixture of the two. In a gel, the migration rate of a macromolecule is influenced by a third factor, its size. This is because the gel comprises a complex network of pores through which the molecules must travel to reach the electrode. The smaller the molecule, the faster it can migrate through the gel. Gel electrophoresis will therefore separate DNA molecules according to size.

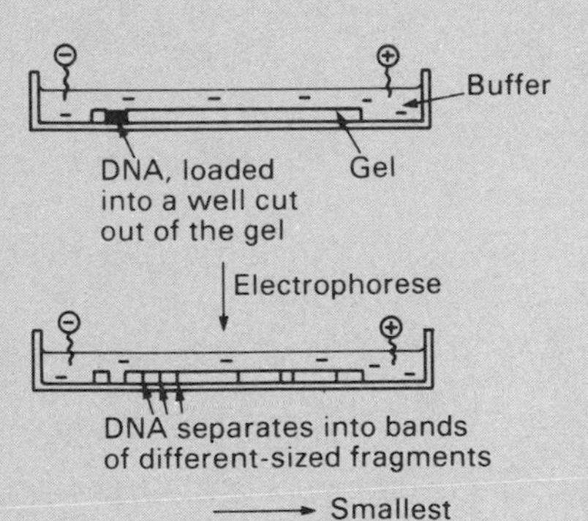

Gel electrophoresis is also routinely used to separate proteins of different molecular masses.

Ultracentrifugation. The ultracentrifuge, invented by Svedberg in 1925, allows samples to be subjected to intense centrifugal forces, up to several hundred thousand × *g*. Cells, cell components and macromolecules sediment during ultracentrifugation at a rate dependent on their size, shape, density and molecular mass.

Two versions of ultracentrifugation are now important in studying DNA. The first is called **velocity sedimentation analysis**, a procedure that involves measuring the rate at which a macromolecule or particle sediments through a dense solution, often of sucrose, whilst subjected to a high centrifugal force. The rate of sedimentation is a measure of the size of the molecule or particle (although shape and density also influence the rate), and is expressed as a sedimentation coefficient (see Section 6.1.1).

The second technique is called **density gradient centrifugation**. A density gradient is produced by centrifuging a dense solution (usually of caesium chloride), as a high centrifugal force will pull the caesium and chloride ions towards the bottom of the tube. Their downward migration will be counterbalanced by diffusion, so a concentration gradient will be set up, with the CsCl density greater towards the bottom of the tube. Macromolecules present in the CsCl solution when it is centrifuged will form bands at distinct points in the gradient, the exact position depending on the buoyant density of the macromolecule.

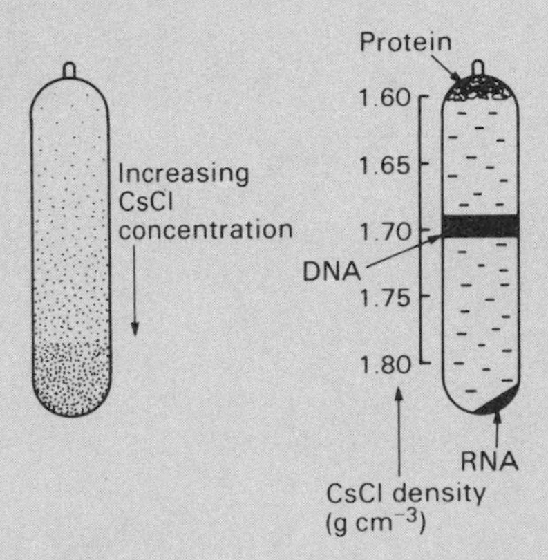

DNA has a buoyant density of about 1.7 g cm^{-3}, and will therefore migrate to the point in the gradient where the CsCl density is also 1.7 g cm^{-3}. Proteins have much lower and RNA somewhat higher buoyant densities.

Ultraviolet spectroscopy. Spectroscopy involves analysis of substances by the spectra they produce. The spectrum of light emitted or absorbed by a substance is characteristic of the substance. For example, DNA strongly absorbs ultraviolet radiation with a wavelength of 260 nm; proteins on the other hand have a strong absorbance at 280 nm.

Ultraviolet spectroscopy is often used to check that samples of DNA obtained from living cells are pure and do not contain protein or other contaminants. As the amount of ultraviolet radiation absorbed by a solution of DNA is directly proportional to the amount of DNA in the solution, the technique can also be used to determine the concentration of DNA in a sample.

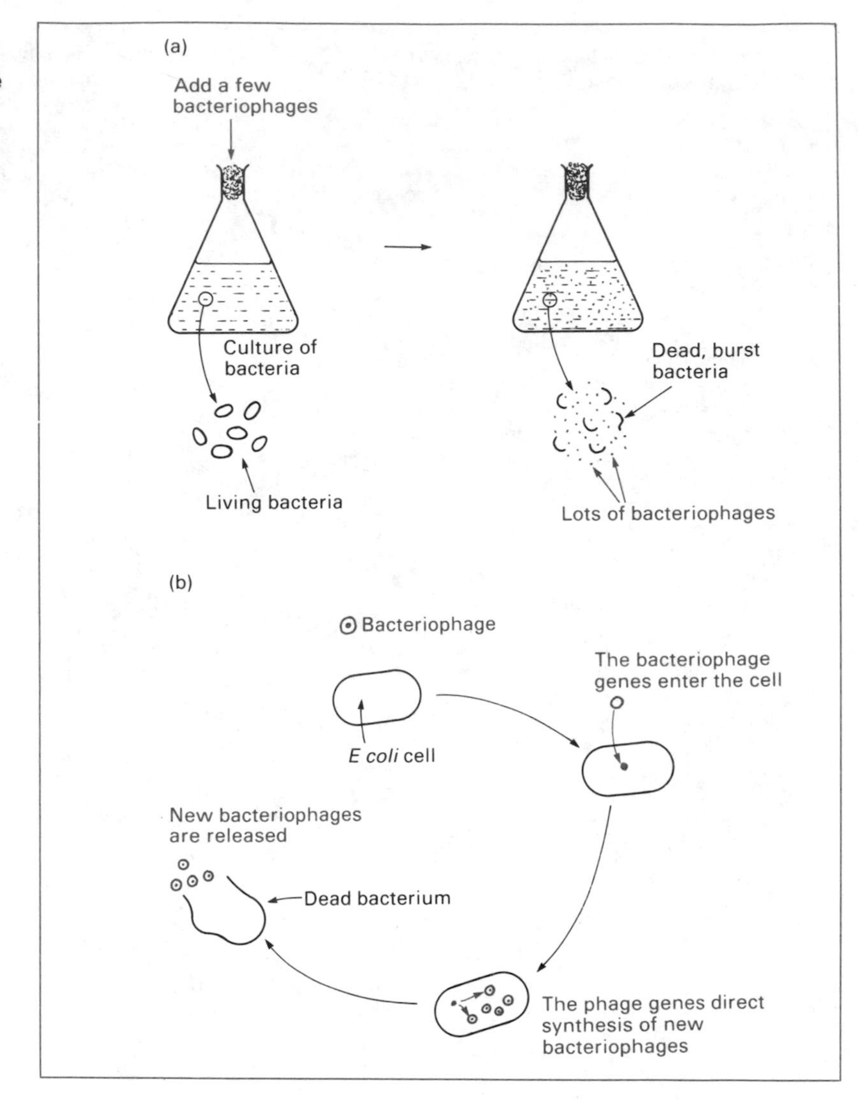

Figure 2.6 Infection of bacteria with bacteriophages. (a) A small number of bacteriophages can infect a culture of bacteria, killing all the cells while lots of new phages are synthesized. (b) During infection of a single bacterium the phage genes must enter the cell in order to direct synthesis of new bacteriophages. Hershey and Chase hypothesized that the scheme must work in a fashion similar to that shown. The details of the infection cycle, as we understand it, are described in Section 13.1.2.

2.2.2 Bacteriophage genes are made of DNA

The second experiment directed at chemical characterization of the genetic material was not carried out until 1951–52, 7 years after Avery's work was published. By then experimental techniques had progressed and new approaches to problems in genetics were available. In particular, the use of bacteriophages as experimental tools for studying molecular genetics had become fully established.

(a) Bacteriophages are viruses that infect bacteria. Bacteriophages, or phages as they are commonly known, are viruses that specifically infect bacteria. Phage T2, for example, is one of several

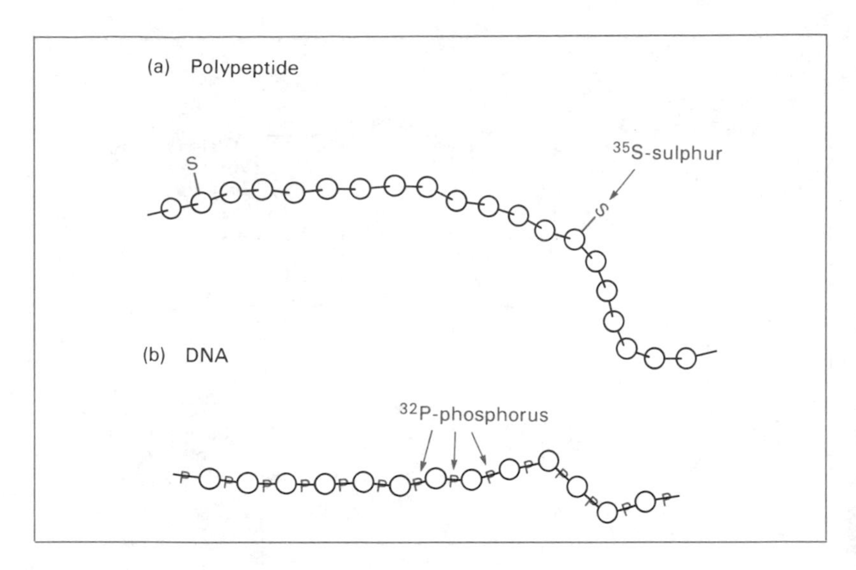

Figure 2.7 Differential labelling of (a) protein and (b) DNA with ^{35}S and ^{32}P. Polypeptides have two sulphur-containing amino acids (cysteine and methionine) but no phosphorus. DNA contains no sulphur, but each bond between individual units of a DNA molecule contains phosphorus.

types of phage that are specific for the bacterium *Escherichia coli*. If T2 bacteriophages are introduced into a culture of *E. coli*, the cells will become infected and will produce large numbers of new phage particles (Fig. 2.6(a)). Details of how the infection cycle proceeds were not worked out until several years later (Section 13.1.2), but it was hypothesized that as phages are not able to replicate on their own the new phage particles must be synthesized by the bacteria. To do this, the bacteria would need to make use of the information carried by the phage genes and probably (so it was argued) these genes would have to enter the bacterial cells for their information to be used (Fig. 2.6(b)). Should it be possible to identify a component of the infecting phage particles that enters the bacterial cell then the chemical nature of the genetic material would be known.

(b) Phage protein and DNA can be labelled with radioactive markers. Identifying the substance injected by phages into bacterial cells is simplified by the fact that phage particles contain only protein and DNA. In 1951 a new method that would allow protein and DNA to be distinguished unequivocally had just been developed. This procedure is called **radiolabelling** and involves attachment of a radioactive atom (or 'marker') to the molecule in question. Protein and DNA can be distinguished because protein can be labelled with ^{35}S, a radioactive isotope of sulphur, which will be incorporated into the sulphur-containing amino acids cysteine and methionine. DNA contains no sulphur and will not be labelled with ^{35}S. Conversely DNA can be labelled with ^{32}P, which will not be incorporated into proteins because proteins do not contain phosphorus (Fig. 2.7). Once labelled, samples of DNA and protein can be distinguished from each other because the two

Isotopes

All elements exist in a number of forms, called isotopes, each differing only in the numbers of neutrons in the atomic nucleus. Different isotopes of the same element therefore have the same atomic number but different atomic masses.

Isotopes may be stable (for example ^{12}C) or unstable (for example ^{14}C). If unstable then the isotope will decay into a stable isotope (possibly not of the same element) with emission of radiation of characteristic energy. An instrument such as a liquid scintillation counter can be used to measure the amount of radioactive disintegration and to identify the radioactive atom involved.

To the biologist the key fact about isotopes is that all the isotopes of a single element have the same number of electrons and therefore have identical chemical properties. Cells are unable to distinguish between different isotopes of the same element; for instance *E. coli* cannot distinguish between ^{14}C-glucose and ^{12}C-glucose and will use each equally efficiently as a carbon source.

Hershey and Chase prepared bacteriophages labelled with ^{32}P and ^{35}S by infecting a culture of *E. coli* grown in the presence of nutrients that contained these radioactive isotopes. Then, after infection of a new bacterial culture with the radioactive phages, they were able to determine if the bacteria contained ^{32}P-DNA or ^{35}S-protein because of the characteristic energies of the radiation emitted by these isotopes. We will encounter isotopic labelling again in Chapter 11.

Figure 2.8 The Hershey–Chase experiment.

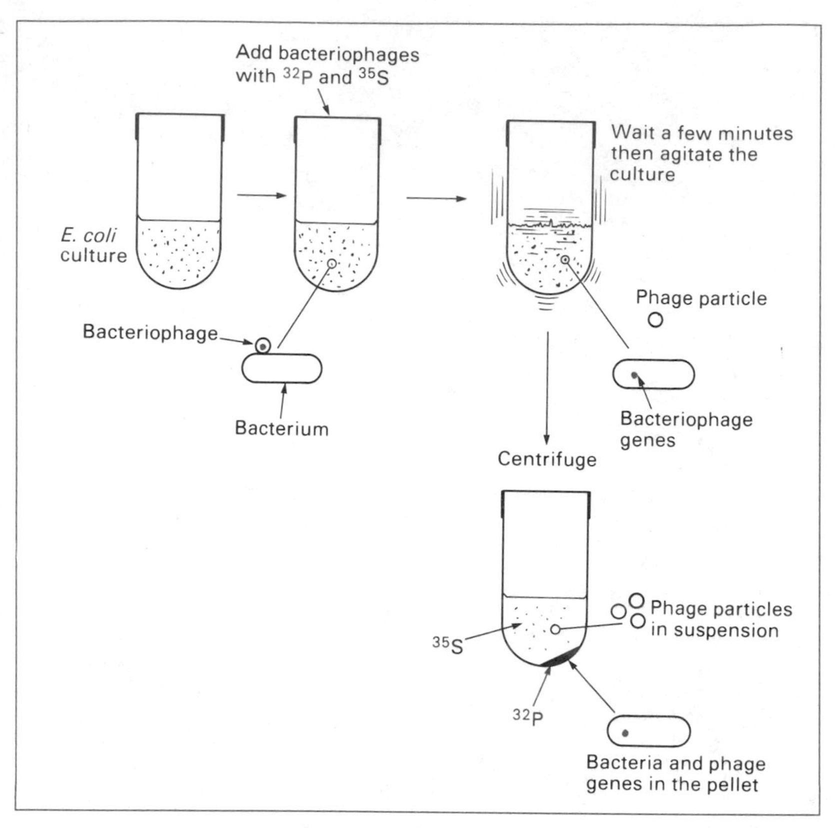

Alfred Hershey b. 4 December 1908, Owosso, Michigan

Alfred Hershey began as a microbiologist, obtaining his PhD from Michigan State College in 1934 for work on the *Brucella* bacteria. He then moved to Washington University, St Louis, where he stayed until 1950, when he joined the Genetics Department of the Carnegie Institute of Washington at Cold Spring Harbor, New York, becoming Director in 1962. He retired in 1974 and now lives near Cold Spring Harbor on Long Island. He was a founder member of Delbrück's 'Phage Group' and was responsible for several discoveries that led to an understanding of the phage infection cycle; the discovery that phage DNA is injected into the host cell was really just a single step in his long-term attack on phages. Delbrück described Hershey as 'a social misfit' and therefore ideally suited to being a molecular biologist. Hershey himself described his idea of heaven as 'To have one experiment that works, and keep doing it all the time'. He shared the 1969 Nobel Prize with Salvador Luria and Max Delbrück.

radioisotopes, ^{32}P and ^{35}S, emit radiation of different characteristic energies.

(c) The Hershey–Chase experiment. The critical experiment was carried out by Alfred Hershey and Martha Chase at the Cold Spring Harbor Laboratory on Long Island, New York. They prepared a radioactive sample of T2 phage, one in which the protein was labelled with ^{35}S and the DNA with ^{32}P, from a T2-infected culture of *E. coli* that had been grown with ^{35}S- and ^{32}P-labelled nutrients. The labelled phages produced by this culture were used to infect a new, non-radioactive culture of *E. coli* (Fig. 2.8). However, this time the infection process was interrupted a few minutes after inoculation by agitating the cells in a Waring blender (hence the popular name of the 'Waring blender experiment'). These few minutes were long enough for the phage genes to enter the bacteria but not enough time for new bacteriophages to be synthesized and the bacteria killed. Hershey and Chase believed that agitation would remove the phage material attached to the outside of the cell so that the only component retained by the bacteria would be the injected substance, the phage genes. The culture was then centrifuged so that the relatively heavy bacterial cells, containing the phage genes, collected at the bottom of the tube, leaving the empty phage particles in suspension. Hershey and Chase discovered that over 80% of the ^{32}P was present in the bacterial pellet. The bacteria were then allowed to continue through the infection process and produce new phages. Almost half of the ^{32}P, but less than 1% of the ^{35}S, was present in these new phage particles. Clearly, the genetic material was DNA and not protein.

2.3 Acceptance of DNA as the genetic material

Although very few doubts could be raised about the interpretation and validity of the Hershey–Chase experiment, several biologists still remained unconvinced that DNA was the genetic material in all organisms. Strictly speaking, all that the Hershey–Chase experiment demonstrated was that genes of bacteriophages are made of DNA. Phages are very unusual organisms, if indeed they can be called organisms at all, so perhaps the results of experiments with phages cannot be extrapolated to real organisms such as man. A biologist seeking rigorous scientific proofs would therefore have to accept that the chemical nature of the gene in higher organisms had still not been established.

Nowadays we do accept that the genetic material is DNA in all organisms except a few types of phage and virus, in which the second type of nucleic acid, RNA, replaces DNA (Section 13.1.1). However, direct experimental evidence that genes are made of DNA in higher organisms has come only surprisingly recently, with the advent of **recombinant DNA technology** and the artificial introduction of pure DNA into bacteria, yeast, fungi, insects, plants and animals (Chapter 20). Until these techniques became available during the 1970s there was only indirect evidence that the genetic material is DNA in higher organisms. This indirect evidence was fairly convincing but on its own would probably not have persuaded all biologists that genes are made of DNA. However, from 1953 onwards DNA was universally accepted as being the genetic material. This was because of the discovery of the structure of the double helix by Watson and Crick, and the realization that the structure was absolutely compatible with the requirements of the genetic material. This breakthrough is the central topic of the next chapter.

Reading

R. Olby (1974) *The Path to the Double Helix*, Macmillan, London – includes a particularly good account of the protein theory of the gene and its eventual overthrow.

P. A. Levene (1921) On the structure of thymus nucleic acid and on its possible bearing on the structure of plant nucleic acid. *Journal of Biological Chemistry*, **48**, 119–25 – an early description of the incorrect, non-variable structure for DNA.

O. T. Avery, C. M. MacLeod and M. McCarty (1944) Studies on the chemical nature of the substance inducing transformation of pneumococcal types. *Journal of Experimental Medicine*, **79**, 137–58; A. D. Hershey and M. Chase (1952) Independent functions of viral protein and nucleic acid in growth of bacteriophage. *Journal of General Physiology*, **36**, 39–56 – the two important papers that demonstrated that genes are made of DNA. Both are reprinted in J. H. Taylor, ed. (1965) *Selected Papers in Molecular Genetics*, Academic Press, New York.

M. McCarty (1980) Reminiscences of the early days of transformation. *Annual Review of Genetics*, **14**, 1–15; M. McCarty (1985) *The Transforming Principle: Discovering that Genes are made of DNA*, Norton, London – two personal accounts.

Problems

1. Define the following terms:

protein	transforming principle
nucleic acid	serotype
DNA	bacterial capsule
nucleoprotein	bacteriophage
macromolecule	radiolabelling

2. Describe the contribution of the following scientists to the understanding of the chemical nature of the gene:

   Griffith
   Avery, MacLeod and McCarty
   Hershey and Chase

3. Explain why biologists in the early 1900s thought that genes must be made of protein.

4. Describe what bacterial transformation is and explain why it is relevant to the chemical nature of the gene.

5. Contrast the experimental methods used by Avery *et al.* and by Hershey and Chase to determine the chemical nature of the gene.

6. Explain how radiolabelling was used in the Hershey–Chase experiment.

7. To what extent is the statement 'Genes are made of DNA' consistent with the results of Avery's work and with the results of the Hershey–Chase experiment?

8. *What properties, in addition to variability, can be predicted for the genetic material from our knowledge of the role of genes in heredity?
9. *Why did 11 years elapse between the discovery of bacterial transformation and the chemical identification of the transforming principle?

3 The structure of DNA

DNA is a polymer – nucleotides – polynucleotides – RNA • The double helix – complementary base pairing – different forms of DNA

Now we have established that genes are made of DNA we must consider exactly what the structure of DNA is and how this relates to the properties and requirements of the genetic material. Some students find molecular structures one of the more tiresome aspects of biology but in the case of DNA the effort must be made to understand and remember the structure. The answer to Schrödinger's conundrum 'What is Life?' lies in the DNA molecule.

3.1 DNA is a polymer

The fact that DNA is a long polymeric molecule was first clearly understood in the late 1930s and led to the realization that DNA, as well as protein, has the potential variability required of the genetic material. A polymer is a long chain-like molecule, comprising numerous individual units, called **monomers**, linked together in a series (Fig. 3.1). Many important biological molecules, including not only proteins and nucleic acids but also polysaccharides and lipids, are polymers of one type or another.

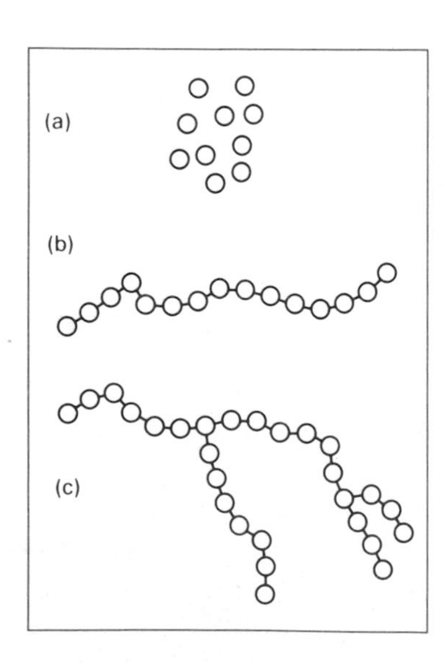

Figure 3.1 (a) Monomers. (b) A linear polymer. (c) A branched polymer. Nucleic acids and polypeptides are linear polymers. Some polysaccharides are branched polymers.

3.1.1 Nucleotides – the monomers in DNA

The basic unit of the DNA molecule is the **nucleotide**. Nucleotides are found in the cell either as components of nucleic acids or as individual molecules, in which form they may play several different roles. For instance, the nucleotides that make up RNA molecules (Section 3.1.3) are also important in the cell as carriers of energy used to power many enzymatic reactions.

The nucleotide is itself quite a complex molecule, being made up of three distinct components: a sugar, a nitrogenous base and phosphoric acid (Fig. 3.2).

(a) The sugar component. In DNA the sugar component of the nucleotide is a pentose (containing five carbon atoms) called 2′-deoxyribose (Fig. 3.2(a)). Pentose sugars can exist in two forms, the straight chain or Fischer structure and the ring or Haworth

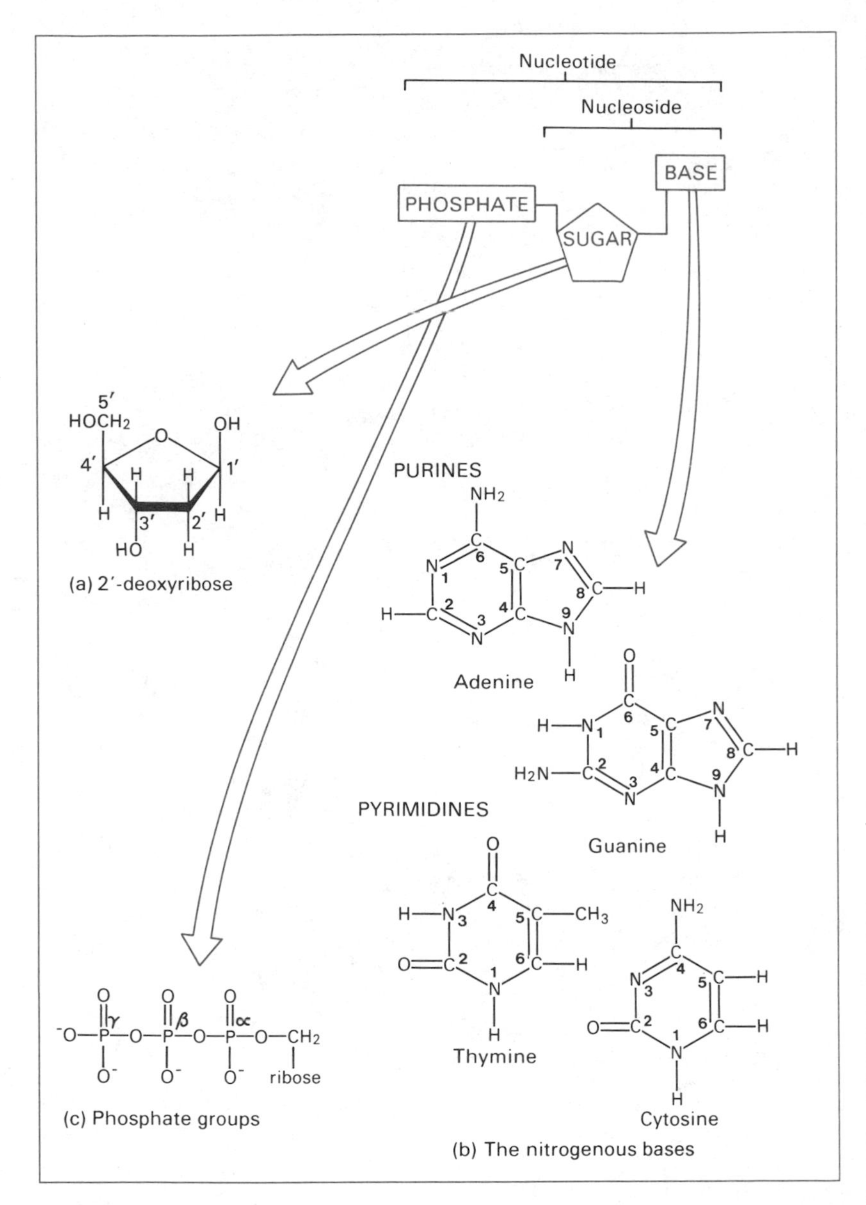

Figure 3.2 The components of a deoxyribonucleotide.

Fischer structure

Haworth structure

Figure 3.3 The two structural forms of 2′-deoxyribose.

structure (Fig. 3.3). It is the ring form of 2′-deoxyribose that occurs in the nucleotide.

The name 2′-deoxyribose indicates that the standard ribose structure has been altered by replacement of the hydroxyl group (—OH) attached to carbon atom number 2′ with a hydrogen group (—H). The carbon atoms are always numbered in the same way, with the carbon of the carbonyl group (—C=O), occurring at one end of the chain form, numbered 1′. It is important to remember the numbering of the carbons because it is used to

Figure 3.4 The four nucleotides found in DNA molecules.

2′-deoxyadenosine 5′-triphosphate

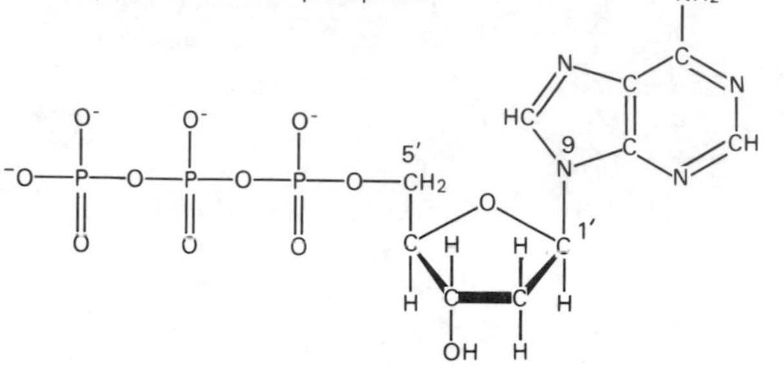

2′-deoxyguanosine 5′-triphosphate

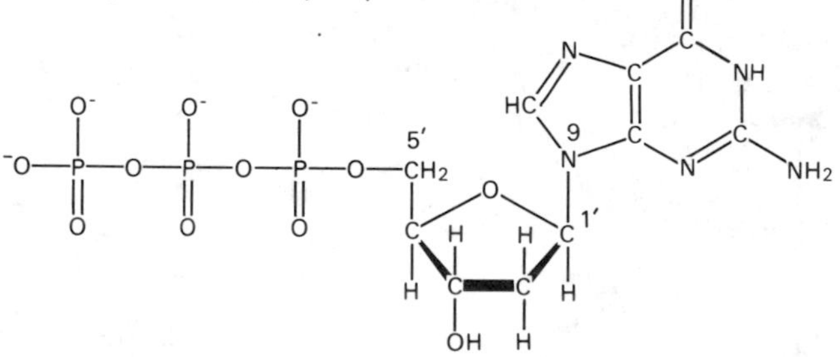

2′-deoxycytidine 5′-triphosphate

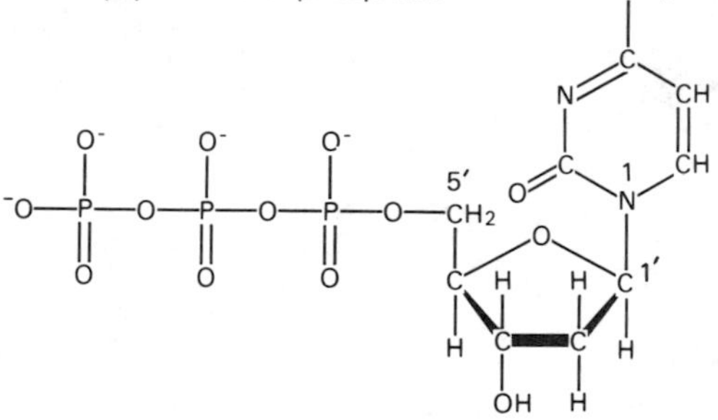

2′-deoxythymidine 5′-triphosphate

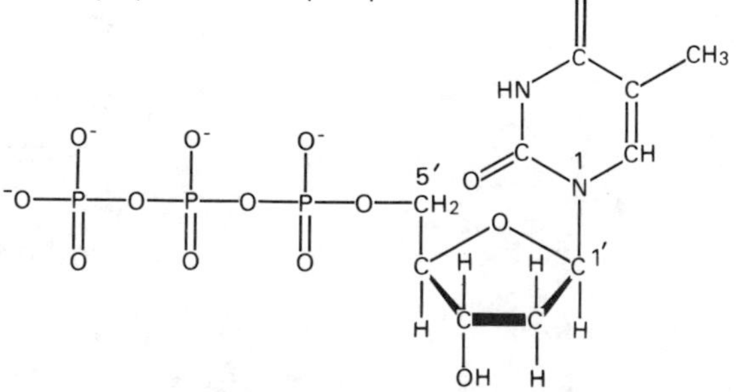

indicate at which positions on the sugar other components of the nucleotide are attached. Note that the number is not just 1, 2, 3, 4, 5 and 6, but 1′, 2′, 3′, 4′, 5′ and 6′ (the dash is called the 'prime' and the numbers are called 'one-prime', 'two-prime', and so on). The prime is used to distinguish the carbon atoms in the sugar from the carbon and nitrogen atoms in the nitrogenous base. The latter are numbered 1, 2, 3 and so on and it is important not to confuse these with the carbon atoms on the sugar.

(b) The nitrogenous bases. These are rather complex single- or double-ring structures that are attached to the 1′-carbon of the sugar. In DNA any one of four different nitrogenous bases can be attached at this position. These are called adenine and guanine, which are double-ring **purines**, and thymine and cytosine, which are single-ring **pyrimidines**. Their structures are shown in Fig. 3.2(b).

(c) The phosphoric acid component. A molecule comprising the sugar joined to a base is called a **nucleoside**. This is converted into a nucleotide by attachment of a phosphoric acid group to the 5′-carbon of the sugar (Fig. 3.2(c)). Up to three individual phosphate groups may be attached in series giving a nucleoside monophosphate (NMP), nucleoside diphosphate (NDP) or nucleoside triphosphate (NTP). The individual phosphate groups are designated α, β, and γ, with the α-phosphate being the one attached directly to the sugar.

(d) The nomenclature of nucleotides. The four different nucleotides that polymerize to form DNA are shown in Fig. 3.4. Their full names are:

2′-deoxyadenosine 5′-triphosphate
2′-deoxyguanosine 5′-triphosphate
2′-deoxycytidine 5′-triphosphate
2′-deoxythymidine 5′-triphosphate

Normally, however, these are abbreviated to dATP, dGTP, dCTP and dTTP, or even to just A, G, C and T, especially when writing out the sequence of nucleotides found in a particular DNA molecule.

3.1.2 Polynucleotides

The next stage in building up the structure of a DNA molecule is to link the individual nucleotides together to form a polymer. This polymer is called a **polynucleotide** and is formed by attaching one nucleotide to another by way of the phosphate groups.

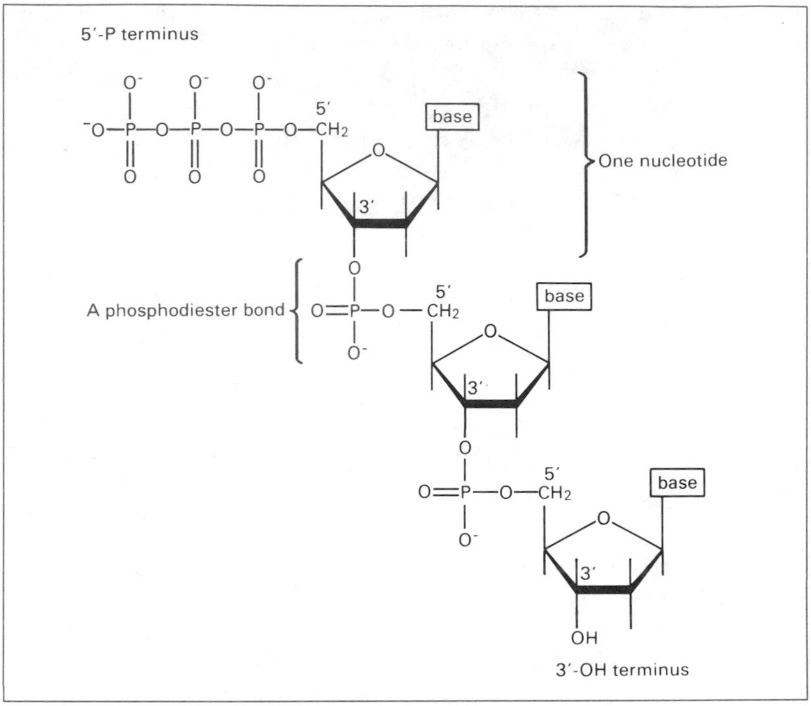

Figure 3.5 The structure of a short polynucleotide.

(a) Nucleotides are joined by phosphodiester bonds. The structure of a trinucleotide, a short DNA molecule comprising three individual nucleotides, is shown in Fig. 3.5. The nucleotide monomers are linked together by joining the α-phosphate group, attached to the 5′-carbon of one nucleotide, to the 3′-carbon of the next nucleotide in the chain. Normally a polynucleotide is built up from nucleoside triphosphate subunits, so during polymerization the β- and γ-phosphates are cleaved off. The hydroxyl group attached to the 3′-carbon of the second nucleotide is also lost.

The linkage between the nucleotides in a polynucleotide is called a **phosphodiester bond**: 'phospho' indicating the presence of a phosphorus atom, and 'diester' referring to the two ester (C—O—P) bonds in each linkage. To be precise we should call this a 3′–5′ phosphodiester bond so that there is no confusion about which carbon atoms in the sugar participate in the bond.

(b) Polynucleotides have chemically distinct ends. An important feature of the polynucleotide is that the two ends of the molecule are not the same. This is clear from an examination of Fig. 3.5. The top of this polynucleotide ends with a nucleotide in which the triphosphate group attached to the 5′-carbon has not participated in a phosphodiester bond and the β- and γ-phosphates are still in place. This end is called the **5′** or **5′-P terminus**. At the

A—T—A—G—A—A—C—A—G—G

A—A—A—G—A—A—C—A—G—G

A—T—A—A—A—A—C—A—G—G

A—T—A—C—A—A—C—A—G—G

A—T—A—G—T—A—C—A—G—G

Figure 3.6 Five of the 1 048 576 different possible sequences for a DNA molecule ten nucleotides in length.

other end of the molecule the unreacted group is not the phosphate but the 3′-hydroxyl. This end is called the **3′** or **3′-OH terminus**. The chemical distinction between the two ends means that polynucleotides have a direction, which can be looked on as 5′→3′ (down in Fig. 3.5) or 3′→5′ (up in Fig. 3.5). The direction of the polynucleotide is very important in molecular genetics, as will become clear in later chapters.

(c) Polynucleotides can be any length and have any sequence. There is apparently no limitation to the number of nucleotides that can be joined together to form an individual DNA polynucleotide. Molecules containing several thousand nucleotides are frequently handled in the laboratory and the DNA molecules in chromosomes are much longer, possibly several million nucleotides in length. In addition, there are no chemical restrictions on the order in which the nucleotides can join together. At any point in the chain the nucleotide could be A, G, C or T. Consider a polynucleotide just ten nucleotides in length: it could have any one of $4^{10} = 1\,048\,576$ different sequences (Fig. 3.6). Now imagine the number of different sequences possible for a polynucleotide 1000 nucleotides in length, or one million. This is the variability of DNA that enables the genetic material to exist in an almost infinite number of forms.

3.1.3 RNA is also a polynucleotide

DNA is not the only type of nucleic acid found in living cells. The closely related substance RNA is also a nucleic acid and, like DNA, plays a vital role in molecular genetics. RNA is also a polynucleotide but with two important differences from DNA.

1. ***The sugar in RNA is ribose.*** In RNA the sugar component of the nucleotide is not 2′-deoxyribose, but ribose itself (Fig. 3.7(a)).
2. ***RNA contains uracil instead of thymine.*** Three of the nitrogenous bases – adenine, guanine and cytosine – are found in both DNA and RNA. However, the fourth base is different. In

Figure 3.7 The components of RNA that are different from those in DNA.

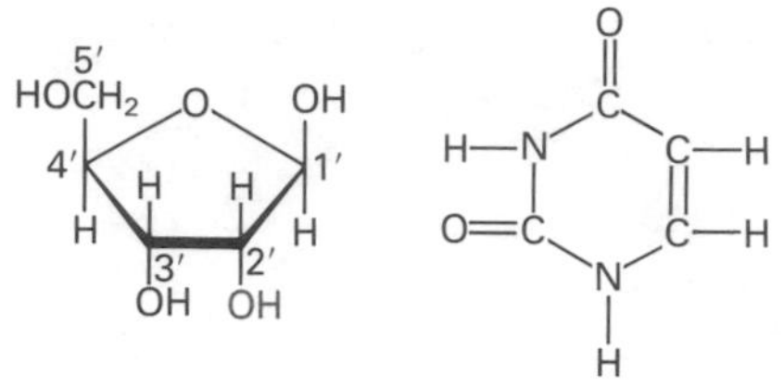

(a) Ribose (b) Uracil

DNA it is thymine but in RNA it is another pyrimidine, **uracil** (Fig. 3.7(b)).

The names of the four nucleotides that polymerize to make RNA are:

adenosine 5′-triphosphate
guanosine 5′-triphosphate
cytidine 5′-triphosphate
uridine 5′-triphosphate

which are abbreviated to ATP, GTP, CTP and UTP, or A, G, C and U.

The polynucleotide structure of RNA is exactly the same as that of DNA, with 3′–5′ phosphodiester bonds linking together the individual nucleotides in the molecule. As with DNA there are no restrictions on the sequence of nucleotides within an RNA molecule. There is, however, one very important difference between RNA and DNA at this higher level of structure. RNA in the cell usually exists as a single polynucleotide whereas DNA is almost invariably in the form of two polynucleotides wrapped round one another. This is the famous **double helix**, the discovery of which convinced molecular geneticists that genes are indeed made of DNA.

3.2 The double helix

The discovery of the double helix by James Watson and Francis Crick at Cambridge in 1953 was one of the great deductive triumphs in the history of science and has influenced every aspect of molecular genetics, indeed of biology as a whole.

The structure of the polynucleotide, as described in Section 3.1.2, was already known when Watson and Crick came on to the scene. The next step was to understand the exact structure that a DNA molecule takes up in a living cell. Is just one polynucleotide involved or is a DNA molecule made up of two or more polynucleotides attached to one another in some way? The answer might reveal how DNA is able to act as the genetic material. To solve the problem Watson and Crick used **model-building** – they literally built a scale model of what they thought a DNA molecule must look like based on all the information that was available at that time. Of course the model had to obey the laws of chemistry, which meant that if a polynucleotide was coiled up or folded in any way then the various atoms must not be placed too close together, and any new chemical bonds that were postulated must be between atoms the appropriate distances apart. In addition, it was vital that the model took account of the results of two other investigations into DNA structure, one carried out by Erwin Chargaff in New York, the other by Rosalind Franklin at King's College, London. Before we examine Watson and Crick's double helix we must, like them, consider the results of these two important pieces of work.

Erwin Chargaff b. 11 August 1905, Czernowitz, Austria (now Chernovtsy, the Ukraine)

'We can only seek what we have already found'.

Erwin Chargaff received his PhD from the University of Vienna in 1928, and then after periods at Yale, Berlin and Paris joined Columbia University, New York City in 1935. He has worked in many fields of biochemistry and his research papers include studies on subjects as diverse as lipid metabolism and blood coagulation. However, his most influential contribution was the demonstration during 1945–1950 that the base ratios in DNA are constant. Although this was one of the seminal discoveries that led to the structure of DNA, Chargaff has never claimed any allegiance to molecular biology, and indeed has become one of its sternest critics (see for example *Nature*, **326**, 199–200, 1987). He is the source of many pungent comments about contemporary science and scientists. He described Watson and Crick as two pitchmen in search of a helix and explains his objection to molecular biology as 'by its claim to be able to explain everything, it actually impedes the flow of scientific explanation'. His autobiography is *Heraclitean Fire: Sketches From a Life Before Nature*, Rockefeller University Press, New York, 1978.

(a) Chargaff's base ratios paved the way for the correct structure. The discovery by Avery, MacLeod and McCarty that the transforming principle is DNA influenced another New York research worker, Erwin Chargaff of Columbia University, to carry out a comprehensive analysis of the chemical composition of DNA. In particular, Chargaff, together with his colleagues E. Vischer and S. Zamenhof, used new and sensitive paper chromatographic techniques to determine the exact amounts of each of the four nitrogenous bases in samples of DNA purified from different tissues and different organisms. Their results were quite startling: they revealed a simple mathematical relationship between the proportions of the bases in any one sample of DNA. The relationship is that the number of adenine residues equals the number of thymines, and the number of guanines equals the number of cytosines, that is A = T and G = C. Implicit in this is that the total purines (A + G) will equal the total pyrimidines (T + C). This rule

Table 3.1 A summary of some of Chargaff's results indicating a consistent pattern to the ratios of the bases in DNA from different organisms

	Ratio of bases in DNA samples	
Organism	*Adenine:Thymine*	*Guanine:Cytosine*
Cow	1.04	1.00
Man	1.00	1.00
Salmon	1.02	1.02
Escherichia coli	1.09	0.99

Data from E. Chargaff (1963) *Essays on Nucleic Acids*, Elsevier, p. 30.

Chromatography

Chromatographic techniques are used to separate compounds on the basis of their relative affinities for absorption on to a solid matrix. They are particularly useful for separating individual proteins, amino acids or nucleotides.

There are several different chromatographic methods, distinguished by the nature of the solid matrix. In paper chromatography, this matrix is a strip of filter paper. A sample of the compounds to be analysed is placed at one end of the paper strip and eluted by soaking an aqueous or organic solution (the solvent) along the strip. As the solvent soaks along the paper it carries the compounds with it, but at different rates depending on the relative affinities of the compounds for absorption on to the matrix.

Chargaff's method for analysing the base composition of DNA involved first breaking the molecule into its component nucleotides by treatment with acid or alkali. The sample was then eluted along a paper strip with any of several solvents (for example, an *n*-butanol–water mixture) and the resulting spots cut out. The pure nucleotides were recovered from the paper by soaking in an aqueous solution, and their concentrations measured.

Chromatography is now more routinely carried out in a glass column packed with tiny beads, composed of substances such as cellulose or agarose, immersed in the solvent. The sample is layered on to the top of the column and eluted by passing through more solvent. The separated compounds are then collected as they drip out of the bottom of the column.

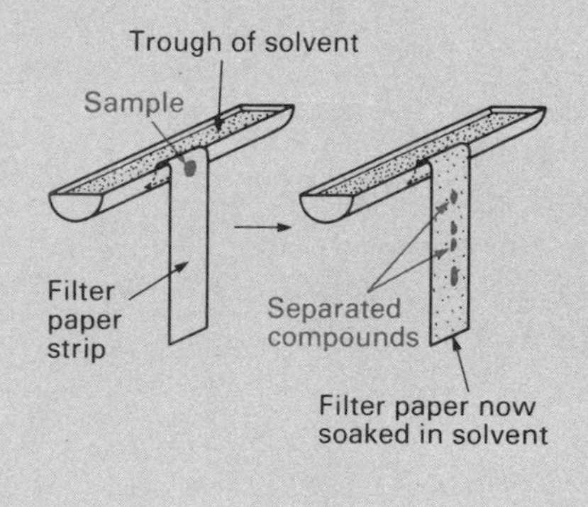

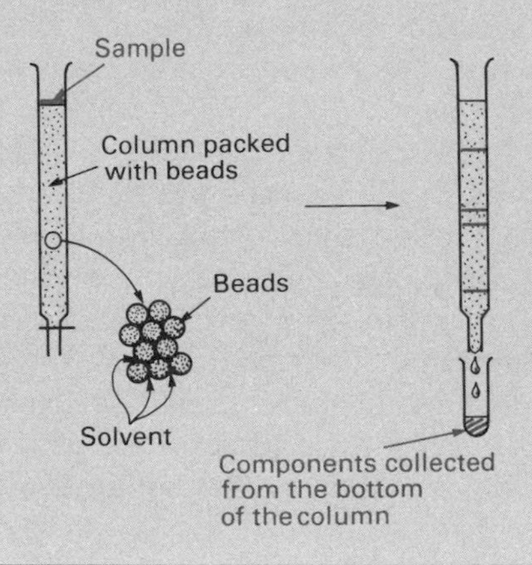

is illustrated by Table 3.1, which shows some of the original results published by Chargaff's group.

Chargaff was a very thorough scientist who worked strictly according to the Sherlock Holmes maxim of never theorizing before being sure of the data. He did not speculate to any great extent, at least not in his publications, about the relevance of the base ratios to the structure of DNA. He believed that to do so before confirming the results, by analysing DNA from many additional sources, would be to run the risk of falling into the same trap that led to the earlier incorrect ideas about DNA being a small and simple molecule. However, an acceptable model for the structure of DNA must explain how these base ratios arise.

Before leaving Chargaff's base ratios, one further point should be noted. It is important to realise that although A = T and G = C, A + T does not equal G + C. In fact, the **GC** (or conversely AT) **content** of an organism's DNA varies considerably from species to species. Why this should be so is not known, but it can have a significant influence on gene structure and the way in which the action of genes is controlled.

(b) X-ray diffraction analysis indicates that DNA is a helical molecule. The second piece of evidence available to Watson and Crick was the **X-ray diffraction pattern** obtained when a crystallized DNA fibre is bombarded with X-rays. The theory behind **X-ray diffraction analysis** is very complex but is based on the fact that the angles at which X-rays are deflected on passage through a crystal will be determined by the three-dimensional structure of the molecules in the crystal. The deflections can be recorded by allowing the X-ray beam, after passage through the crystal, to expose a photographic film (Fig. 3.8). The result is a pattern of spots, the positions and intensities of which may allow the structure of the molecule to be deduced.

X-ray diffraction analysis began in the late nineteenth century and has been successfully applied to crystals of many compounds. Its greatest achievement up to 1953 had been in providing the structure of several natural proteins, an example being keratin, a component of wool and hair. Since 1953 it has been used with increasing sophistication to help elucidate the complex three-dimensional structures of proteins such as haemoglobin. But the most famous X-ray diffraction pictures of all were taken with crystals of DNA by Rosalind Franklin at King's College, London during May 1952, using techniques previously developed by Maurice Wilkins. These pictures show that DNA is a helix with two regular periodicities of 3.4 Å and 34 Å along the axis of the molecule. But how does this relate to Chargaff's base ratios and to the actual structure of DNA?

(c) Watson and Crick solve the structure. The story of how Watson and Crick deduced that DNA exists as a double helix has been told many times, most entertainingly by Watson himself in his book *The Double Helix*. Watson and Crick put together all the experimental data concerning DNA and decided that the only

GC contents of different DNAs

The GC content of an organism's DNA varies considerably from species to species. Here are some representative values:

Organism	*GC content (% total nucleotides)*
Plasmodium falciparum (malaria parasite)	19.0
Slime mould	23.0
Sea urchin	35.4
Man	40.3
Wheat	45.8
Escherichia coli	51.0
Mycobacterium tuberculosis	65.0
Streptomyces griseolus	72.4

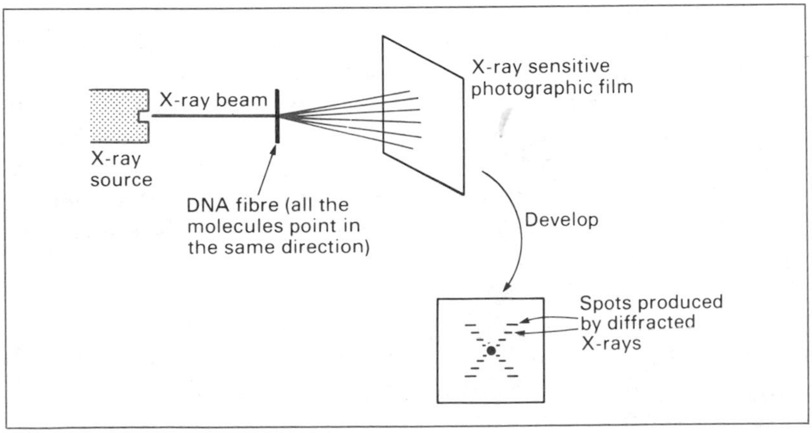

Figure 3.8 X-ray diffraction analysis.

Maurice Wilkins

Rosalind Franklin

Maurice Wilkins b. 15 December 1916, Pongaroa, New Zealand, and **Rosalind Franklin** b. 25 July 1920, London; d. 16 April 1958, London

Maurice Wilkins began as a physicist (BA Cambridge 1938, PhD Birmingham 1940) but his experiences on the atomic bomb project during the War turned him away from physical science and towards biology. He joined King's College, London in 1947 and began studying the structure of chromosomes and genes using physical methods. In May 1950 he obtained from Rudolf Signer of Bern a sample of what was probably the purest DNA available at the time, and from it obtained extremely detailed X-ray diffraction patterns. In January 1951 Rosalind Franklin joined King's from Paris and took over analysis of the Signer DNA, producing pictures that showed clearly that DNA is a helix. Unfortunately there was a personality clash between Franklin and Wilkins and the intellectual discussion needed to solve the structure of DNA (which Watson and Crick enjoyed) never took place at King's. Nevertheless, Franklin was very close to a double helix structure when Watson and Crick announced their results.

Franklin moved to Birkbeck College to work on virus structure in 1953 and remained there until she died of cancer. Her portrayal in *The Double Helix* as 'Rosie' is now considered a fiction and it is clear from accounts of her contemporaries that, had she lived, she would probably have become one of the most eminent British scientists of her day (see *Rosalind Franklin and DNA* by A. Sayre in 'Reading'). Wilkins continued with DNA analysis for a number of years and shared the 1962 Nobel Prize with Watson and Crick. In 1969 he became the founding President of the British Society for Social Responsibility in Science.

Units of length

1 micrometre or micron (μm) = 10^{-6} metres = one-thousandth of a millimetre
1 nanometre (nm) = 10^{-9} metres = one-millionth of a millimetre
1 Ångström unit (Å) = 10^{-10} metres

structure that fitted all the facts was the double helix shown in Fig. 3.9. There are seven important features of this helix.

1. ***The double helix comprises two polynucleotides***. The number of polynucleotides in the helix was a problematical question for some time. Several pieces of experimental evidence indicated that the number was two or three, and in fact the famous American scientist Linus Pauling proposed an incorrect triple helix model for DNA in the months leading up to the Watson–Crick structure. Eventually the measurement of the density of DNA was refined so that a double-stranded molecule became most likely.
2. ***The nitrogenous bases are stacked on the inside of the helix***. The experimental evidence also indicated that the sugar-phosphate 'backbone' of the molecule is on the outside, with the bases inside the helix. In fact the bases are stacked on top of each other rather like a pile of plates.

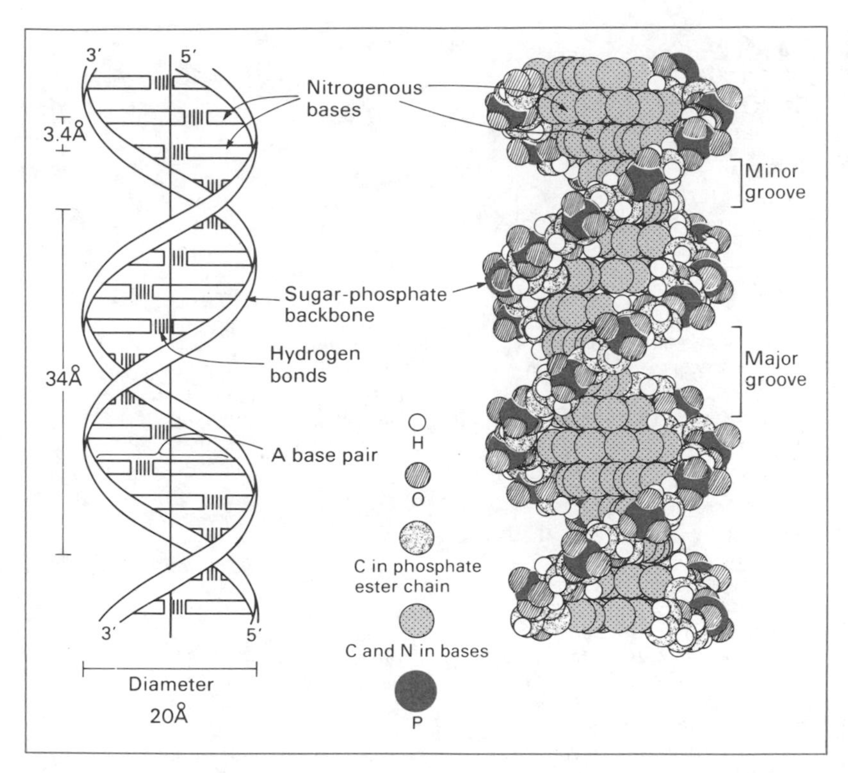

Figure 3.9 Two representations of the double helix structure for DNA. On the left is a simplified model similar to the one originally published by Watson and Crick. On the right is a 'space-filling' model, where an attempt is made to indicate the relative sizes of the atoms in the molecule.

Hydrogen bonds

A hydrogen bond is a weak electrostatic attraction between an electronegative atom (such as oxygen or nitrogen) and a hydrogen atom attached to a second electronegative atom. It arises because the electrons in a covalent bond such as O—H or N—H are displaced towards the electronegative atom so that the hydrogen atom possesses a partial positive charge and can therefore form a weak bond with a second electronegative atom. In effect, the hydrogen atom is shared between the two electronegative atoms.

Hydrogen bonds are longer than covalent bonds (perhaps 2 Å as compared with 1 Å) and are much weaker. A hydrogen bond usually has a bond energy of between 3 and 10 kcal mol^{-1} at 25°C, whereas covalent bond energies are generally greater than 50 kcal mol^{-1}. However, a large number of hydrogen bonds can cooperate to form a relatively strong attractive force.

In biology, hydrogen bonds are important because they contribute to the three-dimensional structures of macromolecules such as nucleic acids and proteins.

3. ***The bases of the two polynucleotides interact by hydrogen bonding.*** This is the explanation of Chargaff's base ratios. An adenine residue in one of the polynucleotides is always adjacent to a thymine in the other strand; similarly guanine is always adjacent to cytosine (Fig. 3.10(a)). These two pairs of bases, and no other combinations, are able to form **hydrogen bonds** between each other, two bonds between A and T, and three between G and C (Fig. 3.10(b)). These hydrogen bonds are the only attractive forces between the two polynucleotides of the

James Watson

Francis Crick

James Watson b. 6 April 1928, Chicago, and **Francis Crick** b. 8 June 1916, Northampton

James Watson was a child prodigy who entered the University of Chicago at the age of fifteen, graduating in 1947 and obtaining his PhD from the University of Indiana in 1950. He was then awarded a fellowship to study in Copenhagen but after a year there he moved to the Cavendish Laboratory, Cambridge, with the specific objective of studying the gene. There he met Francis Crick, who was working on protein structure for his PhD after starting as a physicist but becoming diverted to biology after the War. The famous fusion of minds that led Watson and Crick to deduce that DNA is a double helix has been described many times, notably by Watson in *The Double Helix*. The two shared the 1962 Nobel Prize with Maurice Wilkins.

After the double helix, Watson and Crick went separate ways. Watson returned to the USA in 1953 and eventually took up a professorship at Harvard before moving in 1976 to the Cold Spring Harbor Laboratory on Long Island, where he is Director. He has worked on RNA synthesis, protein synthesis and the role of viruses in cancer. Recently he has been one of the prime motivators behind the Human Genome Project, the ambitious research programme that has as its goal the complete nucleotide sequence of the human genome (Section 16.1.13). Crick stayed in Cambridge until moving to the Salk Institute in Southern California in 1977. During the 1950s and 1960s he was the major influence in molecular biology and many of the great advances in understanding genes and gene expression are due directly to him. His autobiography is *What Mad Pursuit*, Weidenfield and Nicolson, London, 1989.

double helix and serve to hold the structure together.

4. ***Ten base pairs occur per turn of the helix.*** The double helix executes a turn every ten **base pairs** (abbreviated as 10 bp). The pitch of the helix is 34 Å, meaning that the spacing between adjacent base pairs is 3.4 Å; these are the periodicities apparent from the X-ray diffraction pattern. The helix is 20 Å in diameter.
5. ***The two strands of the double helix are antiparallel.*** One polynucleotide runs in the 5′→3′ direction, the other in the 3′→5′ direction. Only antiparallel polynucleotides will form a stable helix.
6. ***The double helix has two different grooves.*** The helix is not absolutely regular: a **major** and a **minor groove** can be distinguished. This feature is important in the interaction between the double helix and the proteins involved in DNA replication and in the expression of genetic information.
7. ***The double helix is right-handed.*** This means that if the double helix was a spiral staircase that you were climbing up, the banister (= the sugar-phosphate backbone) would be on your right-hand side.

Describing the lengths of DNA molecules

The basic monomeric unit of the double helix is the base pair, with ten base pairs occurring per turn of helix. The length of a DNA molecule is usually described as so many base pairs, abbreviated to bp (for example 100 bp, 250 bp). A kilobase pair, usually called a kilobase or kb, is 1000 bp.

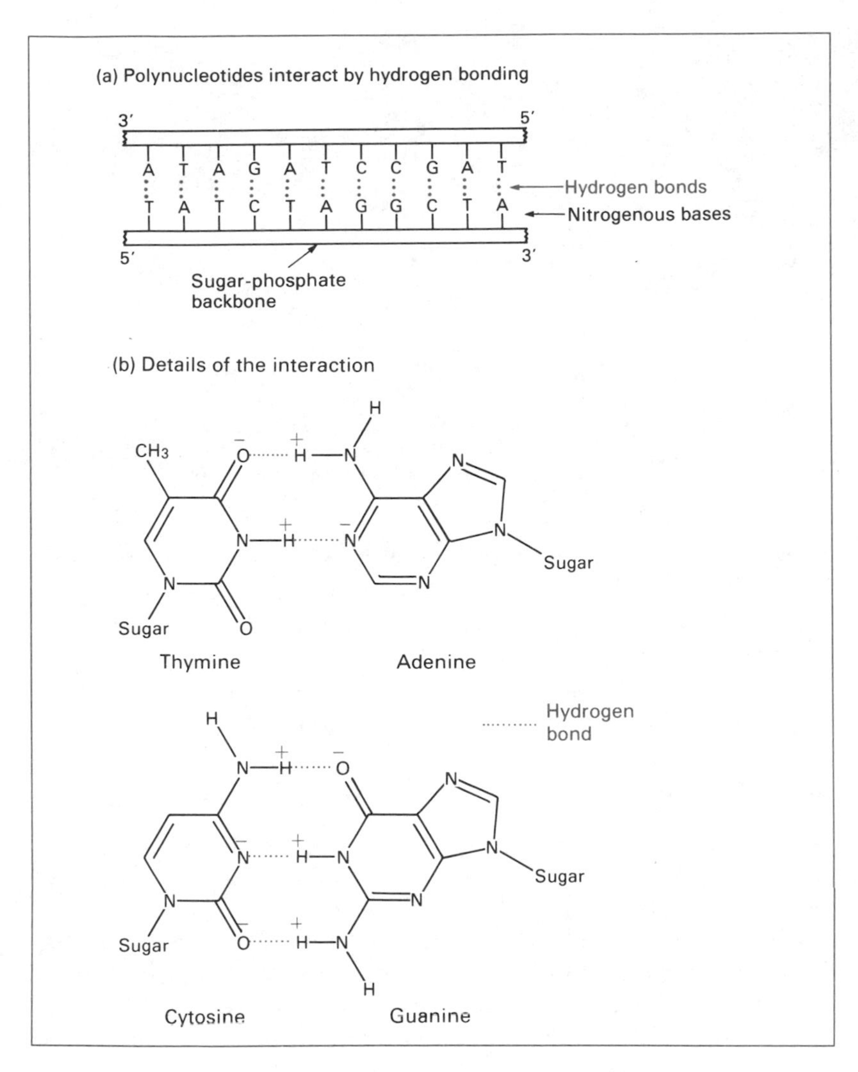

Figure 3.10 Base pairing.

3.2.1 Complementary base pairing is the fundamental fact of molecular genetics

Of these various features it is the base pairing that is most important. The rule is that A base pairs with T, and G base pairs with C. Other base pairs are not normally permitted because they will either be too large to fit in the helix (as is the case with a purine–purine base pair), or too small for the helix (pyrimidine–pyrimidine), or will not align in a manner allowing hydrogen bonding to occur (as with A–C and G–T). The two polynucleotides in a double helix are therefore **complementary**, the sequence of one determining the sequence of the other. Complementary base pairing is of fundamental importance in molecular genetics. As we

Franklin's ideas on the structure of DNA

'The results suggest a helical structure . . . containing probably 2, 3 or 4 co-axial nucleic acid chains per helical unit, and having the phosphate groups near the outside.'

From a report written by Franklin in February 1952, a year before Watson and Crick proposed the double helix structure. (Taken from *Rosalind Franklin and DNA* by A. Sayre in 'Reading')

The different structures of DNA

The double helix described by Watson and Crick is now called the B-form of DNA. In the cell two other forms of DNA called A-DNA and Z-DNA are also thought to occur. Under laboratory conditions, additional helical structures, such as C-, D- and E-DNA, have been produced but these probably never exist in the cell. The different forms of DNA are distinguished by different dimensions to their double helices. The data for the best-studied forms are shown in the table.

A-, B- and C-DNA are formed in fibres of DNA subjected to different relative humidities (75% for A, 92% for B and 66% for C). Z-DNA will form in a double helix that has a sequence made up of alternating purine and pyrimidine nucleotides.

Form	*Direction of helix*	*Base pairs per turn*	*Distance between base pairs (Å)*	*Diameter of helix (Å)*
A	Right-handed	11	2.6	23
B	Right-handed	10	3.4	19
C	Right-handed	9.3	3.3	19
Z	Left-handed	12	3.7	18

shall see in later chapters it provides the means by which the sequence of a DNA molecule is retained during replication of the double helix, something which is crucial if the biological information contained in the gene is not to become scrambled during cell division. Complementary base pairing is also vital for expression of the biological information in a form utilizable by the cell.

3.2.2 The double helix exists in several different forms

By the time that Watson and Crick discovered the double helix it had already been shown by Rosalind Franklin that there are two distinct crystalline forms of DNA, the **A-form** and the **B-form**. The form taken up depends on the amount of water present in the DNA solution from which the crystals form. The Watson–Crick structure refers to the B-form, of which the classic X-ray diffraction pictures were taken. The B-form is in fact the structure that DNA takes up in the cell, at least under most conditions.

The A-form differs from the B-form in relatively minor though important ways. A-DNA is still a double helix, but more compact than B-DNA, with 11 bp per turn of the helix and a diameter of 23 Å. Since 1953, four other forms of DNA have been found: C-, D-, E- and **Z-DNA**, each with a slightly different conformation to the double helix. Z-DNA is the most strikingly different, as in this structure the helix is left-handed, not right-handed as it is with A- to E-DNA. These alternative forms of DNA have received increasing attention in recent years as it has been shown that the nucleotide sequence is one of the factors that influence the form taken by a segment of the double helix. Regions of DNA that are predisposed, by virtue of their nucleotide sequences, to take up non-standard double helices (e.g. Z-DNA regions) have recently been discovered in chromosomes.

Reading

J. D. Watson and F. H. C. Crick (1953) Molecular structure of nucleic acids: a structure for deoxyribose nucleic acid. *Nature*, **171**, 737–8 – the original publication of the double helix.

J. D. Watson and F. H. C. Crick (1953) Genetic implications of the structure of deoxyribose nucleic acid. *Nature*, **171**, 964 – explains why the double helix is perfectly suited to its role as the genetic material.

J. D. Watson (1968) *The Double Helix*, Atheneum, London – Watson's own account of the events leading up to the discovery of the double helix. The reaction (some of it outraged) of the rest of the scientific community to Watson's book is described in Stent, G. S. (ed.) (1981) *The Double Helix: A New Critical Edition*, Weidenfield and Nicolson, London.

A. Sayre (1975) *Rosalind Franklin and DNA*, W. W. Norton, New York; H. R. Wilson (1988) The double helix and all that. *Trends in Biochemical Sciences*, **13**, 275–8 – two other personal accounts of the events surrounding the discovery of the double helix. If you read Watson's book then you should also read Sayre to get a balanced point of view.

R. Olby (1974) *The Path to the Double Helix*, Macmillan, London; F. H. Portugal and J. S. Cohen (1977) *A Century of DNA*, MIT Press, Cambridge; H. F. Judson (1979) *The Eighth Day of Creation*, Jonathan Cape, London – provide less personal accounts of the work that led to the double helix.

E. Chargaff (1950) Chemical specificity of nucleic acids and mechanism of their enzymatic degradation. *Experientia*, **6**, 201–9 – describes the methodology for determining the base composition of DNA, together with some of the original data.

L. Pauling and R. B. Corey (1953) A proposed structure for the nucleic acids. *Proceedings of the National Academy of Sciences USA*, **39**, 84–97 – Linus Pauling's incorrect triple-stranded model for DNA, which appeared just before the Watson–Crick structure.

R. E. Dickerson, H. R. Drew, B. N. Conner *et al.* (1982) The anatomy of A-, B- and Z-DNA. *Science*, **216**, 475–85 – a detailed description of X-ray analysis of different forms of DNA.

The development of knowledge about DNA up to 1953

1869	Miescher discovers DNA in an extract prepared from pus cells
1894	Kossell and Neumann identify three of the bases in DNA
1900	Hammersten shows that DNA contains a pentose sugar
1929	Levene demonstrates that the sugar is 2′-deoxyribose
1934	The tetranucleotide theory of DNA structure (see Fig. 2.1(a)) is overthrown when Caspersson proves that DNA is a macromolecule
1944	Avery, MacLeod and McCarty discover that the transforming principle is DNA, the first indication that DNA is the genetic material
1950	Chargaff publishes his results demonstrating a consistency to the base ratios in DNA from different sources
1952	Hershey and Chase show that phage genes are made of DNA
1952	Franklin takes the X-ray diffraction pictures that indicate that DNA is a helix
1953	Watson and Crick deduce the double helix structure

Problems

1. Define the following terms:

polymer	purine
monomer	pyrimidine
nucleotide	polynucleotide

phosphodiester bond	double helix
5′-terminus	hydrogen bond
3′-terminus	base pair
model-building	major and minor grooves
X-ray diffraction pattern	complementary

2. Describe the contribution of each of the following to the determination of the structure of DNA:

Chargaff	Watson
Franklin	Crick
Wilkins	

3. Draw the structure of a nucleotide.
4. Distinguish between the structures of the nucleotides found in DNA and those found in RNA.
5. Describe how nucleotides are attached together to form a polynucleotide.
6. Explain why the two ends of a polynucleotide are chemically distinct.
7. Why can DNA provide the variability required by the genetic material?
8. Outline the two major types of experimental analysis that laid the foundations for the deduction of the structure of DNA by Watson and Crick.
9. What are the important features of the double helix structure?
10. What is meant by 'complementary base pairing' and why is it important?
11. *The most famous statement in Watson and Crick's paper is 'It has not escaped our notice that the specific pairing we have postulated immediately suggests a possible copying mechanism for the genetic material'. Explain.
12. *To what extent can the double helix structure of DNA be deduced from just a knowledge of Chargaff's base ratios?
13. *In Chapter 1, Question 6 you helped Mendel obtain a research grant. Now do the same for Watson and Crick.
14. *Discuss why the double helix gained immediate universal acceptance as the correct structure for DNA.
15. If the nucleotide sequence of one polynucleotide of a DNA double helix is 5′-ATAGCAATGCAA-3′, what will be the sequence of the complementary polynucleotide?
16. Thirty per cent of the nucleotides in DNA from the locust are

As. What are the percentage values for: (a) T, (b) G + C, (c) G, (d) C?

17. DNA from the fungus *Neurospora crassa* has an AT content of 46%. What is the GC content?

4 Genes and biological information

Genes are segments of DNA • Gene organization – clusters – pseudogenes – discontinuous genes • Gene expression – the Central Dogma – transcription – translation – the importance of proteins

We have seen that genes are made of DNA, and that DNA is a double helix of two intertwined polynucleotides. Now we can look more closely at exactly what a gene is and how it carries biological information. In this chapter we will also consider how the information contained in a gene is read and made use of by the cell. These are topics that will be expanded in the following chapters when the individual aspects of gene expression are described in detail.

4.1 Genes are segments of DNA

A gene is simply a segment of a DNA molecule (Fig. 4.1(a)). This segment may be any length from about 75 nucleotides to over 2300 kb.

(a) The nucleotide sequence is the crucial feature of the gene. The biological information carried by a gene is contained in its nucleotide sequence. This information is in essence a set of instructions for the synthesis of an RNA molecule that may subsequently direct the synthesis of an enzyme or other protein molecule, or may itself be functional in the cell. This process is called **gene expression** and we will return to it later in the chapter. At this stage it is sufficient to appreciate that the nucleotide sequence of the gene is the crucial factor as far as the biological information is concerned.

Genes are discrete segments of DNA molecules, separated from one another by **intergenic DNA**. As we shall see in Part Two of this book, genes are arranged in different ways in different types of organism. In viruses, for instance, genes are closely packed with very little intergenic DNA between them (indeed, some genes may overlap). On the other hand, in higher organisms the genes are

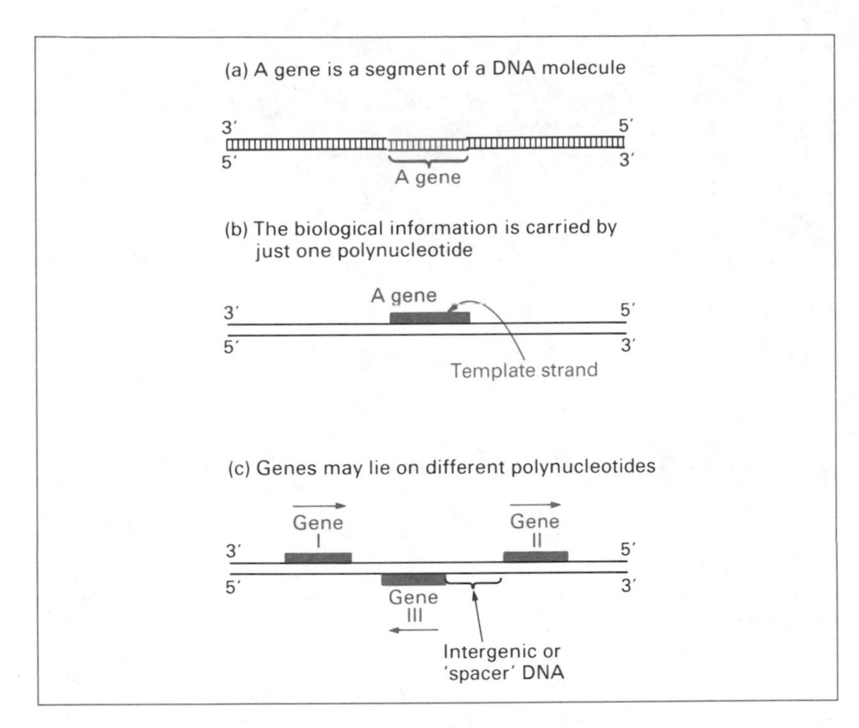

Figure 4.1 Genes are segments of DNA. In (c) the arrows indicate the direction in which the biological information is read during gene expression.

very spread out and are often separated by very long intergenic regions. In humans the genes make up less than 30% of the total DNA in the cell (Section 16.1).

The biological information of a gene is carried by just one of the two polynucleotides of the double helix (Fig. 4.1(b)). This polynucleotide is called the **template strand** because it acts as a template for synthesis of a complementary RNA molecule during the first stage of gene expression. The biological information carried by the template strand is read in the 3′→5′ direction. Two genes on the same DNA molecule do not necessarily have the same polynucleotide as their template strand (Fig. 4.1(c)).

It should be clear to the reader that the information-carrying capability of genes is almost unlimited. Remember that a gene 150 bp in length, which is in fact quite short, could have any one of 4^{150} different nucleotide sequences. In practice, not all of these 'genes' would be meaningful in biological terms because there are certain rules that place limitations on the number of sequences that make sense. Nevertheless, a large proportion of the 4^{150} sequences would contain biological information. There is no difficulty in rationalizing the known nature of the gene with the vastness of biological variability.

Other names for the template strand

When you read scientific papers you may come across the terms 'sense', 'antisense', 'coding' and 'non-coding' to refer to the two polynucleotides in a gene. These terms are not very helpful and can easily lead to confusion.

Most biologists think of the sense strand as being the template strand, with 'antisense' referring to the complementary, non-template strand. However, at least one important textbook has it the other way round. The use of 'coding' and 'non-coding' is even more confusing. You may think that the template strand would logically be the 'coding' strand, with the complement being the 'non-coding' strand. Unfortunately most (not all) textbooks that use this terminology call the template strand the 'non-coding' one.

This is a good example of the way in which jargon confuses science and scientists. Always try to use real words when talking or writing about genetics (try finding 'antisense' in the dictionary) and make sure that the terms you use are not ambiguous.

Chromosomes and genes

This table gives a very rough estimate of the numbers of genes carried by individual DNA molecules in different organisms

Organism	*Number of chromosomes*	*Approximate number of genes*	*Average number of genes per chromosome*
Escherichia coli	1	2 800	2800
Yeast	16	8 750	550
Man	23	50 000	2200

4.2 The organization of genes on DNA molecules

In the cells of a higher organism such as man all the genes are carried by a small number of chromosomes, each of which contains a single DNA molecule. Clearly each of these DNA molecules must carry several hundred, if not thousands, of genes. This holds for lower organisms too. Although a bacterium is much simpler than a human being, and therefore has fewer genes, with most bacterial species there is only one 'chromosome', so again a large number of genes are present on a single DNA molecule. How are these genes organized?

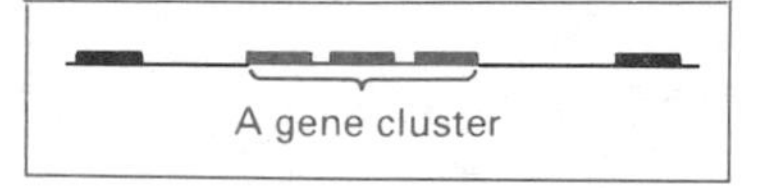

Figure 4.2 Genes may occur in clusters. As is the convention when drawing genes, only the template strand of the double helix is shown.

4.2.1 Genes may occur in clusters

The majority of genes are spaced out more or less randomly along the length of a DNA molecule. In some cases, however, they are grouped into distinct clusters (Fig. 4.2). Sometimes the individual genes in a cluster are unrelated and there is no apparent reason or advantage to having them organized in this way. More frequently clusters are made up of genes that contain related units of biological information. Two examples are the **operon** and the **multigene family**.

(a) Operons. Operons are fairly common features of the organization of genes in bacteria. An operon is a cluster of genes coding for a series of enzymes that work in concert in an integrated biochemical pathway (Fig. 4.3). The first operon to be discovered was the lactose or *lac* operon of *E. coli*, a cluster of three genes each coding for one of the three enzymes involved in conversion of the disaccharide lactose into its monosaccharide units glucose and galactose. These enzymes will not be required all the time, only when lactose is available to the bacterium. But when needed they will be needed together: the three units of biological information carried by the *lac* genes must be read at the same time. As we shall

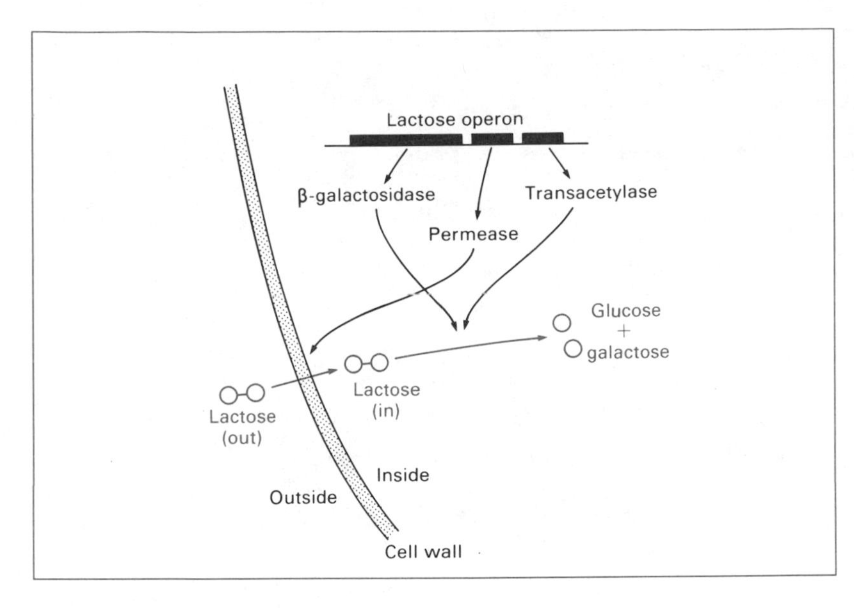

Figure 4.3 The lactose operon of *E. coli*. The three genes specify three enzymes involved in utilization of lactose as a carbon and energy source. The permease transports lactose into the cell, and β-galactosidase and transacetylase split lactose into glucose and galactose.

see in Chapter 10, a sophisticated system enabling the *lac* genes to be expressed together and only when needed has evolved round the fact that the genes are clustered.

(b) Multigene families. Operons have no direct counterparts in organisms other than bacteria. In contrast, multigene families are found in many organisms. A multigene family is also a cluster of related genes, but in multigene families the individual genes have identical or similar nucleotide sequences, and therefore contain identical or similar pieces of biological information. Genes with similar nucleotide sequences are referred to as **homologous genes**, and usually have a figure, such as 90%, attached to them to indicate the degree of sequence similarity between members of the family.

There are two types of multigene family:

1. ***Simple multigene families***, in which all of the genes are exactly or virtually the same. For example, most higher organisms have multiple copies of the gene for the 5S rRNA (a component of ribosomes – Section 6.1), probably because large quantities of 5S rRNA molecules must be synthesized at certain times, creating a demand that just one or a few genes would be unable to satisfy. Rather than being spread randomly throughout the chromosomes, all the 5S rRNA genes are clustered into a single multigene family (Fig. 4.4(a)), which in humans is made up of about 2000 copies of the gene.
2. ***Complex multigene families*** are made up of similar but non-identical genes. A good example is provided by the gene family

Figure 4.4 Multigene families.

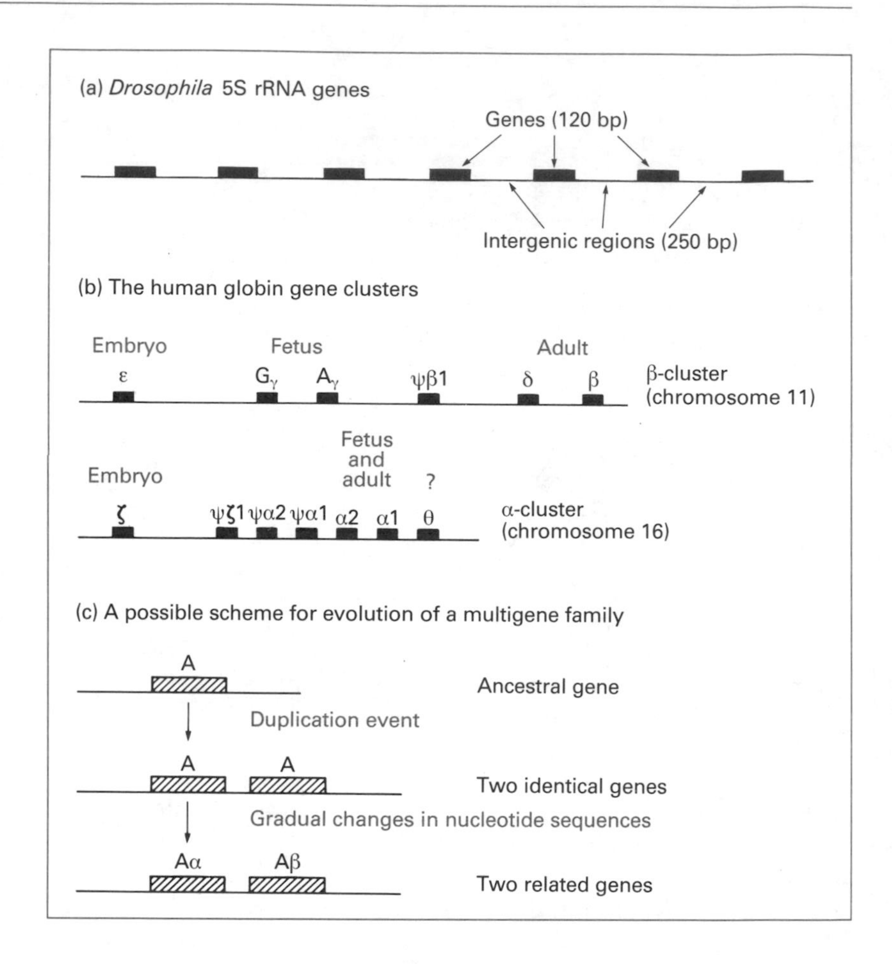

for the globin polypeptides of vertebrates. In humans four globin polypeptides, 2α and 2β, combine with a haem cofactor to make a single molecule of haemoglobin, the protein responsible for transporting oxygen in the bloodstream. Each type of globin polypeptide, α and β, exists as a family of related molecules differing from one another at just a few amino acid positions within the polypeptide chain. In humans there are in fact two α-like globins, α and ζ, and five β-like globins, β itself, δ, ${}^{A}\gamma$, ${}^{G}\gamma$ and ε. The types of globin found in the bloodstream depend on the stage of development of the organism, with ζ and ε, for instance, found only in the early embryo. The genes for the human globin proteins were among the first to be studied by recombinant DNA technology. Tom Maniatis and his colleagues at the California Institute of Technology discovered during the late 1970s that the human globin genes are clustered into an α multigene family on chromosome 16 and a β multigene family on chromosome 11. Subsequent analysis has allowed the

Recombinant DNA technology

Much of what we know about gene structure has been gained by the application of recombinant DNA techniques. These will be described in Chapters 20 and 21, but at this point the reader should note the main types of experimental manipulation and the sort of information that can be obtained:

1. **DNA molecules can be cleaved into defined segments.** DNA molecules can be cut into fragments by enzymes called **restriction endonucleases**, which cleave DNA only at certain nucleotide sequences. A single gene, or group of genes, can therefore be obtained as a small fragment of DNA.
2. **DNA fragments can be purified.** DNA fragments can be separated by gel electrophoresis (p. 19) and individual ones recovered in pure form from the gel.
3. **Cloning provides large quantities of a chosen gene.** DNA fragments can be inserted into a **cloning vector** and large amounts obtained by gene cloning. Usually the entire DNA content of an organism, rather than a purified gene, is used as the starting material for a cloning experiment. If this is the case then many different clones, each containing a different fragment of DNA, will be obtained. This **clone library** can then be screened to identify the clone containing the gene of interest.
4. **PCR can also provide large quantities of a chosen gene.** As an alternative to cloning, a defined segment of DNA can be amplified into multiple, identical copies by the **polymerase chain reaction** (**PCR**). As with cloning, the experiment can be designed so that a selected gene is obtained, even if the starting material is the entire DNA content of a cell.
5. **The nucleotide sequence of a gene can be determined,** by either of two experimental methods that will be described in Chapter 21.

precise positions of the individual genes in the clusters to be mapped (Fig. 4.4(b)).

We now have information on a whole range of multigene families in a variety of organisms, including some families with which the individual genes are not in fact clustered but scattered at different positions on a single or on more than one chromosome. This indicates that clustering of members of a multigene family is not an aid to coordinated expression of the genes (as is the case with operons) but probably is merely a result of the evolutionary processes that gave rise to the family. We can speculate that the homologues in each family arose from an ancestral gene by a series of duplications, followed by events that have slightly altered the sequences of the resulting genes (Fig. 4.4(c)). Large-scale rearrangements within and between DNA molecules (which are known to occur but are poorly understood) could subsequently break the cluster and scatter the members of the family.

4.2.2 Some genes have lost their function

The globin multigene families illustrate a second feature of genes and gene organization: the **pseudogene**. As well as the genes coding for the α-like and β-like globins, the human gene clusters also include additional genes, labelled $\psi\zeta_1$, $\psi\alpha_1$, $\psi\alpha_2$ and $\gamma\beta_1$ (Fig. 4.4(b)). These 'genes' are very similar to the other members of their family but we can tell from their nucleotide sequences that the biological information they contain has become scrambled and is

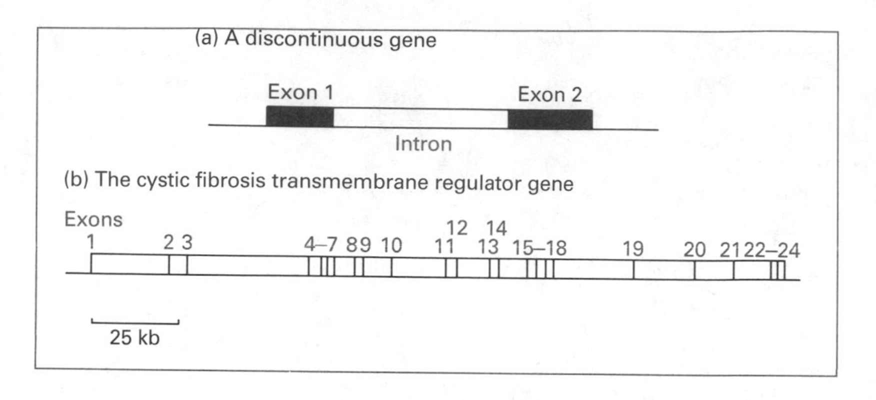

Figure 4.5 (a) The structure of a discontinuous gene. (b) The human gene for the cystic fibrosis transmembrane regulator. The vertical lines indicate the positions of the exons.

Table 4.1 Introns in a selection of human genes

Gene	*Length (kb)*	*Number of introns*	*Intron component of the gene (% length)*
Insulin	1.4	2	67
Serum albumin	18	13	88
Apolipoprotein-B	43	17	67
Phenylalanine hydroxylase	90	25	97
Blood clotting factor VIII	186	23	97
Cystic fibrosis transmembrane regulator	250	26	98
Dystrophin	2300	>100	99

Data from T. Strachan (1992) *The Human Genome*, BIOS, Oxford.

no longer readable. As we will see in later chapters, apparently minor changes in a gene's nucleotide sequence can completely destroy the meaning of the biological information, leaving a pseudogene that can no longer perform its function. It appears that once the biological information is lost a pseudogene will undergo sequence alterations relatively rapidly until it is no longer recognizable as a gene relic. More and more pseudogenes are being found as the DNA molecules of lower and higher organisms are studied in increasing detail with the sophisticated techniques now available to the molecular biologist.

4.2.3 Some genes are discontinuous

One of the most startling findings of the past 20 years was made in 1977 when several investigators discovered that the biological information carried by some genes is split into distinct units separated by intervening segments of DNA (Fig. 4.5(a)). The sections containing biological information are called **exons** and the

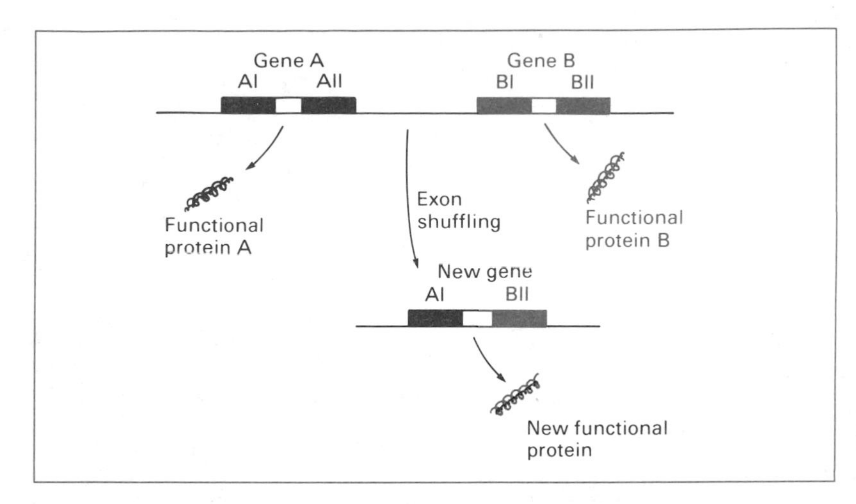

Figure 4.6 The hypothesis of 'exon shuffling' suggests that new functional genes could be formed by rearranging exons of existing genes.

intervening sequences are referred to as **introns**. **Discontinuous genes** (also called split or mosaic genes) are now known to be common in higher organisms and many types of virus, but until recently were thought to be absent in bacteria. This is probably the case with eubacteria (the vast majority of known species) but the latest research has revealed introns in certain genes of the ancient organisms called archaebacteria (Section 14.2.2).

In higher organisms a gene may contain no introns or may have as many as a hundred (Table 4.1). Often the introns are much longer than the exons, a point that is illustrated by the human gene for the cystic fibrosis transmembrane regulator (Fig. 4.5(b)). This gene, which can cause cystic fibrosis when it does not function correctly, is 250 kb from beginning to end and is split into 24 exons and 23 introns. The average length of the exons is 227 bp, so all the exons added together make up only 2.4% of the gene. These exons are scattered throughout the entire length of the gene, separated by introns that range in size from 2 to 35 kb.

Why some genes should be split into exons and introns has been the subject of intense speculation. An interesting, although not necessarily correct, theory has been put forward by Walter Gilbert and associates, who suggest that each exon of a discontinuous gene contains a different subcomponent of the biological information carried by the gene as a whole. Although these subcomponents are not sufficient on their own to code for the complete protein, the information they contain is meaningful in that it specifies a recognizable portion of the protein's function, for example enabling the protein to bind a particular substrate or attach to a specific site in the cell. Gilbert's hypothesis is that, during evolution, exons from different discontinuous genes can be 'shuffled', creating new combinations of biological information (Fig. 4.6). This kind of process would be more likely to produce

Walter Gilbert b. 21 March 1932, Boston

Walter Gilbert obtained his PhD in mathematics in 1957 from Cambridge University (USA) and was a theoretical physicist at Harvard University before joining James Watson's group in 1960. His first important work concerned the identification of mRNA (Chapter 7) in 1961. This led to work on the mechanism of protein synthesis. Later he studied repressor proteins (Chapter 10) and did a technically demanding series of experiments that led to isolation of the lactose repressor in 1966. His most lasting contribution perhaps has been the development in the mid-1970s of one of the two rapid DNA sequencing techniques (Section 21.2) that have provided a major stimulus to the study of gene structure. It was for this work that he shared the 1980 Nobel Prize for chemistry with Frederick Sanger and Paul Berg. During the 1980s Gilbert became one of the first biotechnology entrepreneurs, being chairman of Biogen from 1981 to 1984, before returning to Harvard in 1985.

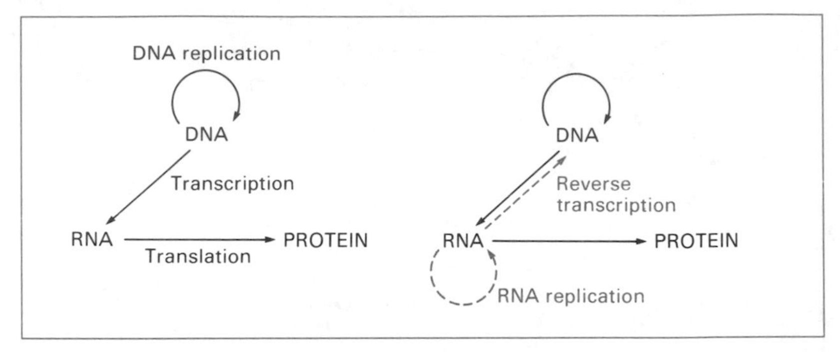

Figure 4.7 Crick's Central Dogma. The diagram on the left shows the transfers of biological information that occur in all cells. On the right additional transfers are included: these occur only in a few organisms, mainly viruses. The underlying concept is that 'once "information" has passed to protein it cannot get out again' (Crick, 1958).

new functional proteins than would an entirely random rearrangement of existing genes.

4.3 Gene expression

The next six chapters of this book will be concerned with different aspects of the same topic: gene expression, the process by which the biological information contained in the gene is made available to the cell. Although gene expression is complex it is nevertheless relatively straightforward in outline. At this stage we will look briefly at the overall process and what it achieves, thereby providing a framework which can be built on when details of the individual events are described in the subsequent chapters.

4.3.1 The Central Dogma

The process that we now call gene expression was first put forward by Francis Crick in 1958 in a lecture called 'On Protein Synthesis' presented to the Society for Experimental Biology. Crick postulated that the biological information contained in the DNA of the gene is transferred first to RNA and then to protein (Fig. 4.7). We now accept that this is, in its simplest form, what happens. Crick also stated that the information flow is unidirectional, that proteins cannot themselves direct synthesis of RNA and that RNA cannot direct synthesis of DNA. This part of the Central Dogma was shaken in 1970 when Howard Temin and David Baltimore independently discovered that certain viruses do transfer biological information from RNA to DNA (Section 13.2.2). However, the **Central Dogma** still remains today as one of the underlying concepts of molecular genetics.

Evolution of the first genes

The discovery of exons has led to many speculations concerning the evolution of genes before the origin of living cells. One theory is that the first 'genes' to form in the prebiotic soup were not continuous sequences of nucleotides, but instead were made up of a number of short mini-genes, each containing a small fragment of biological information. Initially these mini-genes would have been separated by vast regions of meaningless nucleotide sequence, or possibly were even on different molecules, but eventually they became associated together to form a discontinuous gene, with the mini-genes becoming the exons. These earliest genetic molecules may have been DNA, but it now seems more likely that they were RNA, with DNA copies being formed at a subsequent stage in evolution. If genes were formed in this way then bacteria must have originally possessed introns, losing them at a later stage, possibly to streamline their DNA molecules enabling more rapid DNA replication and a shorter generation time between one cell division and the next.

4.3.2 Transcription – the first stage of gene expression

All genes undergo the first stage of gene expression which is called **transcription** (Fig. 4.8). During transcription the template strand of the gene directs synthesis of an RNA molecule, the sequence of this RNA **transcript** being determined by complementary base pairing.

4.3.3 Translation – the second stage of gene expression

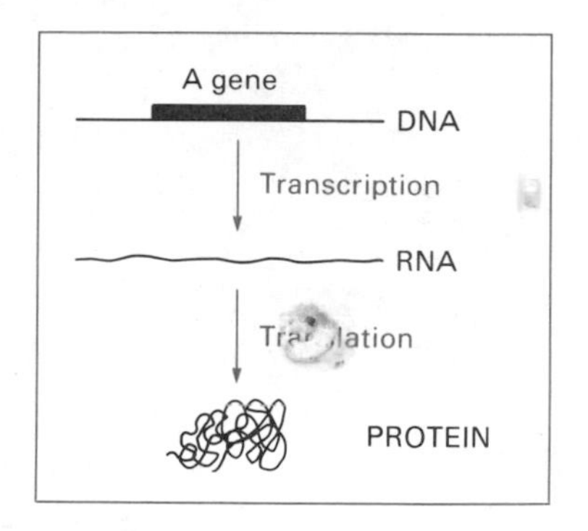

Figure 4.8 The two stages of gene expression.

For some genes the RNA transcript is itself the end-product of gene expression. For others the transcript undergoes the second stage of gene expression, called **translation**. During translation the RNA molecule (called a **messenger** or **mRNA**) directs synthesis of a polypeptide, the amino acid sequence of which is determined by the nucleotide sequence of the mRNA (which is of course itself derived from the nucleotide sequence of the gene). Each triplet of adjacent ribonucleotides specifies a single amino acid of the polypeptide, the identity of the amino acid corresponding to each triplet being set by the **genetic code**.

4.3.4 Protein synthesis is the key to expression of biological information

For all genes the end-point of gene expression is synthesis of either an RNA molecule or a polypeptide. How can this simple process bring about the utilization of the biological information contained in genes and thereby enable construction of a living organism?

The answer lies in the functional flexibility of protein molecules. This is because polypeptides of different amino acid sequences can have quite different chemical properties and can therefore play quite different roles in the cell. Consider a few of the roles played by different proteins:

1. ***Structural proteins*** form parts of the framework of organisms. Examples are collagen which is associated with tendons, bone and cartilage in vertebrates, sclerotin in the exoskeleton of insects, and virus coat proteins.
2. ***Contractile proteins*** enable organisms to move. Examples are actin and myosin in muscles, and dynein in cilia and flagella.
3. ***Enzymes*** catalyse the multitude of biochemical reactions that bring about the release and storage of energy (catabolism) and the synthesis of new compounds (anabolism). Examples are: hexokinase, the enzyme that catalyses the first step in glycolysis, the pathway by which energy is released from glucose; RNA polymerase, which is responsible for transcription of DNA into RNA during gene expression; tryptophan synthetase, which catalyses the last step in the lengthy biosynthetic pathway that converts phosphoenolpyruvate, itself derived from glucose, into the amino acid tryptophan.
4. ***Transport proteins*** carry important molecules around the body.

Examples are haemoglobin, which carries oxygen in the bloodstream of vertebrates; haemocyanin, which performs the same function in some invertebrates; and serum albumin, which transports fatty acids in the blood.

5. ***Regulatory proteins*** control and coordinate biochemical reactions in the cell and in the organism as a whole. Examples are catabolite activator protein, which regulates expression of the genes involved in sugar metabolism in *E. coli*, and hormones such as insulin which controls glucose metabolism in vertebrates.
6. ***Protective proteins*** have evolved to greatest sophistication in vertebrates and protect against infectious agents and injury. Examples are immunoglobulins and other antibodies which form complexes with foreign proteins, and thrombin and other components of the blood clotting mechanism.
7. ***Storage proteins*** store compounds and molecules for future use by the organism. Examples are ovalbumin, which stores amino acids in egg white, and ferritin which stores iron in the liver.

The development and functioning of a living organism can be looked on as nothing more than the coordinated activity of a vast range of different protein molecules. The biological information contained in the genes, the blueprint for life itself, is simply the instructions for synthesizing these proteins at the correct time and in the correct place. It is perhaps worth pointing out that at the present time we probably understand something less than 1% of how all this works.

Reading

T. Strachan (1992) *The Human Genome*, BIOS, Oxford – Chapters 1 and 2 are readable descriptions of gene structure and organization in human genes.

E. F. Fritsch, R. M. Lawn and T. Maniatis (1980) Molecular cloning and characterization of the human α-like globin gene cluster. *Cell*, **19**, 959–72 – details of the initial analysis of this cluster.

P. F. R. Little (1982) Globin pseudogenes. *Cell*, **28**, 683–4 – a useful mini-review on pseudogenes and their origin.

N. Proudfoot (1986) Globin gene monkey business. *Nature*, **321**, 730–1 – the discovery of a new pseudogene.

P. Chambon (1981) Split genes. *Scientific American*, **244**(5), 48–59.

F. H. C. Crick (1979) Split genes and RNA splicing. *Science*, **204**, 264–71 – Crick's first thoughts on introns and the problems they pose.

W. Gilbert (1978) Why genes in pieces? *Nature*, **271**, 501, and W. Gilbert (1985) Genes in pieces revisited. *Science*, **228**, 823–4– Gilbert's initial and later ideas on exon shuffling.

F. H. C. Crick (1958) On protein synthesis. *Symposium of the Society for Experimental Biology*, **12**, 138–63 – the first clear description of the Central Dogma.

H. Temin (1972) RNA-directed DNA synthesis. *Scientific American*, **226**(1), 24–33 – a description of the transfer of biological information from RNA to DNA.

F. H. C. Crick (1970) Central Dogma of molecular biology. *Nature*, **227**, 561–3 – an updated version to take account of the RNA to DNA transfer.

Problems

1. Define the following terms:

gene expression	exon
intergenic DNA	intron
template strand	Central Dogma
operon	transcription
multigene family	transcript
homologous genes	translation
pseudogene	messenger RNA
discontinuous gene	genetic code

2. Distinguish between the roles of the template and non-template strands of a gene.

3. Explain why the information-carrying capability of genes is almost unlimited.

4. Describe the types of relationship that are often displayed by genes that occur in clusters.

5. Provide examples of two types of multigene family.

6. What aspects of gene structure and organization are displayed by the human globin gene families?

7. Explain what is meant by the Central Dogma. How has the original concept of the Central Dogma been changed to account for more recent discoveries?

8. *Discontinuous genes are common in higher organisms but virtually absent in bacteria. Discuss the possible reasons for this.

9. *Introns are now known to be present in the genes of the very ancient organisms called archaebacteria. What are the implications of this discovery for theories on the evolution of genetic systems?

10. *What would be the implications for theories of evolution if it were discovered that the Central Dogma is fundamentally incorrect and that information can flow from proteins to RNA and thence to DNA?

11. *Support or challenge the statement 'The development and functioning of a living organism can be looked on as nothing more than the coordinated activity of a vast range of different protein molecules'.

12. How many possible nucleotide sequences are there for a gene 95 bp in length?

13. Here are the nucleotide sequences of the template strands of two imaginary (and much too short) genes:

3′-TACCCATGATTTCCCGCGATATATATGCCTGCAATC-5′
3′-TACCCTTGGTTTCGCGCGATGTATATACCAGGAATC-5′

What is the percentage sequence similarity of these two genes?

14. Write down the nucleotide sequences of the messenger RNA molecules that would be transcribed from the two genes described in Question 13.

Transcription

5

Eukaryotes and prokaryotes • Nucleotide sequences • RNA synthesis – RNA polymerases • Transcription in *E. coli* – initiation – elongation – termination • Transcription in eukaryotes

During the first stage of gene expression the template strand of the gene directs the synthesis of a complementary strand of RNA. This process is called transcription and the RNA molecule that is synthesized is called the transcript. In this chapter we look at the chemical reaction involved in RNA synthesis, at the enzyme that catalyses the reaction, and at the events involved in the overall process of transcription. First, however, two important general points need to be made concerning the distinctions between higher and lower organisms, and the notation used to describe the nucleotide sequence of a gene.

5.1 Eukaryotes and prokaryotes

So far in this book the terms 'higher' and 'lower' organism have been used to distinguish between vertebrates on the one hand and bacteria on the other. The more correct terminology used from here on is **eukaryote** and **prokaryote**.

Eukaryotes and prokaryotes are distinguished by their fundamentally different cellular organizations (Fig. 5.1). The typical eukaryotic cell is usually larger and more complex than a prokaryotic cell, with a membrane-bound nucleus containing the chromosomes, and with other distinctive membranous organelles such as mitochondria, vesicles and Golgi bodies. Prokaryotes, in contrast, lack an extensive cellular architecture: membranous organelles are absent and the genetic material is not enclosed in a distinct structure.

Eukaryotes can be unicellular or multicellular and comprise all the macroscopic forms of life such as plants, animals and fungi. Prokaryotes are usually unicellular and include bacteria and the blue-green algae, the latter now more properly called cyanobacteria. Eukaryotes and prokaryotes probably diverged during the earliest stages of cellular evolution.

The distinction between eukaryotes and prokaryotes is important because although the basic features of genes were laid down before the two diverged, there are nevertheless important dif-

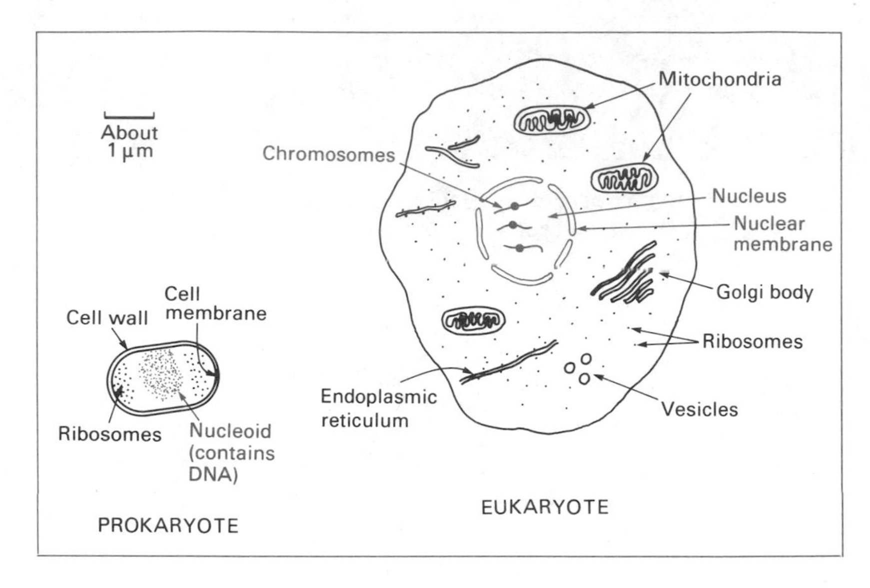

Figure 5.1 A typical prokaryotic and a typical eukaryotic cell, showing the differences in structure. The prokaryote is a rod-shaped bacterium, similar to *E. coli*; the eukaryote is a relatively small, unspecialized animal cell.

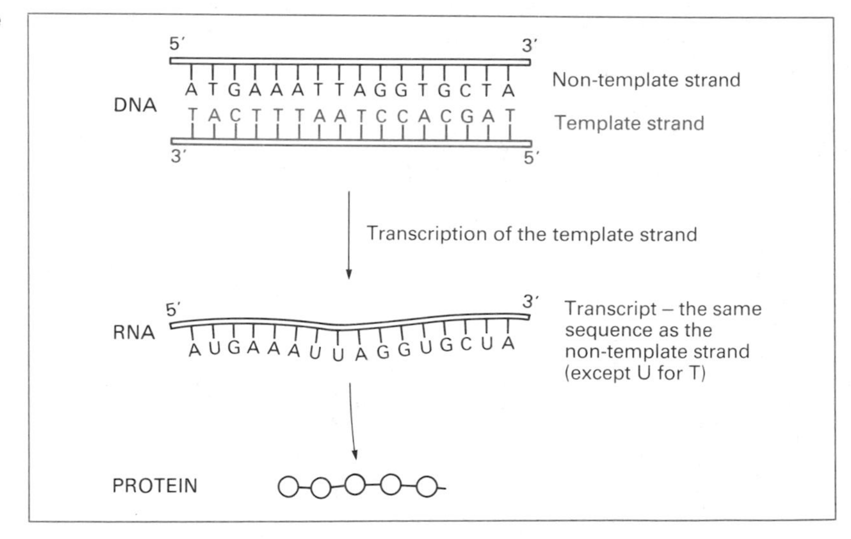

Figure 5.2 When writing a nucleotide sequence, the non-template strand is given as this has the same sequence as the RNA transcript.

ferences between them in gene structure and expression. We have already noted that some eukaryotic genes are discontinuous whereas prokaryotic ones are not. For reasons to do mainly with ease of handling in experiments, most of the pioneering work in molecular genetics has been carried out with a prokaryote, *Escherichia coli*. It is much more difficult to do experiments with mammalian cells and we still know much more about gene expression in *E. coli* than in eukaryotes. Things are changing though, especially as microbial eukaryotes (such as the yeast *Saccharomyces cerevisiae*) have been studied extensively over the past 10 years. Gene expression in yeast is not exactly the same as in man, but it is sufficiently similar for us to be able to use results obtained with

Growing bacteria in culture

There are several reasons why bacteria are popular for experimental work, not only in genetics and molecular biology, but also in biochemistry and other areas of biology. The ease with which bacteria can be grown in culture is an important factor. Bacteria can be grown in a variety of culture media, most of which are easy to prepare. For instance, LB medium, commonly used for the growth of *E. coli*, is made simply by dissolving 10 g of tryptone (supplying amino acids), 5 g of yeast extract (a dried preparation of yeast cells) and 10 g of sodium chloride in one litre of water. The medium is sterilized by heat treatment at a high pressure (autoclaving), to kill any bacteria already present, and then inoculated with a small sample of the organism under study. After a short lag period the bacteria start to grow and divide, with each cell dividing as frequently as once every 20 min if the growth conditions are absolutely right. Division will continue for several hours until a high cell density (2 to 3 × 10^9 ml^{-1}) is reached. A large number of bacterial cells can therefore be obtained fairly quickly. These cells can be used as a source of DNA for study by the recombinant DNA techniques described on p. 49 and in Chapters 20 and 21.

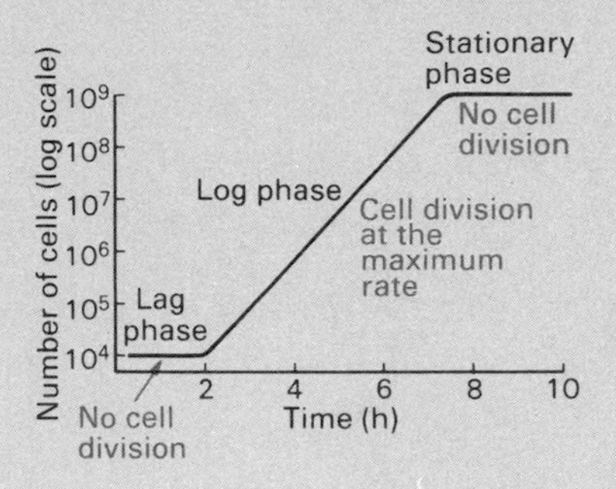

Alternatively, the cells can be broken open and enzymes and other molecules from within purified and analysed. Many of the molecules involved in gene expression (for example, the RNA polymerase enzyme) have been studied in this way. Information can also be obtained from an examination of culture growth. If the culture conditions are kept exactly the same, then the growth of a second culture will precisely parallel that of the first. The effect of changes in the growth conditions (for example, addition of a novel nutrient, or infection with a bacteriophage) on cell metabolism or gene activity can therefore be examined.

In contrast to bacteria, most eukaryotes are much more difficult to grow under controlled conditions. Techniques for growing plant cells in culture media were developed during the 1960s but this is still not possible for all species. Animal cells have been grown in culture for several years but the growth rates and cell yields are generally low. However, microbial eukaryotes such as yeasts (for example, *Saccharomyces cerevisiae*) and fungi (such as *Aspergillus nidulans* and *Neurospora crassa*) are exceptions and can be grown in culture almost as easily as bacteria. Studies with these organisms over the past 10 years or so have opened up our understanding of gene expression in eukaryotes.

yeast to plan informative experiments with animal cells. This is the best way to make progress in biology: start by studying a 'simple' organism and then gradually work up to the more complex ones.

5.2 Nucleotide sequences

From this point on increasing use will be made of nucleotide sequences to illustrate aspects of gene structure and expression. Usually the sequence presented will be that of the non-template strand of the gene in the 5′→3′ direction. This may at first seem illogical as we know that the biological information is carried by the template strand and not by its complement. However, as will become clear in later chapters, it is the sequence of the RNA transcript (which is the same as that of the non-template strand) that the molecular biologist is really interested in, because it is from this sequence that the important conversion to amino acids is made by way of the genetic code (Fig. 5.2). For this reason, when detailing a gene sequence the convention is to describe the non-template strand.

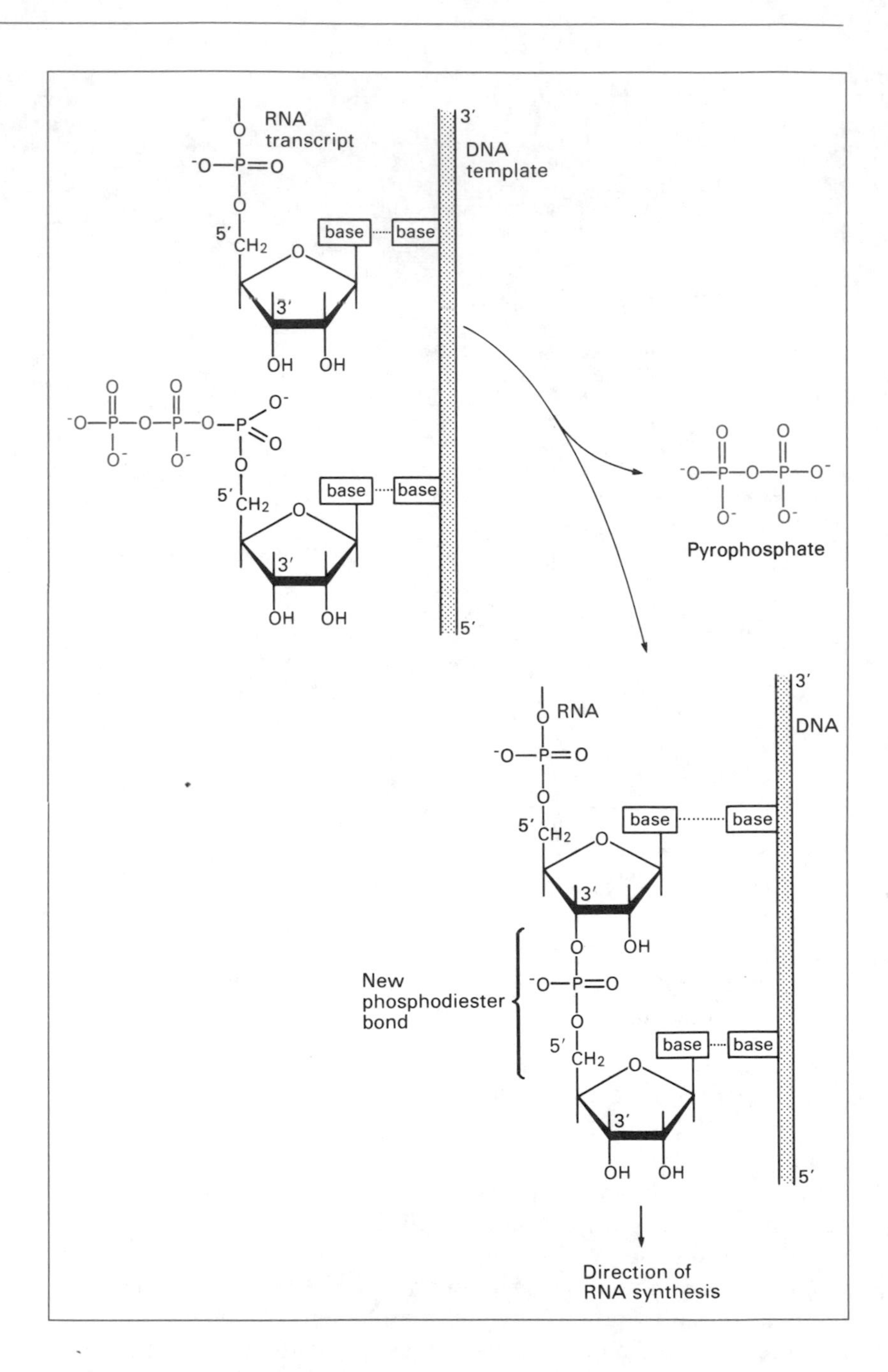

Figure 5.3 Transcription – RNA synthesis on a DNA template.

5.3 RNA synthesis

The underlying chemical reaction in transcription is synthesis of an RNA molecule. This occurs by polymerization of ribonucleotide

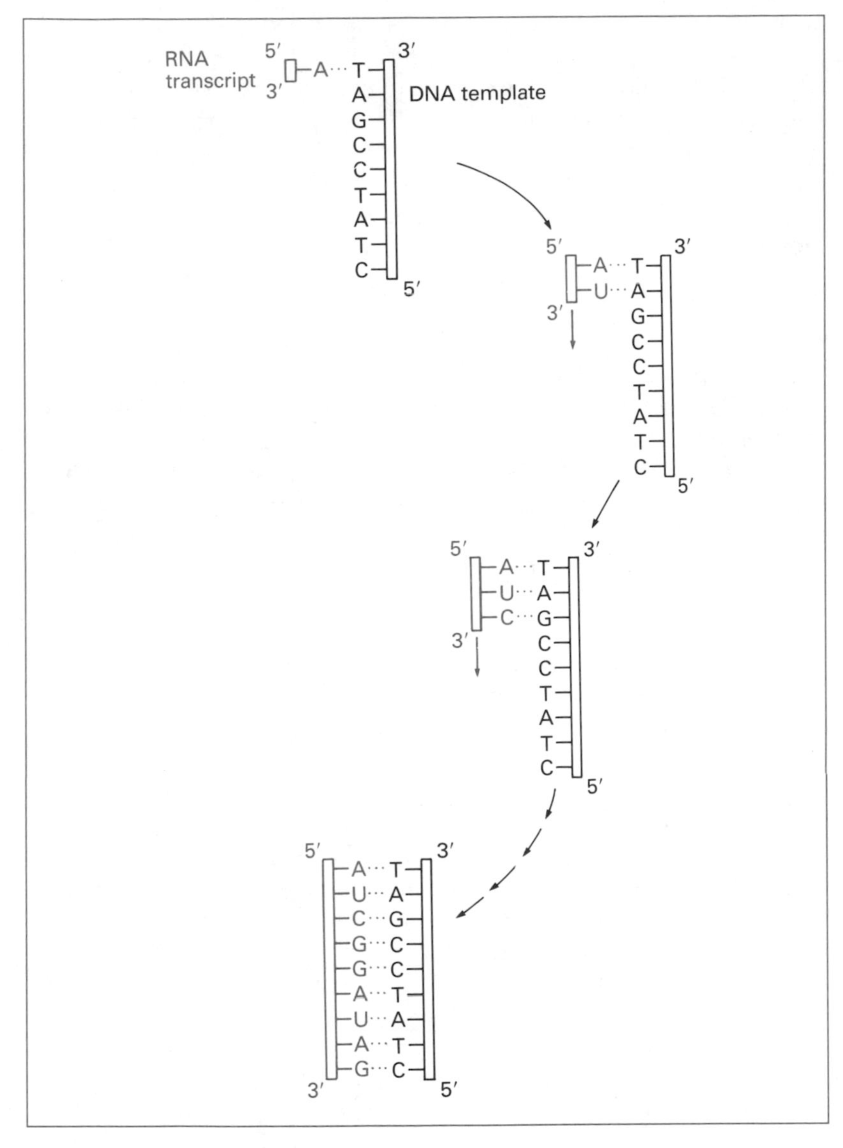

Figure 5.4 During transcription the RNA transcript is built up in a step-by-step fashion on the DNA template. The template is read in the 3′ → 5′ direction and RNA synthesis occurs in the 5′ → 3′ direction.

subunits and can be summaried as:

$$n(\text{NTP}) \rightarrow (\text{NTP})\text{—}(\text{NMP})_{n-1} + n-1(\text{PPi})$$

The reaction is shown in detail in Fig. 5.3.

During polymerization the 3′-OH group of one ribonucleotide reacts with the 5′-P group of the second to form a phosphodiester bond. This results in loss of a pyrophosphate molecule (PPi) for each bond formed. In transcription, the chemical reaction is modulated by the presence of the DNA template, which directs the order in which the individual ribonucleotides are polymerized into RNA (Fig. 5.4). The RNA transcript is built up in a step-by-step fashion, with new ribonucleotides being added to the free 3′ end of the existing polymer. The transcript is therefore synthesized in the 5′ → 3′ direction. Remember that in order to base pair, complementary polynucleotides must be antiparallel: this means that the template strand of the gene must be read in the 3′ → 5′ direction. The enzyme that catalyses the reaction is called DNA-dependent RNA polymerase (usually referred to simply as **RNA polymerase**).

5.3.1 RNA polymerases

The enzyme that catalyses RNA synthesis during transcription in *E. coli* was first discovered in 1958 and has since been subjected to detailed study. Like many enzymes involved in molecular genetic processes, an RNA polymerase has to perform several tasks, as will be seen when the events involved in transcription are described later in this chapter. It is not surprising, therefore, that the *E. coli* RNA polymerase is a large protein made up of several different polypeptide subunits.

(a) The RNA polymerase of *E. coli* comprises five subunits. Each *E. coli* cell contains about 7000 RNA polymerase molecules, between 2000 and 5000 of which may be actively involved in transcription at any one time. The structure of the enzyme is described as $\alpha_2\beta\beta'\sigma$, meaning that each molecule is made up of two α polypeptides plus one each of β, the related β', and σ (Fig. 5.5). This version of the enzyme is called the **holoenzyme** (molecular mass 480 000 daltons) and is distinct from a second form, the **core enzyme** (molecular mass 395 000 daltons), which lacks the σ subunit and is just $\alpha_2\beta\beta'$. The two versions of the enzyme have different roles during transcription, as will be described below.

(b) Eukaryotes possess more complex RNA polymerases. All *E. coli* genes are transcribed by the same type of RNA polymerase enzyme. In contrast, eukaryotes possess three different RNA polymerases (called RNA polymerase I, II and III), each of which

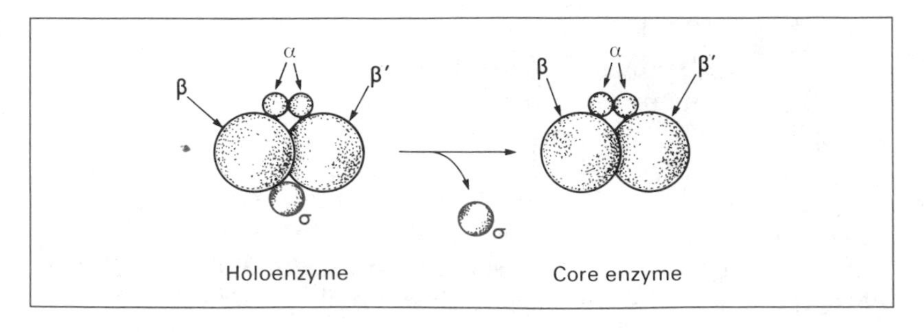

Figure 5.5 The relationship between the holoenzyme and core enzyme versions of *E. coli* RNA polymerase.

Roles of eukaryotic RNA polymerases

The three eukaryotic RNA polymerases have different functions in that each transcribes a different set of genes:

RNA polymerase	*Types of gene transcribed**
I	Ribosomal RNA genes
II	Genes for proteins; most U-RNA genes
III	Genes for small RNAs: transfer RNAs, 5S ribosomal RNA, U6-RNA, plus a few others

*See Chapters 6 and 7 for descriptions of these different types of gene.

transcribes a different set of genes. The eukaryotic enzymes are each larger than the *E. coli* version, with molecular masses in excess of 500 000 daltons, and have more complex structures. RNA polymerase II of yeast is made up of ten different polypeptides called RPB1 to RPB10. The three largest of these polypeptides (RPB1, RPB2 and RPB3) appear to be equivalent to the α, β and β' subunits of the *E. coli* enzyme. This has been deduced by comparing the amino acid sequences of the yeast and *E. coli* polypeptides and by examining the exact role of each subunit during transcription. The functions of the other RPB polypeptides are unknown, and in fact a few of them seem to be dispensable as transcription occurs perfectly well in their absence.

5.4 Transcription in *E. coli*

The actual process by which a gene is transcribed is conveniently divided into three phases: initiation, elongation and termination.

5.4.1 Initiation

The key point about the initiation of transcription is that an RNA polymerase enzyme must transcribe genes rather than random pieces of DNA. This means that the initial binding of an RNA polymerase enzyme to a DNA molecule must occur at a specific position, just in front ('**upstream**') of the gene to be transcribed. Specific attachment points are essential because in all organisms a large proportion of the total DNA is non-genic and efficient transcription of genes will not occur purely by chance.

(a) The transcription initiation site is signalled by the promoter. These attachment points are called **promoters**. A promoter is a short nucleotide sequence that is recognized by an RNA polymerase enzyme as a point at which to bind to DNA in order to

DNase protection experiments

A number of methods have been used to study the interaction between RNA polymerase and promoter sequences. One of the most important of these is called deoxyribonuclease (DNase) protection.

DNase is an enzyme that degrades DNA by cutting phosphodiester bonds. The result of DNase activity on pure DNA is therefore breakdown of the double helix into individual nucleotides.

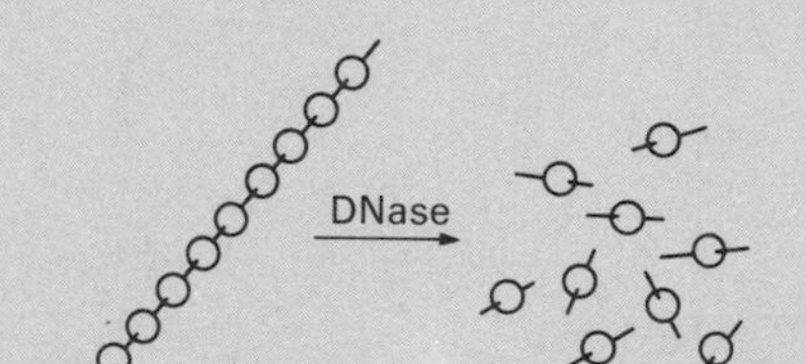

However, if a purified fragment of DNA contains a promoter, to which an RNA polymerase protein is attached, then not all of the phosphodiester bonds will be available to attack by DNase: some will be 'protected' by the bound enzyme. The result of DNase treatment will now be a few mononucleotides and an uncleaved DNA fragment. The size of this DNA fragment will indicate the extent of the interaction between RNA polymerase and DNA: its nucleotide sequence (which could be worked out by DNA sequencing) would show exactly which part of the molecule is protected by the enzyme.

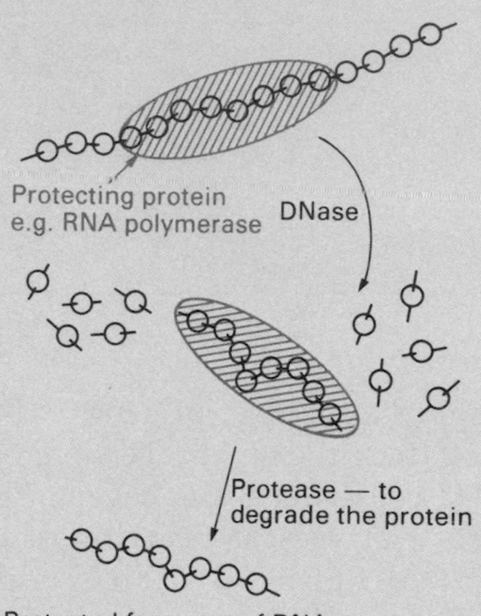

The results of these experiments show that *E. coli* RNA polymerase protects between 41 and 44 bp of DNA including the −10 and −35 boxes and some of the surrounding sequence.

begin transcription. Promoters occur just upstream of genes, and nowhere else.

Clearly all promoters must have the same or very similar sequences if each is to be recognized by the same enzyme. In *E. coli* the promoter sequence has been dissected and found to be made up of two distinct components, the −35 box and the −10 box (Fig. 5.6). The latter is also called the **Pribnow box**, after the scientist who first characterized it. The names refer to the location of the boxes on the DNA molecule relative to the position at which transcription starts. The sequences of the boxes are:

−35 box 5′-TTGACA-3′
−10 box 5′-TATAAT-3′

(Remember that these are the sequences found on the non-template polynucleotide.)

In fact the actual sequences of the components of the promoter vary from gene to gene, but all are related to and recognizable with the **consensus sequences** shown above. Whether there is any importance in the slightly different versions of the promoter sequences found upstream of different genes is not known. It has

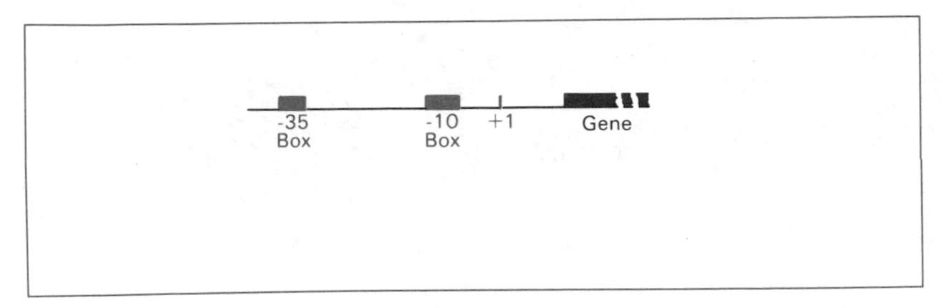

Figure 5.6 The positions of the promoter sequences upstream of a gene in *E. coli*. Transcription begins at the point marked as '+1'. Note that this point is a short distance upstream of the start of the gene itself.

Real promoter sequences

The sequences of a few of the promoters recognized by the *E. coli* RNA polymerase are as follows:

Gene	*−35 Sequence*	*−10 Sequence*
E. coli consensus	TTGACA	TATAAT
lac operon	TTTACA	TATGTT
trp operon	TTGACA	TTAACT
tRNA genes	TTTACA	TATGAT

From these few examples we can see that the actual promoter sequence can differ from the consensus at two or possibly more positions. Nevertheless, it has been shown experimentally that not all variations on the consensus sequence are allowed and some alterations result in a 'promoter' that is no longer recognized by the RNA polymerase enzyme. The reasons for this are not fully understood but are being studied by looking closely at the physical interaction between the holoenzyme and the DNA molecule in the region of the promoter.

been suggested that the actual sequence may affect the efficiency with which RNA polymerase locates and binds to the promoter, thereby influencing the extent to which the gene is expressed. These proposals are complicated by the observation that certain minor alterations in the promoter sequence can completely block recognition by RNA polymerase.

(b) The σ subunit of RNA polymerase recognizes the promoter. In *E. coli* the holoenzyme version of RNA polymerase ($\alpha_2\beta\beta'\sigma$) is required for initiation; in fact it is the σ subunit of the enzyme that is responsible for promoter site recognition. If the σ subunit is absent the core enzyme will still be able to bind to DNA, but in a more random fashion at a variety of sites and not specifically to promoters.

The initial structure formed between RNA polymerase and DNA

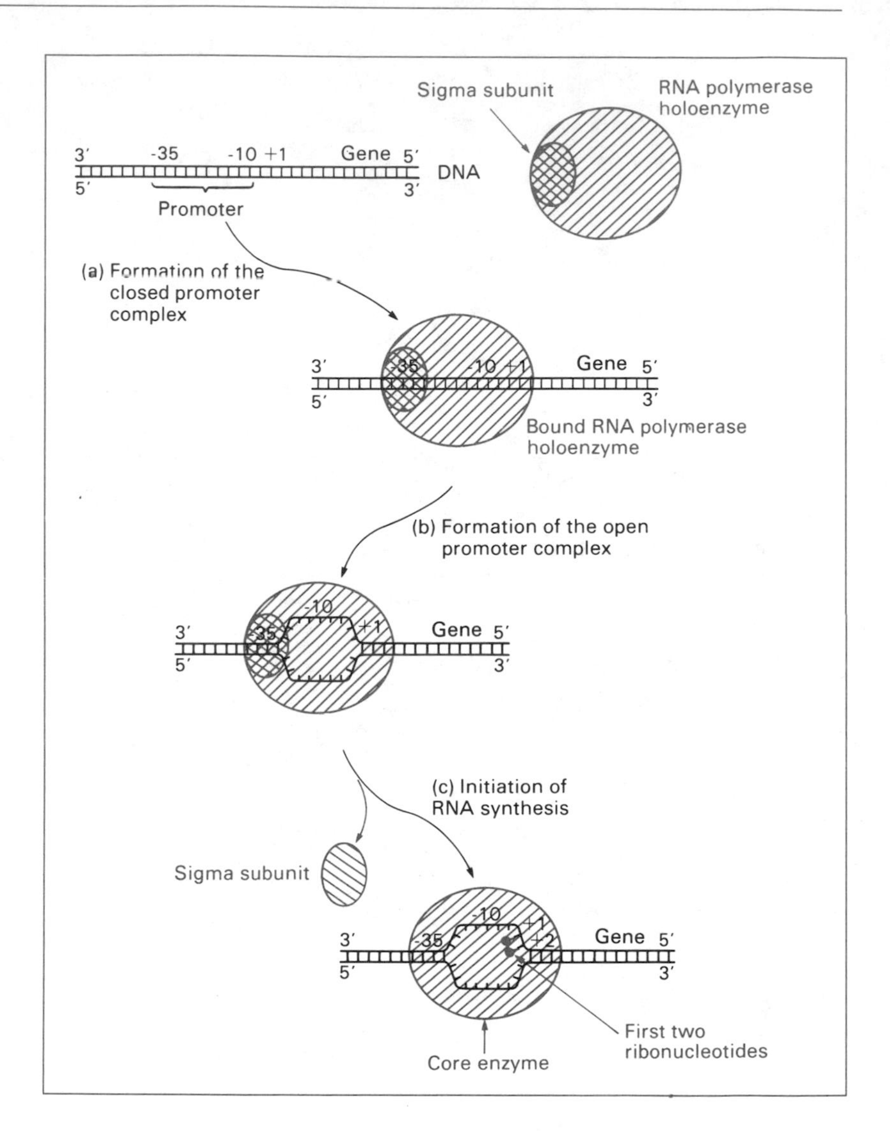

Figure 5.7 Initiation of transcription in *E. coli*.

is called the **closed promoter complex** (Fig. 5.7(a)). In this complex the enzyme covers or 'protects' about 60 bp of the double helix, from just upstream of the −35 box to just downstream of the −10 box. Experimental evidence suggests that the RNA polymerase holoenzyme specifically recognizes the −35 box as the DNA binding site, although these conclusions are still controversial. However, it is clear that the −10 box is the region where breakage of base pairing (called 'melting') and unwinding of the DNA double helix first occurs, forming a second structure, the **open promoter complex** (Fig. 5.7(b)). Melting is, of course, an essential prerequisite for transcription because the bases of the template strand must be exposed in order to direct transcript synthesis. The fact

Functions of the core subunits of *E. coli* RNA polymerase

Subunit	*Function*	*Equivalent subunit in RNA polymerase II*
α	Attachment to the promoter	RPB3
β	Binds the ribonucleotides prior to RNA synthesis	RPB1
β'	Attachment to the DNA template	RPB2

Although the α subunit attaches to the promoter it cannot actually recognize it: that is the function of the σ subunit. The β and β' subunits probably cooperate in catalysing RNA synthesis during transcription.

that the −10 box is made up entirely of A–T base pairs, which comprise just two hydrogen bonds each, compared with three for a G–C pair, is presumed to make melting of the helix easier in this region.

When the open promoter complex has formed, the σ subunit dissociates and the holoenzyme is converted into the core enzyme. At the same time the first two ribonucleotides can be base paired to the template polynucleotide at positions +1 and +2, and the first phosphodiester bond of the RNA molecule synthesized (Fig. 5.7(c)).

5.4.2 Elongation

During the elongation stage of transcription the RNA polymerase enzyme migrates along the DNA molecule, melting and unwinding the double helix as it progresses, while sequentially attaching ribonucleotides to the 3′ end of the growing RNA molecule; base pairing to the template polynucleotide of the gene determines the identity of the ribonucleotide added at each position (Fig. 5.8).

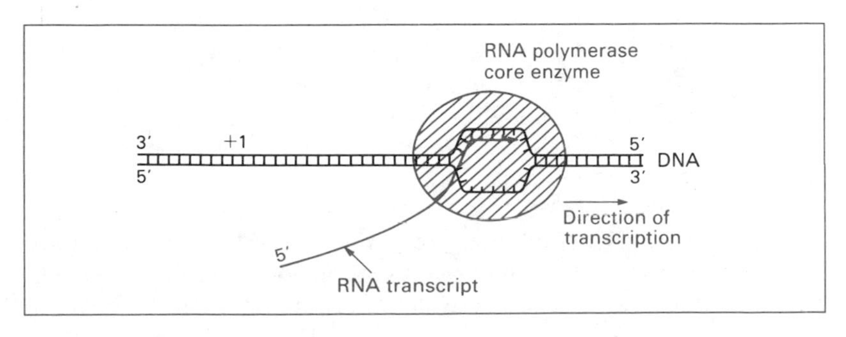

Figure 5.8 The elongation step of transcription.

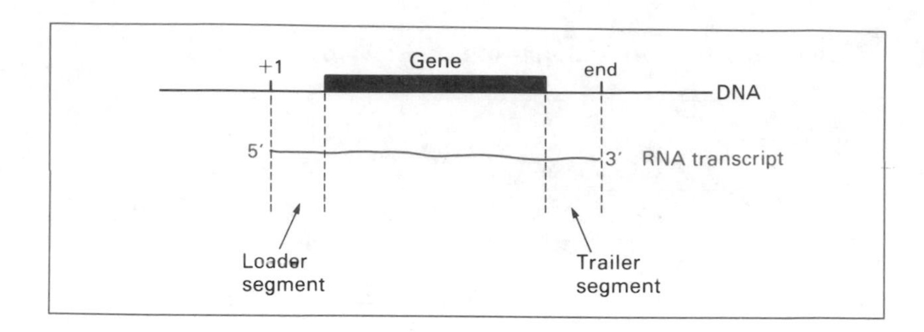

Figure 5.9 Transcripts are usually longer than genes as they include leader and trailer segments.

Three points should be noted:

1. ***The transcript is longer than the gene***. Position +1 in Fig. 5.7, where synthesis of the RNA transcript actually begins, is rarely the start of the gene itself. Almost invariably the RNA polymerase enzyme transcribes a **leader segment** before reaching the gene (Fig. 5.9). The length of this leader varies from gene to gene: in *E. coli* it may be as short as 20 nucleotides or longer than 600 nucleotides. Similarly, when the end of the gene has been reached the enzyme continues to transcribe a **trailer segment** before termination occurs.
2. ***Only a small region of the double helix is unwound at any one time***. Once elongation has begun, the polymerized portion of the RNA molecule can gradually dissociate from the template strand allowing the double helix to return to its original state. This means that only a limited region of the DNA molecule is melted at any one time (see Fig. 5.8). The open region, or transcription 'bubble', contains between 12 and 17 RNA–DNA base pairs and so is only a very short stretch of the gene as a whole. This is important because unwinding a portion of the double helix necessitates overwinding the adjacent areas and there is only limited molecular freedom for this to occur.
3. ***The rate of elongation is not constant***. A feature of transcription that is attracting increasing interest from molecular biologists is the variation in transcription rate that can occur as RNA polymerase moves along a DNA molecule. Occasionally the enzyme will slow down, pause, and then reaccelerate, or even go so far as to reverse over a short stretch, removing ribonucleotides from the end of the newly synthesized transcript before resuming its forward course. Explanations for these phenomena are still being sought.

5.4.3 Termination

As with initiation, termination of transcription is not a random process but must occur only at suitable positions shortly after the ends of genes. Surprisingly, perhaps, termination is not mediated

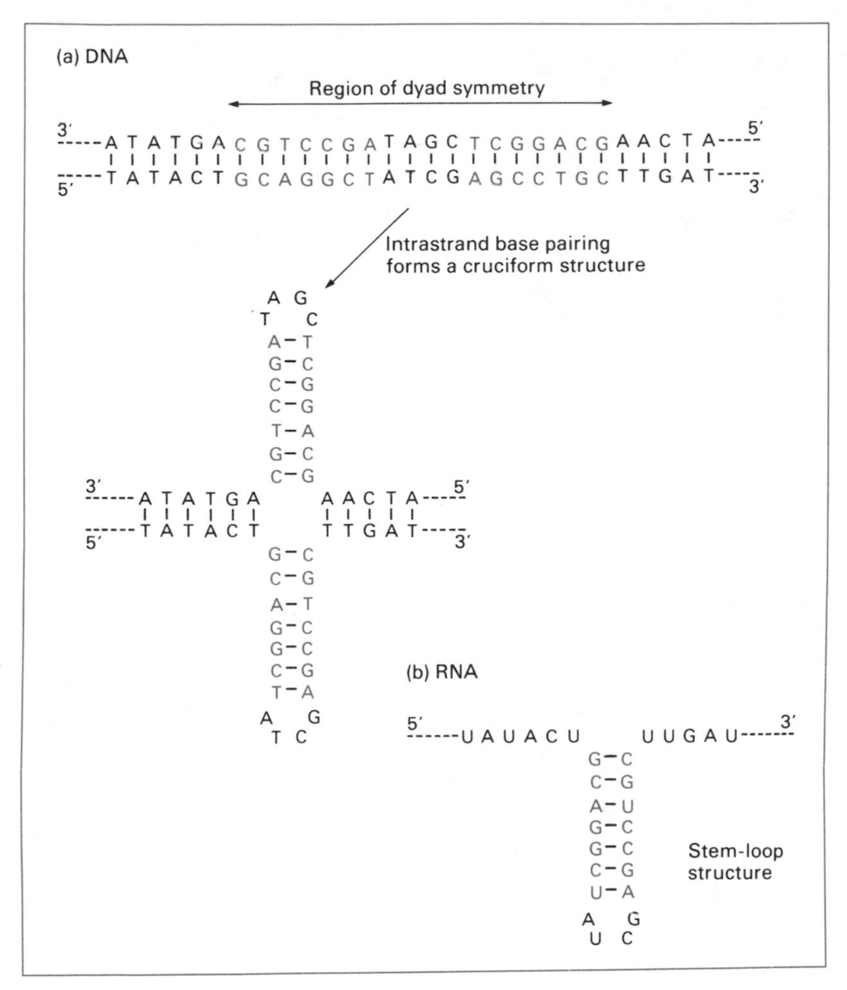

Figure 5.10 Possible consequences of the presence of a complementary palindrome in (a) a DNA double helix, and (b) the RNA molecule transcribed from it. Base pairing between nucleotides in the same strand can result in a cruciform structure in a double-stranded DNA molecule, or a stem-loop structure in the single-stranded RNA.

by a specific sequence analogous to the promoter, but by a more complex signal.

(a) Terminators are complementary palindromes. Termination signals for *E. coli* genes vary immensely in their actual nucleotide sequences, but all possess a common feature: they are complementary palindromes. This means that base pairing can occur not only between the two strands of the double helix, but also within each strand and also within the RNA transcript (Fig. 5.10). The result of intrastrand base pairing is a **cruciform structure** in double-stranded DNA or a **stem-loop structure** in single-stranded RNA.

A cruciform structure is unlikely to form in double-stranded DNA because the interstrand base paired double helix is more energetically stable due to the much larger number of potential

Figure 5.11 Termination of transcription in *E. coli*.

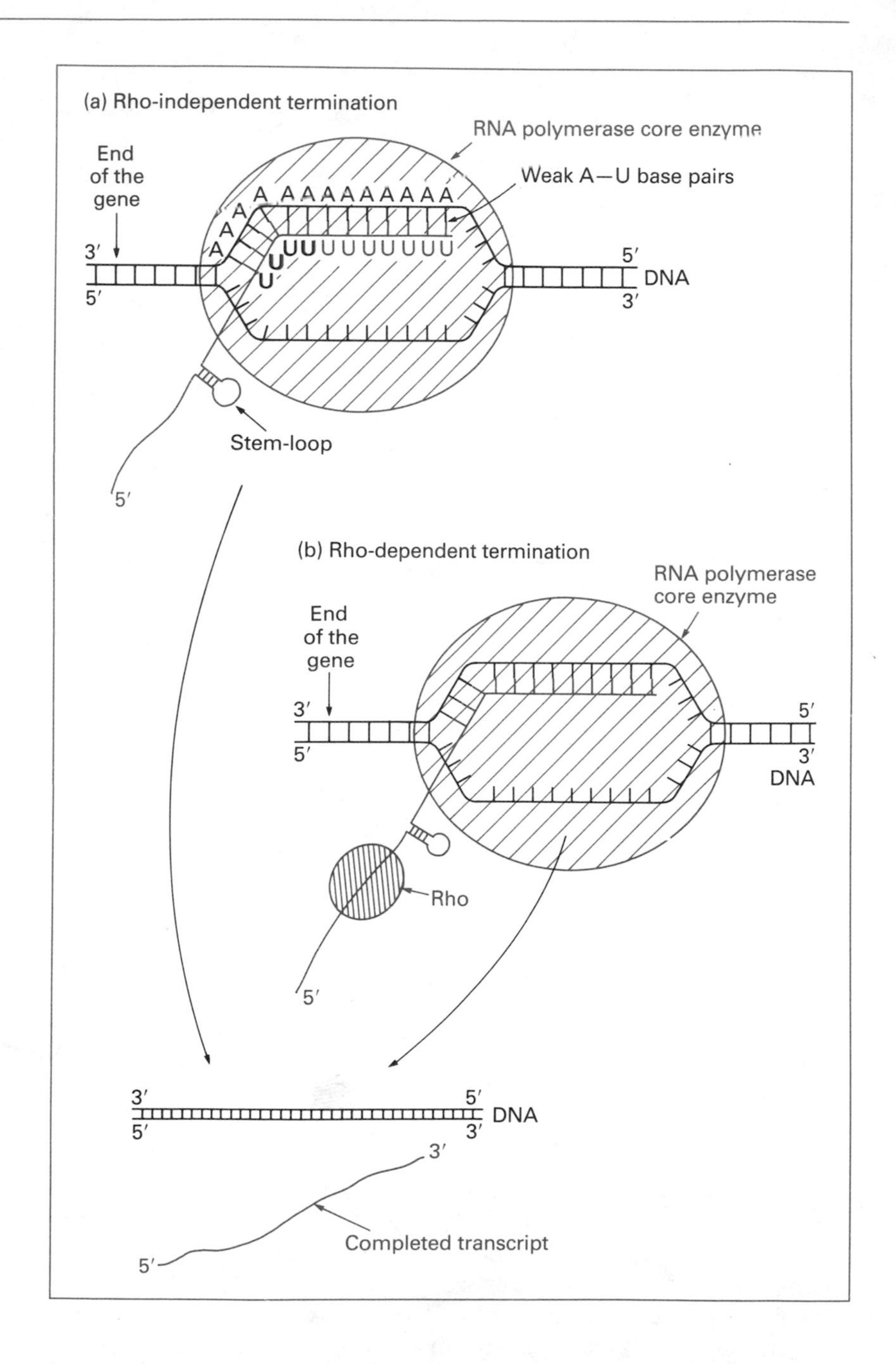

base pairs that can form. In fact, the complementary palindrome exerts its influence on transcription by enabling a stem-loop to form in the growing RNA molecule. Formation of this stem-loop is an essential part of termination, but unfortunately this stage of transcription has proved difficult to study and exactly what happens is still a mystery. Some stem-loop terminators are followed by a run of five to ten As in the DNA template which, after transcription into Us, will leave the transcript attached to the template by a series of weak A–U base pairs. It is thought that the RNA polymerase pauses just after the stem-loop structure and that the A–U base pairs then break, so that the transcript detaches from the template (Fig. 5.11(a)).

Other terminators do not have the run of As and must work in a different way. These are called **Rho-dependent terminators** as they can function only in the presence of a protein called **Rho**. Rho is quite large (molecular mass 276 000 daltons) and is probably not a true subunit of the RNA polymerase. Instead Rho attaches to the growing RNA molecule and, when the polymerase pauses at the stem-loop structure, actively disrupts the base pairing between the DNA template and RNA transcript (Fig. 5.11(b)).

(b) Completion of transcription. Termination of transcription results in dissociation of the RNA polymerase enzyme from the DNA and release of the RNA transcript. The core enzyme is now able to reassociate with a sigma subunit and begin a new round of transcription on the same or a different gene. The RNA transcript is ready to play its role either as the end-product of gene expression or as the message for synthesis of a polypeptide by translation.

5.5 Transcription in eukaryotes

Transcription in eukaryotes occurs by a process very similar to that just described for *E. coli*. The most important differences are that in eukaryotes initiation of transcription is more complicated, and termination does not involve stem-loop structures.

5.5.1 Initiation of transcription by RNA polymerase II

As in *E. coli*, initiation of transcription in eukaryotes depends on the RNA polymerase enzyme recognizing and binding to a promoter sequence. The three different RNA polymerases in eukaryotes each have their own promoters, so each enzyme transcribes only its own set of genes. RNA polymerase II, for example, cannot transcribe transfer RNA genes as these have promoters for RNA polymerase III.

In fact the term 'promoter' is not wholly appropriate when we look at transcription by a eukaryotic RNA polymerase. This is because transcript initiation in eukaryotes depends on a number of different types of nucleotide sequence. The most important is the RNA polymerase attachment site, which is usually just upstream of the gene to be transcribed. This attachment site is equivalent to the *E. coli* promoter, but is not the only nucleotide sequence that

Figure 5.12 Nucleotide sequences upstream of the attachment site can influence the binding of RNA polymerase II to the DNA template.

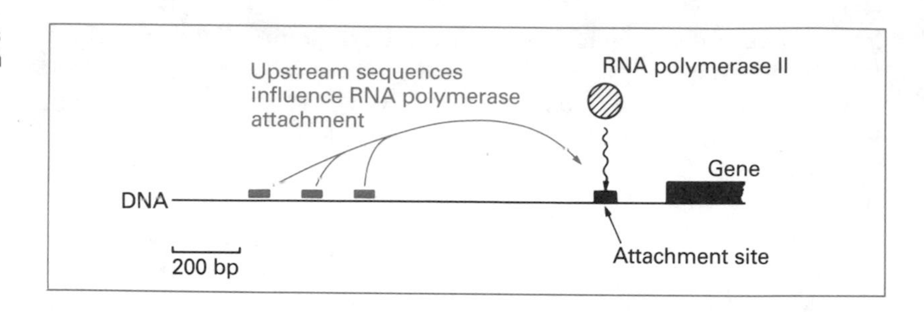

Figure 5.13 The sequence of events thought to occur during attachment of RNA polymerase II to a TATA box. At the moment we know the order of events but do not understand why they should occur in this way.

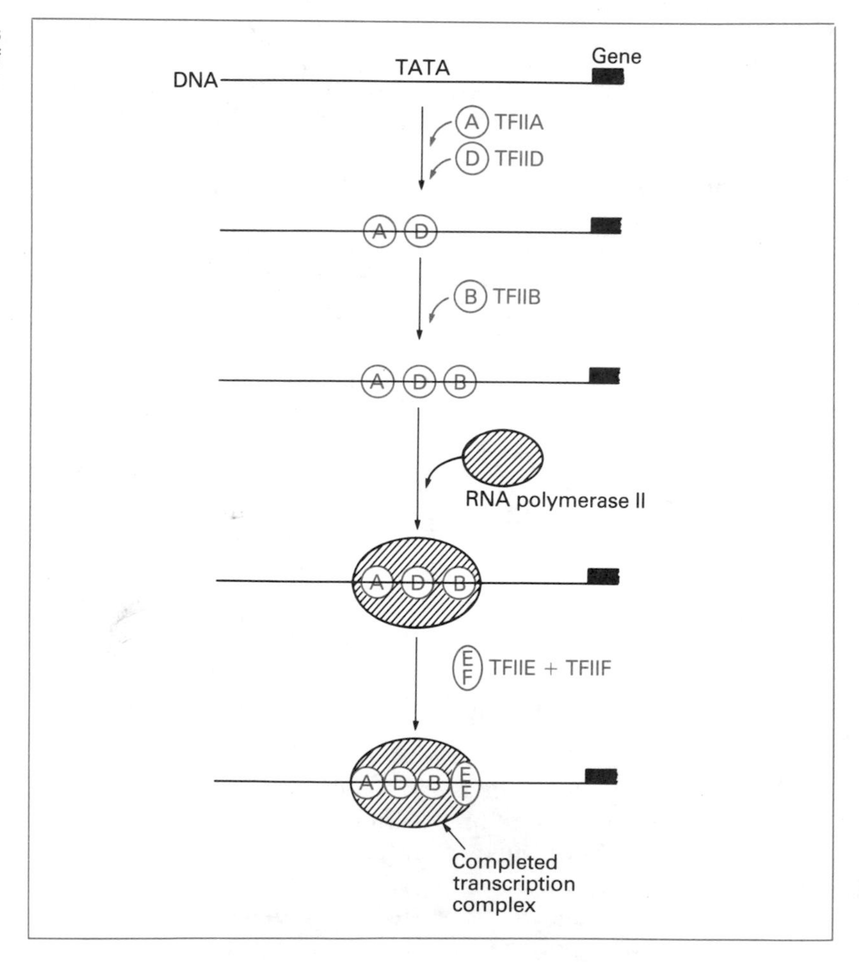

has a role in transcript initiation in eukaryotes. Other sequences, possibly spread out over several hundred base pairs, will also be involved in the binding of the RNA polymerase enzyme to the DNA template (Fig. 5.12).

The main function of many of these additional sequences is to regulate expression of the gene by stimulating or preventing initiation of transcription. This is a topic that we will return to in Chapter 10. For the time being all we will worry about is the attachment site and how RNA polymerase binds to it.

(a) The transcription complex for RNA polymerase II. The attachment site for RNA polymerase II has the following consensus sequence:

5′-TATAAAT-3′ the **TATA box**

The TATA box is usually located about 25 nucleotides upstream of the position at which transcription starts, and so is also called the **−25 box**. RNA polymerase II attaches to the TATA box, but not directly. Instead the attachment is mediated by a set of proteins called **transcription factors**. The transcription factors for RNA polymerase II have been given the names TFIIA, TFIIB and so on, distinguishing them from the TFI and TFIII transcription factors, which work with RNA polymerases I and III respectively.

The events that occur during attachment of RNA polymerase II to the DNA template have only recently been described. The very first step is the binding of TFIID to the TATA box, probably mediated by TFIIA (Fig. 5.13). Once TFIID has attached to the DNA template the transcription complex starts to form. TFIIB binds next, followed by RNA polymerase II itself, and then the two final transcription factors, TFIIE and TFIIF. The transcription complex is now ready to begin synthesizing RNA.

Transcription by RNA polymerases I and III

Each eukaryotic RNA polymerase has a slightly different method for initiating and terminating transcripts. RNA polymerase I in humans uses promoters that are located between nucleotides −45 and +20, spanning the transcription start point. This region corresponds to the attachment site for RNA polymerase I and contains the equivalent structure to the RNA polymerase II TATA box. Other sequences further upstream regulate RNA polymerase I binding and transcript initiation. Termination occurs about 600 bp after the end of the gene, at a signal provided by an 18 bp consensus sequence.

Promoters for RNA polymerase III are more unusual in that they occur within the gene to be transcribed. These are called internal control regions and are typically some 50 bp in length. The first step in transcript initiation is attachment of a transcription factor to the internal control region of the gene. The transcription complex is then built up, with the RNA polymerase III enzyme reaching back to the start of the gene. Termination occurs fairly soon after the end of the gene, apparently at a run of As, though without the formation of a stem-loop structure.

5.5.2 Termination of transcription by RNA polymerase II

Exactly what happens when RNA polymerase II has completed a transcript is not known. Identifying termination signals for RNA polymerase II has proved difficult, partly because the 3′ ends of the transcripts are removed immediately after synthesis by cleavage at an internal phosphodiester bond (see Section 7.3.2). One thing that is clear is that the trailer regions may be very long, possibly extending for 1000 to 2000 bp downstream of the gene. With some genes the termination points seem to vary from transcript to transcript: some transcripts of the mouse β-globin gene have trailer regions about 600 bp in length, whereas others are over 1750 bp. This had led to the suggestion that there are in fact no set termination points downstream of genes transcribed by RNA polymerase II. Possibly one of the transcription factors detaches from the complex at the end of the gene, destabilizing the complex and leading to it falling off the template at some later point. As with termination in *E. coli*, a lot of work needs to be done before we can be sure exactly what happens.

Reading

N. J. Woychik and R. A. Young (1990) RNA polymerase II: subunit structure and function. *Trends in Biochemical Sciences*, **15**, 347–51 – a detailed review on this eukaryotic RNA polymerase.

A. A. Travers and R. R. Burgess (1969) Cyclic re-use of the RNA polymerase sigma factor. *Nature*, **222**, 537–40 – the first demonstration of the role of the σ subunit.

H. Bujord (1980) The interaction of *E. coli* RNA polymerase with promoters. *Trends in Biochemical Sciences*, **5**, 274–8 – a review on the DNA–protein interaction during transcription initiation.

D. G. Bear and D. S. Peabody (1988) The *E. coli* Rho protein. *Trends in Biochemical Sciences*, **13**, 343–7 – some of the latest ideas on how Rho works in transcript termination.

N. J. Proudfoot (1989) How RNA polymerase II terminates transcription in higher eukaryotes. *Trends in Biochemical Sciences*, **14**, 105–10 – some possible answers to this puzzling question.

Problems

1. Define the following terms:

eukaryote
prokaryote
RNA polymerase
holoenzyme
core enzyme
upstream
downstream
promoter
consensus sequence
closed promoter complex
open promoter complex
leader segment
trailer segment
cruciform structure
stem-loop structure
Rho-dependent terminator
transcription factor

2. Draw a detailed diagram of the chemical reaction involved in synthesis of RNA from individual nucleotides.

3. Distinguish between the core and holoenzyme versions of *E. coli* RNA polymerase. What are the roles of each in transcription?

4. Outline the differences between prokaryotic and eukaryotic RNA polymerases.

5. Explain how the *E. coli* RNA polymerase recognizes the correct position within a DNA molecule at which to begin transcription.

6. Describe the general features of a signal for transcription termination in *E. coli*.

7. Explain what is meant by a 'complementary palindrome'.
8. Distinguish between the events that are thought to occur during transcription termination at Rho-dependent and Rho-independent structures.
9. Outline the events involved in initiation and termination of transcription by RNA polymerase II.
10. *Some species of bacteria possess more than one σ factor. For instance, *Bacillus subtilis* (a bacterium able to produce spores) has at least four. What might be the role of these?
11. *With some types of bacteriophage, transcription of the host bacterium's genes ceases shortly after infection. All the cell's RNA polymerase enzymes start transcribing the bacteriophage's genes instead. Suggest events that might underlie this phenomenon.
12. Propose a suitable consensus sequence for the following:

    5′-ATAGACATA-3′
    ATAGCCATT
    ATACACTTA
    AGAGAGAAT
    ATAGACATA
    TTTGACATT
    ATAAATATA

13. Which of the following RNA molecules would be able to form stem-loop structures?

    5′-ACGUUUGGCAGUCCAAACU-3′
    5′-AGCUAGCUACAGCUAGCUUUGUGAGCUAGCUUGUG-3′
    5′-GGCUAGGUCGAAACGACCUAGCC-3′

6 Types of RNA molecule: rRNA and tRNA

Ribosomal RNA – ribosome structure – rRNA synthesis • Transfer RNA – the cloverleaf – processing and modification

The RNA molecules produced by transcription can be grouped into three major classes according to function. These are **ribosomal** or **rRNA**, **transfer** or **tRNA**, and **messenger** or **mRNA**. Ribosomal and transfer RNAs are the end-products of gene expression and perform their roles in the cell as RNA molecules (Fig. 6.1). Messenger RNA, on the other hand, undergoes the second stage of gene expression, translation, and has no function beyond acting as the intermediate between a gene and its final expression product, a polypeptide.

Ribosomal and transfer RNAs are often referred to as **stable RNA**, indicating that these molecules are long-lived in the cell, in contrast to mRNA which has a relatively rapid **turnover rate**, possibly as short as a minute for some bacterial mRNAs. Both types of stable RNA – rRNA and tRNA – are involved in protein synthesis even though they are not themselves translated. In this

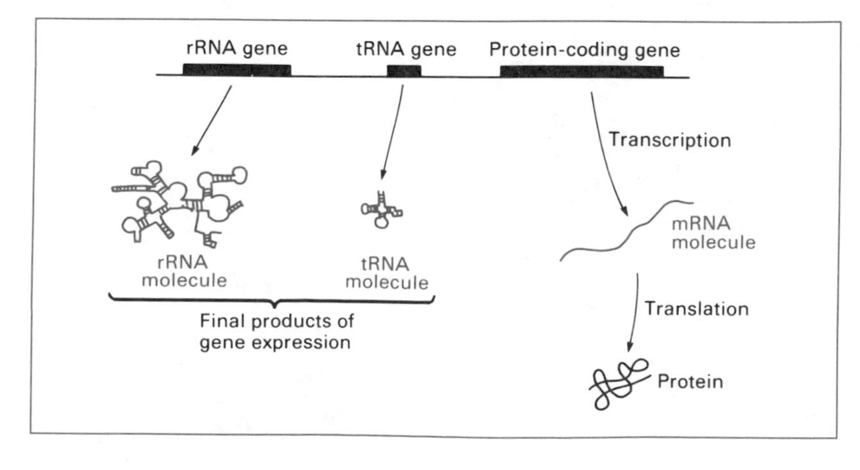

Figure 6.1 The three major types of RNA molecule produced by transcription. rRNA and tRNA are the final products of expression of their genes; mRNA undergoes the second stage of gene expression, called translation.

chapter the structures of these molecules and the special features of their synthesis will be described as a prelude to the examination in Chapters 8 and 9 of the details of their precise roles in protein synthesis.

Other types of RNA in eukaryotes

The three main classes of RNA (rRNA, tRNA and mRNA) are found in both prokaryotes and eukaryotes. Eukaryotes also have several other types of RNA molecule:

Type of RNA	*Function*
U-RNA	mRNA and rRNA processing
7SL RNA	Protein transport in the cell
7SK RNA	Unknown
Alu RNA	Unknown

6.1 Ribosomal RNA

Ribosomal RNA molecules are components of **ribosomes**, the large multimolecular structures that act as factories for protein synthesis. During translation, ribosomes attach to mRNA molecules and migrate along them, synthesizing polypeptides as they go, analogous in a way to the role of RNA polymerase in transcription. Ribosomes are made up of rRNA molecules and proteins, and are extremely numerous in most cells. An actively growing bacterium may contain over 20 000 ribosomes, comprising about 80% of the total cell RNA and 10% of the total protein.

6.1.1 The structure of ribosomes

The prokaryotic ribosome has a total molecular mass of 2 520 000 daltons and is very roughly ovoid with approximate dimensions of 29 nm × 21 nm (1 nm = 10^{-9} m). Eukaryotic ribosomes are larger: 4 220 000 daltons and 32 nm × 22 nm.

(a) Ribosome sizes are determined by sedimentation analysis. Ribosomes, like many macromolecules and multimolecular assemblies, are so large that estimates of their molecular masses are very difficult to obtain. Instead the sizes of these structures are determined by velocity sedimentation analysis (p. 19), a procedure that measures the rate at which a molecule or particle sediments through a dense solution (often of sucrose) while subjected to a very high centrifugal force (700 000 × *g* or more). The sedimentation coefficient is expressed as an **S value** (S = Svedberg unit, after the Swede, The Svedberg, who built the first ultracentrifuge in the early 1920s) and is dependent on several factors, notably molecular mass and the shape of the macromolecule or macromolecular assembly. The S values for several molecules and assemblies of importance in molecular biology are shown in Table 6.1. Prokaryotic ribosomes have a sedimentation coefficient of 70S, somewhat smaller than eukaryotic ribosomes, which are variable but average about 80S.

(b) Ribosomes are made of two subunits. In *E. coli* these subunits have sedimentation coefficients of 50S and 30S. The large subunit contains two rRNA molecules, of 23S and 5S, together

Table 6.1 Sedimentation coefficients of important biological molecules and structures

Molecule or structure	*Sedimentation coefficient*	*Molecular mass (daltons)*
Ribosomes		
Prokaryotic	70S	2 520 000
Eukaryotic	80S	4 220 000
RNA molecules		
rRNA	5S	35 000
	16S	560 000
	18S	700 000
	23S	1 104 000
	28S	1 800 000
tRNA	4S	23 000–30 000
mRNA	5–25S	25 000–1 000 000
Proteins		
Human cytochrome *c*	1.17S	13 370
Human haemoglobin	4.46S	64 500

Variations in rRNA sizes in different eukaryotes

The figures given in Table 6.2 refer to mammalian ribosomes. In other eukaryotes the large (23S in mammals) and small (18S) rRNA molecules are slightly shorter:

Organism	*Large rRNA*	*Small rRNA*
Mammals	4718 nucleotides	1874 nucleotides
Amphibia	4110 nucleotides	1825 nucleotides
Yeast	3392 nucleotides	1799 nucleotides

with 34 different polypeptides. The smaller subunit has just a single 16S rRNA plus 21 polypeptides. This information is summarized in Table 6.2.

Eukaryotic ribosomes are also made of two subunits but in this case the sizes are 60S and 40S (Table 6.2). The large subunit has three rRNAs (28S, 5.8S and 5S) and 49 polypeptides; the small subunit has a single 18S rRNA and 33 polypeptides. The additional rRNA of the eukaryotic large subunit is the 5.8S molecule which in *E. coli* is present as an integral part of the 23S rRNA.

(c) The molecular structure of the ribosome. The traditional view of ribosomal structure is that the rRNA molecules act as scaffolding onto which the proteins, which provide the functional activity of the ribosome, are attached. To fulfil this role the rRNA molecules must be able to take up a stable three-dimensional

Table 6.2 The compositions of typical prokaryotic and eukaryotic ribosomes. With eukaryotes the lengths of the large rRNAs and the numbers of proteins are not exactly the same in all species; the figures given here refer specifically to mammalian ribosomes

	Prokaryotes	*Eukaryotes*
Ribosome		
Sedimentation coefficient	70S	80S
Molecular mass (daltons)	2 520 000	4 220 000
Number of subunits	2	2
Large subunit		
Sedimentation coefficient	50S	60S
Molecular mass (daltons)	1 590 000	2 820 000
RNA molecules		
number	2	3
sizes	23S = 2904 n*	28S = 4718 n
	5S = 120 n	5.8S = 160 n
		5S = 120 n
Number of polypeptides	34	49
Small subunit		
Sedimentation coefficient	30S	40S
Molecular mass (daltons)	930 000	1 400 000
RNA molecules		
number	1	1
sizes	16S = 1541 n	18S = 1874 n
Number of polypeptides	21	33

* n = nucleotides

structure. This is achieved by inter- and intra-molecular base pairs, with different rRNAs of a subunit base pairing in an ordered fashion with each other, and also more importantly with different parts of themselves. A two-dimensional representation of the base paired structure of the *E. coli* 16S rRNA, the component of the small subunit of the prokaryotic ribosome, is shown in Fig. 6.2. This model has been developed by a combination of experimental approaches, the three most important being:

1. An examination of the nucleotide sequence of the *E. coli* 16S rRNA and identification of the regions that could form base pairs with one another.
2. Comparison of the *E. coli* sequence with the slightly different sequences of the equivalent rRNAs of other prokaryotes, as it would be expected that regions of base pairing in the *E. coli* molecule would be conserved in other bacteria.

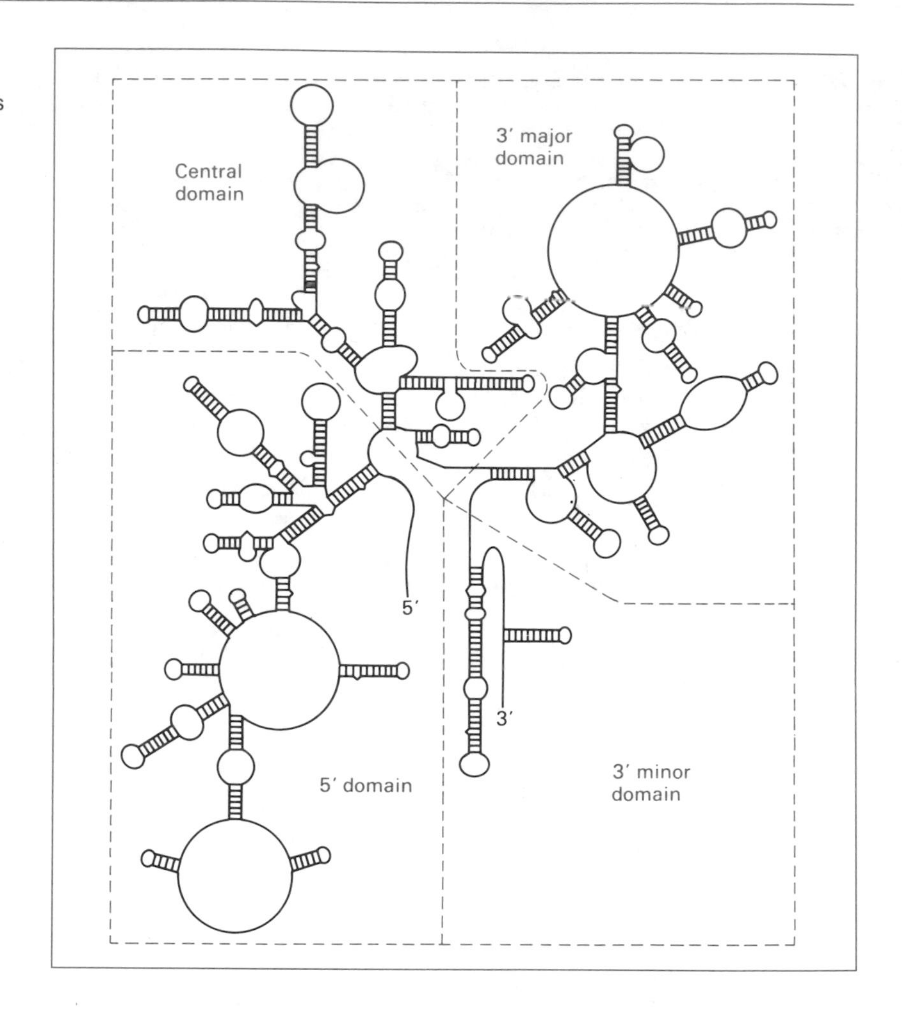

Figure 6.2 The proposed base paired structure for the 16S rRNA molecule of *E. coli*. This molecule is 1541 nucleotides in length.

3. Treatment of the base paired rRNA with ribonucleases that act on single-stranded RNA but have no effect on base paired regions. The double-stranded regions that remain are then analysed to see if they correspond with those predicted by the model (Fig. 6.3(a)).

Additional experiments have located those regions of the 16S rRNA to which the polypeptide components of the *E. coli* small subunit attach. One important technique here is analogous to the DNase protection experiments used to examine the interaction between RNA polymerase and the promoter (p. 64). A complex between an rRNA and an individual ribosomal protein is treated with a non-specific ribonuclease, one that digests both single- and double-stranded RNA. All rRNA, except that protected by the protein, will be digested. The protein can then be degraded and the RNA fragment identified and located on the rRNA structure

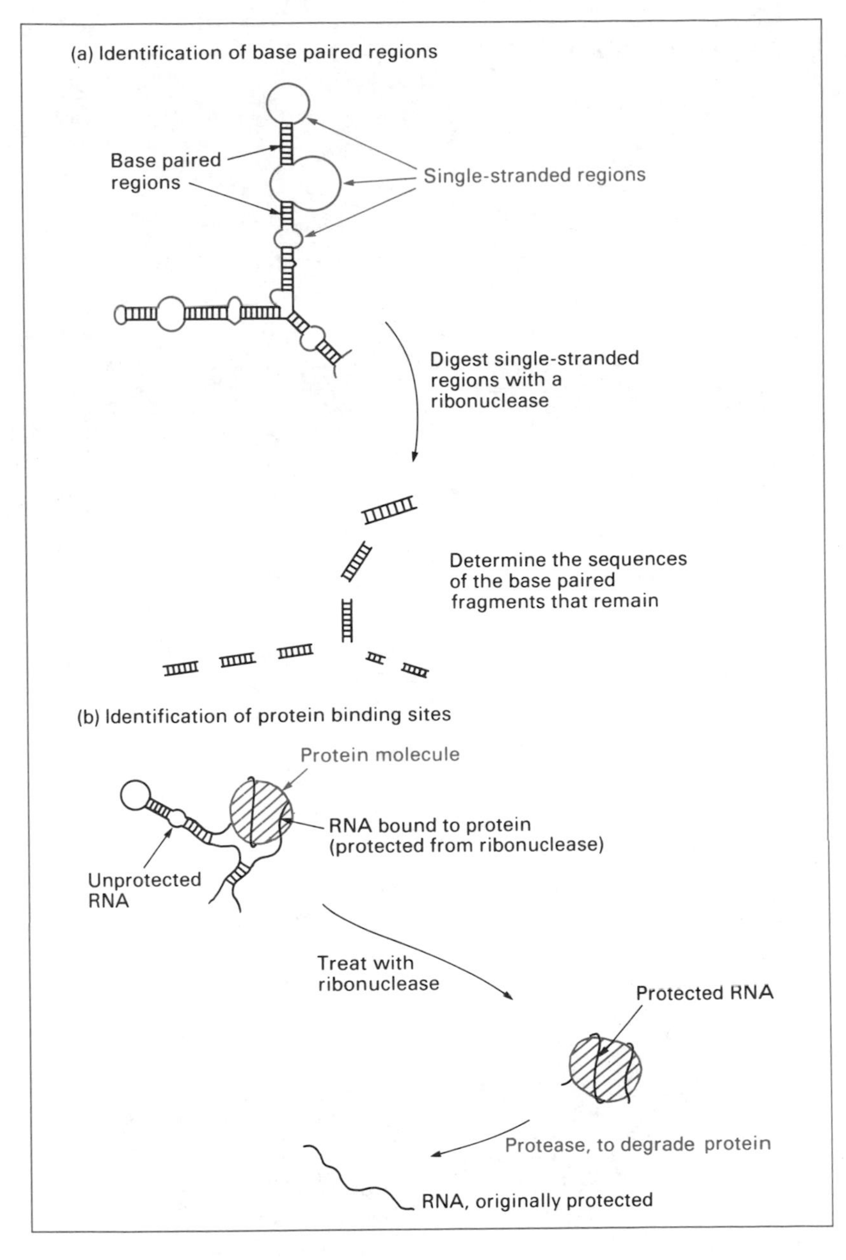

Figure 6.3 Two methods for studying the structure of RNA. (a) Identification of base paired regions by treatment with a single-strand specific ribonuclease. (b) Identification of protein binding sites by nuclease protection.

(Fig. 6.3(b)). Binding sites for several ribosomal proteins have been identified in this way.

(d) The three-dimensional structure. The base paired structures shown in Fig. 6.2 are two-dimensional representations and tell us little about the real three-dimensional structure of the ribosome. In fact this is the problem that is being pursued most actively at

The binding site for a ribosomal protein

The binding sites for several ribosomal proteins have now been located on rRNA molecules. An example is the *E. coli* ribosomal protein called S8, which binds to the following stem-loop in the 16S rRNA:

```
                          U
                         / \
                        A   A                        /
  U-G-C-A-U-C-U-G            C-U-G-G-C-A-A-G-C
 ( •  •  •  •  •  •  •  •           •  •  •  •  •  •  •  •  •
  G-U-G-U-A-G-A-C-U-G-A  U-U-G-U-U-U-C
                          \ /                    \
                           A
```

The nucleotides shown in red are those that are most closely linked to the ribosomal protein. Note that in RNA an additional type of base pair is allowed. As well as the 'Watson–Crick' base pairs (A–U and G–C), G can base pair with U.

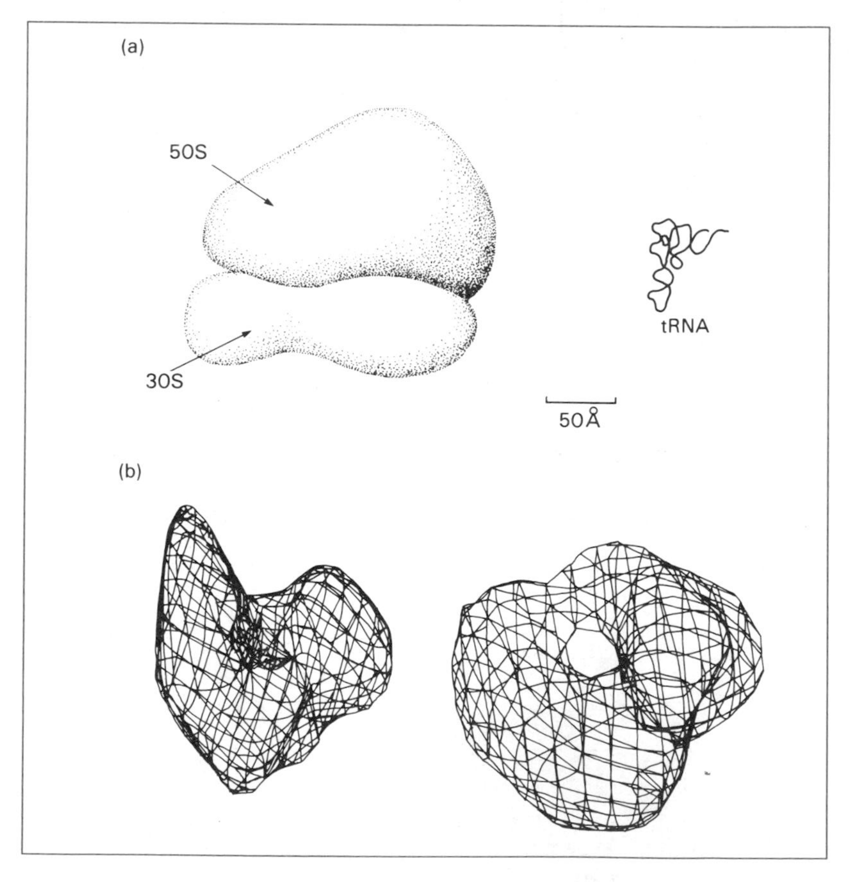

Figure 6.4 The three-dimensional structure of the *E. coli* ribosome. (a) A model of the ribosome drawn from electron microscopic images, with a tRNA molecule drawn to the same scale. (b) Two views of the 50S subunit, derived from X-ray diffraction analysis.

present. Ribosomes can be examined with the electron microscope, albeit towards the limits of resolution, and approximate three-dimensional reconstructions have been built up by analysing the images obtained (Fig. 6.4(a)). These reconstructions have been refined by a variety of techniques that have located the positions of individual ribosomal proteins on the three-dimensional image. But the most significant advance has been the application of X-ray diffraction analysis (Section 3.2) to the ribosome. Diffraction patterns have been obtained and approximate three-dimensional structures derived from them (Fig. 6.4(b)). The complexity of the diffraction patterns makes precise interpretation difficult, but rapid advances are being made in this area, particularly with the help of computer-aided analysis, and the structures are gradually becoming understood in more and more detail.

(e) rRNA molecules may have enzymatic roles during protein synthesis. The traditional view of the rRNA as the scaffolding and the proteins as attachments that provide the real biological activity of the ribosome is now being challenged by the latest exciting ideas about the function of RNA. For many years it was believed that enzymatic catalysis is uniquely a feature of proteins and that RNA molecules cannot act as enzymes. Accordingly, any catalytic activity that the ribosome displays during protein synthesis must be a property of the ribosomal proteins. Recently, however, it has been discovered that certain RNA molecules have enzymatic activity. This was first demonstrated in 1982 with the self-splicing intron of *Tetrahymena* (Section 7.4.2) and has subsequently been proven or implied for other RNA molecules also. As yet there is no proof that rRNA has an enzymatic function in ribosomes, but a discovery along these lines could well be just around the corner.

6.1.2 Synthesis of rRNAs

Each ribosome contains one copy of each of the different rRNA molecules, three rRNAs for the prokaryotic ribosome or four for the eukaryotic version. The most efficient system would be for the cell to synthesize equal numbers of each of these molecules. Of course the cell could make different amounts of each one but this would be wasteful because some copies would be left over when the least abundant rRNA was all used up.

(a) The rRNA transcription unit. Synthesis of equal numbers of each rRNA molecule is assured by having an entire complement of rRNA molecules transcribed together as a single unit. The product of transcription, the **primary transcript**, is therefore a long RNA precursor, the **pre-rRNA**, containing each rRNA separated by short spacers. The spacers are removed by processing events that release the mature rRNAs. The scheme followed during rRNA synthesis in *E. coli* is shown in Fig. 6.5(a). A similar series of events brings about the synthesis of eukaryotic rRNAs, with the exception that only the 28S, 18S and 5.8S genes are transcribed together

Figure 6.5 Transcription and processing of the rRNA genes of (a) *E. coli*, and (b) mammals. Note that in mammals the 5S rRNAs are transcribed independently of the main unit.

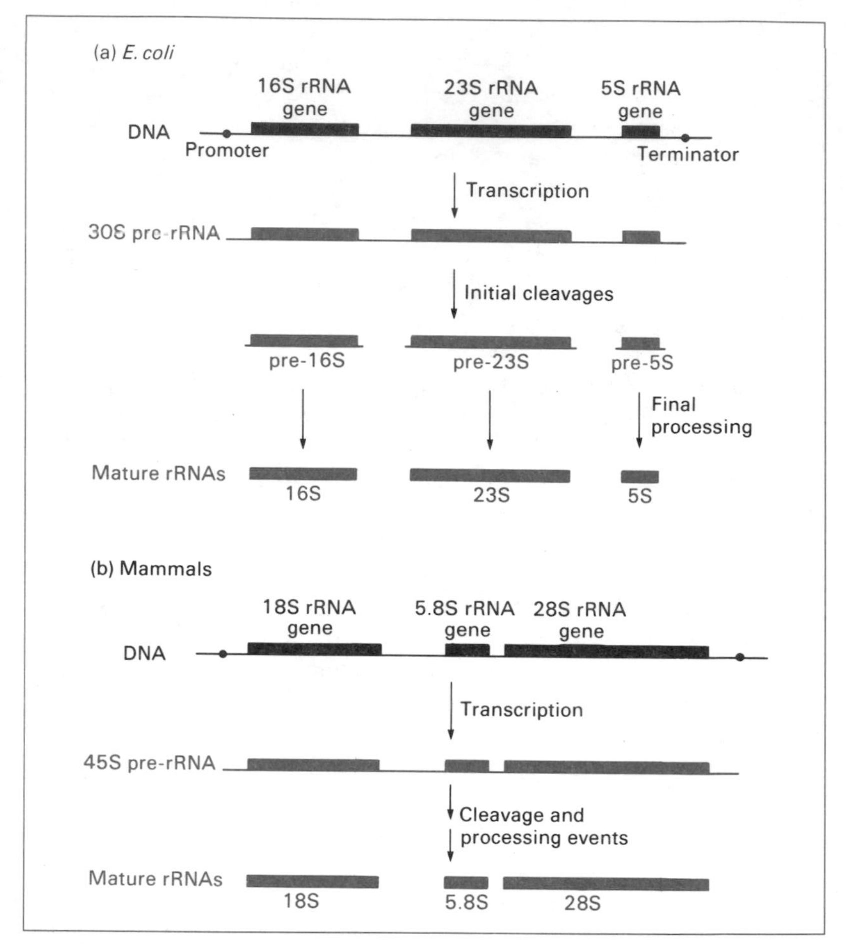

(Fig. 6.5(b)). The 5S genes occur elsewhere on the eukaryotic chromosomes and are transcribed independently of the main unit.

(b) There are multiple copies of the rRNA transcription unit. The actively growing *E. coli* cell that contains 20 000 ribosomes will divide once every 20 minutes or so. Therefore, every 20 minutes it needs to synthesize 20 000 new ribosomes, an entire complement for one of the two daughter cells. This necessitates a considerable amount of rRNA transcription, to such an extent that a single transcription unit would not be able to meet the demand. In fact, the *E. coli* chromosome contains seven copies of the rRNA transcription unit. In eukaryotes there can be an even greater demand for rRNA synthesis and 50 to 5000 identical copies of the rRNA transcription unit may be present depending on species. In eukaryotes these units are usually arranged into multigene families (Section 4.2.1), with large numbers of copies following one after

rRNA and tRNA gene copies

All organisms have multiple copies of their rRNA and tRNA genes. The actual number of copies is different in different species

Species	*Number of genes*		
	Major rRNA transcription unit	*5S rRNA genes**	*tRNA genes*
E. coli	7	7	60
Saccharomyces cerevisiae	140	140	250
Man	280	2 000	1300
Frog	450†	24 000	1150

* In *E. coli* the 5S rRNA gene is part of the major rRNA transcription unit. In eukaryotes it is separate (see text), although in *S. cerevisiae* the number of 5S rRNA genes is linked to the number of the major rRNA transcription units.
† Further increased by gene amplification in certain cells, for example oocytes.

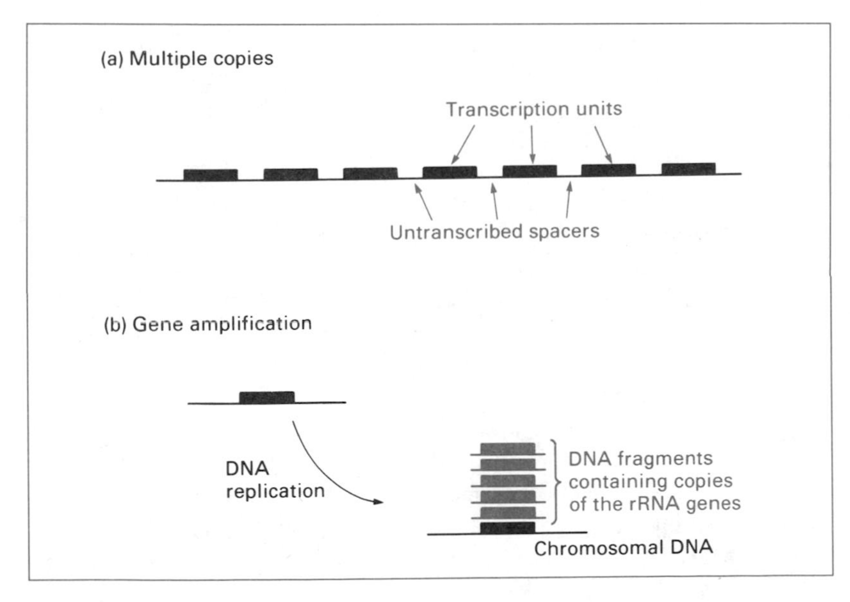

Figure 6.6 Strategies for increasing a cell's capacity for rRNA synthesis. (a) Multiple copies of the rRNA transcription unit, as seen in most eukaryotes. (b) Amplified copies of the transcription unit, a feature of oocytes in amphibians.

the other, separated by non-transcribed spacers (Fig. 6.6(a)). Even this may not satisfy the demand for rRNA synthesis under certain circumstances, and in some eukaryotic cells (for example, amphibian oocytes) an additional strategy known as **gene amplification** may be called upon. This involves replication of rRNA genes into multiple DNA copies which subsequently exist as independent molecules not attached to the chromosomes (Fig. 6.6(b));

Robert Holley b. 28 January 1922, Urbana, Illinois

'It may be as easy to solve an important problem as a trivial one, so study the problems you consider important'.

After obtaining his Bachelor's degree from the University of Illinois, Robert Holley went to Cornell University at Ithaca, New York, gaining his PhD in 1947. From 1944 to 1946 he worked at Cornell University Medical College in New York City on penicillin synthesis. He started working with nucleic acids at the California Institute of Technology in 1955 and continued in this area when he returned to Cornell a year later. He chose to study alanine tRNA because it was relatively easy to purify; nevertheless, it took 3 years to work out new procedures and isolate 1 g of tRNA from 90 kg of yeast. He developed new methods for the isolation and sequencing of large oligonucleotides in order to work out the nucleotide sequence of the tRNA, finally completing the work in March 1965. He states that the cloverleaf structure was not his own idea but was discovered by two of his associates, Betty Keller and John Penswick. In fact the cloverleaf was only one of several possible two-dimensional structures and was not confirmed until other tRNAs were found to fit the same pattern. In 1968 Holley received the Nobel Prize. Since 1968 he has been Professor of Molecular Biology at the Salk Institute in San Diego.

transcription of the amplified copies produces additional rRNA molecules. Gene amplification is not restricted to rRNA genes and occurs with a few other genes whose transcription is required at a greatly enhanced rate in certain situations.

6.2 Transfer RNA

Transfer RNA molecules are also involved in protein synthesis but the part they play is completely different from that of rRNA. Transfer RNAs are in fact the adaptor molecules that read the nucleotide sequence of the mRNA transcript and convert it into a sequence of amino acids. The existence of tRNA was postulated by Crick in the 1950s and individual tRNA molecules were first isolated in 1959 by Robert Holley, who 6 years later presented the complete nucleotide sequence of one of these molecules.

The exact function of tRNA in protein synthesis will be covered in Chapters 8 and 9. Here we will look at the structure of tRNA molecules, the manner in which they are transcribed, and the processing and modification events that they undergo after transcription.

6.2.1 Structure of tRNAs

Transfer RNA molecules are relatively small, mostly between 74 and 95 nucleotides for different molecules in different species.

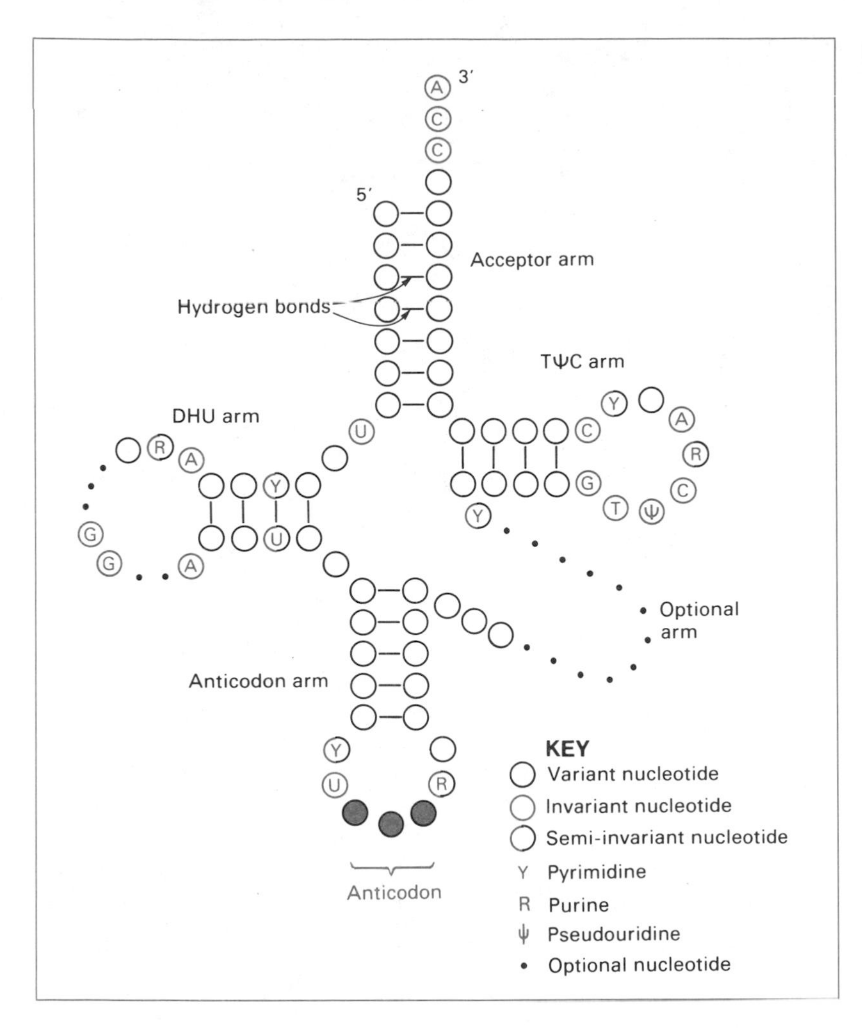

Figure 6.7 The tRNA cloverleaf.

Each organism synthesizes a number of different tRNAs, each in multiple copies. However, virtually all tRNAs take up the same structure after synthesis.

(a) The tRNA cloverleaf. Almost every tRNA molecule in every organism can be folded into a base paired structure referred to as the **cloverleaf** (Fig. 6.7). This structure is made up of the following components:

1. ***The acceptor arm,*** formed by a series of usually seven base pairs between nucleotides at the 5′ and 3′ ends of the molecule. During protein synthesis an amino acid is attached to the acceptor arm of the tRNA (Section 9.1.1).
2. ***The D or DHU arm,*** so named because the loop at its end almost invariably contains the unusual pyrimidine dihydrouracil.

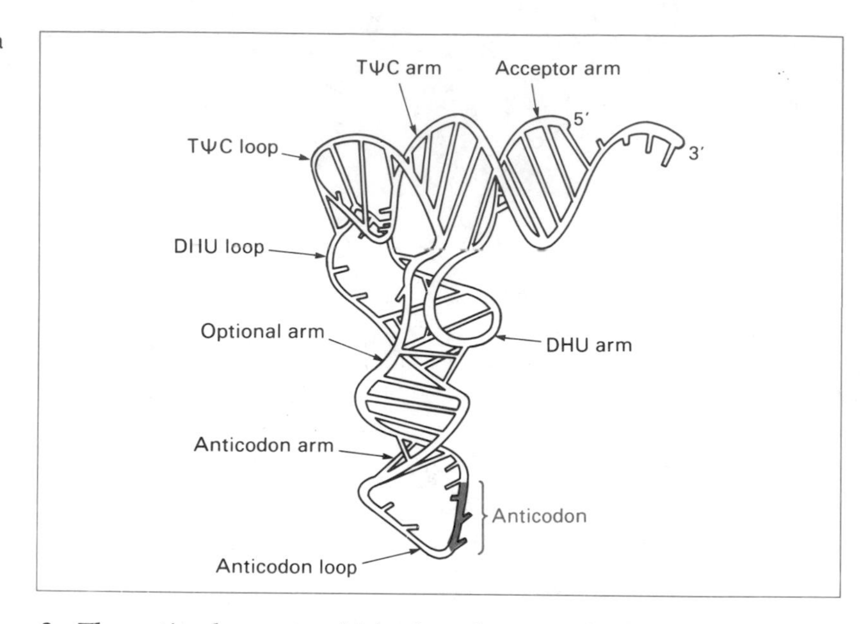

Figure 6.8 The tertiary structure of a tRNA molecule.

3. ***The anticodon arm***, which plays the central role in decoding the biological information carried by the mRNA (Section 9.1.2).
4. ***The extra, optional or variable arm***, which may be a loop of just three to five nucleotides (Class I tRNAs, about 75% of all tRNAs) or a much larger stem-loop of 13–21 nucleotides with up to five base pairs in the stem (Class II tRNAs).
5. ***The TψC arm***, which is named after the nucleotide sequence T, ψ, C (ψ = a nucleotide containing pseudouracil, another unusual pyrimidine base), which its loop virtually always contains.

In addition to having a common secondary structure, different tRNAs also display a certain amount of nucleotide sequence conservation. Some positions are invariant: in all tRNAs such a position will be occupied by the same nucleotide. Others are semi-invariant and will always contain the same type of nucleotide, either one of the pyrimidine nucleotides or one of the purine nucleotides. The invariant and semi-invariant positions are shown in Fig. 6.7.

Unusual tRNA structures

The cloverleaf is not universal. In particular many of the tRNA molecules present in mitochondria (Section 15.5.1) have unusual structures.

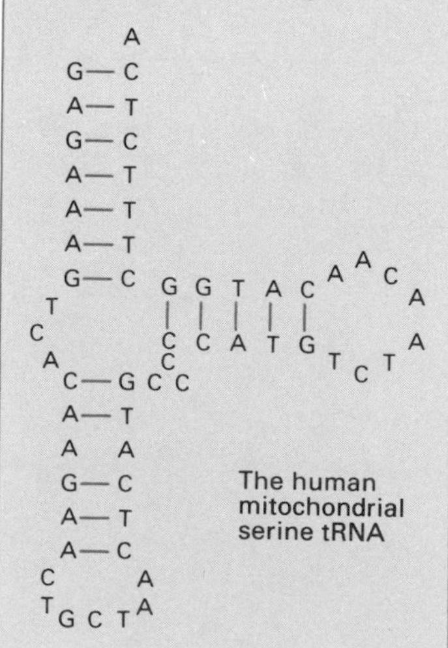

(b) The tertiary structure. It should be appreciated that although the cloverleaf is a convenient way to represent the structure of a tRNA, the molecule itself will have a much less recognizable three-dimensional tertiary structure in the cell. This tertiary structure has been determined by X-ray diffraction analysis and is shown in Fig. 6.8. The base pairings in the stems of the cloverleaf are maintained in the tertiary structure, but several additional base pairings form between nucleotides that appear widely separated in the cloverleaf. This folds the molecule into a compact conformation. Many of the nucleotides involved in the tertiary base pairing are the invariant or semi-invariant ones that are conserved in different tRNAs. The tertiary conformation places the acceptor arm and anticodon loop at opposite ends of the molecule. The precise roles that these parts of the tRNA play during protein synthesis will be described in Chapters 8 and 9.

6.2.2 Processing and modification of tRNA transcripts

In both prokaryotes and eukaryotes tRNAs are transcribed initially as precursor-tRNA which is subsequently processed to release the mature molecules. In *E. coli* there are several separate tRNA transcription units, some containing just one tRNA gene and some with as many as seven different tRNA genes in a cluster. A pre-tRNA molecule is processed by a combination of different ribonucleases that make specific cleavages at the 5′ (**RNase P**) and 3′ (**RNase D**) ends of the mature tRNA sequence (Fig. 6.9). RNase P is particularly interesting because the enzyme is a complex between a small protein (molecular mass 20 000 daltons) and a 377 nucleotide RNA molecule. Recently it has been shown that the RNA molecule in RNase P possesses at least part of the enzymatic activity of the complex and is hence an example of an RNA enzyme or '**ribozyme**'.

Eukaryotic tRNA genes are also clustered and in addition occur in multiple copies, a reflection of the huge demand for tRNA synthesis in the eukaryotic cell. The events involved in processing of eukaryotic pre-tRNA have not been characterized to any great extent as yet but enzymes similar in activity to RNase P and RNase D are implicated.

(a) The three nucleotides at the 3′ end may be added after transcription. All tRNAs have at their 3′ end the trinucleotide sequence 5′-CCA-3′. When the genes coding for tRNAs in eukaryotes are examined it is found that in most cases the CCA sequence does not occur at the expected position in the DNA molecule. Instead the 3′-terminal CCA is added after transcription by another processing enzyme, this one called **tRNA nucleotidyl transferase** (see Fig. 6.9). In prokaryotes the final CCA is more frequently coded by the tRNA gene and is therefore transcribed in the normal manner. Nevertheless, it appears that occasionally this sequence, or part of it, is removed by RNase D during processing of the pre-tRNA and has to be replaced by a prokaryotic nucleotidyl transferase enzyme. This CCA is crucial because it is the point at which the amino acid is attached to the tRNA during protein synthesis.

(b) Certain nucleotides undergo chemical modification. We have already noted the existence of unusual nucleotides in certain

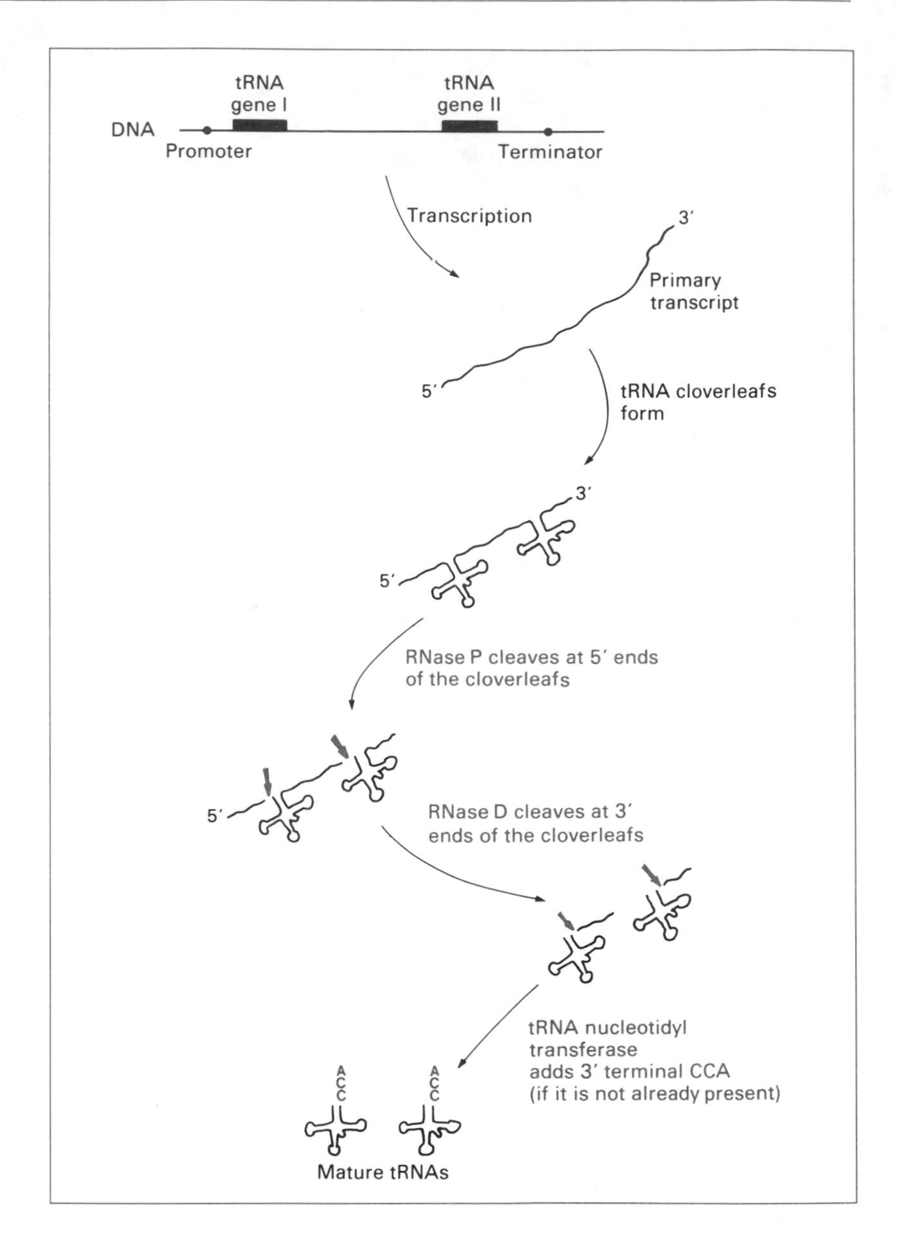

Figure 6.9 Transcription and processing of tRNA molecules in *E. coli*.

stem-loops of the tRNA cloverleaf. In fact a number of different nucleotides in a single tRNA will undergo chemical modification after transcription; each will always be modified into the same unusual nucleotide, though a number of different modifications are known. The most common types are:

1. ***Methylation*** – addition of one or more methyl (—CH_3) groups to the base or sugar component of the nucleoside. Examples: guanosine to 7-methylguanosine (Fig. 6.10(a)) and uridine to ribothymidine (Fig. 6.10(b)).

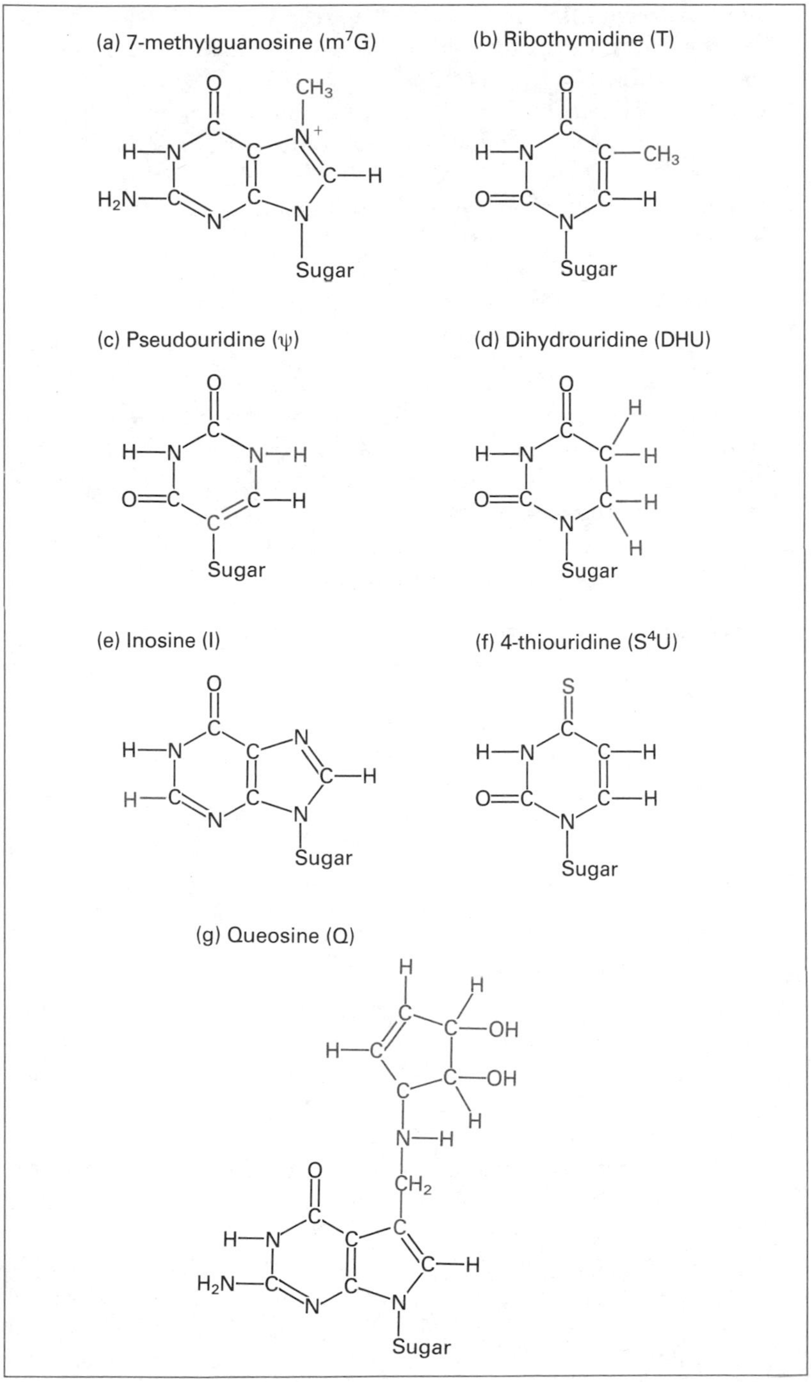

(a) 7-methylguanosine (m^7G)

(b) Ribothymidine (T)

(c) Pseudouridine (ψ)

(d) Dihydrouridine (DHU)

(e) Inosine (I)

(f) 4-thiouridine (S^4U)

(g) Queosine (Q)

Figure 6.10 A few examples illustrating the range of modified nucleosides found in tRNA. You should compare each structure with that of the standard RNA nucleoside from which it is obtained (see Figs 3.2(b) and 3.7(b)).

2. ***Base rearrangements*** – interchanging the position of atoms in the purine or pyrimidine ring. Example: uridine to pseudouridine (Fig. 6.10(c)).
3. ***Double-bond saturation*** – converting a double bond in the base to a single bond. Example: uridine to dihydrouridine (Fig. 6.10(d)).
4. ***Deamination*** – removal of an amino group (—NH_2). Example: guanosine to inosine (Fig. 6.10(e)).
5. ***Sulphur substitution*** – for the oxygen atom of guanosine or uridine. Example: uridine to 4-thiouridine (Fig. 6.10(f)).
6. ***Addition of more complex groups***. Example: guanosine to queosine (Fig. 6.10(g)).

Over 50 types of chemical modification have been discovered so far with tRNA nucleotides, each catalysed by a different tRNA modifying enzyme. The reasons for most of these modifications are unknown, although roles have been assigned to some specific cases, in particular those involving a nucleotide within the anticodon. These will be described in Chapter 9.

Reading

J. A. Lake (1981) The ribosome. *Scientific American*, **245**(2), 56–69– particularly useful for the three-dimensional structure of the ribosome.

A. Yonath (1984) Three-dimensional crystals of ribosomal particles. *Trends in Biochemical Sciences*, **9**, 227–30; H. Eisenberg (1987) Life at the end of the ribosome tunnel. *Trends in Biochemical Sciences*, **12**, 207–8 – describe the application of X-ray diffraction analysis to the ribosome.

R. Brimacombe (1984) Conservation of structure in ribosomal RNA. *Trends in Biochemical Sciences*, **9**, 273–7; R. Brimacombe and W. Stiege (1985) Structure and function of ribosomal RNA. *Biochemical Journal*, **229**, 1–17 – two reviews on the secondary structure and protein contact points for rRNA.

F. H. C. Crick (1958) On protein synthesis. *Symposium of the Society for Experimental Biology*, **12**, 138–63 – includes the first proposal of an adaptor molecule, now known to be tRNA, in protein synthesis.

M. B. Hoagland, M. L. Stephenson, J. F. Scott *et al.* (1958) A soluble ribonucleic acid intermediate in protein synthesis. *Journal of Biological Chemistry*, **231**, 241–57 – the discovery of tRNA.

R. W. Holley, J. Apgar, G. A. Everett *et al.* (1965) Structure of a ribonucleic acid. *Science*, **147**, 1462–5; R. Holley (1966) The nucleotide sequence of a nucleic acid. *Scientific American*, **214**(2), 30–9 – a research paper and review, respectively, on the first tRNA sequence determination.

S. K. Kim, F. L. Suddath, G. J. Quigley *et al.* (1974) Three-dimensional tertiary structure of yeast phenylalanine transfer RNA. *Science*, **185**, 435–40; J. D. Robertus, J. E. Ladner, J. T. Finch *et al.* (1974) Structure of yeast phenylalanine tRNA at 3 Å resolution. *Nature*, **250**, 546–51 – these two papers appeared simultaneously.

A. Rich and S. Kim (1978) The three-dimensional structure of transfer RNA. *Scientific American*, **238**(1), 52–62.

P. Gegenheimer and D. Apirion (1981) Processing of procaryotic ribonucleic acid. *Microbiological Reviews*, **45**, 502–41 – includes details of processing enzymes.

D. Soll (1971) Enzymatic modification of transfer RNA. *Science*, **173**, 293–9 – very detailed review of the chemical modification of tRNA.

Problems

1. Define the following terms:

   ribosomal RNA — primary transcript
   transfer RNA — pre-rRNA
   stable RNA — gene amplification
   turnover rate — tRNA cloverleaf
   ribosome — ribozyme
   sedimentation coefficient

2. Describe the contribution of the following people to our knowledge of RNA molecules:

   Svedberg Crick Holley

3. Distinguish between the compositions of typical prokaryotic and eukaryotic ribosomes.

4. Describe the methods that have been used and are being used to determine the structure of the ribosome.

5. How do prokaryotic and eukaryotic cells satisfy the demand for synthesis of (a) rRNA and (b) tRNA?

6. Draw and annotate the cloverleaf structure of tRNA. To what extent is the cloverleaf a true representation of the actual structure of tRNA?

7. Describe the role of the following enzymes in tRNA synthesis: RNase P, RNase D, tRNA nucleotidyl transferase.

8. *Ribosomal and transfer RNA molecules are relatively long-lived in the cell. In contrast, mRNAs are subject to quite rapid turnover rates. Discuss the possible reasons and consequences.

9. *In *Saccharomyces cerevisiae* the 5S rRNA genes are transcribed separately from the main rRNA transcription units. Nevertheless, the number of 5S rRNA genes is the same as the number of main transcription units. Propose a possible explanation for this.

10. Here is the sequence of the yeast alanine tRNA, first determined by Holley and his colleagues:

5′-GGGCGUGUGGCGUAGUCGGUAGCGCGCUCCCUU-
IGCIUGGGAGAGGUCUCCGGUUCGAUUCCGGACUC-
GUCCACCA-3′

Make a fully annotated drawing of a possible cloverleaf structure for this tRNA.

11. An RNA molecule has the sequence:

5′-AUUAGCAUGUAAAUGCUAUCUGGCAAC-3′

After digestion with a single-strand specific RNase the base paired molecule 5′-AGCAU-3′ / 3′-UCGUA-5′ is left. What was the base pair structure of the original molecule?

12. A second RNA molecule has the sequence:

5′-CUCCUCAAAUGAUUCAGUGCAUCUAACCCUAUU-
UAAACGCACUAGCUCAUGAGAGGAG-3′

After digestion with the same enzyme the following are obtained:

5′-UAA-3′ 5′-AUGA-3′ 5′-AGUGC-3′ 5′-CUCCUC-3′
3′-AUU-5′ 3′-UACU-5′ 3′-UCACG-5′ 3′-GAGGAG-5′

Draw the structure of the original molecule.

Types of RNA molecule: mRNA

7

Most mRNAs are unstable • Modification and processing • End-modifications • Intron splicing • RNA editing

Messenger RNA acts as the intermediate between the gene and the polypeptide translation product. Its existence was, like many of the major features of gene expression, postulated by Crick and his associates during the 1950s. Among the circumstantial evidence for mRNA at that time was the knowledge that in eukaryotic cells the genes reside on the chromosomes in the nucleus, whereas protein synthesis occurs on ribosomes in the cytoplasm. The physical separation of genes and ribosomes means that some sort of messenger molecule must carry the biological information from nucleus to cytoplasm. In bacteria the physical separation of DNA and ribosomes is less distinct but nevertheless a messenger molecule is still necessary.

Evidence that this messenger is in fact RNA came first from Elliot Volkin and Lazarus Astrachan of the Oak Ridge National Laboratory in 1956, but more convincingly from Sol Spiegelman and Benjamin D. Hall of Illinois University 5 years later. Both groups demonstrated that after infection of a culture of bacteria with a bacteriophage, the new RNA that is synthesized is related in sequence to the phage DNA, suggesting that the phage genes are copied into RNA before synthesis of the phage proteins occurs. Shortly afterwards two independent groups – Sydney Brenner, François Jacob and Matthew Meselson, who worked at different universities but did the crucial experiment at the California Institute of Technology, and James Watson's group at Harvard – directly identified mRNA molecules in *E. coli* cells.

Sydney Brenner b. 13 January 1927, Germiston, South Africa

Sydney Brenner first came to the UK in 1952 to work as a doctoral student at Oxford with Sir Cyril Hinshelwood. He met Watson and Crick in May 1953 when he visited Cambridge to see the double helix model of DNA. This meeting led to a long friendship and in 1957 Brenner moved to Cambridge, to what became the Medical Research Council Laboratory of Molecular Biology, where he was Director from 1979 to 1986. He has made numerous important contributions to our understanding of gene expression. As well as being involved in the first direct identification of mRNA in 1961, he also participated in the experiments that proved that the genetic code is triplet (Section 8.2.1), and coined the term 'codon' for the triplet codeword. In the 1960s he initiated research on the nervous system of the nematode *Caenorhabditis*, which he has used as a simple model for studying the genetics of development and other complex functions in eukaryotes.

7.1 Most mRNA molecules are unstable

Messenger RNA molecules are not generally long-lived in the cell. Most bacterial mRNAs have a **half-life** of only a few minutes and so are turned over very rapidly. In eukaryotic cells, mRNA molecules have longer half-lives (e.g. 6 hours for many mammalian

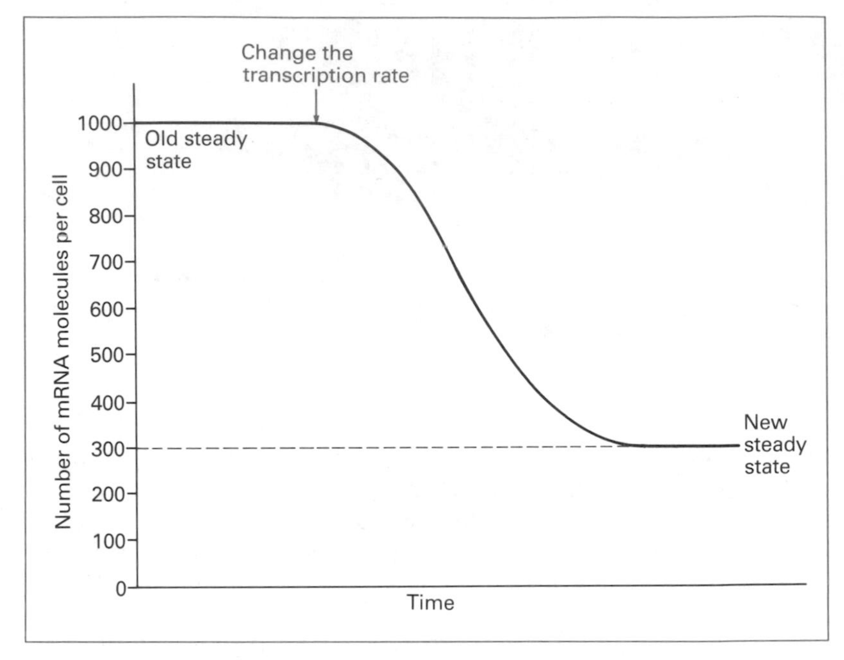

Figure 7.1 Effect of transcription rate on the mRNA steady state in the cell. In this example, the half-life of the mRNA molecule is 5 minutes. The initial transcription rate is 100 molecules of mRNA synthesized per minute, leading to a steady state of 1000 molecules per cell. The transcription rate then decreases to 30 molecules of mRNA synthesized per minute; the steady state gradually drops until the new value of 300 molecules per cell is reached.

mRNAs) but are still subject to turnover. Turnover of mRNA is important because it means that the absolute amount of a particular mRNA in the cell can be controlled by adjusting the rate at which the relevant gene is transcribed. If the transcription rate for the gene decreases then the level of the mRNA in the cell will also decrease until a new steady state is reached (Fig. 7.1). As we shall see in Chapter 10, regulation of gene expression primarily involves control over transcription, a system that can work only if mRNAs are turned over relatively rapidly.

There are a few exceptions and examples of long-lived mRNAs are known. For instance, globin mRNA in reticulocytes is almost fully stable. In these cells regulation of globin gene expression is not important because the maximum rate of globin synthesis is required virtually all the time. Other long-lived mRNAs are being discovered as different types of specialized cell are studied.

7.2 Modification and processing of mRNA

In bacteria the mRNA molecules that are translated are direct copies of the genes. In eukaryotes the situation is different and most mRNAs undergo a fairly complicated series of modification and processing events before translation occurs. These include:

1. Chemical modifications to the two ends of the mRNA molecule.
2. Removal of introns.

RNA modification and processing in the context of the cell

The difference between prokaryotic and eukaryotic mRNA with respect to modification and processing possibly reflects the greater structural complexity of the eukaryotic cell. In prokaryotes there is only one cell compartment and transcription and translation can occur together; indeed, the two processes may be coupled so that translation of an mRNA starts before transcription of the molecule has been completed:

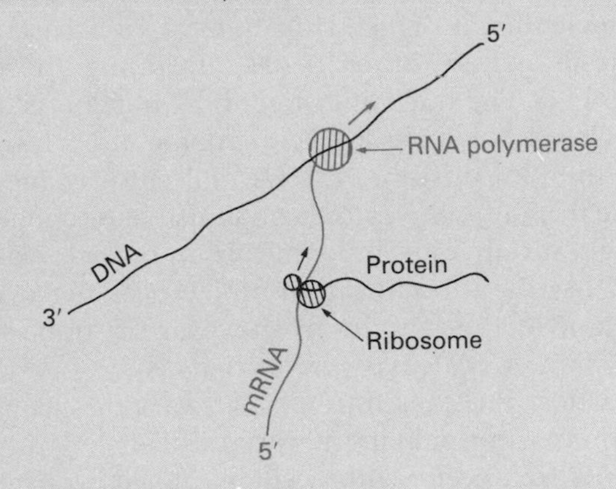

In eukaryotes the transcribed mRNA must be transported to the cytoplasm before translation can occur:

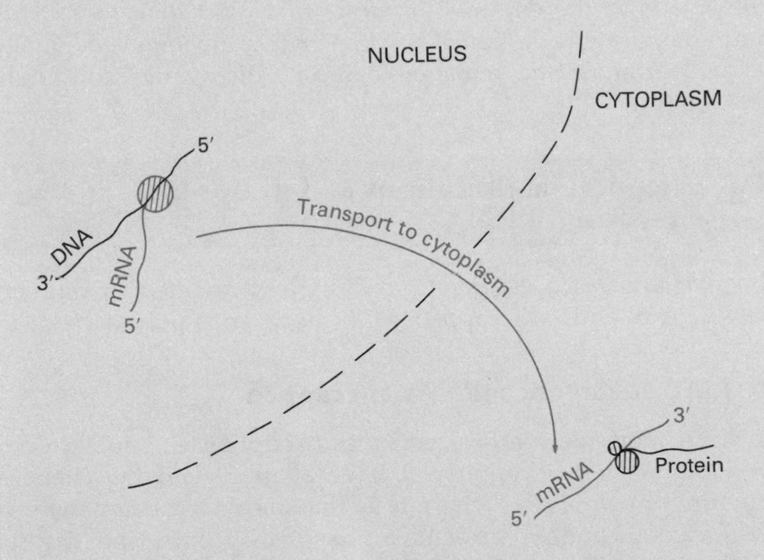

Modification and processing events take place in the nucleus and are almost invariably completed before the mRNA is transported into the cytoplasm. Links have been sought between the modification/processing events and the subsequent ability of the mRNA to be transported to its correct site in the cytoplasm, but as yet no definite connection can be made.

3. In a few special cases, alterations to the nucleotide sequence of the mRNA (**RNA editing**).

Remember, these events occur only in eukaryotes.

The first inkling that eukaryotic mRNAs undergo modification and processing came when RNA fractions present in the nucleus and cytoplasm were compared. The nucleus can be divided into two regions: the **nucleolus**, in which rRNA genes are transcribed, and the **nucleoplasm**, where other genes, including those for mRNA, are transcribed. The nucleoplasmic RNA fraction is called **heterogeneous nuclear RNA** or **hnRNA**, the name indicating that it is made up of a complex mixture of RNA molecules, some over 20 kb in length. The mRNA in the cytoplasm is also heterogeneous, but its average length is only 2 kb. If the mRNA in the cytoplasm is derived from hnRNA, then modification and processing events, including a reduction in the lengths of the primary transcripts, must occur before the mRNA leaves the nucleus.

The hnRNA fraction includes mRNAs at varying stages of modification and processing, making it very difficult to study the exact order of events that occur with a single molecule. There is evidence that the end-modifications are completed before all the introns are removed, but this may not be true for all mRNAs. We will deal with the end-modifications first, and then look at the more complicated question of how the introns are removed. Finally, we will examine the remarkable recent discoveries concerning RNA editing.

7.3 Chemical modifications at the ends of eukaryotic mRNAs

The two ends of a eukaryotic mRNA are modified in different ways. The 5′ ends are 'capped', and the 3′ ends polyadenylated.

7.3.1 All eukaryotic mRNAs are capped

An RNA molecule synthesized by transcription and subjected to no additional modification will have at its 5′ end the chemical structure pppNpN . . . , where N is the sugar-base component of the nucleotide and p represents a phosphate group. Note that the 5′ terminus will be a triphosphate (see Fig. 3.5). With mature eukaryotic mRNA this is not the case as the 5′ terminus has a more complex chemical structure described as m^7GpppNpN . . . (Fig. 7.2), where m^7G is the nucleotide carrying the modified base 7-methylguanine. The m^7G nucleotide is added to the mRNA molecule after transcription by a two-step process, with methylation occurring only after a standard G has been added. As can be seen from Fig. 7.2 the phosphate linkage between the m^7G and

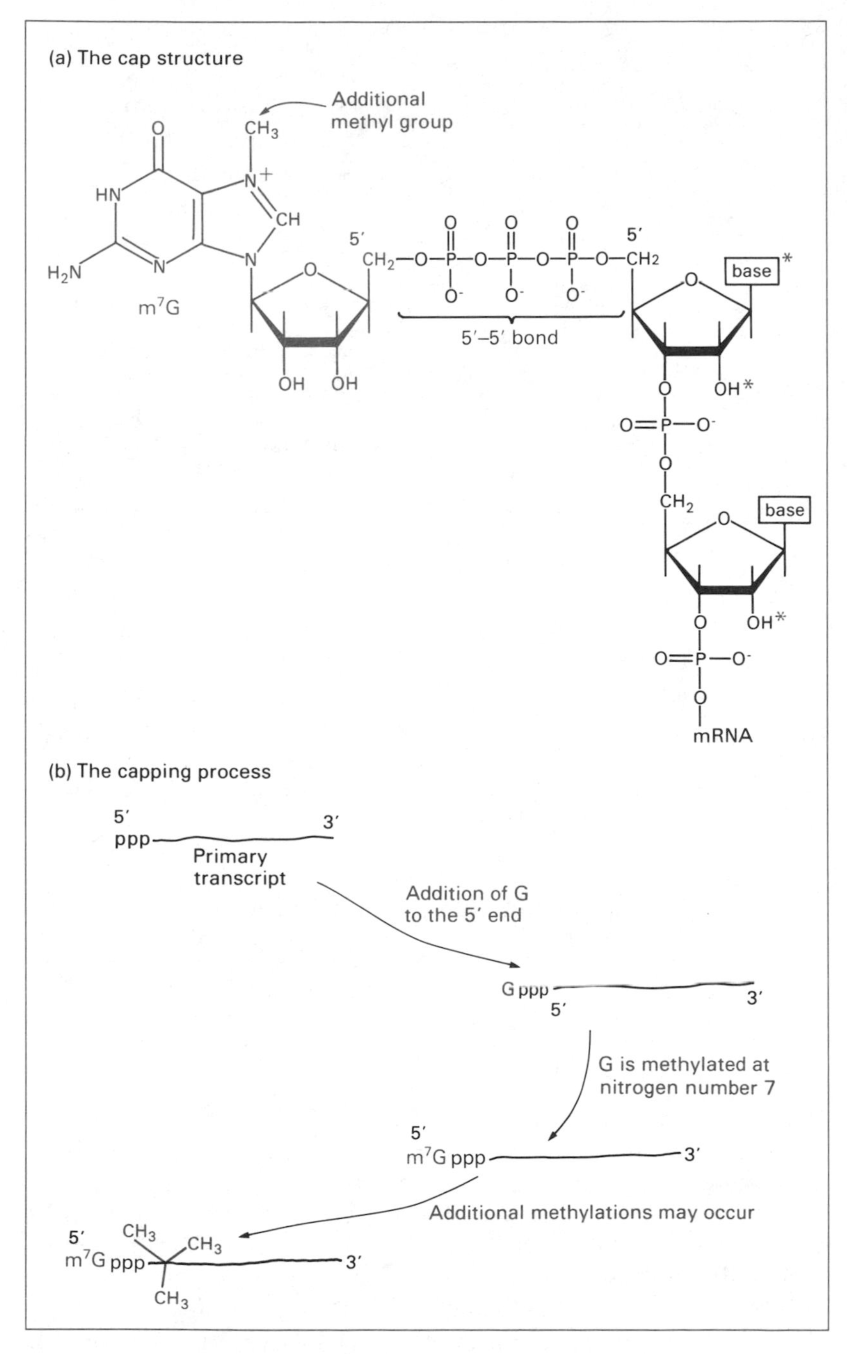

Figure 7.2 Capping of eukaryotic mRNA. (a) The cap structure, showing the modified G attached to the 5′ end of the mRNA molecule by the unusual 5′–5′ bond (*positions of possible additional methylations). (b) The steps in the capping process.

the first nucleotide of the transcript is an unusual one: rather than the normal 5′–3′ bond found in a polynucleotide, the bond in the **cap structure** is between the two 5′-carbons of the adjacent nucleotides.

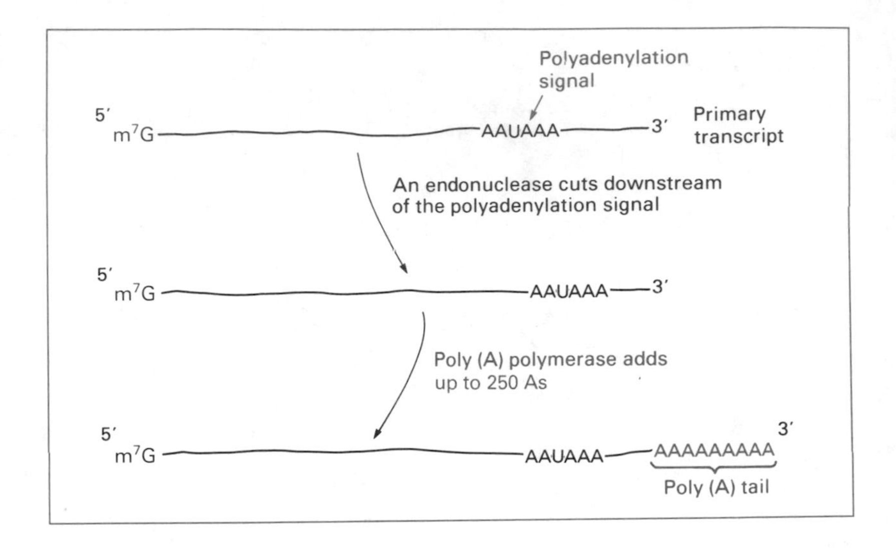

Figure 7.3 Polyadenylation of eukaryotic mRNA.

As well as this basic cap structure, some eukaryotic mRNAs may undergo further modification of the 5′ end with the addition of further methyl groups to one or both of the next two nucleotides of the molecule. No definite function has yet been assigned to capping, but the cap structure is thought to play a role during translation of the mRNA (Section 9.3).

7.3.2 Most eukaryotic mRNAs are polyadenylated

A second modification of most eukaryotic RNAs is the addition to the 3′ end of the molecule of a long stretch of up to 250 A nucleotides, producing a poly(A) tail:

$$\ldots \mathrm{pNpNpA(pA)_{n}pA}$$

Polyadenylation does not occur at the 3′ end of the primary transcript. Instead, the final few nucleotides are removed by a cleavage event that occurs between 10 and 30 nucleotides downstream of a specific polyadenylation signal (consensus sequences 5′-AAUAAA-3′; Fig. 7.3) to produce an intermediate 3′ end to which the poly(A) tail is subsequently added by the enzyme **poly(A) polymerase**.

The reason for polyadenylation is not known although several hypotheses have been built round the idea that the length of the poly(A) tail determines the time the mRNA survives in the cytoplasm before being degraded. All theories must take into account the fact that certain eukaryotic mRNAs, notably those coding for histones (Section 15.2.1), are not polyadenylated.

Purification of mRNA

The poly(A) tails present on most eukaryotic mRNAs provide a convenient means for purification of these molecules. The technique is called affinity chromatography. This is a special type of chromatography (p. 34) where the solid matrix carries reactive groups to which the molecules to be purified bind selectively. For purification of mRNA the solid matrix carries poly(U) or poly(dT) polynucleotides. When a sample of total cell RNA (mRNA, rRNA and tRNA) is applied, the mRNA molecules attach to the matrix via their poly(A) tails. RNA molecules that are not polyadenylated do not base pair with the matrix and so can be removed. The mRNA can then be eluted in pure form by washing with a solution containing a high concentration of salt (usually 1.5 M NaCl), which destabilizes the base pairing.

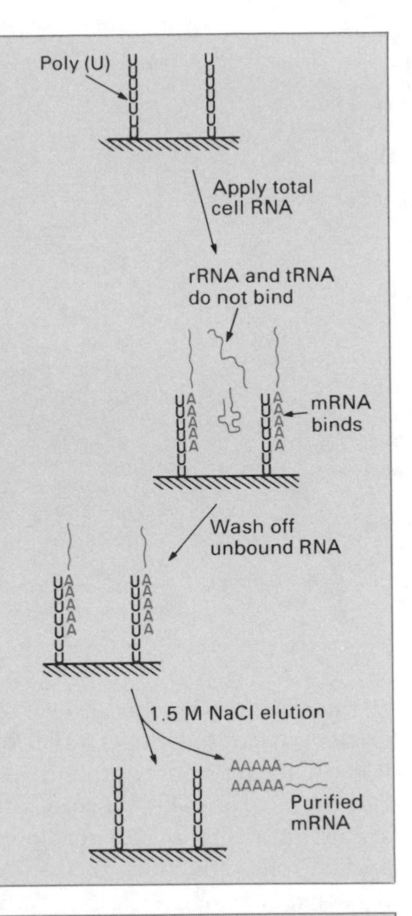

7.4 Intron splicing

Transcription produces a faithful copy of the template strand of the gene. Therefore, if the gene contains introns then the primary transcript will include copies of these. However, the introns must be removed and the exon regions of the transcript attached to one another before translation can occur. This process is called **splicing**; it is very complex and by no means fully understood.

One complication in studying splicing is that there are at least four different types of intron, each with their own special features and each with a different splicing mechanism. Similarities between the four classes of intron are beginning to emerge, but there are still important distinctions. We will devote most of our attention to the 'GT–AG' class of introns, as all the introns found in nuclear genes that are transcribed into mRNA fall into this category.

7.4.1 Splicing of GT–AG introns

The designation 'GT–AG' refers to the fact that virtually all members of this class of intron possess the same dinucleotide sequences

Examining a gene for introns

Introns are easily identified when the nucleotide sequence of a gene is compared with the nucleotide sequence of its mRNA or the amino acid sequence of the protein it specifies. Introns can also be detected experimentally by a method that depends on the ability of the gene to base pair ('hybridize') to its RNA transcript. If the gene does not contain introns then the mRNA will be a direct and continuous copy of the DNA and an uninterrupted heteroduplex will form:

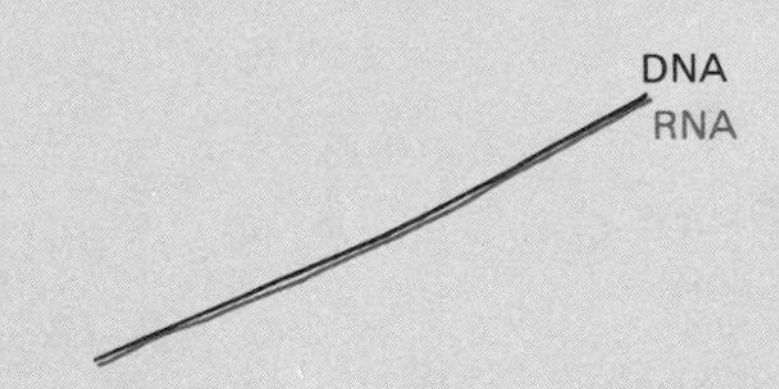

If on the other hand the gene contains one or more introns then regions of the DNA will not be able to base pair with the spliced RNA, as the introns will have been removed from the latter. The result will be a series of loops within the heteroduplex:

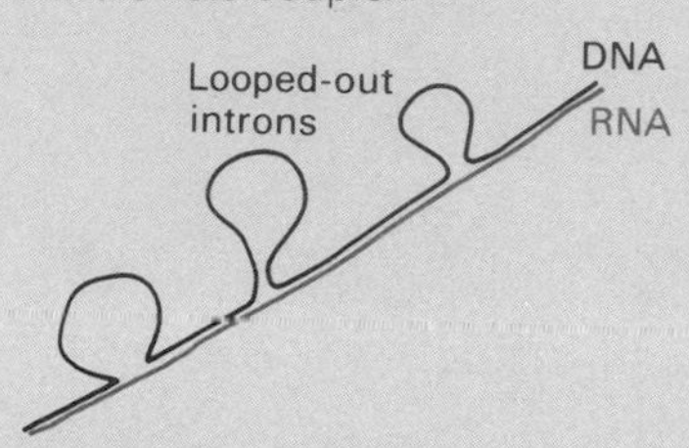

Two methods can be used to determine whether or not the heteroduplex contains loops. Polynucleotides can be seen with the electron microscope, as long as the molecules have been treated with proteins that bind to the nucleic acids and increase their effective diameter. Alternatively, a single-strand specific deoxyribonuclease (an example is S1 nuclease, Section 21.3.1) can be used to digest the loops in the heteroduplex. The DNA fragments that remain are examined by electrophoresis: the number and sizes of the fragments indicate the number and relative positions of the introns.

at their 5′ and 3′ ends. These sequences are GT for the first two nucleotides of the intron (GT is of course the sequence in the DNA; it is GU in the transcript) and AG for the last two nucleotides (Fig. 7.4). In fact the **GT–AG rule** represents only a component of what are actually slightly longer consensus sequences found at the 5′ and 3′ splice junctions of GT–AG introns. In vertebrates the full consensus sequences are:

5′ splice site	5′-AG↓GTAAGT-3′
3′ splice site	5′-PyPyPyPyPyPyNCAG↓-3′

These are slightly more complicated consensus sequences than we have encountered before. 'Py' indicates a position where either of the two pyrimidine nucleotides (C or U) can occur, and 'N' represents any nucleotide. The consensus sequence for the 3′ splice site is therefore six pyrimidines in a row, followed by any nucleotide, followed by a C and then the AG. Remember that as

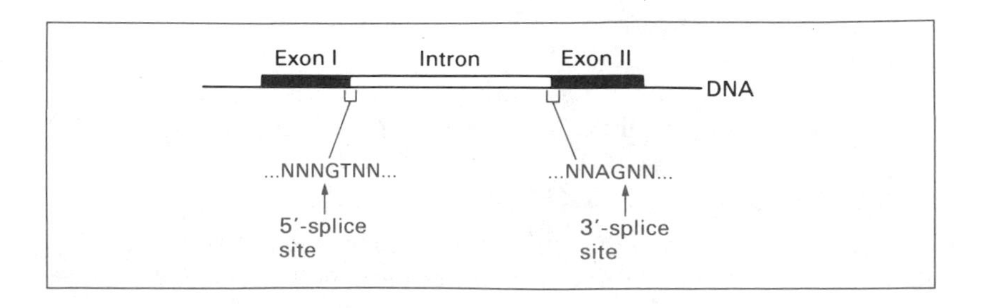

Figure 7.4 The GT–AG rule.

these are consensus sequences the actual sequences seen with an individual intron may be slightly different, although the GT and AG components are almost always present.

The sequences around the two splice sites were originally thought to be the only regions of nucleotide similarity in GT–AG introns. However, in the yeast *Saccharomyces cerevisiae* there is an additional conserved sequence:

The **TACTAAC box** 5′-TACTAAC-3′

This sequence is exactly the same in all yeast introns that have been sequenced so far and is positioned within the intron, between 18 and 140 bp upstream of the 3′ splice site. Other lower eukaryotes (for instance, fungi) have sequences similar to the TACTAAC box, but the motif is not present in vertebrate introns.

(a) U-RNAs are involved in splicing GT–AG introns. It has been known for some time that a special type of RNA molecule is involved in intron splicing. These molecules are called **U-RNAs** (for 'uracil-rich RNAs') or **small nuclear RNAs** (**snRNAs**). There are a number of different U-RNAs in vertebrate nuclei, the most abundant being U1 to U6. They are quite short, varying in size from 106 nucleotides for U6 to 217 nucleotides for U3. The sequence of each type of U-RNA is very similar in all higher eukaryotes, and analogous molecules are also present in lower eukaryotes such as yeast. U-RNAs are in fact another class of stable RNA and are just as important in eukaryotes as rRNA and tRNA. In the cell U-RNAs exist within particles called **small nuclear ribonucleoproteins** (**snRNPs**), each of which contains one U-RNA (except for U4 and U6 which are present in the same snRNP) and several proteins. SnRNPs can be purified from nuclear extracts as particles with sedimentation coefficients of 10S to 12S.

(b) The splicing pathway for GT–AG introns. Our understanding of the splicing process for GT–AG introns has advanced immensely during recent years, mainly thanks to the development of experimental procedures for purifying unspliced transcripts and subjecting them to splicing under controlled conditions in the test-tube. In outline, the splicing reaction involves the following three steps (Fig. 7.5):

1. Cleavage occurs at the 5′ splice site.
2. The resulting free 5′ end is attached to an internal site within the intron to form a lariat structure. In yeast the branch site is the last A in the TACTAAC box. In vertebrates the branch site is also an A, but as vertebrate introns do not have TACTAAC boxes it is not clear how this A is selected.
3. The 3′ splice site is cleaved and the two exons joined together.

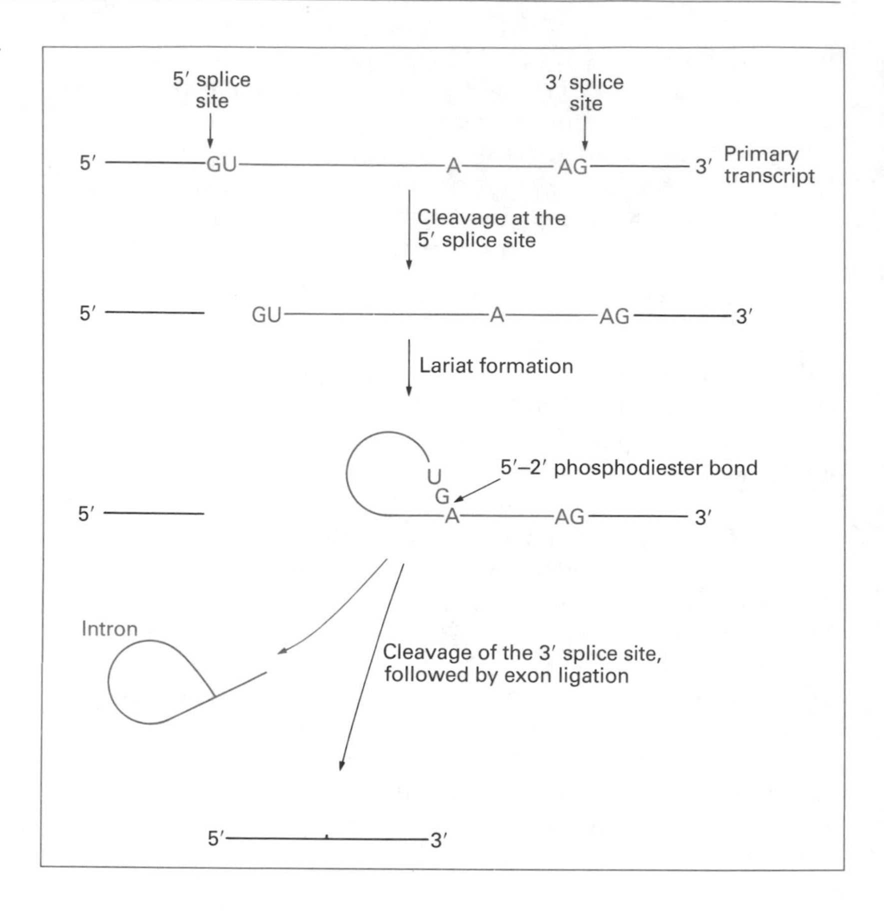

Figure 7.5 The splicing pathway for GT–AG introns.

How are the snRNPs involved in this process? A lot of research effort has been put into answering this question but several pieces of the jigsaw are still missing. For some time it has been known that U1-RNA is able to base pair with the consensus sequence around the 5′ splice site. Attachment of U1-snRNP to this region of the intron is thought to be one of the first steps in the splicing reaction (Fig. 7.6(a)). It is also known that U2-snRNP attaches to the branch site, and again base pairing may be involved, although this is less certain. The interaction could be between the branch site and one of the proteins in U2-snRNP, rather than with U2-RNA itself.

Other U-RNAs

Not all U-RNAs are involved in intron splicing. U3 has an undetermined role in processing of pre-rRNA (Section 6.1.2), U11 assists in mRNA polyadenylation (Section 7.3.2), and U7 is involved in processing of the 3′ ends of histone mRNAs (these are unusual pre-mRNAs in that they are not polyadenylated). Functions have not yet been assigned for U8, U9 and U10.

What happens next? It has been suggested that the U1- and U2-snRNPs have an affinity for each other, and that this draws the 5′ splice site towards the branch point (Fig. 7.6(b)). The other snRNPs involved in splicing (U5-snRNP and U4/U6-snRNP) also attach themselves to the intron at this stage, resulting in a large complex (sedimentation coefficient of 35S) which has been called the **spliceosome** (Fig. 7.6(c)). The cutting and joining reactions that excise the intron and ligate the exons take place within the

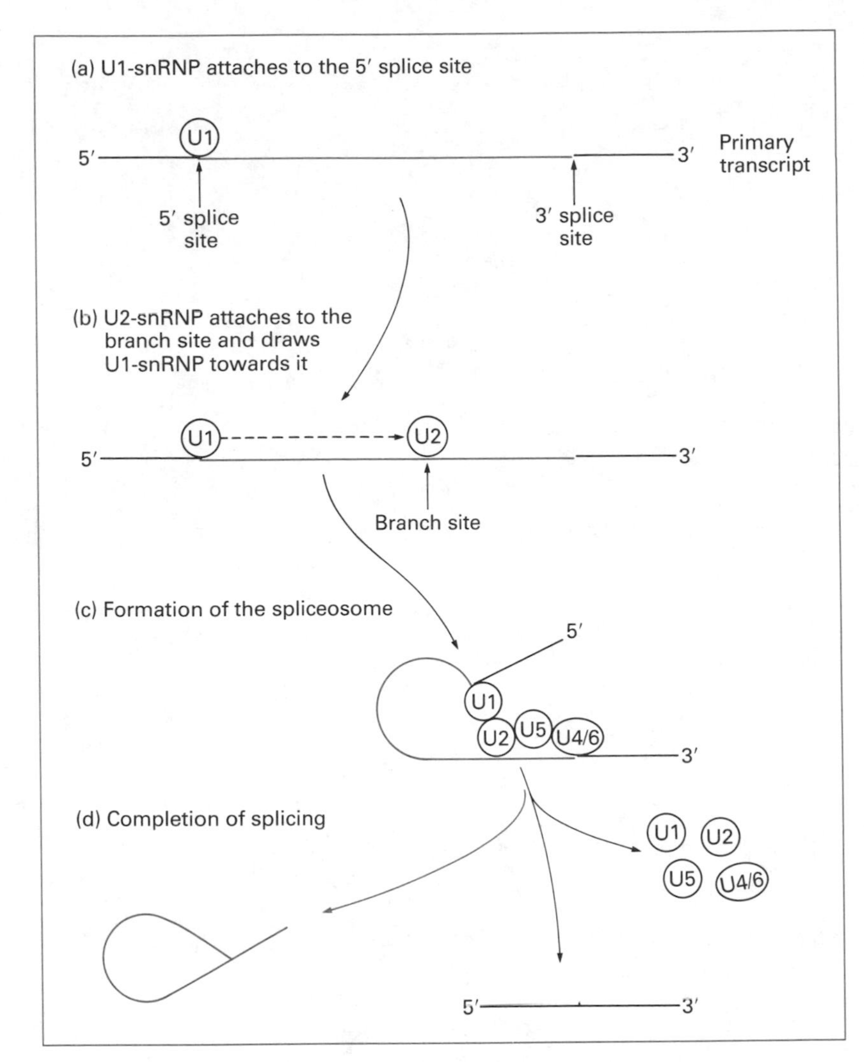

Figure 7.6 The involvement of snRNPs in the splicing of GT–AG introns.

spliceosome. Once completed the splicesome dissociates into its component snRNPs and the spliced mRNA is released (Fig. 7.6(d)).

The splicing pathway for GT–AG introns still presents a number of problems. The spliceosome also contains additional proteins, not components of the snRNPs, and the functions of these are still largely unknown. The cutting and joining reactions must be catalysed by one or more enzymes but none of the proteins in the spliceosome have been shown to have the appropriate enzymatic activities. This leads to the possibility that the intron RNA is itself enzymatic, another example of a ribozyme. This is still speculation with GT–AG introns, but is relevant as another type of intron definitely does have ribozyme activity. This is the famous self-splicing intron of *Tetrahymena*, to which we now turn our attention.

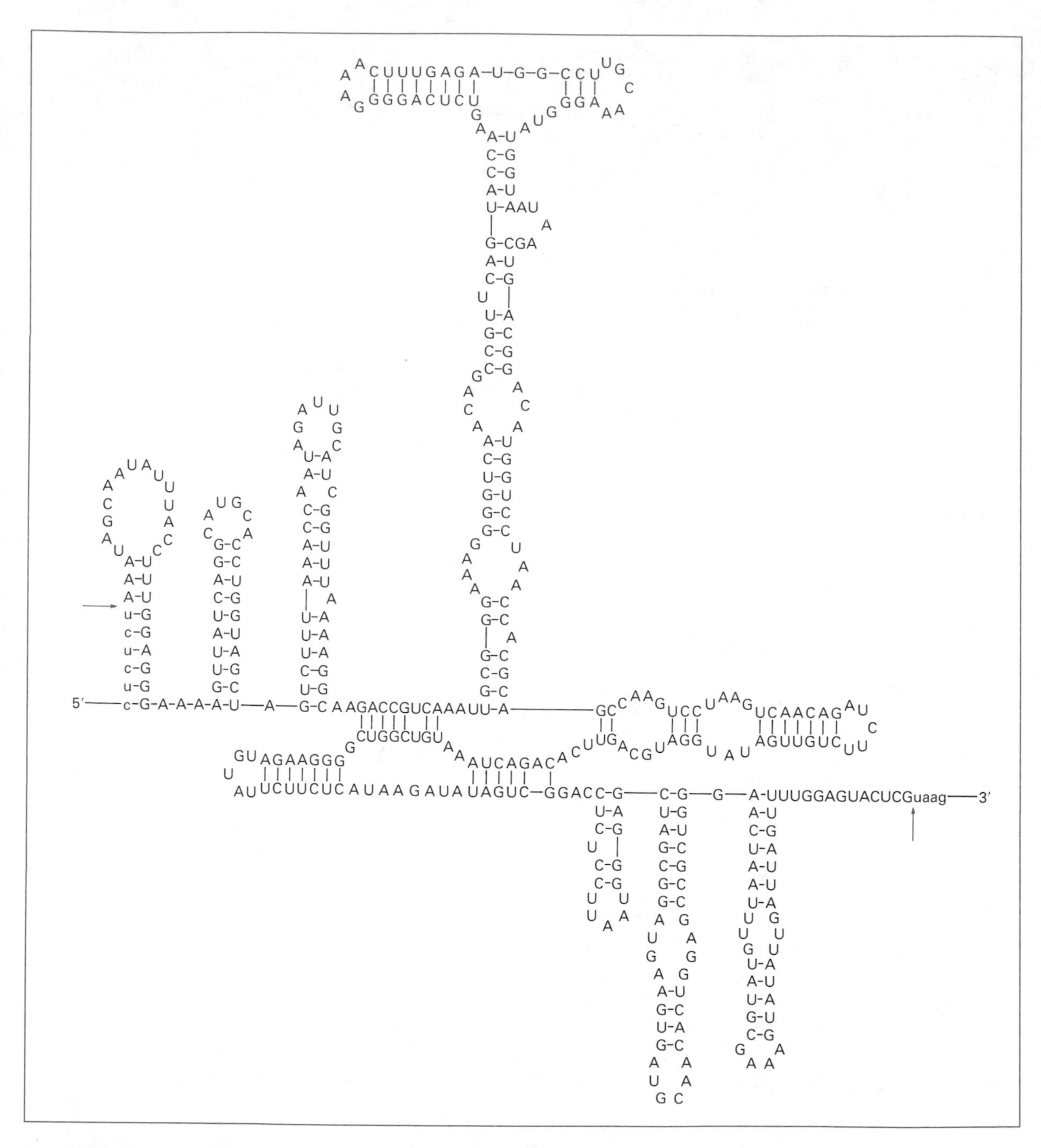

Figure 7.7 Two-dimensional representation of the base paired structure taken up by the *Tetrahymena* self-splicing intron. In this drawing nucleotides in the intron are shown in capital letters, with small letters used for the exons. The two arrows point to the splice sites. The actual three-dimensional structure is more complex, with the intron folded up more tightly so the splice sites are brought close together. Redrawn from J. M. Burke *et al.* (1987) *Nucleic Acids Research*, **15**, 7217–21.

Thomas Cech b. 8 December 1947, Chicago

Thomas Cech received his PhD from the University of California, Berkeley and then moved to the Massachusetts Institute of Technology in 1975. In 1978 he joined the faculty of the University of Colorado at Boulder and began the research that led to the discovery of self-splicing RNA. The first suggestion that the *Tetrahymena* intron might mediate its own splicing was published in 1981, and was followed the next year by the demonstration that the intron RNA excises itself from the adjoining exons in the complete absence of any proteins. In 1986 Cech and his group engineered the intron RNA into an RNA enzyme, capable of cleaving other RNA molecules without being altered in the process. As well as the importance of these discoveries in understanding introns and how they are spliced, Cech's work has implications regarding the origins of life, as ribozymes may have served as both templates and catalysts to produce a primordial, self-replicating biochemical system. Cech shared the 1989 Nobel Prize for Chemistry with Sydney Altman of Yale University.

'Scientists frequently get strange and unexpected experimental results and typically get very excited about them. The trick is deciding if such results are meaningful and worth pursuing, or if one is being misled by some flaw in the experimental method.'

7.4.2 The self-splicing intron

Introns are not only found in genes that are transcribed into mRNA. In particular, the rRNA genes of certain protozoa, notably the ciliate *Tetrahymena*, contain introns. These introns are not members of the GT–AG class as they lack the characteristic consensus sequences and they are not associated with spliceosomes. Instead, the intron folds up by intramolecular base pairing into a complex tertiary structure in which the two splice sites are brought close together (Fig. 7.7). This base paired intron is a ribozyme that catalyses its own splicing reaction: the intron is cut out and the two exons joined together in the complete absence of any protein molecules.

The *Tetrahymena* self-splicing intron was the first ribozyme to be discovered and caused quite a stir. To convince the doubters, Thomas Cech and his collaborators devised an elegant series of experiments to demonstrate first, that the splicing reaction takes place in the complete absence of proteins, and second, that the intron is able to act as an enzyme according to the strictest biochemical definitions of the term. Cech has also investigated the base paired structure of the intron and determined how the catalytic reaction occurs.

The self-splicing intron is a member of the 'Group I' class of introns. Group I introns are also found in the rRNA genes of other protozoa and in mitochondrial genes of fungi and yeast (Section 15.5.2). They are characterized by their ability to take up the base

Ribozymes

We have so far encountered three known or possible ribozymes in this book. The *Tetrahymena* self-splicing intron was the first ribozyme to be discovered, with RNase P (Section 6.2.2) following soon after. Whether the RNA components of ribosomes possess any enzymatic activity (Section 6.1.1) is still open to debate.

In addition to these examples, several unusual viruses have ribozymes. These are called satellite viruses or virusoids, and are probably the most extreme form of 'life' on our planet. All they consist of is an RNA molecule of about 350 nucleotides, without any obvious genes, but capable of being replicated within the plant cells that they infect. In the final stage of replication the two daughter RNAs, which are joined together head-to-tail, are separated by a self-catalysed cleavage:

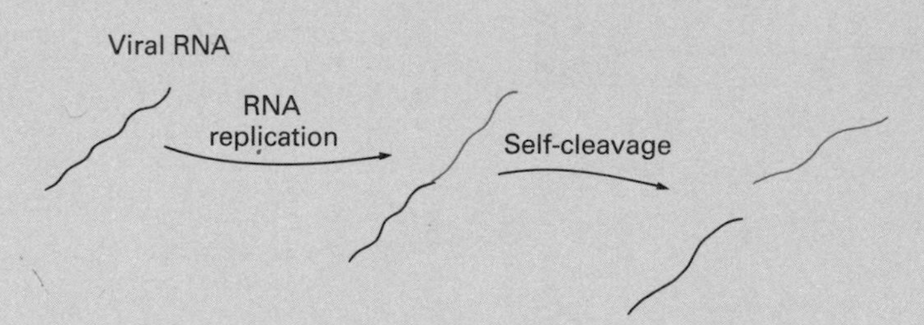

The discovery of ribozymes underlines an important general point in biology. In 1980 the idea that RNA could act as an enzyme was totally outrageous. Now we accept it as an important facet in our understanding of biological processes. In biology you must always be prepared to throw away your preconceptions and embrace totally new ideas. Do not fall into the trap of believing that we already understand most of what there is to know.

paired structure displayed by the *Tetrahymena* intron, but only the *Tetrahymena* intron has been shown to splice efficiently in the absence of proteins. This does not mean that the other Group I introns are not ribozymes. The splicing reaction may be catalysed by the intron itself, with the proteins known to be required in the cell acting only to stabilize the base paired structure.

7.4.3 Other types of intron

So far we have dealt with two of the four classes of intron. The remaining types are:

1. 'Group II' introns, also found predominantly in mitochondrial genes. These introns are interesting as they seem to be intermediate between the Group I and GT–AG classes. Like Group I introns, they take up a base paired structure (although not the same as the Group I structure) and have some self-splicing activity. On the other hand, they resemble GT–AG introns as their splicing reaction involves a lariat intermediate.
2. tRNA introns, which are relatively short and are found in tRNA genes, almost invariably in the anticodon loop (Section 6.2.1).

The intron is removed by a ribonuclease, after the tRNA has taken up its cloverleaf structure. The steps involved in splicing these introns are similar to those responsible for cutting the tRNAs out of their precursor-tRNA transcripts (Section 6.2.2).

7.5 RNA editing

Now we return to the processing of mRNA, to examine the unusual event called RNA editing. At the moment only a few isolated instances of RNA editing are known in different eukaryotes, and it is by no means clear if the different examples represent the same process, or a series of unrelated phenomena.

In RNA editing the nucleotide sequence of an mRNA is altered by inserting new nucleotides, or by deleting or changing existing ones. RNA editing was discovered in 1986 during studies of mitochondrial genes in the parasitic protozoan *Trypanosoma brucei*. It was realized that the primary transcripts of many mitochondrial genes in trypanosomes are edited by the insertion of Us. The editing is quite extensive (Fig. 7.8(a)), with one or more Us being added at a number of different positions in each mRNA. Since this initial discovery, other examples of RNA editing have been found in a variety of organisms, including the conversion of a C to a U in the mRNA for human apolipoprotein-B (Fig. 7.8(b)).

RNA editing is currently a major talking point in biology. The editing has to be carried out with absolute precision otherwise the genetic message will be scrambled: you will appreciate this point when we encounter the genetic code in the next chapter. Molecular

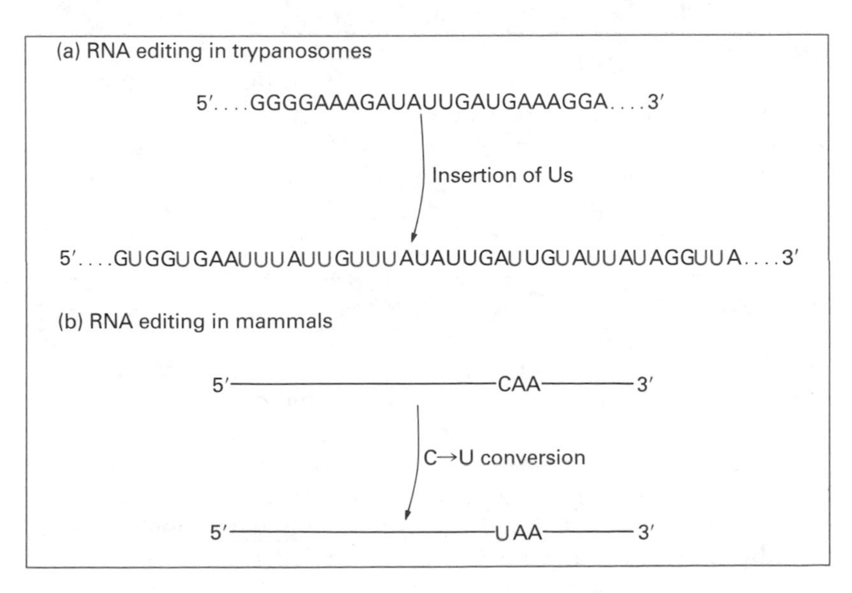

Figure 7.8 RNA editing. (a) An example of RNA editing with a mitochondrial mRNA from *Trypanosoma brucei*. (b) RNA editing of the mRNA for apolipoprotein-B.

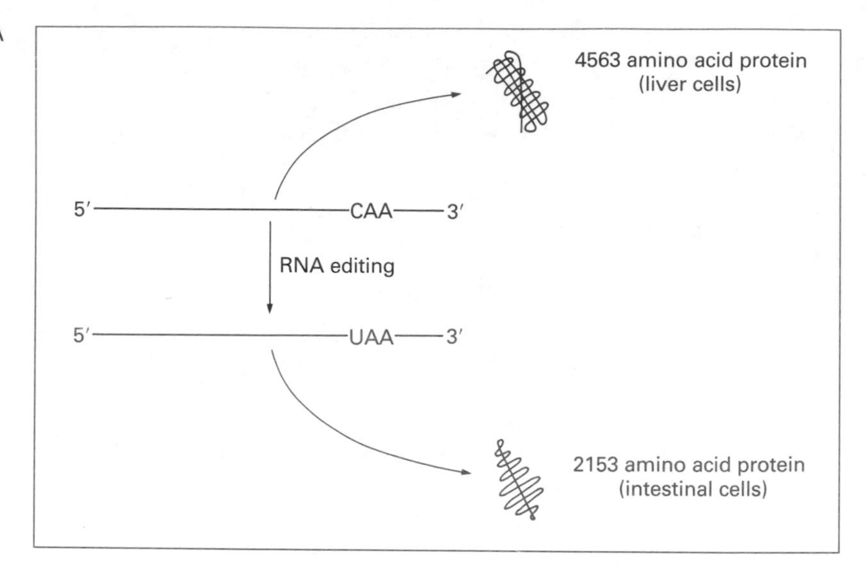

Figure 7.9 RNA editing of the mRNA for human apolipoprotein-B results in a shorter polypeptide being synthesized in intestinal cells.

biologists accept that a few nucleotides in an mRNA could be inserted or modified with precision, perhaps by mechanisms similar to those responsible for tRNA modification (Section 6.2.2). It is much more difficult to imagine how the necessary precision can be achieved with the extensive editing seen with trypanosomal mRNAs. Elaborate hypotheses are currently being tested.

Even more perplexing is why RNA editing exists in the first place. In a few instances we do have answers; for instance, with the apolipoprotein-B mRNA the editing occurs specifically in intestinal tissue and results in a modified protein that functions differently from the apolipoprotein-B produced elsewhere in the body (Fig. 7.9). This is an example of an existing mRNA being given a slightly different function by RNA editing. In contrast, the unedited mRNAs of trypanosomes are non-functional: only after editing does the mRNA have a meaningful genetic message. Why not simply have the complete message in the gene? Nobody knows: if you have any bright ideas then tell me about them.

Reading

S. Brenner, F. Jacob and M. Meselson (1961) An unstable intermediate carrying information from genes to ribosomes for protein synthesis. *Nature*, **190**, 576–81; F. Gros, H. Hiatt, W. Gilbert, C. G. Kurland, R. W. Riseborough and J. D. Watson (1961) Unstable ribonucleic acid revealed by pulse labelling of *Escherichia coli*. *Nature*, **190**, 581–5 – the two demonstrations of the existence of mRNA.

J. M. Adams and S. Cory (1975) Modified nucleosides and bizarre 5′-termini in mouse myeloma mRNA. *Nature*, **255**, 28–33 – an early identification of capping.

C. S. McLaughlin, J. R. Warner, M. Edmonds *et al.* (1973) Polyadenylic acid sequences in yeast messenger ribonucleic acid. *Journal of Biological Chemistry*, **248**, 1466–71 – the clearest early identification of polyadenylation.

M. Wickens (1990) How the messenger got its tail: addition of poly(A) in the nucleus. *Trends in Biochemical Sciences*, **15**, 277–81 – a clear description of our current knowledge on polyadenylation.

J. A. Witkowski (1988) The discovery of 'split' genes: a scientific revolution. *Trends in Biochemical Sciences*, **13**, 110–13 – this is mainly a historical account but it is worth reading if you are interested in the impact that new ideas have on molecular biologists.

R. A. Padgett, P. J. Grabowski, M. M. Konarska and P. A. Sharp (1985) Splicing messenger RNA precursors: branch sites and lariat RNAs. *Trends in Biochemical Sciences*, **10**, 154–7; D. A. Wassarman and J. A. Steitz (1991) Alive with DEAD proteins. *Nature*, **349**, 463–4 – the first review is a good summary of the basic details of splicing of GT–AG introns; the second paper will show you how complicated the process really is.

R. Benne (1990) RNA editing in trypanosomes: is there a message? *Trends in Genetics*, **6**, 177–81 – a well-written review that describes ideas on how RNA editing takes place.

P. Hodges and J. Scott (1992) Apolipoprotein B mRNA editing: a new tier for the control of gene expression. *Trends in Biochemical Sciences*, **17**, 77–81.

Problems

1. Define the following terms:

half-life	poly(A) polymerase
nucleolus	splicing
nucleoplasm	TACTAAC box
heterogeneous nuclear RNA	U-RNA
	small nuclear RNA
cap structure	small nuclear ribonucleoprotein
polyadenylation	spliceosome

2. Explain why the existence of mRNA was suspected before it was actually discovered.

3. Describe the modification events that occur at the 5′ and 3′ ends of a eukaryotic mRNA molecule.

4. Why is it necessary to remove introns from pre-mRNA before translation occurs?
5. Explain what the GT–AG rule refers to.
6. Draw the series of events that are believed to result in removal of a typical GT–AG intron from a primary transcript.
7. What are the differences between the splicing mechanisms for GT–AG and Group I introns?
8. What are the similarities between GT–AG, Group I and Group II introns?
9. Explain what is meant by RNA editing. Give examples of editing in lower and higher eukaryotes.
10. *How might the length of the poly(A) tail influence the half-life of a eukaryotic mRNA?
11. *Eukaryotic cells expend a lot of energy in removing introns from primary transcripts. Yet introns seem to play no role of their own. Discuss.
12. *Discuss the questions raised by the discovery of RNA editing.

8 The genetic code

Polypeptides are polymers – amino acids – levels of protein structure – the importance of the amino acid sequence • The genetic code – first principles – elucidation – features of the code

The three major types of RNA molecule that are produced by transcription – messenger, ribosomal and transfer RNA – work together to synthesize proteins by the process of translation. The central fact about translation is that the sequence of amino acids in the polypeptide being synthesized is specified by the sequence of nucleotides in the mRNA molecule being translated (Fig. 8.1). The rules that determine which sequence of nucleotides specifies which sequence of amino acids are embodied in the genetic code.

The genetic code is usually looked on as a topic independent of protein synthesis itself. This view goes back to the 1950s when the concept of the genetic code and the experiments designed to examine it were very much the province of molecular biologists, whereas unravelling the intricacies of how proteins are made was a problem for biochemists. This distinction between molecular biology and biochemistry is regrettable, but we will nonetheless examine the genetic code in this chapter and the mechanisms of protein synthesis in the next. Before doing either we must remind ourselves about the basic facts concerning protein structure.

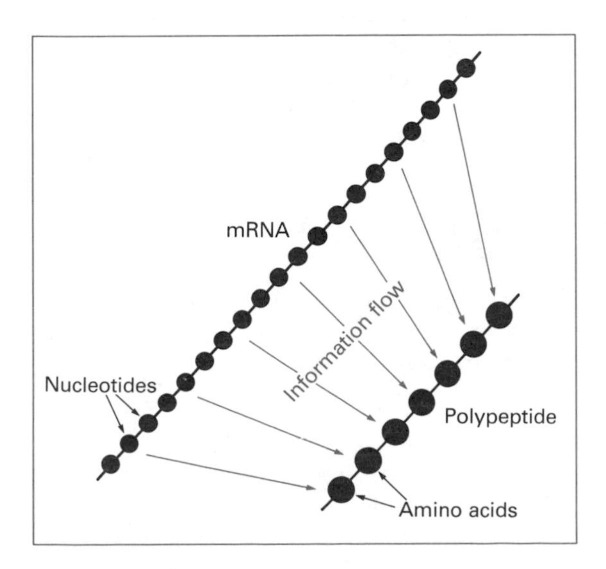

Figure 8.1 During translation the sequence of nucleotides in an mRNA molecule specifies the sequence of amino acids in a polypeptide.

Table 8.1 The 20 amino acids that occur in proteins

Amino acid	Abbreviation* 3-letter	Abbreviation* 1-letter	Molecular weight	Polarity of R group
Alanine	ala	A	89.1	Non-polar
Arginine	arg	R	174.2	Positive polar
Asparagine	asn	N	132.1	Uncharged polar
Aspartic acid	asp	D	133.1	Negative polar
Cysteine	cys	C	121.2	Uncharged polar
Glutamic acid	glu	E	147.1	Negative polar
Glutamine	gln	Q	146.2	Uncharged polar
Glycine	gly	G	75.1	Uncharged polar
Histidine	his	H	155.2	Positive polar
Isoleucine	ile	I	131.2	Non-polar
Leucine	leu	L	131.2	Non-polar
Lysine	lys	K	146.2	Positive polar
Methionine	met	M	149.2	Non-polar
Phenylalanine	phe	F	165.2	Non-polar
Proline	pro	P	115.1	Non-polar
Serine	ser	S	105.1	Uncharged polar
Threonine	thr	T	119.1	Uncharged polar
Tryptophan	trp	W	204.2	Non-polar
Tyrosine	tyr	Y	181.2	Uncharged polar
Valine	val	V	117.2	Non-polar

* The standard abbreviations that are used most frequently are the 3-letter ones. The 1-letter abbreviations should only be used to save space when listing the amino acid sequence of a polypeptide.

8.1 Polypeptides are polymers

At the most fundamental level the structures of polypeptides and nucleic acids are the same: both are polymers made up of a series of linked monomers. In a polypeptide the monomers are called amino acids and the polymeric chain is generally less than 2000 units in length, much shorter than most naturally occurring nucleic acid molecules.

8.1.1 Amino acids

Twenty different amino acids are found in protein molecules (Table 8.1). Each has the general structure shown in Fig. 8.2, comprising a central α-carbon atom to which the following four groups are attached:

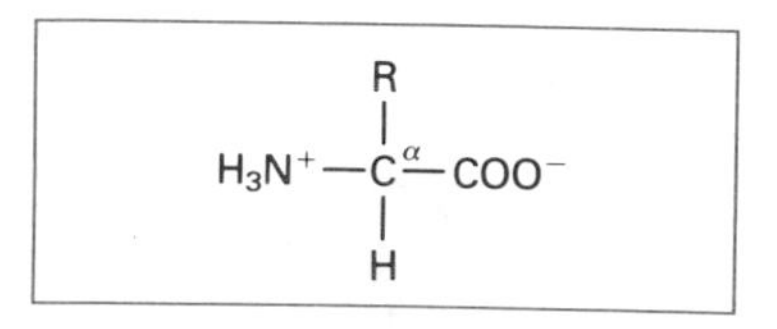

Figure 8.2 The structure of an amino acid.

1. A hydrogen atom.
2. A carboxyl (—COO^-) group.
3. An amino (—NH_3^+) group.
4. The **R group**, which is different for each amino acid (Fig. 8.3).

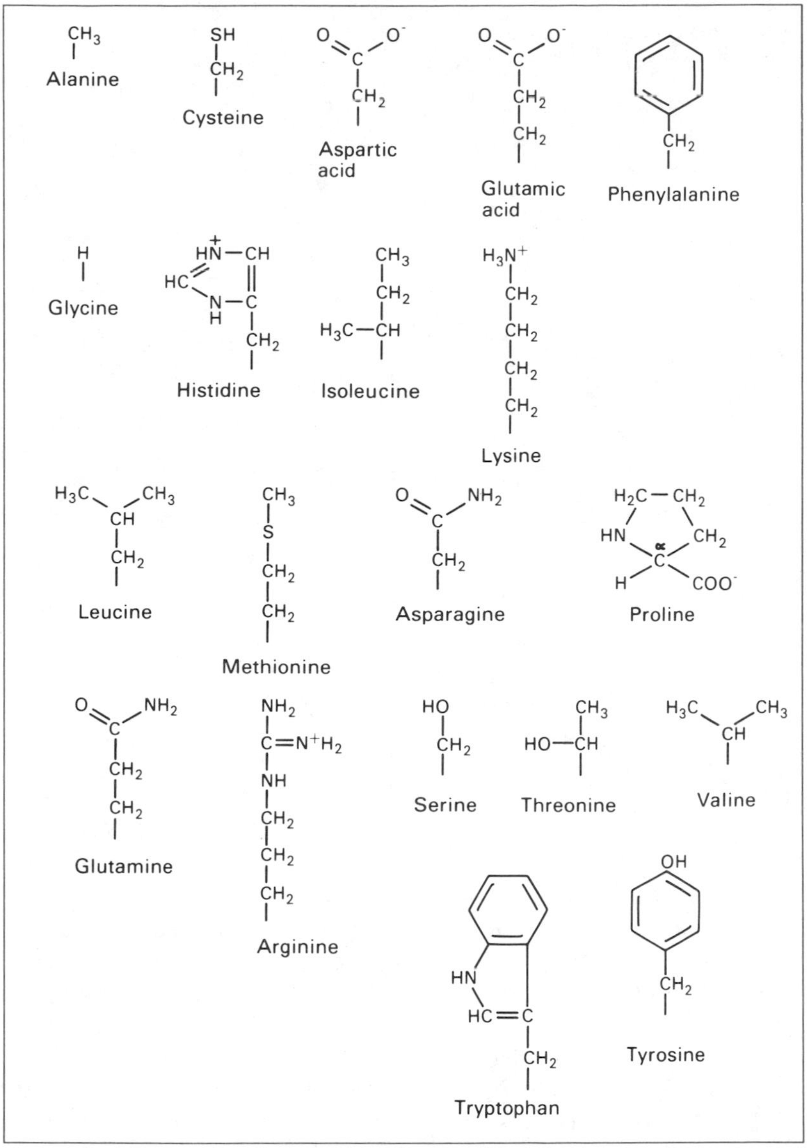

Alanine
Cysteine
Aspartic acid
Glutamic acid
Phenylalanine
Glycine
Histidine
Isoleucine
Lysine
Leucine
Methionine
Asparagine
Proline
Glutamine
Arginine
Serine
Threonine
Valine
Tryptophan
Tyrosine

Figure 8.3 The R groups of each of the 20 amino acids involved in translation. Note that the entire structure of proline is shown, as with this molecule the R group involves the amino group attached to the α-carbon. Proline is in fact an imino acid.

The R groups vary considerably in chemical complexity: for glycine the R group is simply a hydrogen atom, whereas for tyrosine, phenylalanine and tryptophan the R groups are complex aromatic side-chains. The majority of R groups are uncharged, though two amino acids have negatively charged R groups (aspartic acid and glutamic acid) and three have positively charged R groups (lysine, arginine and histidine). Some R groups are **polar** (e.g. glycine, serine and threonine), others are **non-polar** (e.g. alanine, valine and leucine). These differences mean that although all amino acids are closely related, each has its own specific chemical properties.

Figure 8.4 Peptide bonds and polypeptide structure.

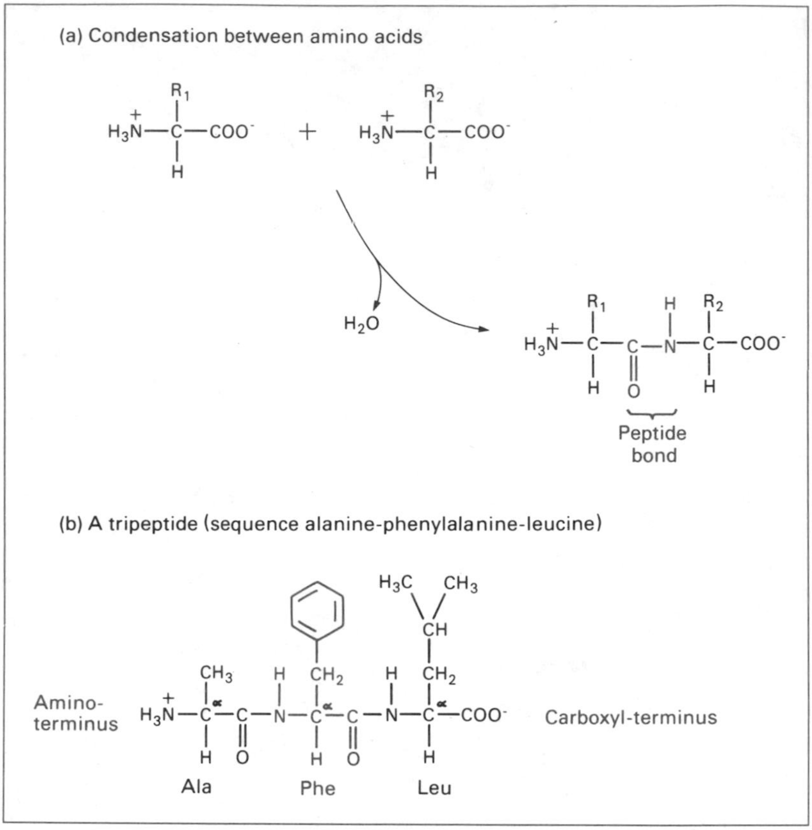

(a) Condensation between amino acids

H_2O

Peptide bond

(b) A tripeptide (sequence alanine-phenylalanine-leucine)

Amino-terminus

Carboxyl-terminus

Ala Phe Leu

(a) Amino acids are linked by peptide bonds. The polymeric structure of a polypeptide is built by linking a series of amino acids by **peptide bonds**, formed by condensation between the carboxyl group of one amino acid and the amino group of a second amino acid (Fig. 8.4(a)). The structure of a tripeptide, comprising three amino acids, is shown in Fig. 8.4(b). Note that, as with polynucleotides, the two ends of a polypeptide are chemically distinct: one has a free amino group and is called the amino-, NH_2-, or N-terminus; the other has a free carboxyl group and is called the carboxyl-, COOH-, or C-terminus.

8.1.2 Different levels of protein structure

Four levels of structure are recognized in protein molecules.

1. ***The primary structure***, which is the amino acid sequence itself.
2. ***The secondary structure***, which is the regular or repeating configuration taken up by the amino acid chain. The two most important secondary structures are the **α-helix** (Fig. 8.5(a)) and

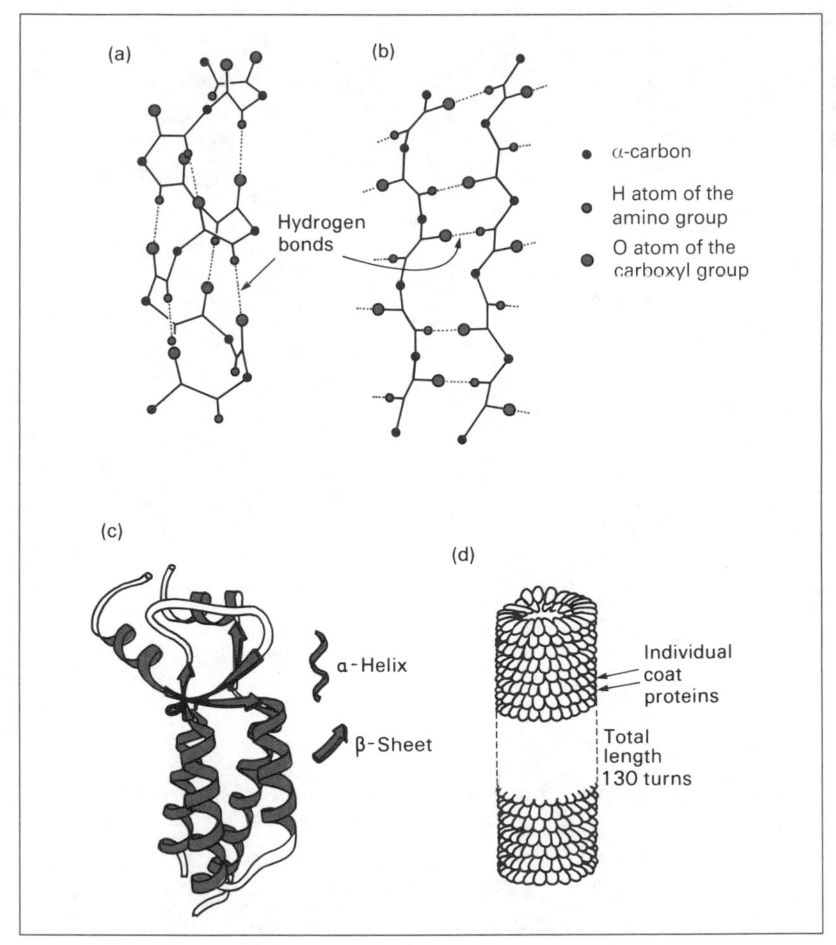

Figure 8.5 The different levels of protein structure. The most important secondary structures are (a) the α-helix, and (b) the β-sheet. (c) The tertiary structure of the coat protein of tobacco mosaic virus. (d) The complex quaternary structure of the tobacco mosaic virus capsid, which is made up of 2130 coat protein subunits.

β-sheet (Fig. 8.5(b)). Both are stabilized primarily by hydrogen bonding between the carboxyl and amino groups of different amino acids. Usually different regions of a polypeptide will take up different secondary structures, so that the whole is made up of a number of α-helices and β-sheets, together with less organized regions.

3. ***The tertiary structure***, which is the three-dimensional conformation formed by folding together the secondary structural components of the polypeptide (Fig. 8.5(c)). The tertiary structure is held together by a variety of interactions between different amino acids, including hydrogen bonds, covalent linkages called **disulphydryl bridges** which can form between two cysteine residues (Fig. 8.6), and the natural tendency of the polypeptide chain to fold up so that non-polar R groups are shielded from water.
4. ***The quaternary structure***, which refers to they way in which two or more polypeptides are orientated together to form a multi-

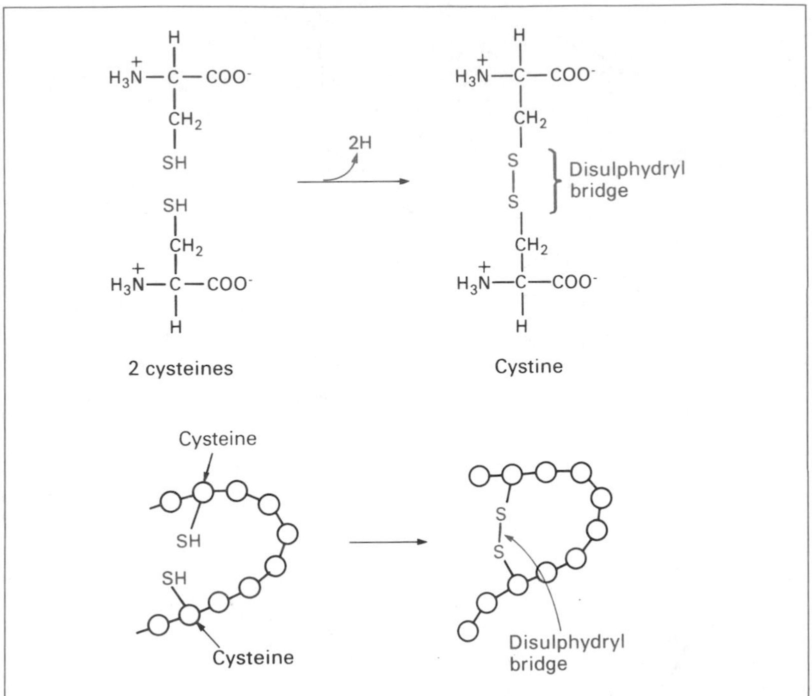

Figure 8.6 Disulphydryl bridges. The reaction between two cysteine amino acid molecules to produce a single cystine is shown (top), together with the consequence of disulphydryl bridge formation between cysteines in the same polypeptide (bottom).

Studying protein structure

Several kinds of experiment can be carried out in order to study the structure of a protein. The following are important examples:

1. The **amino acid composition** can be determined by breaking the protein into its constituent amino acids, usually by placing a sample in hot, concentrated acid for several hours. The free amino acids can then be separated and quantified by chromatography.
2. The **amino acid sequence** can be worked out by any of several related procedures. These involve a stepwise degradation process, with a single amino acid at a time being cleaved from the amino-terminus of the polypeptide. This amino acid is identified by its chromatographic properties and the cycle then repeated. The procedure continues until the entire amino acid sequence has been determined. Automatic polypeptide sequencers have been available since 1967.
3. The **tertiary structure** can be investigated by, for example, X-ray diffraction analysis (Section 3.2). Recently other techniques have also been developed, including nuclear magnetic resonance (NMR) spectroscopy, which allows the environment of different chemical nuclei to be determined. This technique was first developed for structural studies of small molecules, but has gradually been refined so that it can now provide information on molecules as large as proteins.

subunit structure. The quaternary structure may involve several molecules of the same polypeptide or may comprise different polypeptides, as with RNA polymerase ($\alpha_2\beta\beta'\sigma$, see Section 5.3.1). In some cases the quaternary structure is built up from a very large number of polypeptide subunits, to give a complex array; the best examples are the protein coats of viruses (Chapter 13), such as that of tobacco mosaic virus which is made up of 2130 identical protein subunits (Fig. 8.5(d)).

8.1.3 The amino acid sequence is the key to protein structure and function

Each of the higher levels of protein structure – secondary, tertiary and quaternary – is specified by the primary structure, the amino acid sequence itself. This is most clearly understood at the secondary level, where it is recognized that certain amino acids, because of the chemical and physical properties of their R groups, stimulate the formation of an α-helix, whereas others promote formation of a β-sheet. Conversely, certain amino acids more frequently occur outside regular structures and may act to determine the end-point of a helix or sheet. These factors are now so well understood that rules to predict the secondary structures taken up by amino acid sequences have been developed.

Although less well characterized, it is nonetheless clear that the tertiary and quaternary structures of a protein also depend on the amino acid sequence. The interactions between individual amino acids at these levels are so complex that predictive rules, although attempted, are still unreliable. However, it has been established for some years that if a protein is **denatured**, for instance by mild heat treatment or adding a chemical denaturant such as urea, so that it loses its higher levels of structure and takes up a non-organized conformation, it still retains the innate ability upon **renaturation** (by cooling down again, for example) to reform spontaneously the correct tertiary structure (Fig. 8.7). Once the tertiary structure has formed, subunit assembly into a multimeric protein again occurs spontaneously. This shows that the instructions for the tertiary and quaternary structures must reside in the amino acid sequence.

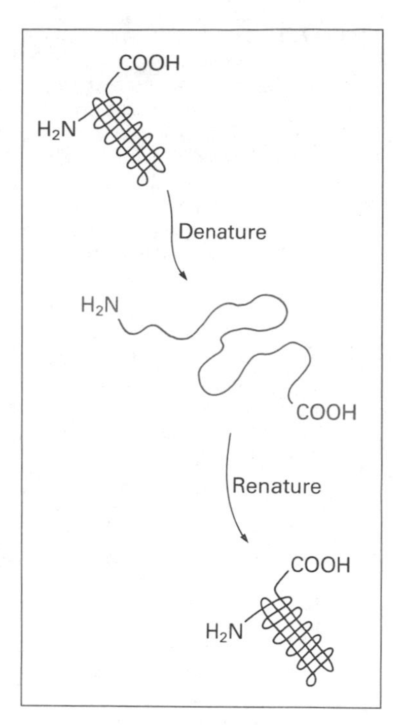

Figure 8.7 A denatured protein will reform its tertiary structure spontaneously on renaturation. The denaturant can be mild heating or treatment with a chemical denaturant such as urea. Renaturation occurs when the temperature is lowered or the urea removed. If the protein contains disulphydryl bridges then these can be broken with a reducing agent such as β-mercaptoethanol. When the reductant is removed the correct disulphydryl bridges will also reform spontaneously.

(a) Function depends on amino acid sequence. From these considerations it is only a small step to appreciate that the function of a protein is also determined by its amino acid sequence. As an illustration we will consider those proteins that must attach themselves to a DNA molecule in order to perform their function in the cell. These DNA-binding proteins form a large and diverse group that includes, for instance, RNA polymerase (Chapter 5) and a number of important regulatory proteins which modulate, and at times block, transcription of individual genes (Chapter 10). An example is provided by Cro, a regulatory protein that controls expression of a number of genes of the bacteriophage called λ (Section 13.1.3). Cro is a dimer of identical polypeptides, each made up of three α-helices and a region of β-sheet. In the active form of the protein, the two α_3 helices (one from each polypeptide in the dimer) are exactly 34 Å apart and therefore fit into two adjacent sections of the major groove of a DNA molecule (Fig. 8.8). If the α_3 helices are absent, or orientated incorrectly, the DNA-binding ability of the protein will be lost. DNA binding depends therefore on the protein assuming the correct secondary, tertiary and quaternary structures, which is, in turn, dependent on the amino acid sequence.

This is a crucial fact in molecular genetics: the function of a protein depends on its amino acid sequence which, in turn, is specified by the sequence of nucleotides in the mRNA, which is itself a copy of the gene. The biological information carried by the gene codes, in essence, for a protein function. Having grasped this important point we can now return to the question of the genetic code.

8.2 The genetic code

The problem of exactly how the genetic code works was the major intellectual preoccupation of molecular biologists from 1953, when

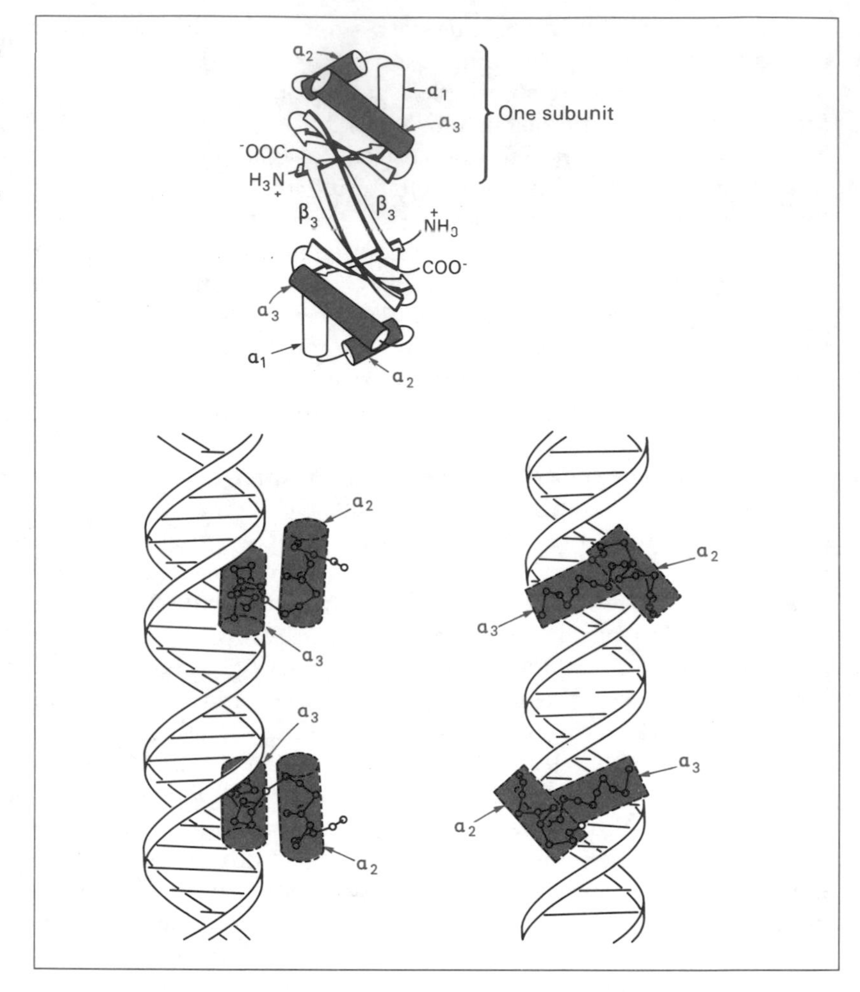

Figure 8.8 The interaction between a typical DNA-binding protein and the DNA double helix. The top part of the figure shows the dimer of the Cro protein, with the three α-helices in each subunit drawn as barrels. The two helices labelled α_3 fit into adjacent sections of the major groove of the double helix, as viewed from two different angles in the lower drawings. Redrawn from R. G. Brennan and B. W. Matthews (1989) *Trends in Biochemical Sciences*, **14**, 286–90.

the double helix was postulated, to 1966 when a complete picture of the genetic code finally emerged. Several of the more outlandish speculations made in the early days seem incredibly farfetched today, but it must be appreciated that until 1960 the available techniques made it very difficult to tackle directly the question of the code. The frustration this caused meant that there was plenty of time for imaginations to dream up sophisticated coding systems.

8.2.1 The code from first principles

During the 1950s the most enlightened molecular biologists, centred as usual around Crick, strove to prevent the ideas about the code becoming too complex. Gradually the simplest system that was compatible with the established facts emerged as a working hypothesis. This system was based on two assumptions, both of which were subsequently confirmed by experimental analysis.

(a) Colinearity between gene and protein. It was more or less assumed right from 1953 that genes are **colinear** with the proteins they code for, meaning that the order of nucleotides in the gene correlates directly with the order of amino acids in the corresponding polypeptide. This is clearly the most straightforward way in which genes could code for proteins but was not in fact proven experimentally until 1964.

The type of experiment needed was fairly obvious: alter the nucleotide sequence of a gene at a specific point and determine whether the resulting change in the amino acid sequence of the corresponding polypeptide occurs at the same relative position or elsewhere (Fig. 8.9). The problem was that although at that time alterations could be introduced into genes without too much trouble (p. 159), determining exactly where in the gene an alteration had occurred was much more difficult. Eventually Charles Yanofsky and his colleagues managed to carry out the necessary analysis, with the *E. coli* gene coding for the enzyme called tryptophan synthase (which is responsible for catalysing the final step in the biosynthetic pathway that results in synthesis of the amino acid tryptophan). The result was a clear demonstration of colinearity between gene and protein. Very shortly afterwards, Sydney Brenner's group completed a similar analysis of a gene and its protein from the phage called T4, confirming Yanofsky's result. Later it was shown that the amino-terminus of the protein corresponds to the 5′ end of the gene.

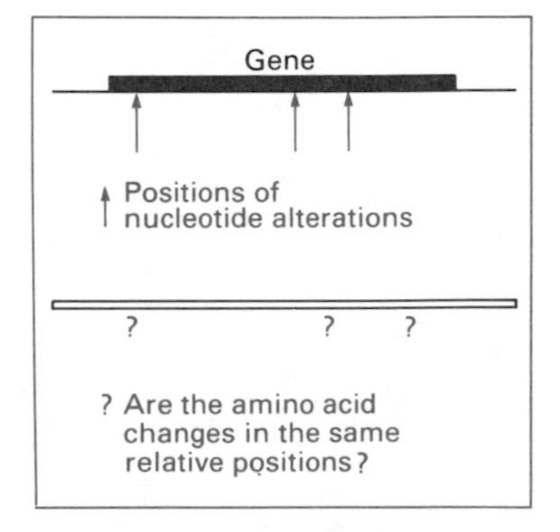

Figure 8.9 The type of experiment needed to determine if a gene is colinear with the polypeptide it codes for.

(b) Each codeword is a triplet of nucleotides. The second fundamental assumption about the genetic code concerned the size of the codeword, or **codon**, the group of nucleotides that code for a single amino acid. Clearly codons cannot be just single nucleotides (A, T, G or C) because that would allow only four different codewords when 20 are required, one for each of the 20 amino acids found in proteins. Similarly, a doublet code (codons such as AT, TA, TT, GC, etc.) can be ruled out as this would contain only $4^2 = 16$ different codons. However, the next stage up, a triplet code (codons AAA, AAT, TAT, GCA, etc.) would be feasible as this would yield $4^3 = 64$ codewords, which would be more than enough.

As with colinearity, experiments to prove the triplet nature of the code were straightforward in concept but difficult in practice. It was known that a group of chemicals called the **acridine dyes** (Section 12.1.3), of which **proflavin** is an example, cause single

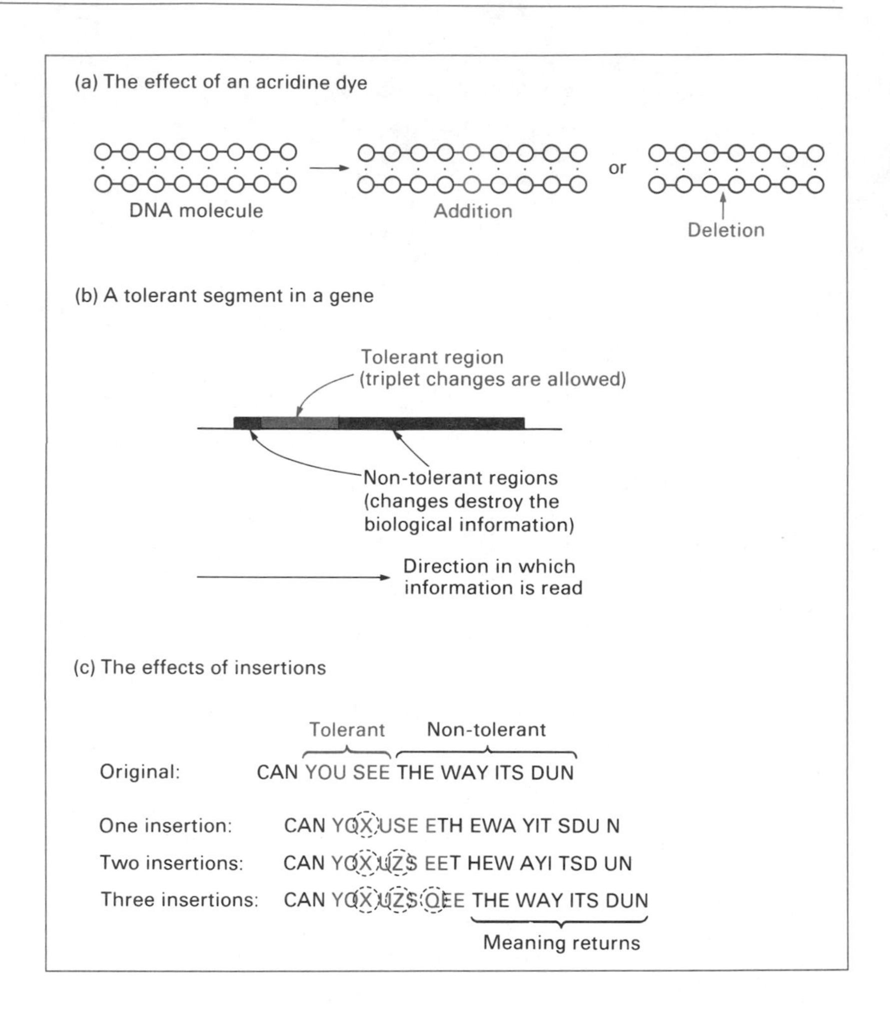

Figure 8.10 The basis to the experiments that established the triplet nature of the genetic code.

base pair deletions or additions in double-stranded DNA molecules (Fig. 8.10(a)). It was also known that certain proteins contain segments where the amino acid sequence can be changed without altering the function of the protein (Fig. 8.10(b)), although other regions of the same protein will not tolerate such changes. What if a series of insertions and/or deletions were introduced into the region of a gene coding for a tolerant segment of a protein? If the code is triplet then a single insertion or deletion would give rise to a non-functional protein, because all the codewords downstream of the mutation would be altered, including those in the non-tolerant segment following the tolerant region (Fig. 8.10(c)). Two insertions or deletions (although not one of each) would have the same effect. But three insertions or deletions would restore the correct **reading frame** in the non-tolerant region and would be predicted to have no effect on protein function. An elegant experiment of this kind was eventually carried out successfully by Crick,

Brenner and others, using the gene called *rIIb* from T4 phage as the target, and determining the effect of proflavin-induced alterations by observing whether or not the treated phage were able to infect *E. coli*. This work established the triplet nature of the code; however, by the time it was published in 1961 the first direct experimental attack on the code itself had begun.

8.2.2 Elucidation of the code

Gradually, as the 1950s progressed, technical advances were made that would eventually allow the question of which triplet codon specifies which amino acid to be answered. The two major advances were:

1. *Synthesis of artificial RNA molecules*. If translation could be directed by an RNA molecule of known nucleotide sequence, then it would be possible to assign individual codons by looking at the amino acid sequence of the protein that is synthesized. To do this it would be necessary to make artificial RNA molecules of known or predictable sequence. This became possible with the discovery by Severo Ochoa in 1955 of **polynucleotide phosphorylase**, an enzyme which in the cell degrades RNA but which, in the test-tube, can be 'forced' to catalyse the reverse reaction and synthesize RNA. This reaction does not require a DNA template and is unrelated to transcription.
2. *Cell-free protein synthesis*. To use the artificial RNA molecule as a message for translation, a cell extract able to synthesize proteins is needed. Such a **cell-free**, or *in vitro*, system must contain all the cellular components necessary for protein synthesis (i.e. ribosomes, tRNAs, amino acids, and so on) but must lack endogenous mRNA so that protein synthesis occurs only when the artificial message is added.

(a) Cell-free protein synthesis with homopolymers. Marshall Nirenberg and Heinrich Matthaei at the National Institutes of Health laboratories in the USA eventually perfected a cell-free system, prepared from *E. coli*, able to synthesize polypeptides specific to an added artificial RNA message. At about 6 a.m. in the morning of Saturday 27 May 1961 Matthaei discovered that when a **homopolymer** made up entirely of uridine nucleotides – poly(U) – was added to the cell-free system, polyphenylalanine was synthesized. The first codon assignation could be made: 5′-UUU-3′ codes for phenylalanine (Fig. 8.11).

Poly(U) was used in this first experiment because homopolymeric RNAs are relatively simple to synthesize. If the reaction mixture from which polynucleotide phosphorylase builds up the RNA molecule contains only uridine mononucleotides then only

Severo Ochoa b. 24 September 1905, Luarca, Spain

'My advice to students of science is that if they have an urge to do research they should do it by all means. Nothing should stand in the way of a strong wish to devote Life to Science.'

'If you have the urge to do scientific research get the proper training and by all means do it; nothing else is likely to give you so much satisfaction and, above all, such a sense of fulfilment.'

Severo Ochoa was educated in his home country, graduating from Malaga University in 1921 and then obtaining his MD from Madrid in 1929. He worked in Germany with Otto Meyerhof on muscle biochemistry and then moved via England to the USA, joining New York University in 1942. During the 1950s his research centred on the enzymes that utilize the energy held in the energy-rich phosphate bonds of ATP. This led in 1955 to his discovery, with Marianne Grunberg-Manago, of polynucleotide phosphorylase, the enzyme that was subsequently used to prepare synthetic mRNA molecules for elucidation of the genetic code. Ochoa received the 1959 Nobel Prize. In 1985 he returned to Madrid University as Professor of Biology.

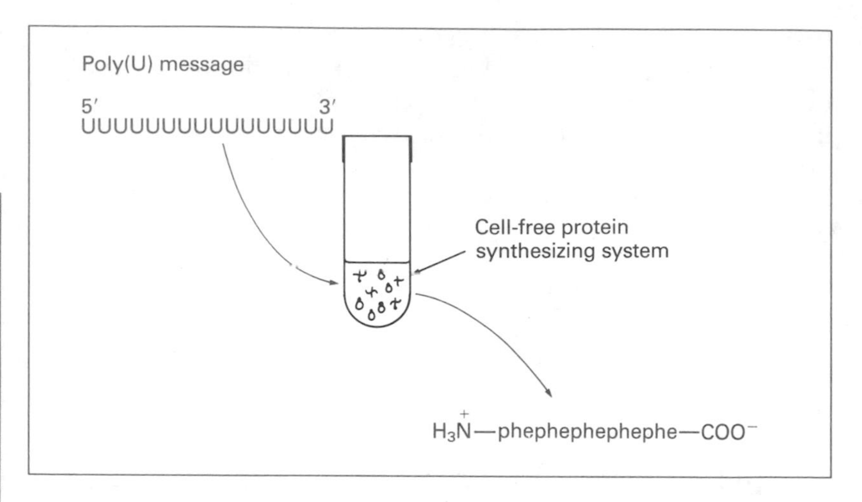

Figure 8.11 The first successful identification of the meaning of a codon.

Marshall Nirenberg b. 10 April 1927, New York City

Nirenberg's career took a circuitous route to the genetic code. His Master's degree was awarded in 1948 by the University of Florida for an ecological and taxonomic study of a group of flies, work that was followed by research into sugar transport by tumour cells, at the University of Michigan (PhD, 1957). He then moved to the National Institutes of Health laboratories at Bethesda, Maryland and began the difficult task of preparing a cell-free extract from *E. coli* that would retain the ability to synthesize proteins when new mRNA was added. The first codon assignations in 1961 followed naturally from this work. Nirenberg's achievements came as a surprise to Crick and his associates, who apparently were not aware of the progress until Nirenberg's first results were announced. After some initial scepticism the value of the work was fully appreciated and Nirenberg remained the central influence in the genetic code studies until their completion in 1966. He shared the 1968 Nobel Prize with Khorana and Holley.

poly(U) will be made. The three other RNA homopolymers were also synthesized, added individually to samples of the cell-free system, and the assignations AAA = lysine and CCC = proline made. For reasons that have never been entirely clear poly(G) would not work and so the amino acid coded by GGG could not at this stage be identified.

(b) Random heteropolymers. The next step was to construct **heteropolymers**, artificial RNA molecules containing more than just one nucleotide. This can be achieved by polymerizing mixtures of nucleotides with polynucleotide phosphorylase. There is a problem, however, as the random nature of the polymerization means that the absolute sequence of the resulting RNA molecule will not be known. For instance, a heteropolymer of A and C will contain eight different codons (AAA, AAC, ACA, CAA, ACC, CCA, CAC, CCC) in a random order and will code for a polypeptide containing six different amino acids (proline, histidine, threonine, asparagine, glutamine and lysine). Clearly these six amino acids are coded by the eight possible codons, but which is coded by which?

An answer is at least partially provided by using a known amount of each nucleotide in the reaction mixture for RNA synthesis. For example, if the ratio of C to A is 5:1 then the probability of a CCC codon occurring is much higher than the probability of an AAA. In fact the frequencies of each of the eight possible codons can be worked out and compared with the amounts of each amino acid in the resulting polypeptide (Fig. 8.12). The codon allocations provided by this method are not definite but are statistically probable and can be cross-checked by changing the composition of the heteropolymer. These techniques allowed Nirenberg, Matthaei and Ochoa to propose meanings for most of the 64 codons of the genetic code.

Random heteropolymer comprising 5C:1A

Probability of C being included in a triplet = $5/6$
Probability of A being included in a triplet = $1/6$

Possible triplets	*Probability*
CCC	$(5/6)^3 = 57.9\%$
CCA, CAC, ACC	$(1/6)(5/6)^2 = 11.6\%$
CAA, ACA, AAC	$(1/6)^2(5/6) = 2.3\%$
AAA	$(1/6)^3 = 0.4\%$

Results of cell-free protein synthesis

Amino Acid	*Amount in Polypeptide*	*Interpretation*
Proline	69%	CCC + one of CCA, CAC, ACC
Threonine	14%	One of CCA, CAC, ACC + one of CAA, ACA, AAC
Histidine	12%	One of CCA, CAC, ACC
Asparagine	2%	One of CAA, ACA, AAC for each
Glutamine	2%	
Lysine	1%	AAA

Figure 8.12 The statistical analysis of the results of a random heteropolymer experiment.

(c) Completion of the code. Although homopolymers and random heteropolymers allowed most of the genetic code to be worked out, unambiguous determination of each codon required two additional types of experiment.

1. ***Ordered heteropolymers*** were synthesized by H. Gobind Khorana. These are made by polymerizing not mononucleotides, but known dinucleotides such as AC (producing ACACACAC, which contains two codons, ACA and CAC) or trinucleotides such as UGU (producing UGUUGUUGU, codons UGU, GUU, UUG). The simpler polynucleotides produced by these messages allowed the meaning of several difficult codons to be determined.
2. ***The triplet binding assay*** was developed by Nirenberg and Philip Leder in 1964 as a modification of the cell-free protein synthesizing system. It was discovered that purified ribosomes will attach to an mRNA molecule of only three nucleotides (a single codon, in fact) and bind the correct amino acid-tRNA molecule (Fig. 8.13). As triplets of known sequence could be constructed in the laboratory, it was possible with the binding assay to check the previously assigned codons and to allocate virtually all of the remaining ones.

Nomenclature of heteropolymers

It is important that the nomenclature used to describe heteropolymers should allow random and ordered molecules to be distinguished. The convention is as follows:

poly(A, C) = random heteropolymer (CACCCAACCCACAC)
poly(AC) = ordered heteropolymer (ACACACACACACA)

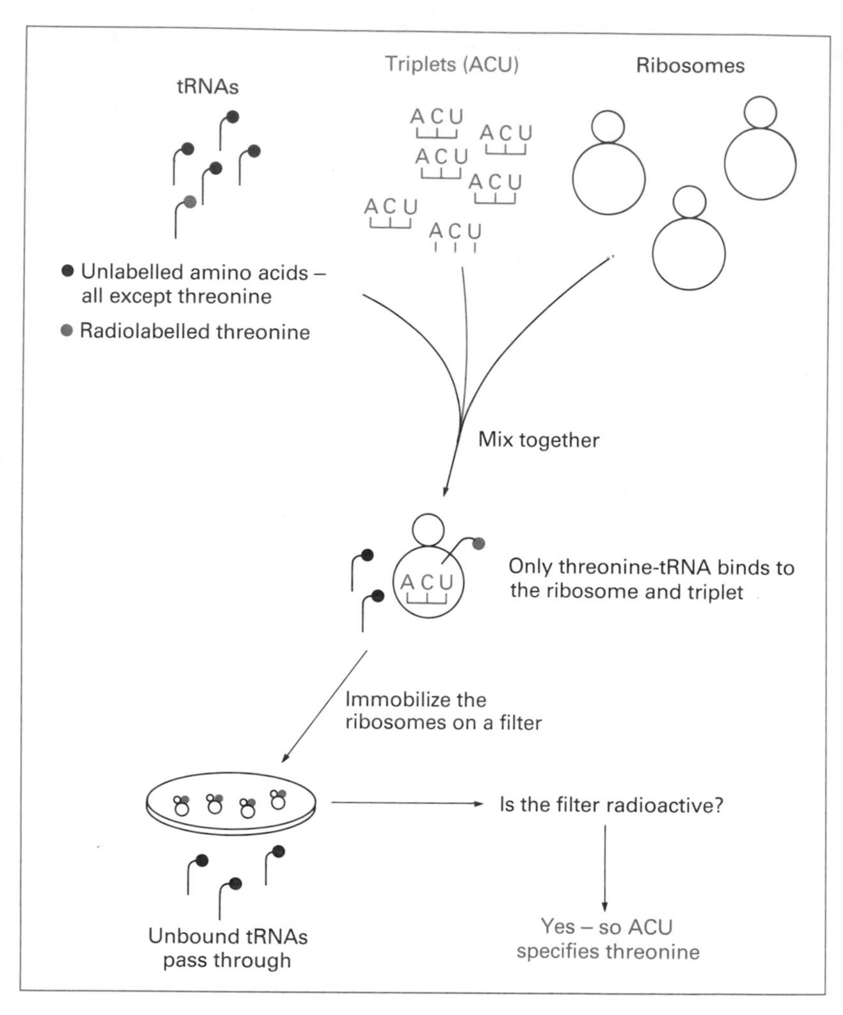

Figure 8.13 The triplet binding assay.

H. Gobind Khorana b. 9 January 1922, Raipur, Pakistan

Khorana has had a major and continuing influence on the development of methods for the artificial synthesis of nucleic acids. He attended the University of the Punjab and then obtained his PhD from Liverpool University. His first experience of work with nucleic acids was gained in Sir Alexander Todd's laboratory at Cambridge in the 1950s. Khorana then moved to the USA, first to the University of Wisconsin (1960) and then the Massachusetts Institute of Technology (1971). He discovered how to synthesize triplet RNA molecules of known sequence, thereby allowing the final assignations of the genetic code to be made. Later he devoted his efforts to synthesis of lengthy DNA molecules that would act as artificial genes. He shared the 1968 Nobel Prize with Marshall Nirenberg and Robert Holley.

Finally in 1966, 5 years after Matthaei's first experiments, the meaning of the final codon was confirmed and the genetic code was complete.

8.2.3 Features of the code

The code is shown in Table 8.2. It has three important features.

1. ***The code is degenerate***. All amino acids except methionine and tryptophan have more than one codon, so that all the possible triplets have a meaning, despite there being 64 triplets and only 20 amino acids. Most synonymous codons are grouped into families (GGA, GGU, GGG and GGC, for example, all code for glycine), a fact that is relevant to the way the code is deciphered during protein synthesis (Section 9.1.2).

Table 8.2 The genetic code

UUU, UUC phe*; UUA, UUG leu	UCU, UCC, UCA, UCG ser	UAU, UAC tyr; UAA, UAG 'stop'	UGU, UGC cys; UGA 'stop'; UGG trp
CUU, CUC, CUA, CUG leu	CCU, CCC, CCA, CCG pro	CAU, CAC his; CAA, CAG gln	CGU, CGC, CGA, CGG arg
AUU, AUC, AUA ile; AUG met	ACU, ACC, ACA, ACG thr	AAU, AAC asn; AAA, AAG lys	AGU, AGC ser; AGA, AGG arg
GUU, GUC, GUA, GUG val	GCU, GCC, GCA, GCG ala	GAU, GAC asp; GAA, GAG glu	GGU, GGC, GGA, GGG gly

* See Table 8.1 for full names of amino acids.

2. ***The code contains punctuation codons***. Three codons, UAA, UGA and UAG, do not code for an amino acid and if present in the middle of a heteropolymer cause protein synthesis to stop, resulting in shorter polypeptides than expected. These are called **termination codons** and one of the three always occurs at the end of a gene at the point where translation must stop.
   Similarly, AUG virtually always occurs at the start of a gene and marks the position where translation should begin. AUG is therefore the **initiation codon** and because it codes for methionine most newly synthesized polypeptides will have this amino acid at the amino-terminus, though it may subsequently be removed by post-translational processing of the protein. Note that AUG is the only codon for methionine, so AUGs that are not initiation codons may be found in the internal region of a gene.
3. ***The code is not universal***. When the genetic code was completed in 1966 it was assumed to be universal. It was postulated that the genetic code must be frozen and unable to evolve because a change in a codon meaning would cause almost every protein in the cell to be altered. The cell-free system used to elucidate the code was prepared from *E. coli* but the experiments were soon repeated with mammalian extracts and essentially the same results obtained. Gradually the assumption of universality became accepted dogma.

Table 8.3 Examples of non-standard codons

Organism	*Genes*	*Codon*	*Universal meaning*	*Actual meaning*
Mammals	All mitochondrial	UGA	Termination	Trp
		AGA, AGG	Arg	Termination
		AUA	Ile	Met
Drosophila	All mitochondrial	UGA	Termination	Trp
		AGA	Arg	Ser
		AUA	Ile	Met
Saccharomyces cerevisiae	All mitochondrial	UGA	Termination	Trp
		CUN*	Leu	Thr
		AUA	Ile	Met
Protozoa	All nuclear	UAA, UAG	Termination	Gln
Candida cylindracea	All nuclear	CUG	Leu	Ser
Mammals	Glutathione peroxidase	UGA	Termination	SeCys
Escherichia coli	Formate dehydrogenase	UGA	Termination	SeCys

*N = any nucleotide.

It was therefore a shock when Frederick Sanger's group at Cambridge discovered in 1979 that human mitochondrial genes (Section 15.5.2) use a slightly different genetic code (Table 8.3). In these genes, UGA (normally a termination codon) codes for tryptophan, AGA and AGG (normally arginine) are termination codons, and AUA (normally isoleucine) specifies methionine. These changes were later confirmed for mitochondrial genes in other mammals, and slightly different non-standard codes were discovered in mitochondrial genes of *Drosophila*, fungi and yeast.

As we will see in Chapter 15, mitochondrial genes are expressed by an RNA polymerase, ribosomes and tRNAs that are different from those responsible for expressing nuclear genes. It is therefore quite acceptable that mitochondrial and nuclear genes should follow slightly different genetic codes. It is difficult to understand how the mitochondrial genetic code could have evolved, but as there are only a few mitochondrial protein-coding genes (13 in humans), a change in a codon meaning would have had a less drastic effect than expected if the nuclear genetic code became altered. Recently though it has been shown that unusual codes are not confined to mitochondria as a few examples have been uncovered where an organism uses a non-standard genetic code for its nuclear genes. This seems to be commonest amongst the protozoa, including *Tetrahymena* and *Paramecium*, in which the two termination codons UAA and UAG both specify glutamine.

Still more surprising is the discovery that the meaning of a codon can vary from gene to gene in the same organism. In human nuclear genes UGA is frequently used as a termination codon, in accordance with its standard meaning. However, in at least two human genes, for the enzymes glutathione peroxidase and Type I iodothyronine 5′-deiodinase, UGA specifies the unusual amino acid called selenocysteine (a cysteine in which the sulphur atom is relaced by selenium). It appears that the mRNAs transcribed from these genes have special stem-loop structures in their trailer regions, and that these stem-loops play some role during translation, ensuring that the UGA triplet is recognized as a selenocysteine codon rather than a termination signal. Further research is being carried out to clarify exactly how this works, and to see if there are other examples of codons having different meanings in different genes.

Unusual codes and RNA editing

A non-standard codon meaning is usually discovered by comparing the nucleotide sequence of a gene with the amino acid sequence of the corresponding protein. Is it possible that the unusual codon is in fact 'corrected' by editing of the mRNA (Section 7.5)? This almost certainly does not happen with most unusual codons, but does appear to be the case in plant mitochondria. In maize, for example, many mitochondrial genes have the triplet CGG at positions where tryptophan codons would be expected. After transcription, RNA editing of the mRNA converts these codons into UGG, which are then read as 'tryptophan', in accordance with the standard genetic code.

Reading

R. F. Doolittle (1985) The proteins. *Scientific American*, **216**(5), 80–94 – a review on protein structure.

F. M. Richards (1991) The protein folding problem. *Scientific American*, **264**(1), 34–41 – an interesting article on the way in which a protein folds up into its correct tertiary structure.

C. Yanofsky, B. C. Carlton, J. R. Guest, D. R. Helsinki and U. Henning (1964) On the colinearity of gene structure and protein structure. *Proceedings of the National Academy of Sciences USA*, **51**, 266–72; C. Yanofsky (1967) Gene structure and protein structure. *Scientific American*, **216**(5), 80–94 – the first research report and a review on colinearity between gene and protein.

F. H. C. Crick, F. R. S. L. Barrett, S. Brenner and R. J. Watts-Tobin (1961) General nature of the genetic code for proteins. *Nature*, **192**, 1227–32 – the experimental proof of the triplet nature of the code.

M. W. Nirenberg and H. Matthaei (1961) The dependence of cell-free protein synthesis in *E. coli* upon naturally occurring or synthetic polyribonucleic acids. *Proceedings of the National Academy of Sciences USA*, **47**, 1588–602; M. W. Nirenberg and P. Leder (1964) RNA codewords and protein synthesis. *Science*, **145**, 1399–407 – two of the primary publications from Nirenberg's laboratory.

F. H. C. Crick (1962) The genetic code. *Scientific American*, **207**(4), 66–74; M. W. Nirenberg (1963) The genetic code II. *Scientific American*, **208**(3), 80–94; F. H. C. Crick (1966) The genetic code

III. *Scientific American*, **215**(4), 55–62 – three excellent reviews that chart the progress in elucidating the code.

Cold Spring Harbor Symposium on Quantitative Biology, Vol. 31 (1966) – captures the excitement of a meeting held just after the code was solved.

B. D. Hall (1979) Mitochondria spring surprises. *Nature*, **282**, 129–30 – reviews the first reports of unusual genetic codes in mitochondrial genes.

T. D. Fox (1985) Diverged genetic codes in protozoans and a bacterium. *Nature*, **314**, 132–3; L. A. Grivell (1986) Deciphering divergent codes. *Nature*, **324**, 109–10 – more recent discoveries concerning unusual codes.

Problems

1. Define the following terms:

   R group
   peptide bond
   disulphydryl bridge
   denaturation
   renaturation
   reading frame
   codon
   homopolymer
   random heteropolymer
   ordered heteropolymer
   termination codon
   initiation codon

2. Describe the contribution made by the following scientists to our understanding of the genetic code:

   Crick
   Yanofsky
   Brenner
   Ochoa
   Nirenberg and Matthaei
   Khorana
   Nirenberg and Leder

3. Draw the structure of an amino acid and indicate which parts of the molecule are the same for each amino acid and which parts are variable.

4. Distinguish between the four levels of protein structure.

5. What experimental evidence supports the contention that the instructions for the higher levels of protein structure reside in the amino acid sequence?

6. Expand upon the statement 'The biological information carried by the gene codes, in essence, for a protein function'.

7. Explain what is meant by colinearity between gene and protein.

8. What considerations suggested that each codon would comprise three nucleotides? How was this fact proven experimentally?

9. Distinguish between the use of random and ordered heteropolymers in elucidation of the genetic code.
10. Explain what is meant by degeneracy with respect to the genetic code.
11. *Discuss the reasons why polypeptides can take up a variety of structures whereas polynucleotides cannot.
12. *During the 1950s it was suggested that adjacent codons may overlap, so that ACAUG might contain two codons: ACA and AUG. Design an experiment to test this proposition.
13. *What result would have been obtained if the first tests for colinearity between genes and proteins had made use of a gene that contains introns?
14. *How can the genetic code change if an alteration in a codon assignation is likely to cause an alteration to every protein in the cell?
15. Calculate the number of codons that would be possible if each codon contained four nucleotides.
16. List the codons that would be contained in a random heteropolymer comprising A and G nucleotides. What amino acids would a polypeptide synthesized from this heteropolymer contain?
17. The non-template strand of a gene has the following sequence:

    5′-ATGTTAGCTGATCCGGAAATGATGTTATATATAATATATGCCCAATAG-3′

    What would be the amino acid sequence of the gene product?

9 Translation

The role of tRNA – aminoacylation – codon recognition • Mechanics of protein synthesis in *E. coli* – initiation – elongation – termination • Translation in eukaryotes

The genetic code is utilized during the final stage of gene expression, when mRNA molecules are translated into polypeptides. Translation is very complex and involves many different components of the cell. However, compliance with the genetic code is central to the entire process, and is ensured by tRNA molecules which act as adaptors forming a physical and informational link between the nucleotide sequence of the mRNA and the amino acid sequence of the polypeptide (Fig. 9.1). Later in this chapter we will look at the detailed mechanism by which proteins are synthesized in the cell. First, however, we must tackle the more fundamental question of exactly how tRNAs are able to ensure that translation proceeds in accordance with the rules of the genetic code.

9.1 The role of tRNA in translation

Each cell contains a number of different types of tRNA molecule, distinguished from one another by their different sequences, although retaining the invariant and semi-invariant nucleotides described in Chapter 6. Each tRNA is distinguished in functional terms by its specificity for one of the 20 amino acids involved in protein synthesis: for example, the tRNA molecule designated $tRNA^{tyr}$ is specific for tyrosine, $tRNA^{gly}$ is specific for glycine. A tRNA molecule forms a covalent linkage with its amino acid (and no other) and can recognize and attach to a codon specifying that amino acid (Fig. 9.2). There may be more than one type of tRNA molecule for a single amino acid, reflecting the fact that the genetic code is degenerate and that most amino acids are coded by more than one codon. Two tRNAs that bind the same amino acid are called **isoacceptors**.

9.1.1 Aminoacylation of tRNA

Each tRNA molecule is able to form a covalent linkage with its specific amino acid by a process called **aminoacylation** or **charging**.

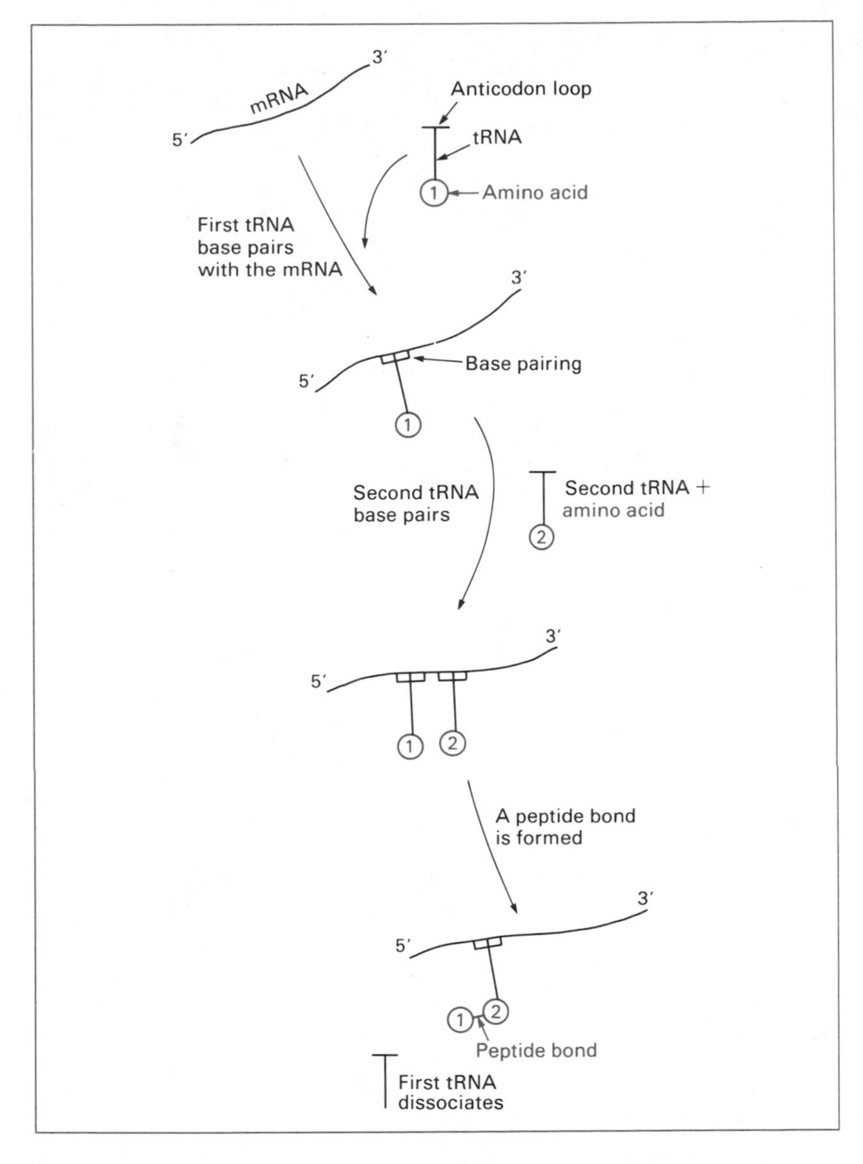

Figure 9.1 The role of tRNA in translation. The tRNA is represented by a 'T'; the bar across the top of the T is the anticodon loop.

The amino acid becomes attached to the end of the acceptor arm of the tRNA cloverleaf. The linkage forms between the carboxyl group of the amino acid and the 3′-OH group of the terminal nucleotide of the tRNA (Fig. 9.3). Remember that this terminal nucleotide is always A because all tRNAs have the sequence 5′-CCA-3′ at their 3′ ends.

(a) Aminoacyl-tRNA synthetases control charging. Aminoacylation is catalysed by a group of enzymes called the **aminoacyl-tRNA synthetases**. In most cells there is a single aminoacyl-tRNA

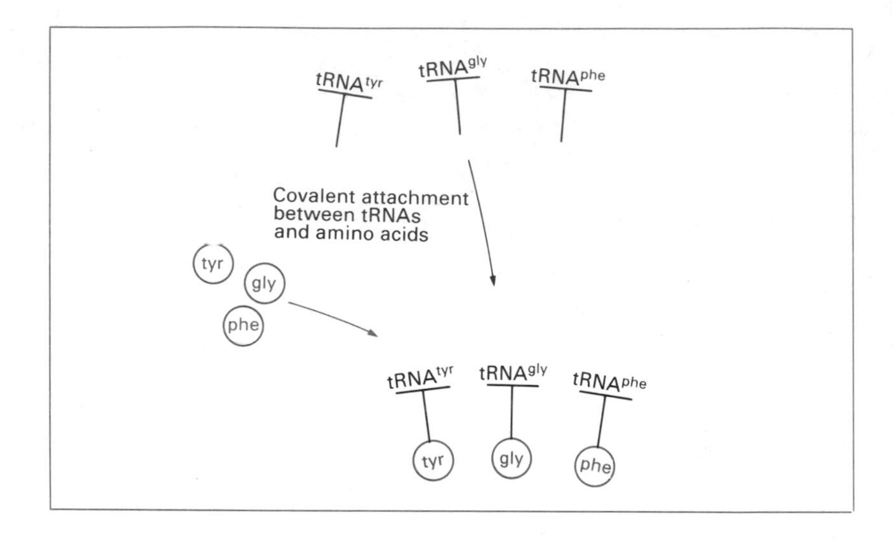

Figure 9.2 Individual tRNA molecules are specific for different amino acids.

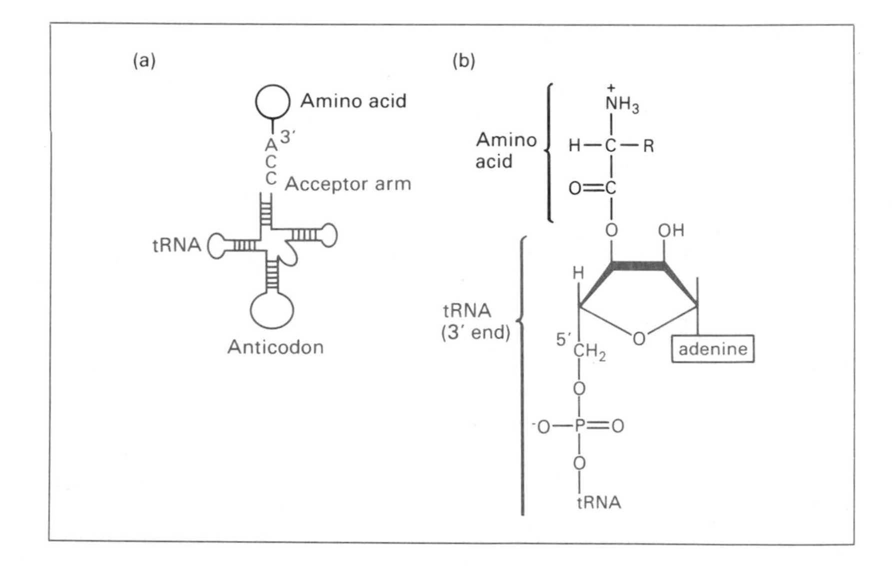

Figure 9.3 The attachment of an amino acid to a tRNA molecule. (a) The amino acid attaches to the 3′ end of the tRNA acceptor arm. (b) The linkage between the amino acid and the 3′ terminal nucleotide of the tRNA.

synthetase for each amino acid, meaning that one enzyme can charge each member of a series of isoaccepting tRNAs.

Although aminoacyl-tRNA synthetases form a fairly heterogeneous group of enzymes, each catalyses the same reaction (Fig. 9.4), with the energy required to attach the amino acid to the tRNA being provided by cleavage of ATP to AMP and pyrophosphate. The reaction takes place in two distinct steps, the first resulting in an activated amino acid intermediate in which the carboxyl group has formed a link with AMP. This intermediate is highly reactive and remains bound to the enzyme until the AMP is replaced in the second stage of the reaction by the tRNA molecule, producing aminoacyl-tRNA and free AMP.

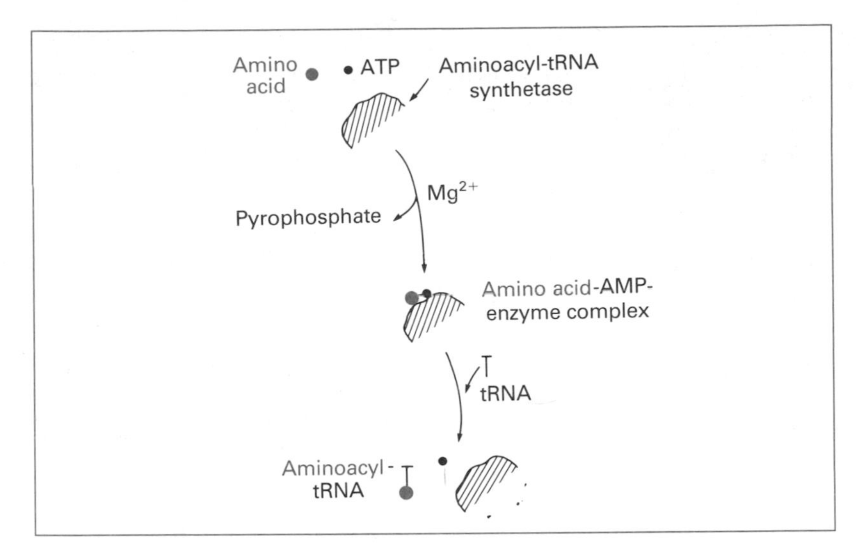

Figure 9.4 The reaction involved in aminoacylation of a tRNA.

The specificity of charging (ensuring that the correct amino acid is attached to the correct tRNA) is a function of the aminoacyl-tRNA synthetase, which is able to recognize both the correct amino acid, primarily on the basis of its unique R-group, and the appropriate tRNA. How the tRNA is recognized is not understood in detail, although it is clear that certain of the variant nucleotides of the cloverleaf are characteristic for an individual tRNA and that one or more of these are distinguished by the aminoacyl-tRNA synthetase.

Understanding how a tRNA is recognized by its aminoacyl-tRNA synthetase

Two complementary experimental approaches are being used to determine how an aminoacyl-tRNA synthetase recognizes its tRNA. The first involves X-ray diffraction analysis (p. 35), which can be used not only with individual proteins (or DNA molecules) but also with molecular complexes such as an aminoacyl-tRNA synthetase bound to its tRNA. In this way the contacts between the two molecules can be visualized. This analysis has been carried out with $tRNA^{gln}$ and its aminoacyl-tRNA synthetase from *E. coli*, revealing that the enzyme makes contact with the acceptor arm and anticodon loop of the tRNA (see Fig. 6.7).

The second approach is to make alterations in the nucleotide sequence of a tRNA and see if the aminoacyl-tRNA synthetase is still able to recognize it. If a nucleotide that is critical to the interaction is changed then the enzyme and tRNA will no longer be able to form a complex. Experiments along these lines with *E. coli* $tRNA^{ala}$ suggest that a single base pair in the acceptor arm is the most important feature recognized by the aminoacyl-tRNA synthetase.

9.1.2 Codon recognition

Once the correct amino acid has been attached to the acceptor arm of the tRNA, the aminoacylated molecule must complete the link between mRNA and polypeptide by recognizing and attaching to the correct codon: that is, one coding for the amino acid that it carries. Codon recognition is a function of the anticodon loop of the tRNA, specifically of the trinucleotide called the **anticodon** (see Fig. 6.7). This trinucleotide is complementary to the codon and can therefore attach to it by base pairing (Fig. 9.5). The specificity of the genetic code is therefore ensured because the anticodon present on a particular tRNA is one that is complementary to a codon for the amino acid with which the tRNA is charged.

(a) Codons and anticodons can wobble. From what has been said so far it might be imagined that there are 61 different types of tRNA molecule in each cell, one for each of the codons that specify an amino acid. In fact it has been known since the early 1960s that there are substantially fewer than 61 different tRNA molecules,

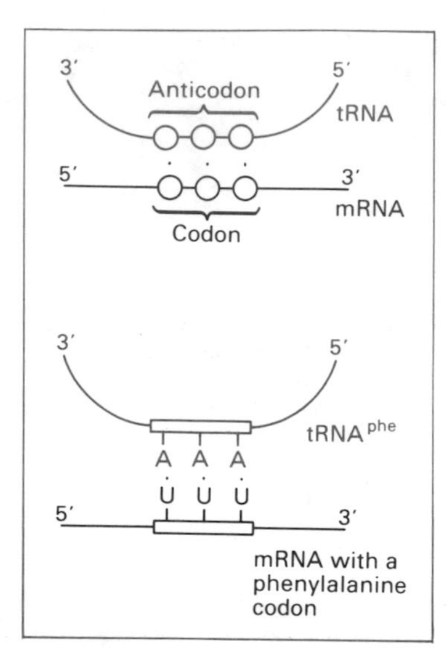

Figure 9.5 Codon recognition. On the top, base pairing between an anticodon on a tRNA molecule and a codon in an mRNA molecule is shown. On the bottom the specific situation for decoding a phenylalanine codon is illustrated.

usually between 31 and 40 depending on the organism. The explanation for this is provided by the **wobble hypothesis** which was first expounded by Crick in 1966.

This hypothesis is based on the fact that the anticodon loop is just that, a loop, and the anticodon itself is not a perfectly linear trinucleotide. The short double helix formed by base pairing between the codon and anticodon does not therefore have the precise configuration of a standard RNA helix: instead its dimensions are slightly altered. As a result, non-standard base pairs can form at the wobble position (between the first nucleotide of the anticodon and the third nucleotide of the codon). So a single anticodon may be able to base pair with more than one codon and a single tRNA may decode more than one member of a codon family.

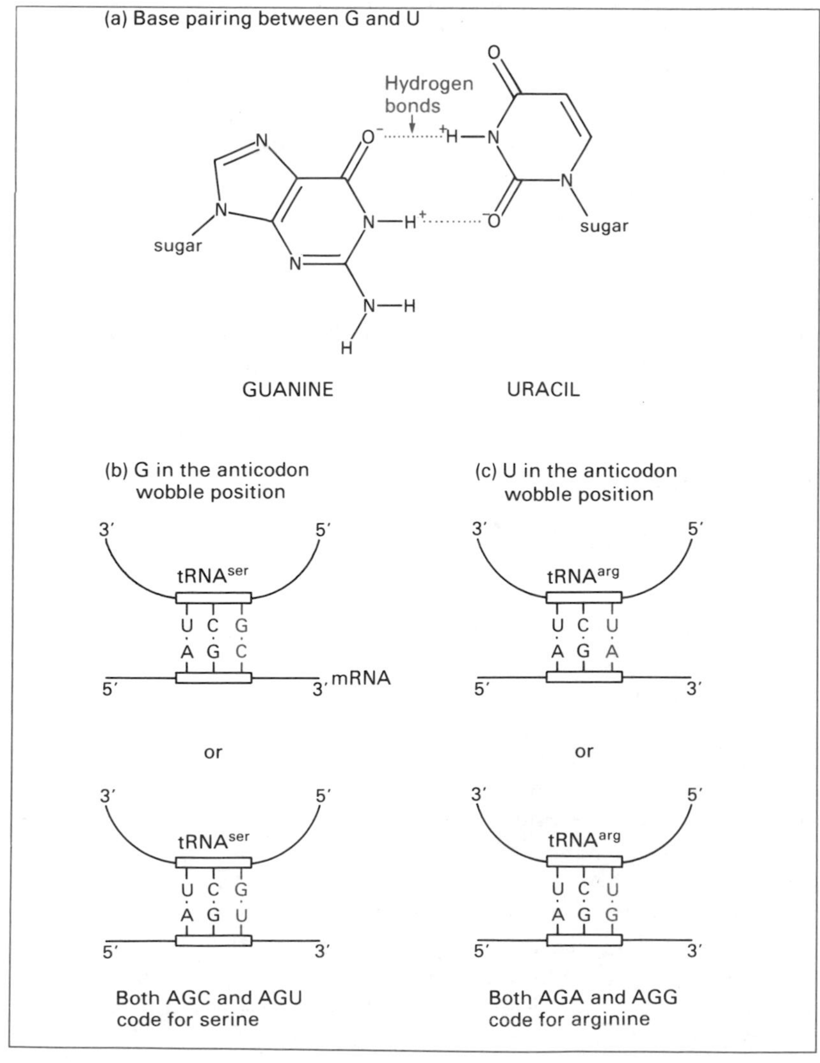

Figure 9.6 Wobble with G–U base pairs. (a) Base pairing between G and U; compare with the standard base pairs shown in Fig. 3.10(b). (b) and (c) Two consequences of G–U in the wobble position.

However, the base pairing rules do not become totally flexible at the wobble position, and only a few types of unusual base pairs are allowed. Common examples are:

1. ***G–U base pairs.*** A G–U base pair (p. 82) is allowable at the wobble position (Fig. 9.6(a)). Two possible consequences of this are shown in Fig. 9.6(b) and (c).

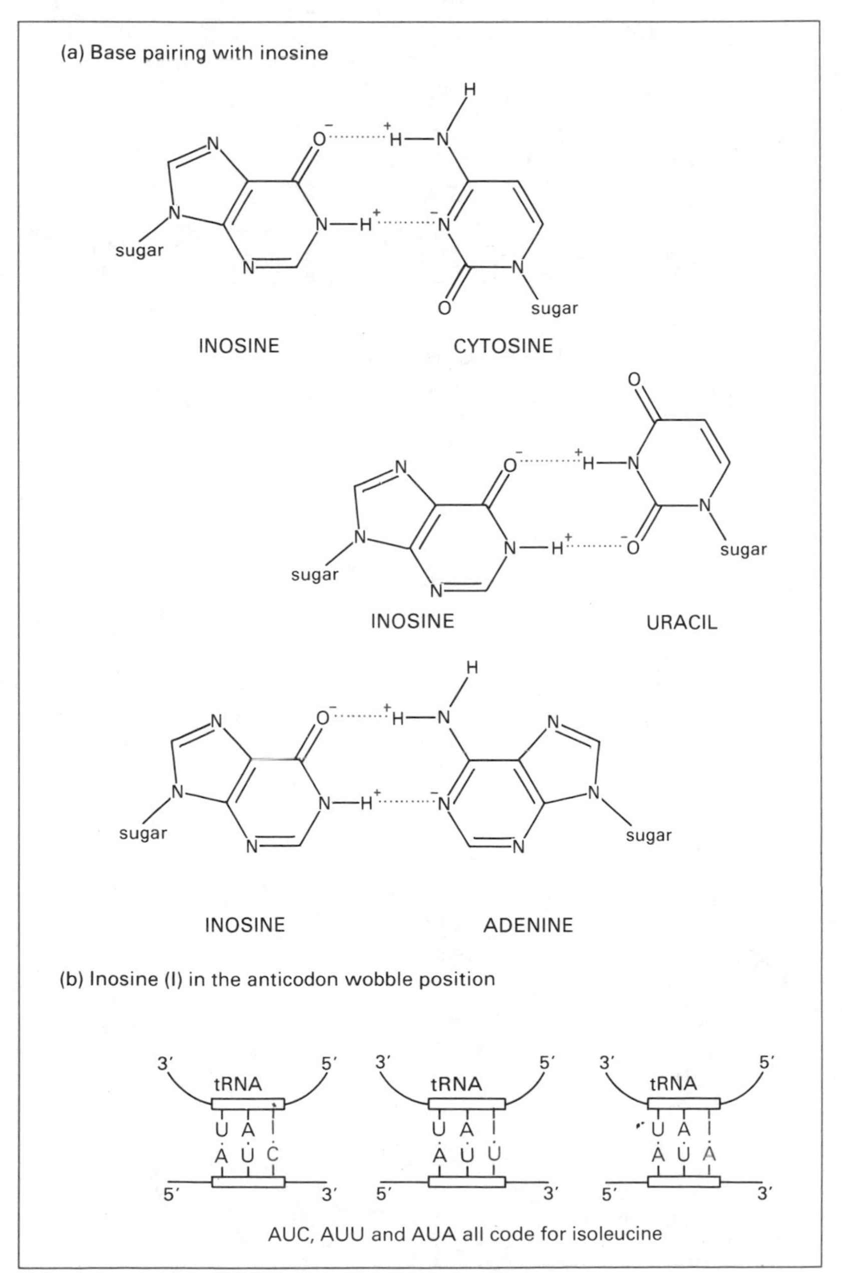

Figure 9.7 Wobble with inosine (I). (a) The base pairs that are possible with this nucleotide. (b) The consequence of having I at the wobble position.

2. ***Inosine can base pair with C, A and U.*** In some tRNAs the wobble nucleotide of the anticodon is inosine (I), a deaminated form of G (p. 92). I can base pair not only with C but also with A and U (Fig. 9.7(a)). A possible consequence of this is shown in Fig. 9.7(b).

Wobble therefore decreases the number of tRNAs needed to decode the genetic code so that a single cell can get by with as few as 31 different tRNAs. Nevertheless, the rules of the genetic code still remain inviolate, and a polypeptide synthesized during translation is polymerized strictly in accordance with the nucleotide sequence of the relevant mRNA. We will now turn our attention to the mechanics of this process.

9.2 The mechanics of protein synthesis in *E. coli*

Once the role of tRNA has been appreciated the exact mechanism by which translation occurs is much easier to follow. Traditionally the process is split into three stages – initiation, elongation and termination – although in practice the events are continuous.

9.2.1 Initiation of translation

The very first step in translation in *E. coli* is attachment of the small (30S) subunit of a ribosome to an mRNA molecule. When

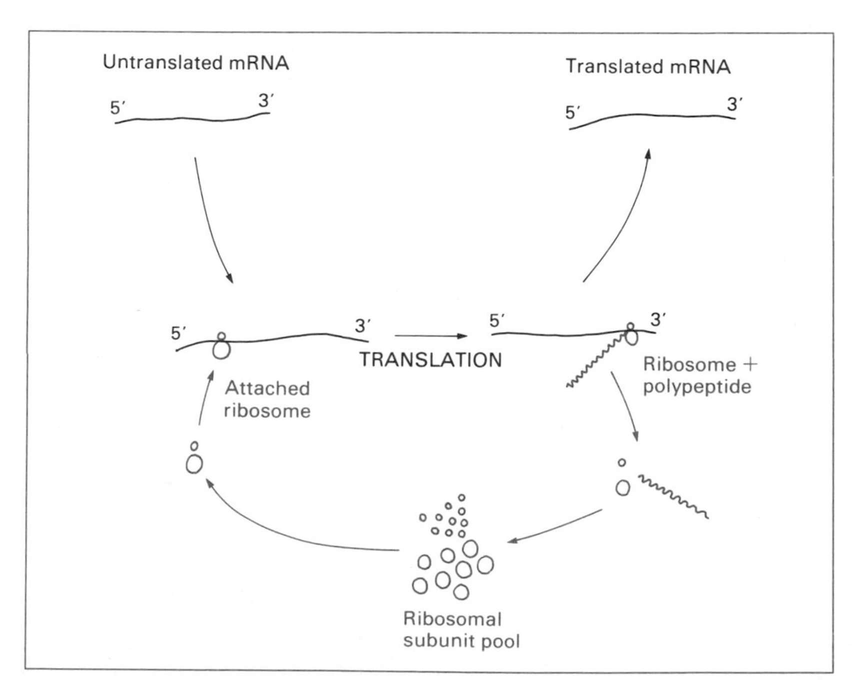

Figure 9.8 The ribosomal pool consists of dissociated subunits not involved in translation.

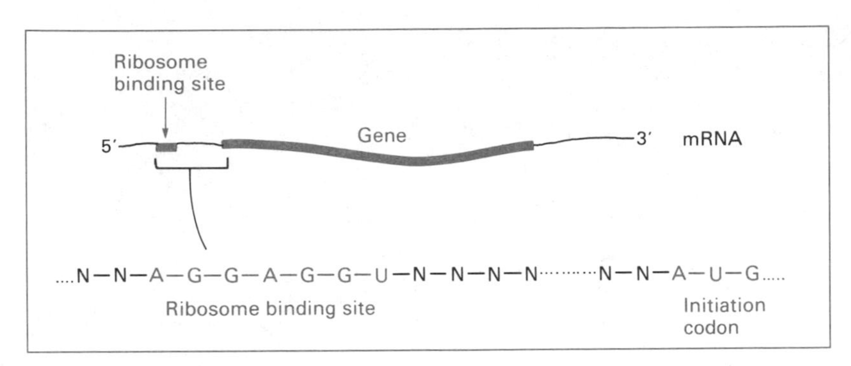

Figure 9.9 The ribosome binding site of *E. coli* and its position relative to the initiation codon for the gene. 'N' indicates any nucleotide.

not actually involved in protein synthesis, ribosomes dissociate into their component subunits so that the ribosome 'pool' in the cell is made up of large numbers of individual 30S and 50S subunits (Fig. 9.8).

(a) The ribosome binding site ensures that translation starts at the correct position. If you recall the transcription process you will remember that it is important to ensure that the RNA polymerase enzyme starts transcription at the correct point so that genes rather than random pieces of DNA are transcribed. An analogous situation exists with translation: the 30S ribosomal subunit must attach to the mRNA molecule to be translated not at random, but at a specific point just upstream of the initiation codon of the gene (Fig. 9.9). The correct attachment point is signalled by the **ribosome binding site** which in *E. coli* has the consensus sequence:

5′-AGGAGGU-3′

This sequence, known as the **Shine–Dalgarno sequence**, is believed to form a transient base paired attachment to a part of the 16S rRNA polynucleotide, allowing the 30S subunit to bind to the mRNA.

Ribosome binding sites

Details of some real ribosome binding sites for *E. coli* genes:

Gene	*Gene product*	*Nucleotide sequence*	*Distance to AUG codon (nucleotides)*
E. coli consensus	–	AGGAGGU	10
lacZ	β-Galactosidase	AGGA	7
galE	Hexose-1-P uridyltransferase	GGAG	6
galT	Galactose-1-P uridyltransferase	AGGA	6
rplJ	Ribosomal protein L10	AGGAG	8
cro	Phage λ regulatory protein	AGGAGGU	3
B	Phage φX174 coat protein	AGGAG	7

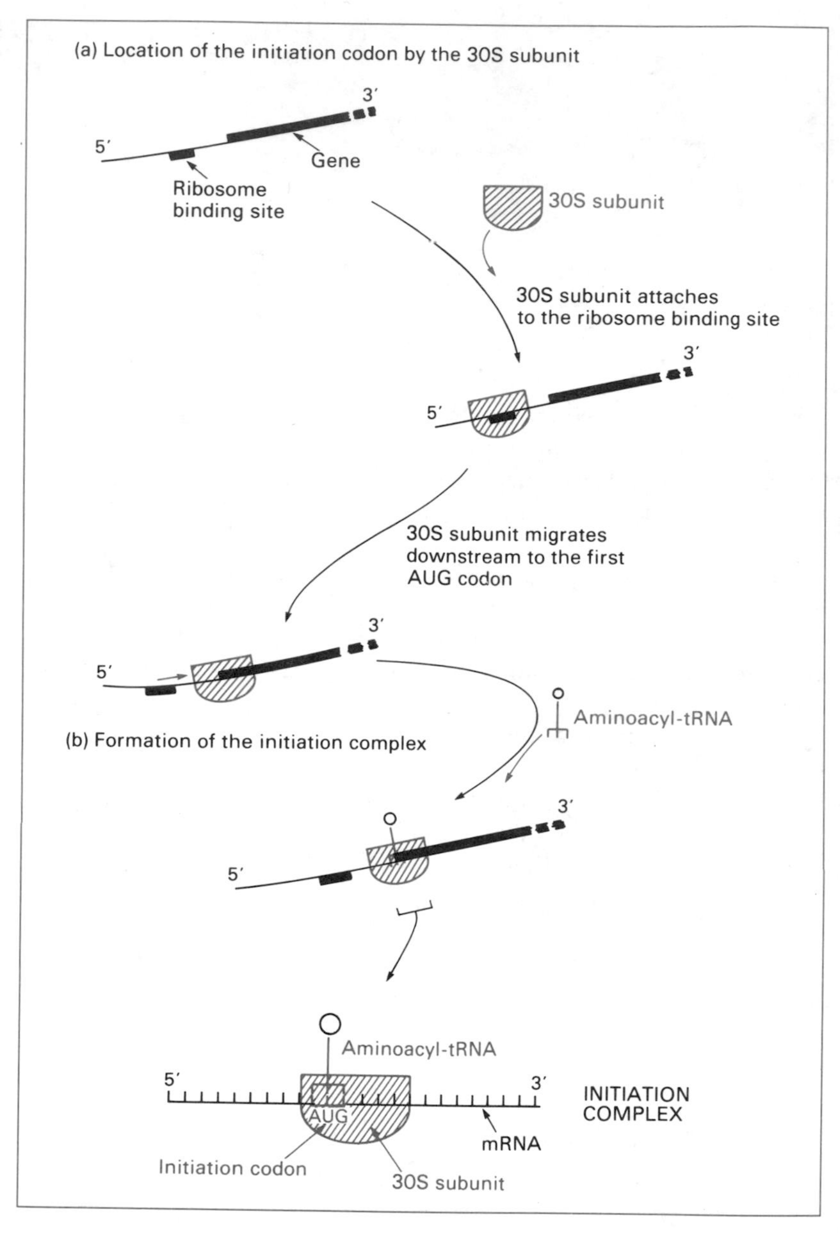

Figure 9.10 Initiation of translation in *E. coli*. For the purposes of clarity the initiation factors and the GTP molecule are left out.

Once the ribosome binding site has been located, the 30S subunit migrates downstream until it encounters an AUG codon, which it will usually find within 10 nucleotides of the binding site (Fig. 9.10(a)). This AUG triplet will be the initiation codon of the gene and will mark the position at which translation must begin.

(b) Formation of the initiation complex. The translation process itself starts when an aminoacylated tRNA base pairs with the

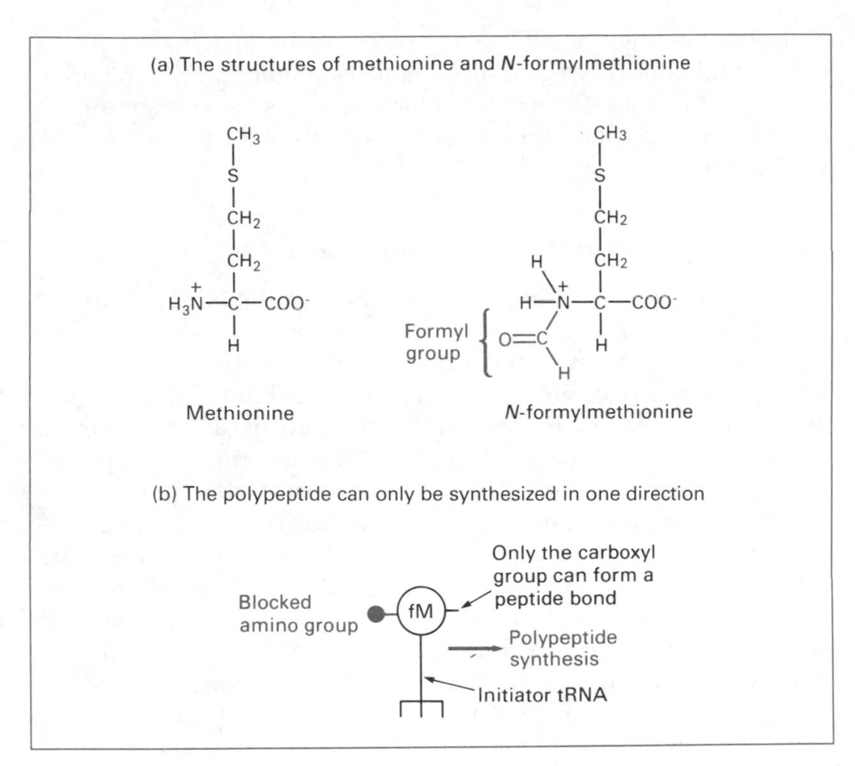

Figure 9.11 *N*-formylmethionine and its role in the initiation of translation in *E. coli*.

initiation codon that has been located by the 30S subunit of the ribosome (Fig. 9.10(b)). This initiator tRNA is charged with methionine, because methionine is the amino acid coded by AUG. However, in bacteria (although not eukaryotes) this methionine is modified after charging by substitution of a formyl group (—COH) for one of the hydrogen atoms of the amino group, to produce *N*-formylmethionine or **fmet** (Fig. 9.11(a)). This substitution blocks the amino group so that it cannot participate in the formation of a peptide bond; polymerization of the polypeptide can therefore take place only in the amino to carboxyl direction (Fig. 9.11(b)).

The resulting structure, comprising primarily the mRNA, 30S subunit and aminoacylated $tRNA^{fmet}$ is called the **initiation complex** and its formation marks the end of the initiation stage of translation.

(c) Initiation factors. The main outstanding area of translation that is not fully understood is the role of various non-ribosomal protein factors. For example, initiation in *E. coli* requires three proteins called **initiation factors**. IF1 and IF3 appear to be responsible for dissociation of the ribosome into 30S and 50S subunits (which is important as the complete ribosome is unable to initiate translation), and IF3 is probably involved in recognition of the ribosome binding site. IF2 participates in the attachment of the charged initiator tRNA and also mediates binding to the initiation complex of a molecule of GTP, which provides the energy for the next step in translation. However, the details of these roles are still very vague.

9.2.2 Elongation of the polypeptide chain

Once the initiation complex has been formed the large subunit of the ribosome can attach. This requires hydrolysis of the GTP molecule associated with the initiation complex and results in two separate and distinct sites to which tRNAs can bind (Fig. 9.12). The first of these, called the **peptidyl-** or **P-site** is at the moment occupied by the aminoacylated $tRNA^{fmet}$, which is still base paired with the initiation codon. The **aminoacyl-** or **A-site** is positioned over the second codon of the gene and at first is empty. Elongation begins when the correct aminoacylated tRNA enters the A-site and base pairs with the second codon (Fig. 9.13(a)). This requires two **elongation factors**, EF-Tu and EF-Ts. EF-Tu is directly involved in entry of the aminoacyl-tRNA into the A-site, with a second molecule of GTP being hydrolysed to provide the required energy. This results in a deactivated form of EF-Tu, which has to be reactivated by EF-Ts before it can participate in a later round of elongation.

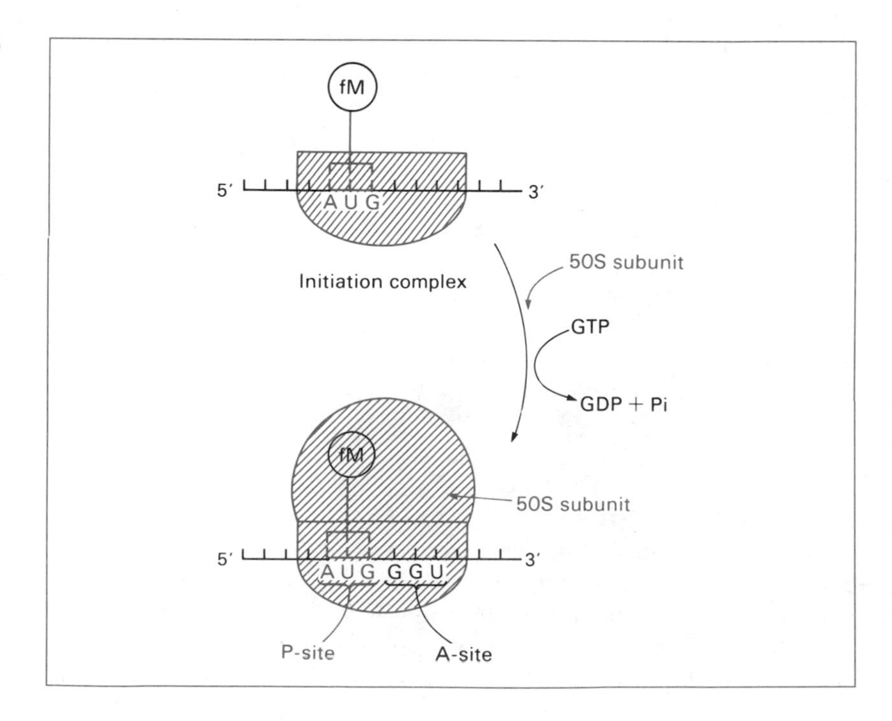

Figure 9.12 Attachment of the 50S subunit to the initiation complex results in two distinct tRNA binding sites.

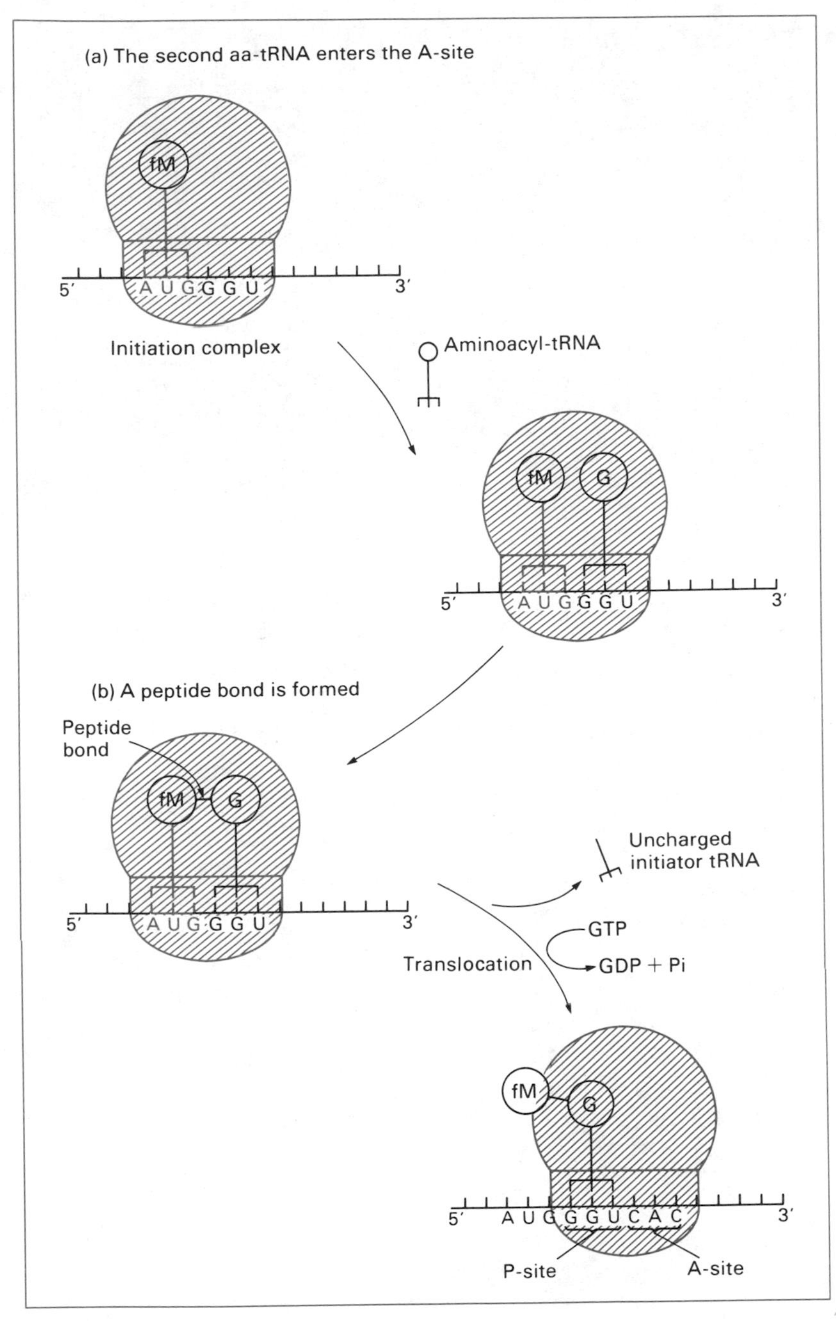

Figure 9.13 The elongation step of translation in *E. coli*.

(a) Peptide bond formation and translocation. Now both sites of the ribosome are occupied by aminoacylated tRNAs and the two amino acids are in close contact. The next step is formation of a peptide bond between the carboxyl group of the fmet and the

Studying translation in *E. coli*

Experiments aimed at understanding translation often make use of inhibitors that block a particular step in the process. This is important in determining the order in which individual events take place.

Inhibitor	*Activity*
Streptomycin	Prevents the initiator tRNA entering the P-site
Kirromycin	Binds to EF-Tu
Chloramphenicol	Inhibits peptidyl transferase
Thiostrepton	Binds to EF-G and prevents translocation

Kirromycin provides a good example of the use of an inhibitor. Kirromycin binds to EF-Tu and prevents this factor from being released from the ribosome after it has delivered the aminoacyl-tRNA to the A-site. As a result, peptide bond formation does not occur. This demonstrates quite clearly that under normal circumstances release of EF-Tu takes place before the new peptide bond is formed.

amino group of the second amino acid. The reaction is catalysed by a complicated enzyme called **peptidyl transferase**, which is possibly a combination of several different ribosomal proteins. Peptidyl transferase acts in conjunction with a second ribosomal enzyme, **tRNA deacylase**, which breaks the fmet–tRNA link. The result is a dipeptide attached to the tRNA present in the A-site (Fig. 9.13(b)).

Translocation now occurs. The ribosome slips along the mRNA by three nucleotides so that the aa–aa–tRNA enters the P-site, expelling the now uncharged $tRNA^{fmet}$ and making the A-site vacant again. The third aminoacylated tRNA now enters the A-site and the elongation cycle is repeated. Each cycle requires hydrolysis of a molecule of GTP and is controlled by a third elongation factor, EF-G.

(b) Each mRNA can be translated by several ribosomes at once. After several cycles of elongation the start of the mRNA molecule is no longer associated with the ribosome, and a second

Figure 9.14 A polysome.

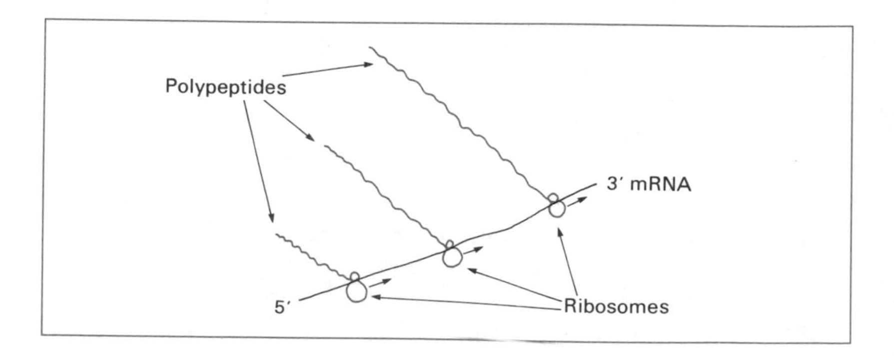

round of translation can begin. A second 30S subunit can attach to the ribosome binding site and form a new initiation complex. The end result is a **polysome** (Fig. 9.14), an mRNA that is being translated by several ribosomes at once. Polysomes have been seen in electron microscopic images of both prokaryotic and eukaryotic cells.

9.2.3 Chain termination

Termination of translation occurs when a termination codon (UAA, UAG or UGA) enters the A-site. There are no tRNA molecules with anticodons able to base pair with any of the termination codons; instead one of two **release factors** – RF1 or RF2 – enters the A-site and cleaves the completed polypeptide from the final tRNA (Fig. 9.15). A third release factor, RF3, plays an ancillary role in the process. The ribosome releases the polypeptide

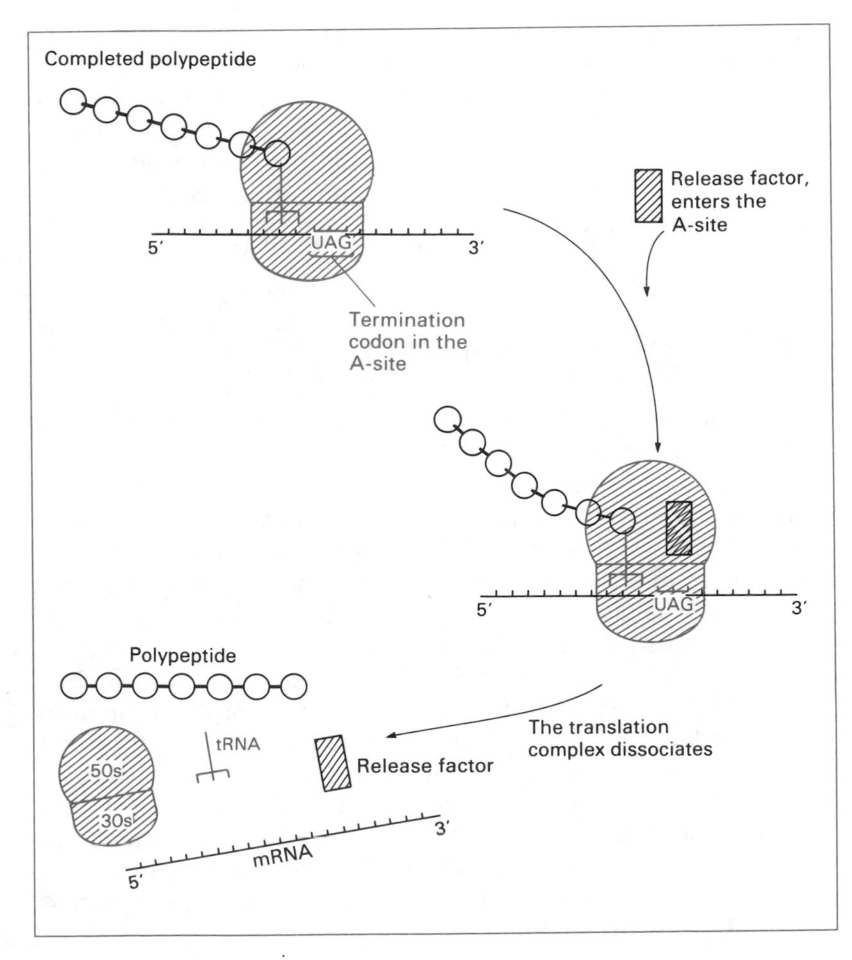

Figure 9.15 Termination of translation in *E. coli*.

E. coli translation factors – a summary

Factor	Molecular mass (daltons)	Function
Initiation		
IF1	9000	Dissociation of ribosomes to subunits
IF2	97 000	Attachment of $tRNA^{fmet}$ to the initiation complex
IF3	23 000	Ribosome dissociation, binding of mRNA to the initiation complex
Elongation		
EF-Tu	43 000	Binding of aminoacyl-tRNA to the A-site
EF-Ts	74 000	Generation of active EF-Tu
EF-G	77 000	Translocation
Termination		
RF1	36 000	Chain release – UAA, UAG codons
RF2	41 000	Chain release – UGA, UAA codons
RF3	46 000	Cooperates with RF1 and RF2

and mRNA, and subsequently dissociates into 30S and 50S subunits which enter the cellular pool before becoming involved in a new round of translation. The polypeptide may be processed by cleavage of a portion of its amino acid sequence, and possibly modified by attachment of chemical groups to specific amino acids. In conjunction with these events the polypeptide chain folds up into its tertiary structure and begins its functional life within the cell.

9.3 Translation in eukaryotes

Translation in eukaryotes occurs in essentially the same fashion as in *E. coli*. The only substantial difference concerns the way in which the small subunit of the ribosome (40S in eukaryotes) binds to the mRNA and locates the AUG codon at which translation should begin. A eukaryotic mRNA does not have an internal ribosome binding site equivalent to the Shine–Dalgarno sequence of *E. coli*. Instead the small subunit of the ribosome recognizes the cap structure (Section 7.3.1) as its binding site and therefore attaches to the extreme 5′ end of the mRNA. The subunit then moves along the mRNA until it reaches an initiation codon and can begin translation (Fig. 9.16). Usually the initiation codon is the first AUG triplet that the small subunit encounters as it **scans** along the mRNA. This is not always the case though, and with some mRNAs initiation of translation begins at a later AUG triplet. To account for this it has been proposed that the nucleotide sequence surrounding the AUG is important in determining whether the triplet is used as an initiation codon.

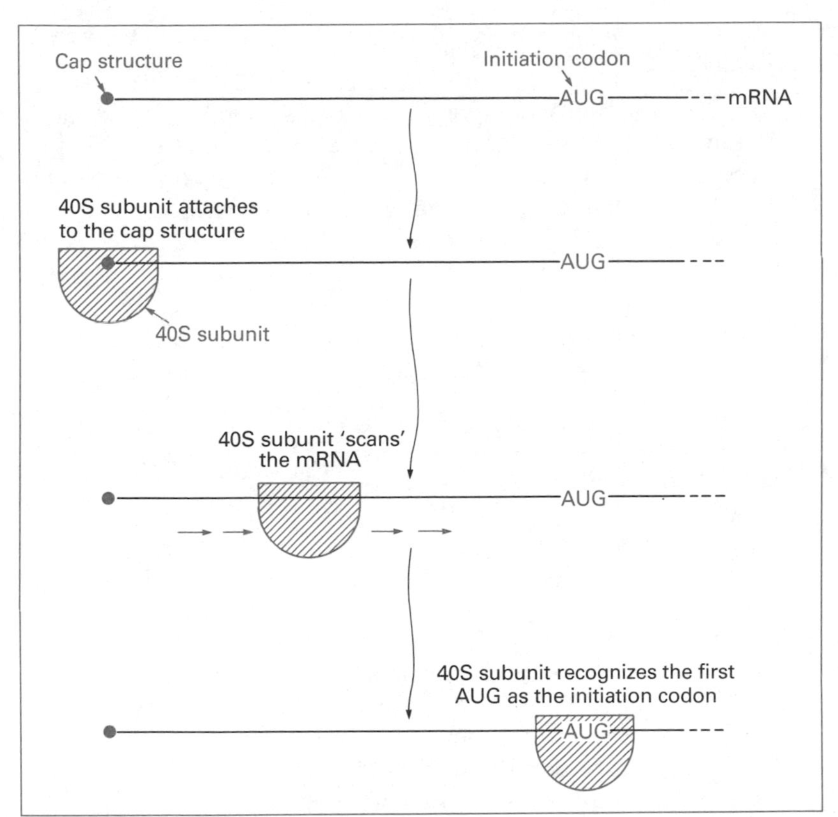

Figure 9.16 Initiation of translation in eukaryotes.

Eukaryotic translation factors – a summary

Factor	*Function*
Initiation	
eIF3, eIF4C	Attach to the 40S subunit before mRNA binding
eIF4A, eIF4E, p220	Attach to the cap structure and assist binding of the 40S subunit
eIF4B	Possibly breaks mRNA stem-loops during scanning
eIF2	Binds to the initiator mRNA
eIF2B	Regeneration of active eIF2
eIF5	Releases eIF2, 3 and 4C from the growing initiation complex
eIF6	Dissociation of ribosome subunits
eIF1	Uncertain
Elongation	
eEF1	Binding of aminoacyl-tRNA to the A-site
eEF2	Translocation
Termination	
eRF	Recognition of all termination codons

Other distinctive features of translation in eukaryotes are as follows:

1. The initiator tRNA in eukaryotes carries an unmodified methionine, not the fmet amino acid used in *E. coli*.
2. Initiation of translation in eukaryotes requires many more initiation factors than the three that are sufficient in *E. coli*.
3. Initiation in eukaryotes requires hydrolysis of ATP molecules to provide energy for constructing the initiation complex and scanning along the mRNA.
4. Termination of translation in eukaryotes requires hydrolysis of a GTP molecule, whereas termination in *E. coli* does not.

Reading

M. B. Hoagland, E. B. Keller and P. C. Zamecnik (1956) Enzymatic carboxyl activation of amino acids. *Journal of Biological Chemistry*, **218**, 345–58 – the discovery of aminoacyl-tRNA synthetases.

P. Schimmel (1991) Classes of aminoacyl-tRNA synthetases and the establishment of the genetic code. *Trends in Biochemical Sciences*, **16**, 1–3 – recent details on the similarities and distinctions between different aminoacyl-tRNA synthetases.

Y.-M. Hou, C. Francklyn and P. Schimmel (1989) Molecular dissection of a transfer RNA and the basis for its identity. *Trends in Biochemical Sciences*, **14**, 233–7 – an investigation into how an aminoacyl-tRNA synthetase recognizes its tRNA.

J. D. Watson (1963) Involvement of RNA in the synthesis of proteins. *Science*, **140**, 17–26 – an account of how ideas developed between 1953 and 1963 about the roles of RNA in protein synthesis.

F. H. C. Crick (1966) Codon–anticodon pairing: the wobble hypothesis. *Journal of Molecular Biology*, **19**, 548–55.

T. Hunt (1980) The initiation of protein synthesis. *Trends in Biochemical Sciences*, **5**, 178–81; B. Clark (1980) The elongation step of protein biosynthesis. *Trends in Biochemical Sciences*, **5**, 207–10; C. T. Caskey (1980) Peptide chain termination. *Trends in Biochemical Sciences*, **5**, 234–7 – three clear reviews on the steps in protein synthesis.

J. Shine and L. Dalgarno (1974) The 3′-terminal sequence of *Escherichia coli* 16S ribosomal RNA: complementarity to nonsense triplets and ribosome binding sites. *Proceedings of the National Academy of Sciences USA*, **71**, 1342–6.

J. M. Adams and M. R. Capecchi (1966) *N*-formylmethionyl-tRNA as the initiator of protein synthesis. *Proceedings of the National Academy of Sciences USA*, **55**, 147–55.

R. A. Garrett and P. Woolley (1982) Identifying the peptidyl transferase centre. *Trends in Biochemical Sciences*, **7**, 385–6 – a short report of this complex enzyme.

R. E. Rhoads (1988) Cap recognition and the entry of mRNA into the protein synthesis initiation cycle. *Trends in Biochemical Sciences*, **13**, 52–6 – a review that illustrates the complexity of translation initiation in eukaryotes.

Problems

1. Define the following terms:

isoaccepting tRNAs	aminoacyl site
aminoacylation	elongation factor
anticodon	translocation
ribosome binding site	polysome
initiation complex	release factor
initiation factor	scanning
peptidyl site	

2. Outline the roles in protein synthesis of the following enzymes:

aminoacyl-tRNA synthetase
peptidyl transferase
tRNA deacylase

3. Describe how an amino acid is attached to a tRNA molecule.

4. Explain the role of tRNA molecules in ensuring that translation occurs in accordance with the rules laid down by the genetic code.

5. Explain why less than 61 tRNAs are sufficient to decode the the entire genetic code.

6. Outline the events involved in formation of the initiation complex during protein synthesis in *E. coli*.

7. What effect does formylation of the initiator methionine have during initiation of translation in *E. coli*?

8. Sketch the events that occur during the elongation and termination stages of translation in *E. coli*.

9. Describe how the 40S subunit of a eukaryotic ribosome locates an initiation codon on an mRNA molecule.

10. *Discuss the connection between the wobble hypothesis and the degeneracy of the genetic code.

11. *In human mitochondria, protein synthesis requires only 22 different tRNAs (Section 15.5.2). What implications does this have for the rules governing codon–anticodon interactions in this system?

12. *Most organisms display a distinct codon bias in their genes. For instance, of the four codons for valine, only two (GUC and GUU) appear at all frequently in genes of *Saccharomyces cerevisiae*: GUG and GUA are less common. It has been suggested that a gene that contains a relatively high number of unfavoured codons might be expressed at a relatively slow rate. Explain the thinking behind this hypothesis and discuss its ramifications.

Control of gene expression

10

Why control gene expression? • Possible strategies • Control of gene expression in bacteria – regulation of lactose utilization • Upstream sites and DNA-binding proteins • Gene regulation during development – *Drosophila* – the homeobox

The previous five chapters have outlined the process of gene expression and described how the biological information contained in the DNA sequence of a gene is released and utilized by the cell. Now we must look at the way in which this process is controlled.

10.1 Why control gene expression?

The entire complement of genes in a single cell represents a staggering amount of biological information. Some of this information is needed by the cell at all times: for instance, most cells continually synthesize ribosomes and so have a continuous requirement for transcription of the rRNA and ribosomal protein genes. Similarly, genes coding for enzymes such as RNA polymerase or those involved in the basic metabolic pathways will be active in all cells all of the time. These genes are sometimes called **housekeeping genes**, reflecting the role in the cell of the biological information they carry.

On the other hand, many genes have a more specialized role and their biological information is needed by the cell only under certain circumstances. All organisms are therefore able to regulate expression of their genes, so that those genes whose RNA or protein products are not needed at a particular time are switched off. This is a straightforward concept but it has extensive ramifications. To illustrate the point we will briefly consider what gene regulation achieves in a bacterium and in a multicellular organism such as man.

10.1.1 Gene regulation in *E. coli*

Gene regulation enables a bacterium to respond to changes in its environment. An example is provided by those bacterial genes

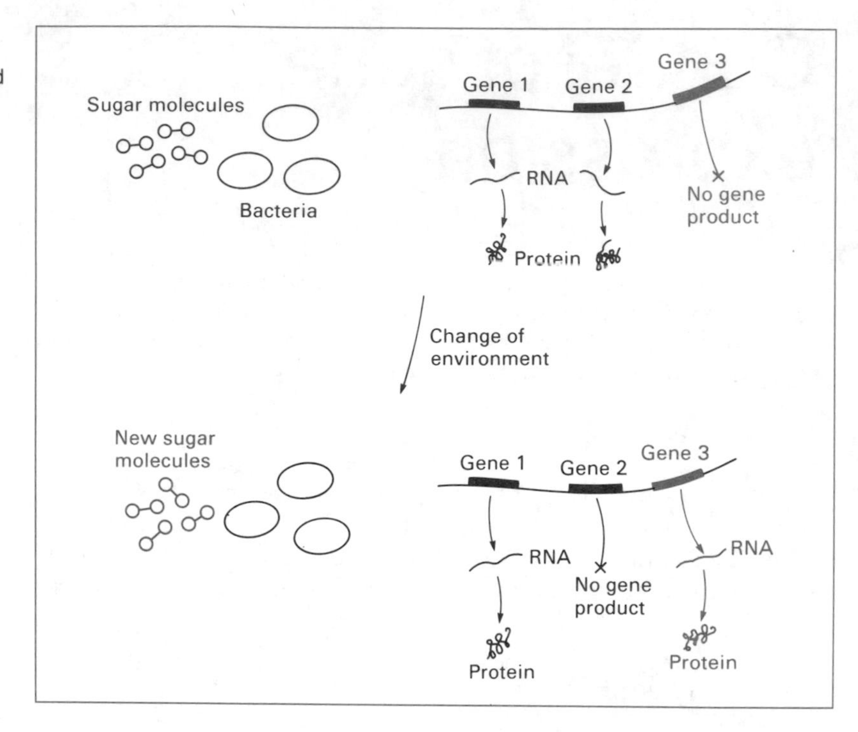

Figure 10.1 Genes can be switched on and off. In this example, gene 1 is a housekeeping gene and is switched on at all times. Genes 2 and 3 code for enzymes involved in metabolism of two different sugars. They are subject to regulatory control and are switched on only when their gene products are required by the cell.

involved in utilization of different sugar compounds as sources of carbon and energy. *E. coli* possesses a number of genes coding for a variety of enzymes that enable the cell to take up and metabolize any of several different sugars. Exactly which enzymes are required by an individual bacterium at any one time depends on which sugars are present in the environment. Sometimes there will be a plentiful supply of glucose, at other times the glucose levels will be low and the bacterium will be utilizing a different sugar, lactose for instance. *E. coli* could continuously express all its sugar utilizing genes, and so have molecules of each enzyme available all the time, but this would waste energy, something all organisms try to avoid. Why synthesize enzymes for lactose utilization if there is no lactose available? Instead, a bacterium expresses only those genes coding for the enzymes needed to metabolize the sugars immediately available; the genes for all the other enzymes are inactive, switched off as their gene products are not required at the present time (Fig. 10.1). If the environment changes, and one sugar is replaced by another, the bacterium can quickly switch on expression of the genes whose products are now needed, and switch off the ones that have become redundant. By regulating expression of their genes, *E. coli* and other bacteria are able to respond quickly to changes in the environment without wasting energy by maintaining in an active state genes whose products are of no immediate use.

10.1.2 Gene regulation in multicellular organisms

Eukaryotic cells can also respond to changes in the environment by altering their gene expression patterns. For example, yeast regulates its genes for sugar-utilization enzymes in a manner analogous to *E. coli*, and plant cells switch on genes for photosynthetic proteins in response to light (Fig. 10.2(a)). In a multicellular organism, individual cells, or groups of cells, must also respond to stimuli that originate from elsewhere within the organism. Hormones, growth factors and other regulatory molecules are produced by one type of cell and cause changes in gene expression in other cells (Fig. 10.2(b)). In this way the activities of groups of cells can be coordinated in a manner that is beneficial to the organism as a whole.

In multicellular organisms we also see another ramification of gene regulation: the existence of specialized cells. In man there are about 250 cell types, each with a different morphology, biochemistry and role to play in the organism. The distinctions between cell types are the result of differences in gene expression patterns. A liver cell is very different from a muscle cell because it expresses a different set of genes (Fig. 10.2(c)). These gene

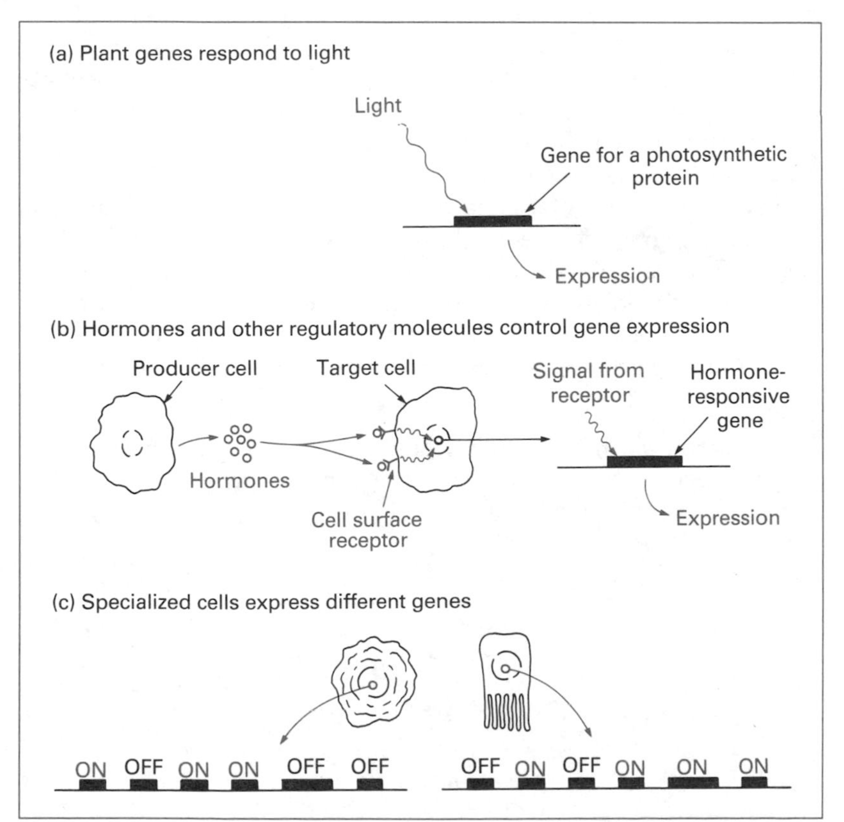

Figure 10.2 Three aspects of gene regulation in multicellular eukaryotes.

expression patterns are permanent – a liver cell will never become a muscle cell – and are laid down very early in development. In the early embryo, cells have already made 'decisions' about which tissue or organ they will give rise to in the fetus. These cells must be at the correct locations within the embryo: if they are in the wrong places then the fetus will not be constructed correctly. How can this spatial control over gene expression be brought about? Some genes are expressed only at certain stages in development. For example, in man different globin genes are expressed in the embryo, fetus and adult (Section 4.2.1). How are these genes switched on and off at the correct times?

These questions all centre on gene regulation. Finding the answers is one of the major challenges of molecular genetics.

10.2 Possible strategies for controlling gene expression

Control of gene expression is, in essence, control over the amount of gene product present in the cell. This amount is a balance between two factors (Fig. 10.3): the rate of synthesis (how many molecules of the gene product are made per unit time) and the degradation rate (how many molecules are broken down per unit time). The result of this balance is a different steady state concentration for each gene product in the cell. If either the synthesis rate or the degradation rate changes then the steady state concentration will also change, although, in practice, the critical variations in steady state of a gene product arise because of changes in synthesis: degradation seems to be relatively constant. This is true both of RNA gene products (rRNA, tRNA, etc.) and proteins.

How can the synthesis rate of a gene product be regulated? The answer is by exerting control over any one, or a combination, of the various steps in the gene expression pathway. Possibilities include (Fig. 10.4):

1. ***Transcription***. If the number of transcripts synthesized per unit time decreases, the amount of the gene product in the cell will also decrease.

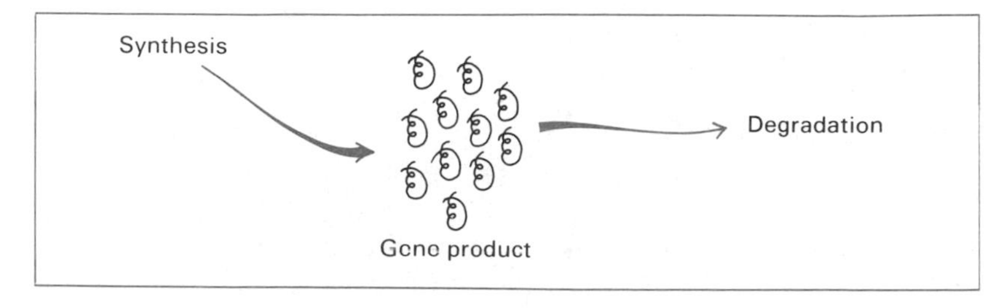

Figure 10.3 The amount of a gene product in the cell is a balance between the rate of synthesis and the rate of degradation.

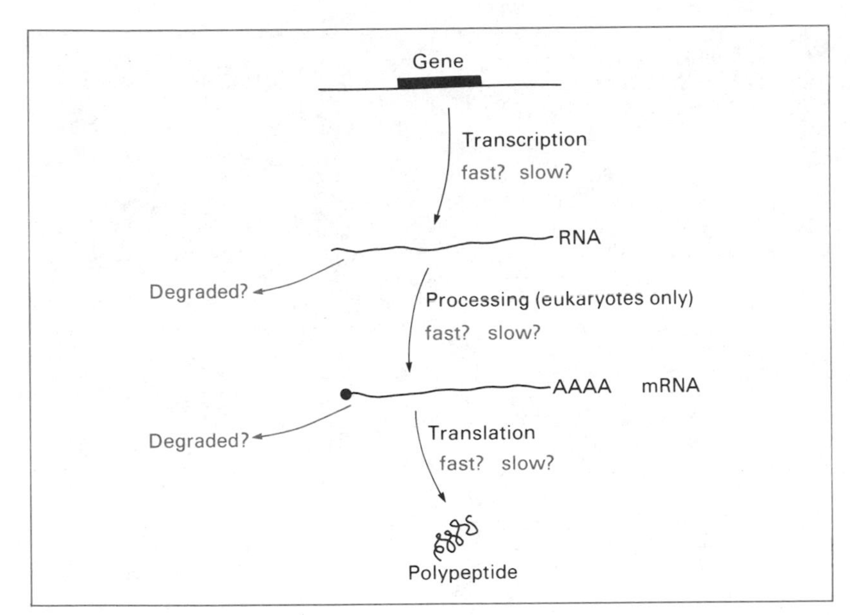

Figure 10.4 Factors that influence the rate of synthesis of a gene product.

2. ***mRNA turnover***. If mRNA molecules are degraded before translation can occur, synthesis of the gene product will be limited.
3. ***mRNA processing***. Events such as capping, polyadenylation and splicing of eukaryotic mRNAs are in most cases prerequisites for translation. If these processing events are slowed, product synthesis will fall.
4. ***Translation***. Control could be exerted over the number of ribosomes that can attach to a single mRNA, or over the rate at which individual ribosomes translate a message.

It is now becoming clear to molecular biologists that control of gene expression is a highly complex business that probably involves each of the above strategies. However, the best understood mechanisms, in both prokaryotes and eukaryotes, depend solely on the first possibility: regulatory control over transcription. We will look first at bacteria and then at eukaryotes and discover how much is known about the way in which transcription is regulated.

10.3 Control of gene expression in bacteria

The foundation to our understanding of how bacteria regulate expression of their genes was laid by François Jacob and Jacques Monod in a paper, published in 1961, that is now considered a classic example of experimental analysis and deductive reasoning. Jacob and Monod based their propositions on an intricate genetic

François Jacob

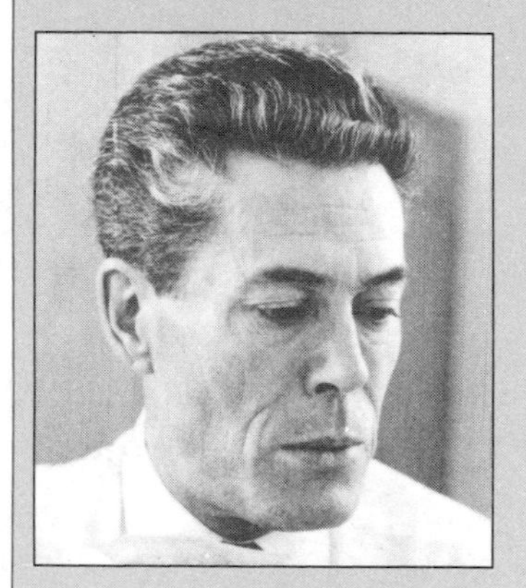

Jacques Monod

François Jacob b. 17 June 1920, Nancy, and **Jacques Monod** b. 9 February 1910, Paris; d. 31 May 1976, Cannes

Both Jacob and Monod served heroically during World War II, Jacob as a member of the Free French Forces, injured seriously in Normandy during August 1944, and Monod as a leader of the Paris Resistance. Monod had already obtained his PhD, Jacob had to wait until 1947 to gain his MD from the University of Paris. Both spent the major parts of their careers at the Pasteur Institute, the famous Paris research centre set up around Louis Pasteur in the late nineteenth century and still one of the most influential European laboratories. Jacob contributed to some of the early work on bacteriophages, with André Lwoff and Elie Wollman, before beginning the collaboration with Monod that led to the operon theory and the concept of messenger RNA in 1961. The two shared the 1965 Nobel Prize with Lwoff. Both have published important books: *Chance and Necessity* by Monod is a powerful vindication of natural selection, *The Logic of Life* by Jacob an illuminating account of changes in scientific thinking over the past three centuries. Jacob's autobiography, *The Statue Within*, was published in 1987.

analysis of lactose utilization by *E. coli* and described an elegant regulatory system that has subsequently been confirmed in virtually every detail. We will use the lactose system as our central example of how gene regulation works in bacteria.

10.3.1 Regulation of lactose utilization

Lactose is a disaccharide composed of a single glucose unit attached to a single galactose unit (Fig. 10.5(a)). In order to utilize lactose as a carbon and energy source, an *E. coli* cell must first transport lactose molecules from the extracellular environment into the cell and then split the molecules, by hydrolysis, into glucose and galactose (Fig. 10.5(b)). These reactions are catalysed by three enzymes, each of which has no function other than lactose utilization: lactose permease, which transports lactose into the cell; β-galactosidase, which is responsible for the splitting reaction; and

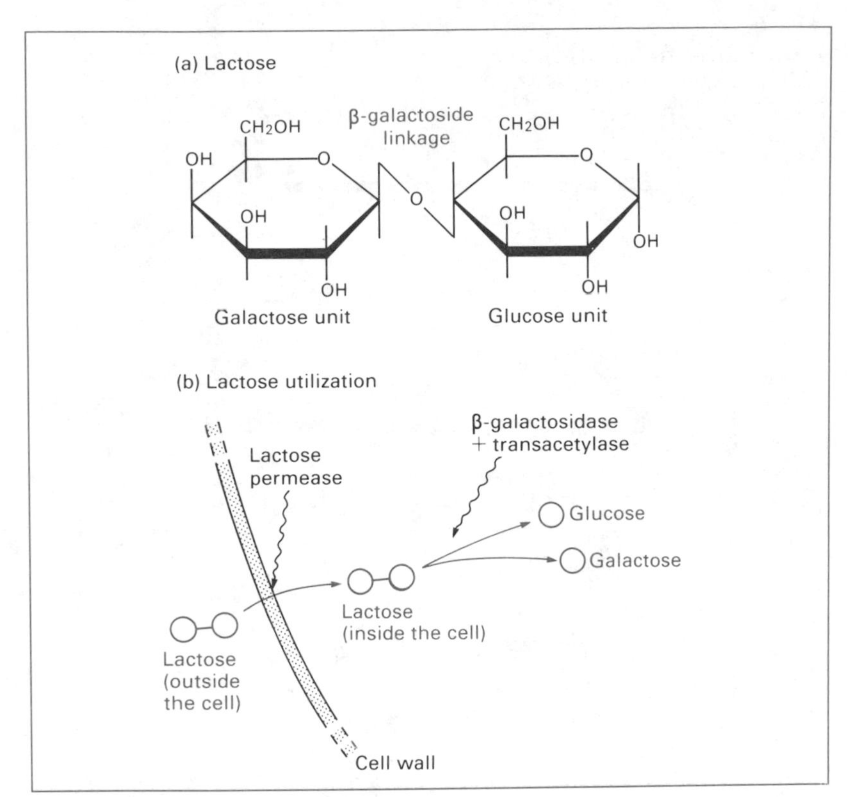

Figure 10.5 Lactose and its utilization by *E. coli*.

β-galactoside transacetylase, which plays a secondary role in the hydrolysis of lactose. In the absence of lactose only a small number of molecules of each enzyme are present in the *E. coli* cell, probably less than five of each. When the bacterium encounters lactose, enzyme synthesis is rapidly induced and within a few minutes levels of up to 5000 molecules of each enzyme per cell are reached. Induction of the three enzymes is coordinate, meaning that each is induced at the same time and to the same extent. This provides a clue to the arrangement of the relevant genes on the *E. coli* DNA molecule.

(a) The lactose utilization genes form an operon. The three genes involved in lactose utilization by *E. coli* are called *lacZ* (*β*-galactosidase), *lacY* (permease) and *lacA* (transacetylase). These genes lie in a cluster with only a very short distance between the end of one gene and the start of the next (Fig. 10.6(a)). In fact the three genes form an operon (Section 4.2.1): each is transcribed on to the same mRNA molecule, delineated by a single promoter upstream of *lacZ* and a single terminator downstream of *lacA*.

Jacob and Monod used genetic analysis techniques to identify *lacZ*, *lacY* and *lacA*, and to determine their relative positions on the

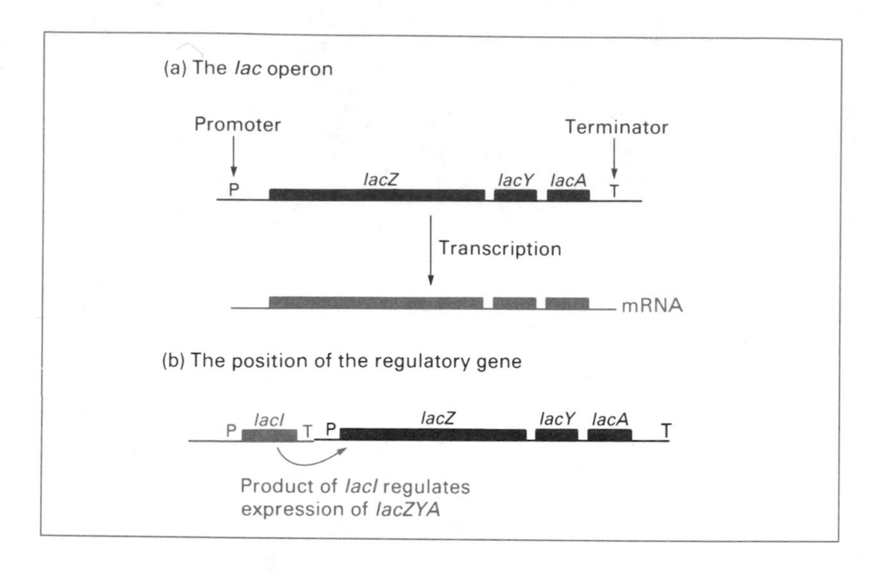

Figure 10.6 The *lac* operon and its regulatory gene *lacI*.

Gene designations

Individual genes are distinguished by a very simple nomenclature, illustrated by the system used with *E. coli*:

lacZ

A three-letter abbreviation indicating the function of the gene

One or more letters to distinguish between genes of related function

The convention is to italicize the name when it is printed, or to underline it when written or typed. Some examples:

Gene	*Function*	*Gene product*
lacZ	LACtose utilization	β-Galactosidase
trpA	TRyPtophan biosynthesis	Tryptophan synthase
rplA	Ribosomal Protein, Large subunit	Ribosomal protein L1
polA	DNA POLymerase	DNA polymerase I
leuA	LEUcine biosynthesis	β-Isopropylmalate synthase
pyrG	PYRimidine biosynthesis	CTP synthetase

E. coli chromosome. They also discovered an additional gene, which they designated *lacI*. This gene lies just upstream of the *lac* gene cluster but is not itself a part of the operon, because it is transcribed from its own promoter and has its own terminator (Fig. 10.6(b)). The gene product of *lacI* is intimately involved in lactose utilization but is not an enzyme directly required for the uptake or hydrolysis of the sugar. Instead the *lacI* product regulates the expression of the other three genes: if *lacI* is inactivated by a mutation, the *lac* operon becomes switched on continuously, even in the absence of lactose. The terminology used by Jacob and

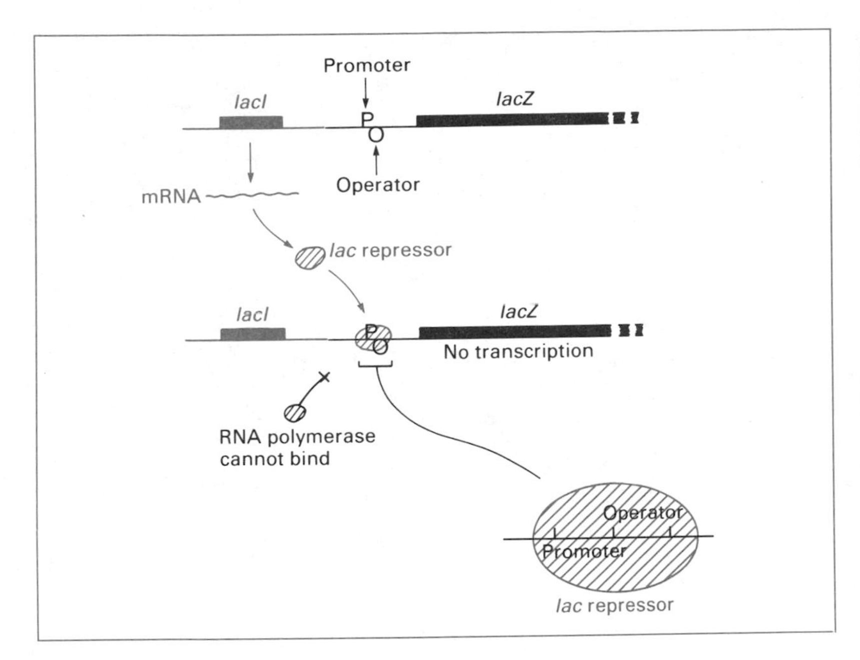

Figure 10.7 In the absence of lactose, the *lac* repressor attaches to the operator site. RNA polymerase cannot bind to the promoter, so transcription does not occur.

Monod is still important today: *lacZ*, *lacY* and *lacA* are called **structural genes** because their products play enzymatic or structural roles in the cell; *lacI* is called a **regulatory gene** because its function is to control the expression of other genes.

(b) The lactose repressor. The gene product of *lacI* is a protein that Jacob and Monod called the lactose **repressor**. This protein is able to attach to the *E. coli* DNA molecule at a site between the promoter for the *lac* operon and the start of *lacZ*, the first gene in the cluster (Fig. 10.7). This attachment site is called the **operator**, and was also located by Jacob and Monod by genetic means. The operator in fact overlaps the promoter so that when the repressor is bound, access to the promoter is blocked so that RNA polymerase cannot attach to the DNA and transcription of the *lac* operon cannot occur. This is what happens if lactose is not available to the cell: **if there is no lactose, transcription of the *lac* operon does not occur because the promoter is blocked by the repressor.**

Allolactose induces transcription. As well as being able to attach to the operator, the *lac* repressor can also bind to allolactose. Allolactose is an isomer of lactose and is synthesized as an intermediate during the splitting of lactose into glucose and galactose. It can be synthesized, albeit at extremely low levels, even when the operon is switched off, as a few molecules of the lactose permease, β-galactosidase and transacetylase enzymes are always present, as mentioned earlier.

When the bacterium encounters a new supply of lactose it takes

The effects of mutations on genes

Jacob and Monod's analysis of the *lac* operon provides an excellent example of the way in which **mutations** are used in genetic research.

A mutation is an alteration in the nucleotide sequence of a DNA molecule. Mutations occur naturally or in response to exposure to agents such as ultraviolet radiation (Section 12.1.3). If a mutation occurs within a gene then it may change one or more codons, resulting in an alteration in the amino acid sequence of the protein that the gene codes for. This alteration may cause the protein to become inactive, or to function in an atypical manner. Mutations can therefore be used to help deduce the function of a gene. For instance, a mutation in the *lacY* gene might result in an *E. coli* cell that is unable to take up lactose, providing a clear indication that the *lacY* gene codes for the lactose permease.

The fact that *lacI* is a regulatory gene became apparent when the effects of mutations in this gene were studied. In a *lacI* cell (that is, one in which the *lacI* gene product is inactive) the genes of the *lac* operon are expressed all the time, even in the absence of lactose. This is called **constitutive expression.** It shows that *lacI* is somehow involved in regulating the response of the *lac* operon to the presence or absence of lactose in the environment.

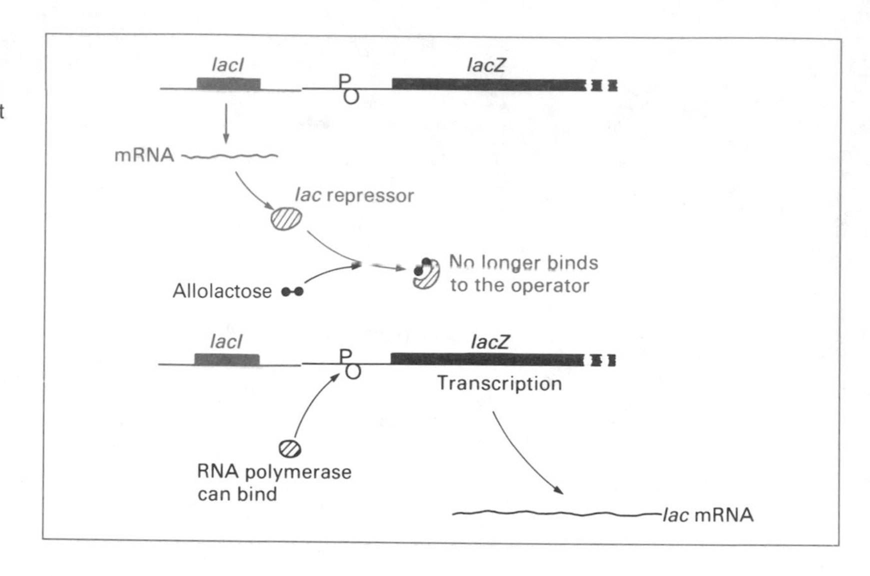

Figure 10.8 In the presence of lactose, the *lac* repressor binds to allolactose. This causes a change in conformation of the repressor so that it is no longer able to bind the operator, and transcription occurs.

up a few molecules and converts them into allolactose. Allolactose then binds to the repressor, causing a change in the conformation of the protein in such a way that the repressor is no longer able to attach to the operator (Fig. 10.8). The repressor–allolactose complex dissociates from the DNA molecule, enabling RNA polymerase to locate the promoter and begin transcription of the operon. This results in synthesis of the much larger numbers of enzyme molecules needed to take up and metabolize the rest of the lactose.

Allolactose therefore acts as the **inducer** of the operon. **If lactose is present, transcription of the *lac* operon occurs as the repressor–inducer complex does not bind to the operator**.

Eventually the lactose utilization enzymes will exhaust the available supply of lactose. Repressor–inducer binding is an equilibrium event (Fig. 10.9), so when the free lactose concentration decreases the number of repressor–allolactose complexes will also decrease and free repressor molecules will start to predominate. These free repressors will have regained their original conformation and so can attach once again to the operator. **When the lactose supply is used up the lactose operon is switched off as the repressor re-attaches to the operator**.

Why is lactose not an inducer of the *lac* operon?

The inability of lactose to induce its own operon is puzzling; especially as a number of other related sugars, not just allolactose, do act as inducers. This has led to the suggestion that in reality lactose utilization is not the main function of the *lac* operon. In fact, in their natural habitat (the intestines of adult mammals) *E. coli* bacteria probably do not encounter lactose to any great extent. The available sugars may, however, include β-galactosyl glycerol, which is a breakdown product of fatty acids and a strong inducer of the *lac* operon. The primary role of the operon may therefore be in the utilization of sugars such as β-galactosyl glycerol. We have come to look on it as the 'lactose' operon simply because this is the sugar used in studies of the operon in bacteria grown in artificial culture media.

(c) Glucose represses the *lac* operon. The presence or absence of lactose is not the only factor that influences expression of the *lac* operon. If the cell has a sufficient source of glucose (one of the breakdown products of lactose) for its energy needs it will not need to metabolize lactose even if lactose is also available in the environ-

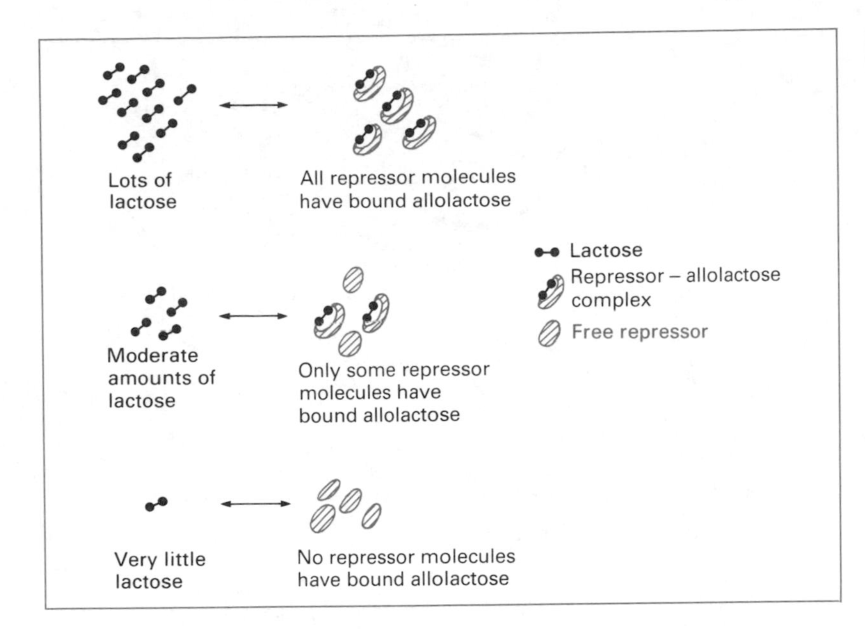

Figure 10.9 Repressor–inducer binding is an equilibrium event.

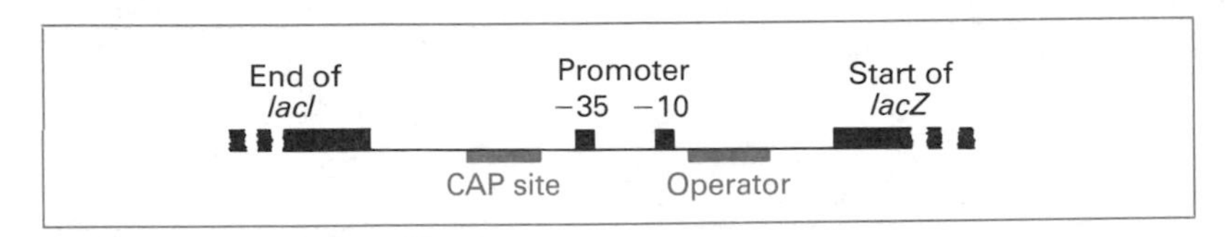

Figure 10.10 The position of the CAP site upstream of the *lac* operon.

ment. A system is therefore needed whereby glucose can override the lactose system and keep the operon switched off if need be.

This process is called **catabolite repression** and involves a second regulatory protein, the **catabolite activator protein** (CAP) and a second upstream binding locality, the **CAP site** (Fig. 10.10). CAP attaches to the CAP site only in the presence of **cyclic AMP** (cAMP), a modified nucleotide derived from ATP by a reaction catalysed by **adenylate cyclase**. The amount of cAMP in the cell is influenced by glucose, which inhibits adenylate cyclase (Fig. 10.11(a)). If the glucose levels are high, then there is relatively little cAMP; if glucose is low, then there is more cAMP.

By controlling the amount of cAMP in the cell, glucose can indirectly regulate attachment of the CAP–cAMP complex to the CAP site. This is important as the CAP–cAMP complex, when attached to the CAP site, stimulates binding of RNA polymerase to the promoter, and therefore also stimulates transcription of the *lac* operon (Fig. 10.11(b)). When the CAP site is vacant, the operon is

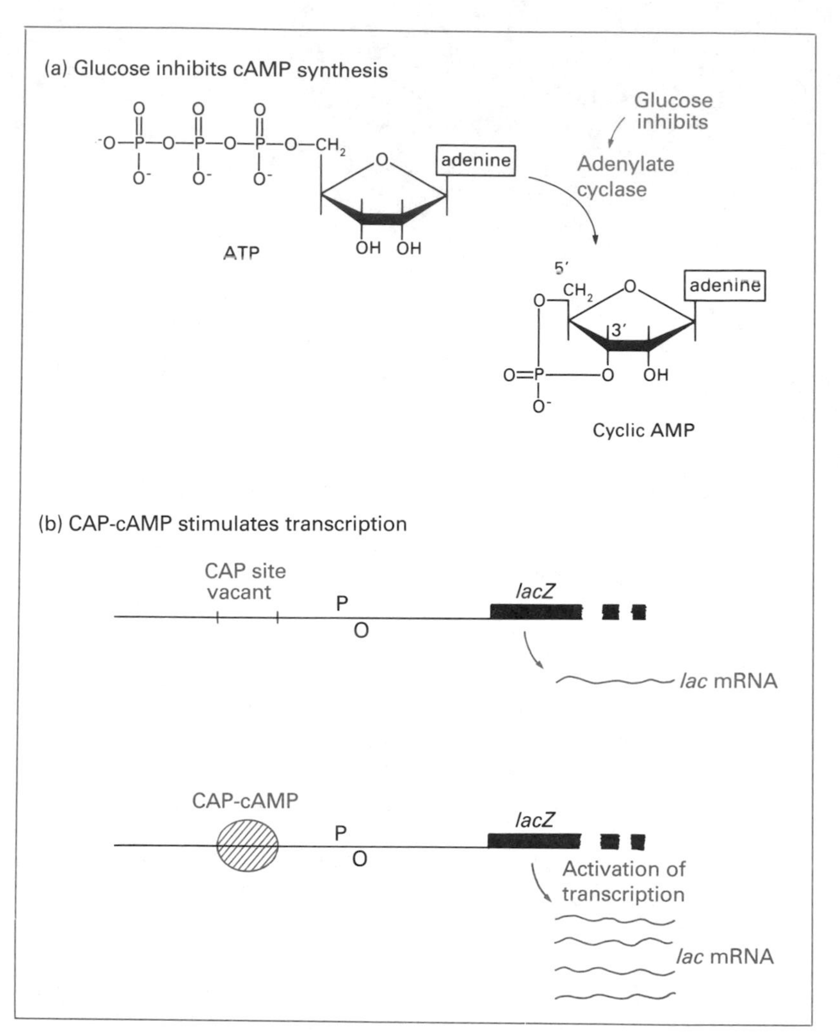

Figure 10.11 Regulation of the *lac* operon by glucose. (a) Glucose controls the amount of cAMP in the cell by inhibiting adenylate cyclase. (b) When cAMP is present (low glucose amounts), CAP–cAMP attaches to the CAP site and stimulates transcription of the *lac* operon.

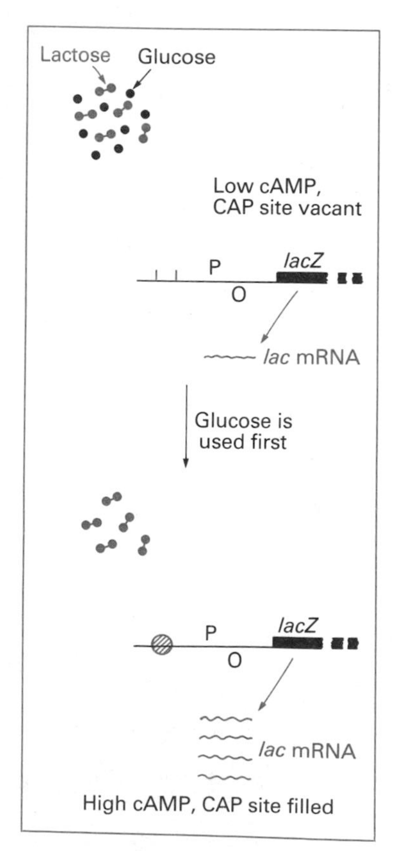

Figure 10.12 Catabolite repression enables *E. coli* to utilize a source of glucose in preference to lactose.

transcribed at only a low rate, even when lactose is present and the *lac* repressor is not attached to the operator. In summary:

> **Glucose present – adenylate cyclase inhibited – cAMP levels low – CAP site vacant – *lac* operon is transcribed at only a low rate.**
>
> **Glucose absent – adenylate cyclase active – cAMP levels high – CAP–cAMP attaches to the CAP site – *lac* operon is transcribed at a high rate.**

So, if glucose and lactose are both available the *lac* operon will not be transcribed to any great extent because the CAP site will be vacant. Once the glucose supply is used up the cAMP levels rise, the CAP site becomes filled, transcription of the *lac* operon is activated, and the cell can start to utilize the lactose (Fig. 10.12).

Features of the tryptophan operon

Other strategies for gene regulation in *E. coli* are illustrated by a second operon. The *trp* operon consists of five genes coding for enzymes involved in synthesis of the amino acid tryptophan. Expression of the operon is controlled by the *trp* repressor, which attaches to the *trp* operator and prevents transcription. In this case though, the repressor on its own cannot attach to the operator. Repression of the operon occurs only when the *trp* repressor binds tryptophan:

NO TRYPTOPHAN

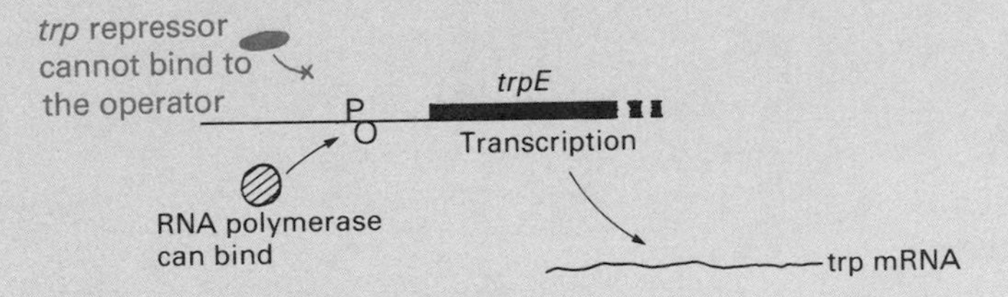

TRYPTOPHAN PRESENT

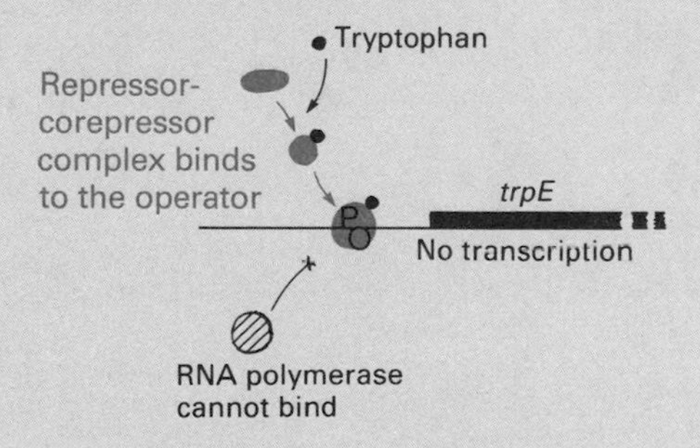

This is of course completely logical, as tryptophan is the *product* of the biochemical pathway controlled by the operon. If tryptophan is absent, then the enzymes for its synthesis are needed, and the operon must be transcribed. So, in the absence of tryptophan the repressor does not attach to the operator. On the other hand, when tryptophan is present the genes must be switched off, so the repressor–tryptophan complex binds to the operator and prevents transcription.

Tryptophan is called the **co-repressor**. This type of operon is called a repressible operon, to distinguish it from the *lac* operon and others like it, which are inducible operons.

Attenuation

The *trp* operon also makes use of a completely different gene regulation strategy, called **attenuation**, which serves to finely tune expression of the operon. When we examine the nucleotide sequence between the promoter and *trpE*, the first gene in the *trp* operon, we see that two stem-loop structures could form in the transcript. The relative positions of these two stem-loops prevent both from forming at once: it must be one or the other. The larger and more stable structure would have no influence on transcription, but the smaller stem-loop has all of the features of a terminator structure (Section 5.4.3). This stem-loop, if it formed, would terminate transcription before the operon is reached: in effect there would be no gene expression.

Attenuation depends on the fact that transcription and translation are linked in bacteria, so that an mRNA molecule still being transcribed may already have one or more ribosomes attached to it (see p. 97). If a ribosome is attached to the *trp* transcript, then it could influence which of the two stem-loops forms, and therefore could determine whether termination of transcription occurs.

The system works as follows. Immediately upstream of the stem-loop region is a short 'gene' that the ribosome must translate. This gene consists of only 14 codons, but two of them are for tryptophan. If there are adequate levels of tryptophan in the cell then the ribosome will have no difficulty translating the gene, and will continue to follow closely behind the RNA polymerase enzyme. Under these circumstances the ribosome will disrupt the larger stem-loop, the terminator will form, and transcription will stop:

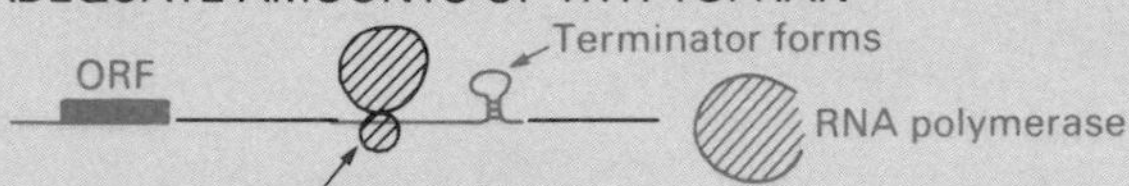

On the other hand, when the tryptophan level is low, the ribosome will stall during translation of the gene as there will be a lack of tryptophan for it to insert into the peptide. The RNA polymerase enzyme will move ahead, the larger stem-loop will not be disrupted by the ribosome, and the terminator will not form. Transcription will continue:

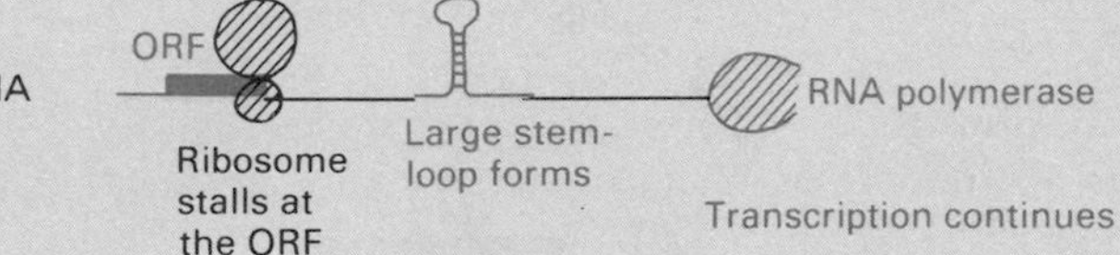

The important point is that the speed at which the ribosome translates the gene will not be exactly the same for each transcript. When tryptophan is present in medium amounts (not abundant but not totally absent either) then some transcripts will terminate, but others will not. Attenuation therefore modulates synthesis of tryptophan. The presence or absence of the repressor determines whether the operon is switched on or off; if the operon is on then attenuation determines how efficiently it is transcribed. Expression of the operon can therefore be tuned to the precise requirements of the cell.

10.4 Upstream sites and DNA-binding proteins

Let us review what the last few pages have told us about regulation of the *lac* operon:

1. There are two regulatory sites immediately upstream of the *lac* operon. These are called the operator and the CAP site.
2. The operator and the CAP site are both binding sites for regulatory proteins. The operator is the binding site for the *lac*

repressor. The CAP site is the binding site for the CAP–cAMP complex.

3. The *lac* repressor, when bound to the operator, prevents transcription by blocking access to the promoter so RNA polymerase cannot bind.
4. The CAP–cAMP complex, when bound to the CAP site, activates transcription by stimulating attachment of RNA polymerase to the promoter.

The general point to emerge from our study of the *lac* operon is that transcription of a gene can be regulated by a DNA-binding protein that attaches to a site upstream of the gene. The regulatory protein may repress transcription, or activate it. We have seen how, with the *lac* operon, just two regulatory proteins enable the cell to respond in an appropriate fashion to the presence or absence of lactose in the environment, and to modify this response to take account of whether glucose is also present. Do similar things happen in eukaryotes?

10.4.1 Upstream sites for eukaryotic genes

In general terms, the basic principles of gene regulation, as illustrated by the *lac* operon, also apply to eukaryotic genes. Expression of a eukaryotic gene is also regulated by DNA-binding proteins that attach to sites upstream of the gene and influence the binding of RNA polymerase to the promoter. However, as might be expected, the situation is rather more complicated than in bacteria.

(a) The human metallothionein gene. We will use the human metallothionein gene as an example to illustrate the complexity of gene regulation in eukaryotes. Metallothionein is a protein that protects cells from the toxic effects of heavy metals, such as cadmium. Small amounts of metallothionein are always present in the cell, but its synthesis is induced by various stimuli, notably the presence of heavy metals. As with all eukaryotic genes that code for proteins, the metallothionein gene is transcribed by RNA polymerase II (Section 5.3.1).

Nine upstream sites are involved in expression of the metallothionein gene (Fig. 10.13). These nine sites can be divided into three groups according to their function:

1. ***The TATA box***. This is the site at which RNA polymerase II,

Operons

Over 75 operons have so far been identified in *E. coli*. These can be broadly divided into the two categories exemplified by the *lac* and *trp* operons:

Inducible operons code for enzymes involved in a metabolic pathway, and are controlled by a substrate to that pathway. The substrate is an inducer of the operon. Examples: lactose operon, galactose operon and arabinose operon.

Repressible operons code for enzymes involved in a biosynthetic pathway, and are controlled by the product of that pathway. The product may be a co-repressor, or may regulate expression via attenuation, or both. Examples: tryptophan operon and phenylalanine operon, which have both repressors and attenuators; histidine operon, leucine operon and threonine operon, which do not have repressors and so depend entirely on attenuation control.

Figure 10.13 The upstream sites involved in expression of the human metallothionein gene. The numbers indicate the nucleotide positions relative to position +1, the point at which transcript synthesis begins. Abbreviations: GRE, glucocorticoid response element; MRE, metal response element; GC, GC box; TATA, TATA box.

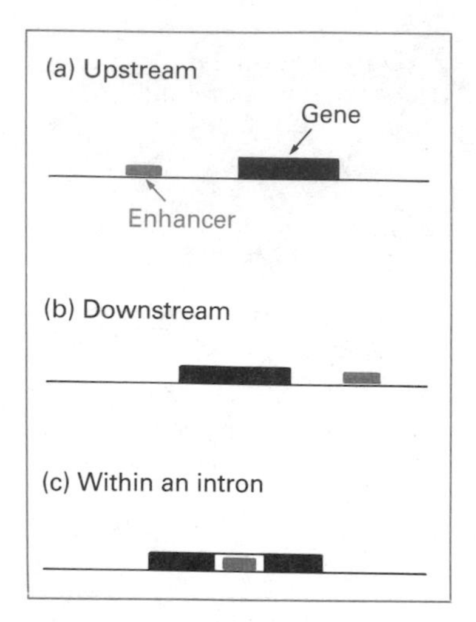

Figure 10.14 Positions where enhancers may be found.

assisted by its transcription factors TFIIA, TFIIB, etc. attaches to the transcript (Section 5.5.1).

2. *'Activation' sites*. The **GC box** and the two **enhancers** are thought to play a general role in activating transcription by RNA polymerase II. Their exact functions are not known and the available information is slightly confusing. If one or other of these sites is disrupted (e.g. by a mutation) then the transcription rate is substantially reduced. This suggests they have an important role in transcriptional activation, and we might therefore expect all genes to have GC boxes and enhancers. In fact many genes lack one or other, but may instead have a different activation site, such as a **CAAT box**. Enhancers are particularly puzzling as a single enhancer may control two or more genes, even ones that are several kilobases away. Also, enhancers are not limited to the regions upstream of their target genes: an enhancer may be positioned downstream of the gene it acts on, or even in an intron (Fig. 10.14).

Enhancers are believed to control tissue-specific gene expression, meaning that they are responsible for ensuring that the correct genes are expressed in each specialized cell type. The evidence for this is a bit shaky but reasonably sound. GC boxes and CAAT boxes are sometimes looked on as integral parts of the promoter, with their binding proteins thought of as components of the RNA polymerase II initiation complex, but this is

Figure 10.15 The role of the transient response elements in regulating expression of the human metallothionein gene.

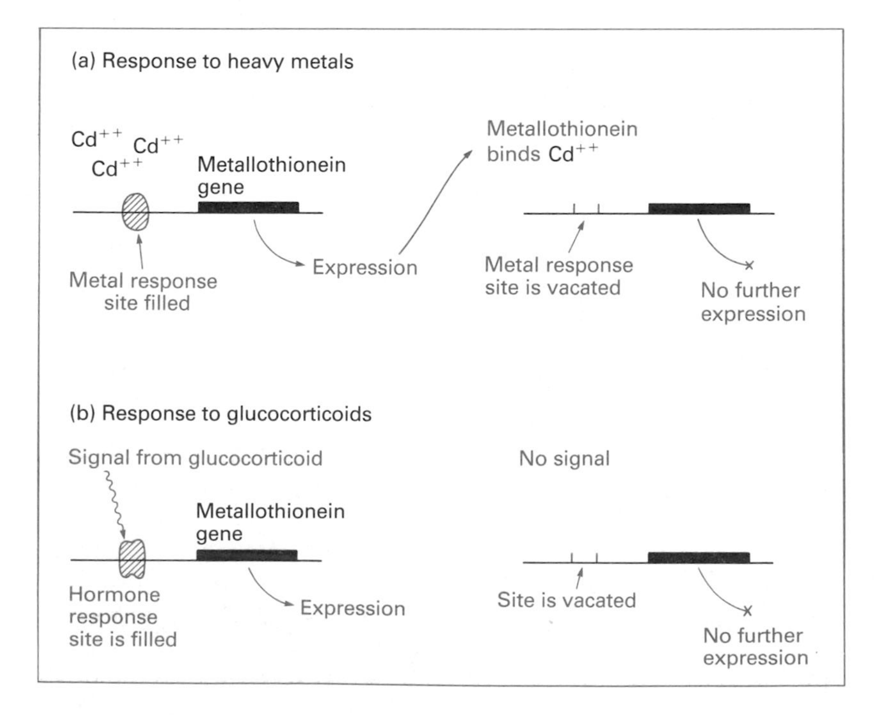

now less certain. We will return to the problem of what the binding proteins actually do in Section 10.4.2.

3. ***'Transient response' elements.*** Activation sites appear to have a continuing and unchanging influence on transcription of their target genes. In contrast a 'transient response' site activates transcription in a temporary fashion, in response to a stimulus from inside or outside of the cell. The transient response elements of the metallothionein gene are the metal response elements, which enable transcription to respond to the presence or absence of heavy metals in the cell, and the glucocorticoid response element, which activates transcription in response to an external signal provided by a glucocorticoid hormone.

   Our experiences with the *lac* operon in *E. coli* point to the way in which the transient response elements work. When heavy metals are present the metal response sites are filled by regulatory proteins that activate RNA polymerase II. The metallothionein gene is switched on and metallothionein protein is synthesized. Once the metallothionein has dealt with the heavy metals, the regulatory proteins vacate the metal response sites, and the gene is switched off again (Fig. 10.15(a)). Equivalent events could occur at the glucocorticoid site in response to the signal provided by this hormone (Fig. 10.15(b)). It is presumed that by having four metal response sites the activity of the gene can be varied with greater precision.

Determining the function of an upstream binding site

The positions of DNA-binding sites upstream of a eukaryotic gene can be mapped by DNase protection experiments (p. 64). The functions of the sites can then be determined by the technique called **deletion analysis**. This involves removing the segment of DNA containing the binding site and seeing the effect the deletion has on gene expression. If an enhancer is removed then an approximately 200-fold decrease in the transcription rate would be seen. Alternatively, removing the glucocorticoid response element from the upstream region of the metallothionein gene (see Fig. 10.13) would mean that transcription of the gene would no longer respond to the presence of the hormone.

There are still many unanswered questions regarding the role of upstream sites in activating and regulating expression of eukaryotic genes. The scheme just described for the metallothionein gene probably holds for many other eukaryotic genes as well, although it may be an over-simplification. Nonetheless, it is a 'good working hypothesis'.

10.4.2 DNA-binding proteins

It has proved easier to identify the sites upstream of a gene than to study the DNA-binding proteins that attach to them. As a result we have very little definite information about these proteins. One thing that is clear is that there are structural similarities between different DNA-binding proteins, reflecting the fact that they perform similar functions. The ability to bind to DNA is often conferred by the **helix-turn-helix** structure (Fig. 10.16(a)), typified by the Cro protein (see Fig. 8.8). 'Helix-turn-helix' refers to the fact that in this structure two α-helices are separated by four amino acids that cause the polypeptide chain to execute a sharp turn. The structure is found in many bacterial repressor proteins, including the *trp* repressor, and is also believed to be present in a group of important DNA-binding proteins in eukaryotes (Section 10.5.2).

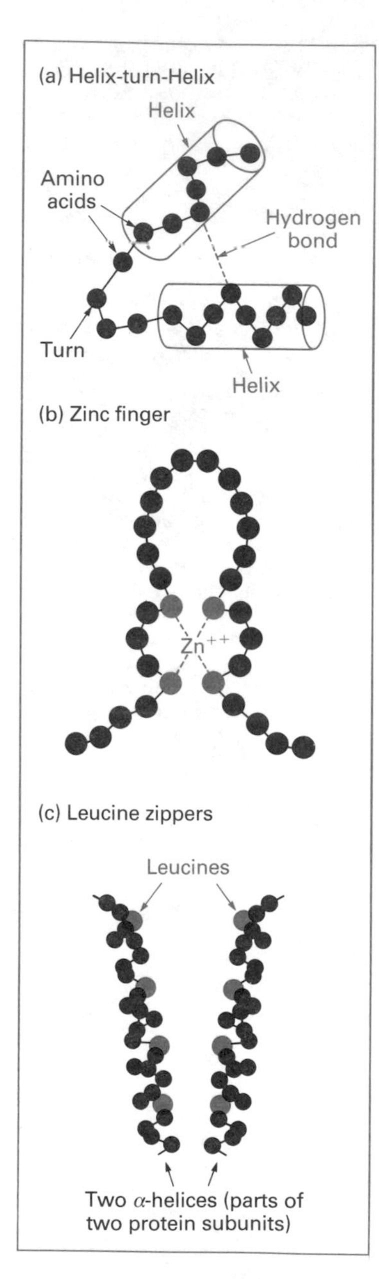

Figure 10.16 Structures found in DNA binding proteins: (a) helix-turn-helix; (b) zinc finger; (c) leucine zipper. In (c) two leucine zippers are shown; these can attach to each other (just like a zip) to hold together two subunits of a dimeric protein.

Amino acids in the two helices make contact with nucleotides in the double helix, ensuring that a particular protein binds to its correct attachment site and nowhere else.

Other DNA-binding proteins possess structures called **zinc fingers**, the name referring to the fact that a zinc atom forms a complex with the polypeptide chain in the DNA-binding region (Fig. 10.16(b)). The protein that binds to the GC box, called SP1, is an example of a zinc finger protein. Still other DNA-binding proteins have **leucine zippers**, which consist of a series of leucines, spaced so they will be in a line on the surface of an α-helix (Fig. 10.16(c)). Leucine zippers do not attach directly to DNA, but instead hold together subunits of a dimeric protein, making sure that other parts of the subunits are orientated so they can make a suitable contact with the double helix. One interesting possibility is that the protein may unzip, so one subunit can be replaced with another of slightly different type, changing the function of the protein in some subtle way.

(a) How do DNA-binding proteins activate transcription? In eukaryotes, as we have seen, most proteins that bind to upstream sites activate rather than repress transcription. This means that a protein attached to a site that is possibly several hundred base pairs away must influence events occurring at the TATA box (see Fig. 5.12). The separation between the activation site and the TATA box does not mean that the DNA-binding protein cannot make direct contact with the transcription complex. DNA is flexible and so can bend to enable contact to be made (Fig. 10.17). Exactly how construction of the transcription complex is activated has been the subject of some debate, but currently the attachment of TFIIB (see Fig. 5.13) is thought to be the critical step. Beyond this, details are difficult to obtain. Many activator proteins are rich in acidic amino acids (aspartic acid and glutamic acid), and this is thought to underlie their function, but exactly how is not known.

10.5 Gene regulation during development

The problems posed by development have occupied biologists for over a century. In the late 1800s embryology was looked on as the key to understanding development. In the twentieth century cell biology, biochemistry and neurobiology have all tried, and to a great extent failed, to explain how the programmed series of changes that lead from fertilized egg cell to adult are controlled. It has to be said that genetics and molecular biology have not yet had much more success than any other approach, but progress has been made.

A molecular biologist looks on development as a series of events

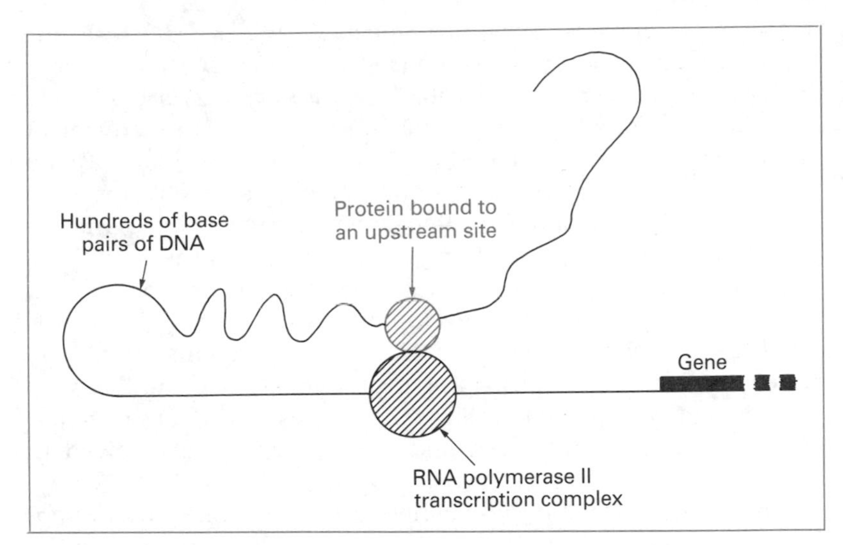

Figure 10.17 DNA bending enables a protein bound to a position several hundred base pairs upstream of a gene to make contact with the RNA polymerase II transcription complex attached to the TATA box.

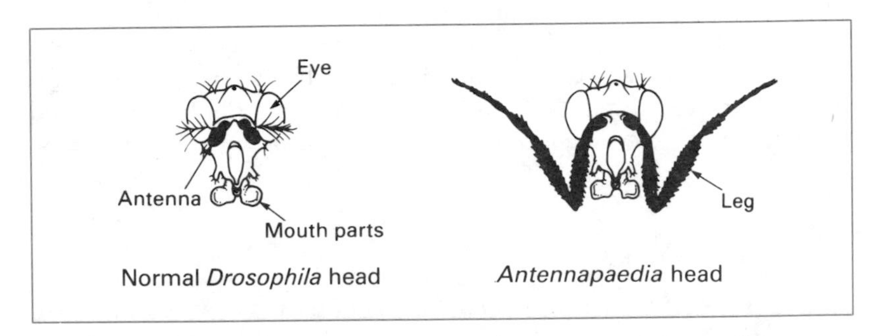

Figure 10.18 The *antennapaedia* mutation.

that result from switching specified genes on and off in the correct cells at the correct times. These changes in gene expression patterns are responsible for 'development' in the morphological sense. Clearly, to understand development we must study gene regulation.

10.5.1 Developmental mutants in *Drosophila*

The main problem in studying gene regulation during development is knowing which genes are important. It is relatively easy to identify a gene whose biological information is 'β-galactosidase enzyme' or 'metallothionein protein', but how can we find, and understand, a gene that says 'grow a leg here, start doing it now?'

The progress that has been made has centred on mutant forms of *Drosophila melanogaster*. In Chapter 1 we learnt how Thomas Hunt Morgan and other pioneers of genetics studied variant characteristics, such as unusual eye colours, in the fruit fly. Among the many mutant forms of *Drosophila* that have been discovered are several in which the development of the fly has gone badly wrong.

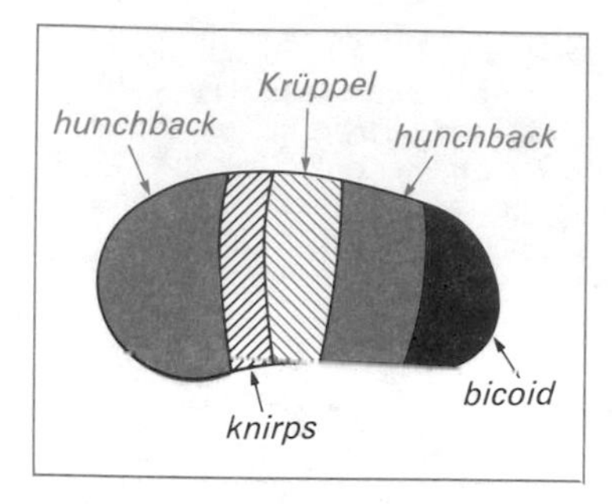

Figure 10.19 Genes involved in development are expressed in specific portions of the *Drosophila* embryo. The drawing shows an embryo a few hours after fertilization. Different portions of the embryo are expressing different genes, as indicated.

A striking example is provided by *antennapaedia*, which causes the fly to grow legs instead of antennae (Fig. 10.18). Other developmental mutations include *deformed, hunchback* and *sex combs reduced*.

During the late 1970s it became possible, with the advent of recombinant DNA techniques (p. 49 and Chapters 20 and 21) to isolate the genes which, when mutated, are responsible for these developmental aberrations. Often when molecular biologists are unsure how to study something they will start by isolating and sequencing the relevant genes, just to see what they look like. It was found that many of the developmental genes code for proteins with features characteristic of DNA-binding proteins (*hunchback*, for example, codes for a zinc finger protein), suggesting that they are regulatory genes controlling the expression of other genes: precisely the function we would expect of a gene involved in development.

Further evidence that these genes really are the key to development in *Drosophila* has come from studying when and where the genes are expressed. This has been done by visualizing the protein products by staining. The results reveal that each gene is expressed only in a certain part of the embryo at a certain time (Fig. 10.19). It is becoming clear that the gene products act together in a concerted fashion, with one developmental gene controlling the expression of another, and so on.

10.5.2 The homeobox

Drosophila is a fly, which means that its body plan is very different from that of humans. So how relevant is the work on *Drosophila* to development in man? The answer is 'very much so'.

Many developmental genes in *Drosophila* are characterized by the presence of a nucleotide sequence, 180 bp in length, that is very similar in different genes (Fig. 10.20). This is called the **homeobox**, after the genes themselves, which are called **homeotic genes** ('homeotic' loosely means 'identity'). Over 20 *Drosophila* genes, all involved in development, contain homeoboxes. The homeobox

Figure 10.20 The homeobox.

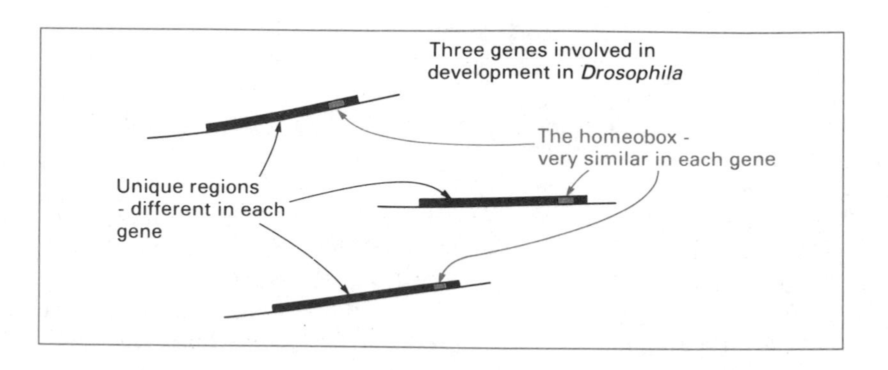

sequence is believed to code for a DNA-binding structure (very similar to helix-turn-helix – Section 10.4.2). Why these homeotic genes should have this particular type of DNA-binding structure is not understood, but the apparent absence of the homeobox from DNA-binding proteins that are not involved in development indicates that the answer, when known, could be very informative.

The big surprise concerning homeoboxes came in 1984 with the discovery of a homeobox, very similar to the *Drosophila* ones, in a vertebrate, the toad *Xenopus laevis*. Soon afterwards the first mammalian homeoboxes were located, and we now know that humans possess a whole array of homeobox genes. Like their counterparts in *Drosophila*, the mammalian homeobox genes are involved in development, as mutations within them result in aberrations in the body plan.

There is a still a long way to go before the role of homeobox genes in mammalian development is understood. But a link has already been made with one type of cancer and rapid progress can be expected in the near future. The role of *Drosophila* in opening up this area of research is a powerful argument for studying the genetics of less complicated organisms in order to understand equivalent processes in man.

Reading

F. Jacob and J. Monod (1961) Genetic regulatory mechanisms in the synthesis of proteins. *Journal of Molecular Biology*, **3**, 318–56 – the classic paper in which the operon theory was first proposed.

M. Ptashne and W. Gilbert (1970) Genetic repressors. *Scientific American*, **222**(6), 36–44 – the mode of action of repressors and the methods used in isolation of the proteins.

G. Zubay, D. Schwartz and J. Beckwith (1970) Mechanisms of activation of catabolite-sensitive genes: a positive control system. *Proceedings of the National Academy of Sciences USA*, **66**, 104–10 – an early description of the CAP system.

D. Latchman (1990) *Gene Regulation: A Eukaryotic Perspective*, Unwin Hyman, London – an excellent book with which to gain a fuller understanding of gene regulation in eukaryotes.

W. Dynan and R. Tjian (1982) Transcription enhancer sequences: a novel regulatory element. *Trends in Biochemical Sciences*, **7**, 124–5.

K. Struhl (1989) Helix-turn-helix, zinc-finger, and leucine zipper motifs for eukaryotic transcriptional regulatory proteins. *Trends in Biochemical Sciences*, **14**, 137–40; S. L. McKnight (1991) Molecular zippers in gene regulation. *Scientific American*, **264**(4), 32–9 – details on DNA-binding proteins.

T. Beardsley (1991) Smart genes. *Scientific American*, **265**(2), 72–81;

W. J. Gehring *et al*. (1990) The structure of the homeodomain and its functional implications. *Trends in Genetics*, **6**, 323–9 – two articles on the study of gene regulation during development.

P. A. Lawrence (1992) *The Making of a Fly*, Blackwell Scientific Publishers, Oxford – an advanced but readable account of the development genetics of *Drosophila*.

Problems

1. Define the following terms:

   housekeeping gene — operator
   structural gene — catabolite repression
   regulatory gene — homeotic gene

2. Carefully distinguish between the meaning of the following terms with respect to the control of bacterial gene expression:

   inducer repressor co-repressor

3. Describe the role of the following in gene regulation in eukaryotes:

   GC box enhancer CAAT box homeobox

4. Explain what is meant by the following terms:

   helix-turn-helix zinc finger leucine zipper

5. Explain why it is desirable that organisms are able to regulate expression of their genes.

6. Explain the nature of the balancing factors that influence gene expression and list all the ways that you can think of by which the amount of a gene product in the cell could be altered.

7. Describe the (a) similarities and (b) differences between the ways in which gene expression is regulated at the *E. coli lac* and *trp* operons. Confine your answer to a consideration of just the repressor–operator systems.

8. How does glucose influence the expression of a range of different sugar utilization operons?

9. Describe the structure of the region upstream of the human metallothionein gene.

10. Outline what is known about the way in which DNA-binding proteins activate transcription.

11. What has the study of mutations such as *antennapaedia* told us about development in *Drosophila*?

12. To what extent have studies with *Drosophila* provided information relevant to development in man?

13. *From Fig. 10.4 it is clear that there are numerous ways in which the synthesis of a gene product could be regulated. Many different strategies are known in individual systems. In view of this, can any real understanding of the control of gene expression be gained by studying a model system such as the *lac* operon?

14. *Operons are very convenient systems for achieving co-ordinated regulation of expression of related genes. Operons are common in bacteria yet they are absent in eukaryotes. Discuss.

15. *Sidney Brenner likens DNA to a computer program containing the instructions for building an organism. Is this a realistic view?

16. In *E. coli*, the galactose operon comprises three structural genes – *galK*, *galT* and *galE* – whose gene products are involved in metabolism of the sugar galactose. There is also a regulatory gene, *galR*, which codes for a repressor protein that binds to an operator upstream of the *gal* operon. Galactose acts as an inducer. Describe the events that you would expect to occur in: (a) the absence of galactose, and (b) the presence of galactose.

17. As we will see in Chapter 12, a mutation is an alteration in nucleotide sequence that may alter the function of a gene, promoter or regulatory site. Deduce and explain the effects of each of the following mutations on expression of the *lac* operon:

   (a) A mutation in the operator, so the repressor is no longer able to bind.
   (b) A mutation in *lacI*, so that the repressor is no longer synthesized.
   (c) A mutation in the promoter, so that RNA polymerase no longer binds.
   (d) A mutation in *lacI*, so that the repressor no longer binds lactose.
   (e) A mutation in *lacY* that causes transcription to terminate in the middle of the gene.
   (f) A mutation, location unknown, which prevents the *lac* mRNA from being degraded.

11 Replication of DNA molecules

Semi-conservative replication • Replication in *E. coli* – DNA polymerase – events at the replication fork – the topological problem • Replication in eukaryotes • Replication of molecules

Every time a cell divides it must make a complete copy of all its genes. This is of course essential if the products of division, the two daughter cells, are each to receive a full complement of the biological information possessed by the parent. A dividing cell therefore has to carry out extensive DNA replication.

A considerable degree of accuracy must be achieved and maintained during DNA replication. Even an error rate of 0.01% (one mistake per 10 000 nucleotides) will cause a significant accumulation of alterations in the genes of a rapidly dividing organism such as a bacterium, quickly leading to the vital DNA sequences becoming meaningless.

Three aspects of DNA replication must be considered by the molecular geneticist. The first concerns the overall pattern of replication and the question of how a single DNA double helix can give rise to two daughter molecules. This problem was solved by a single elegant experiment carried out in 1958. The second aspect is the biochemistry and enzymology of the process: which proteins are involved and what reactions do they participate in during DNA replication? The details are now fairly well understood in *E. coli*, and appear to be very similar in eukaryotes. The final aspect concerns the precise way in which DNA replication is handled in different organisms, as replication of a circular bacterial DNA molecule presents a completely different set of problems to replication of a eukaryotic chromosome.

11.1 The overall pattern of DNA replication

One of the final sentences of Watson and Crick's 1953 paper on the double helix reads 'It has not escaped our notice that the specific pairing we have postulated immediately suggests a possible copying mechanism for the genetic material'. They were referring to the fact that complementary base pairing between the two polynucleotides of the double helix would enable each strand to act as a template for synthesis of its complement (Fig. 11.1). Thus, the double helix structure possesses an obvious means of replicating itself.

Nevertheless, molecular biologists in the early 1950s were uncertain about the overall pattern of the process. Three different strategies for replication of the double helix seemed possible (Fig. 11.2).

1. **Semi-conservative replication**, in which the daughter molecules each contain one polynucleotide derived from the original molecule and one newly synthesized strand.
2. **Conservative replication**, in which one daughter molecule contains both parent polynucleotides and the other daughter contains both newly synthesized strands.
3. **Dispersive replication**, in which each strand of each daughter molecule is composed partly of the original polynucleotide and partly of newly synthesized polynucleotide.

Although the semi-conservative scheme seems most likely purely on intuitive grounds, and in fact was favoured by Watson and Crick, it was necessary to devise an experiment which would distinguish between these three modes of replication and confirm which scheme actually operates. The experiment that settled this

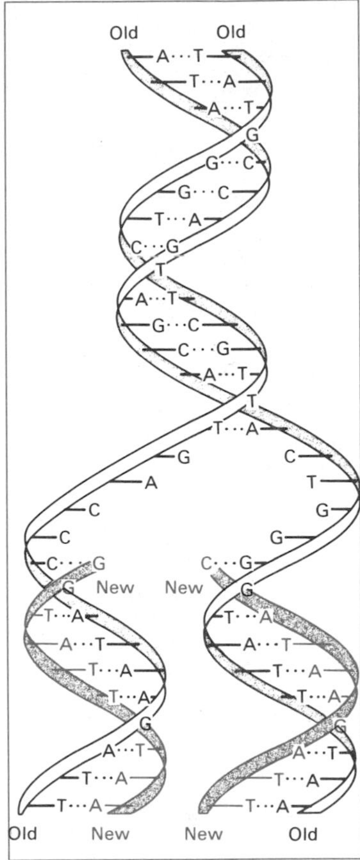

Figure 11.1 Complementary base pairing underlies the replication of DNA.

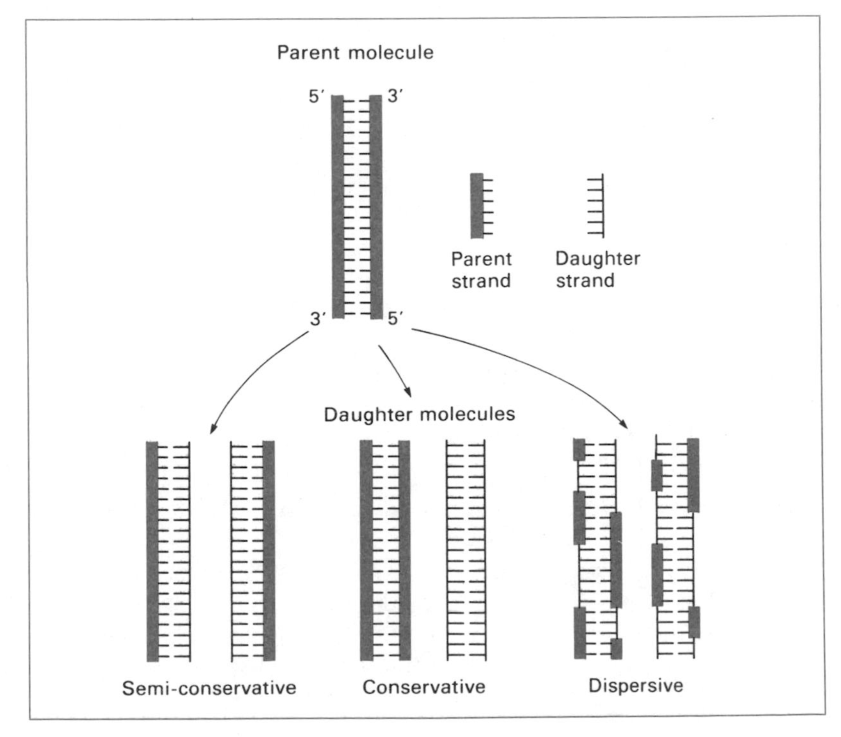

Figure 11.2 Three possible schemes for DNA replication.

question was carried out in 1958 at the California Institute of Technology by Matthew Meselson and Franklin Stahl.

11.1.1 The Meselson–Stahl experiment

Like many advances in biochemistry and molecular biology, the Meselson–Stahl experiment depended on the use of chemical isotopes (p. 21). Nitrogen exists in several isotopic forms. as well as the 'normal' isotope, ^{14}N, which predominates in the environment and has an atomic weight of 14.008, there are a number of other isotopes that occur in much smaller amounts. These include ^{15}N, which because of its greater atomic weight is called 'heavy nitrogen'. Different isotopes can be distinguished from one another by a variety of means, so if a molecule containing an uncommon isotope is provided to a cell as a nutrient, the fate of the

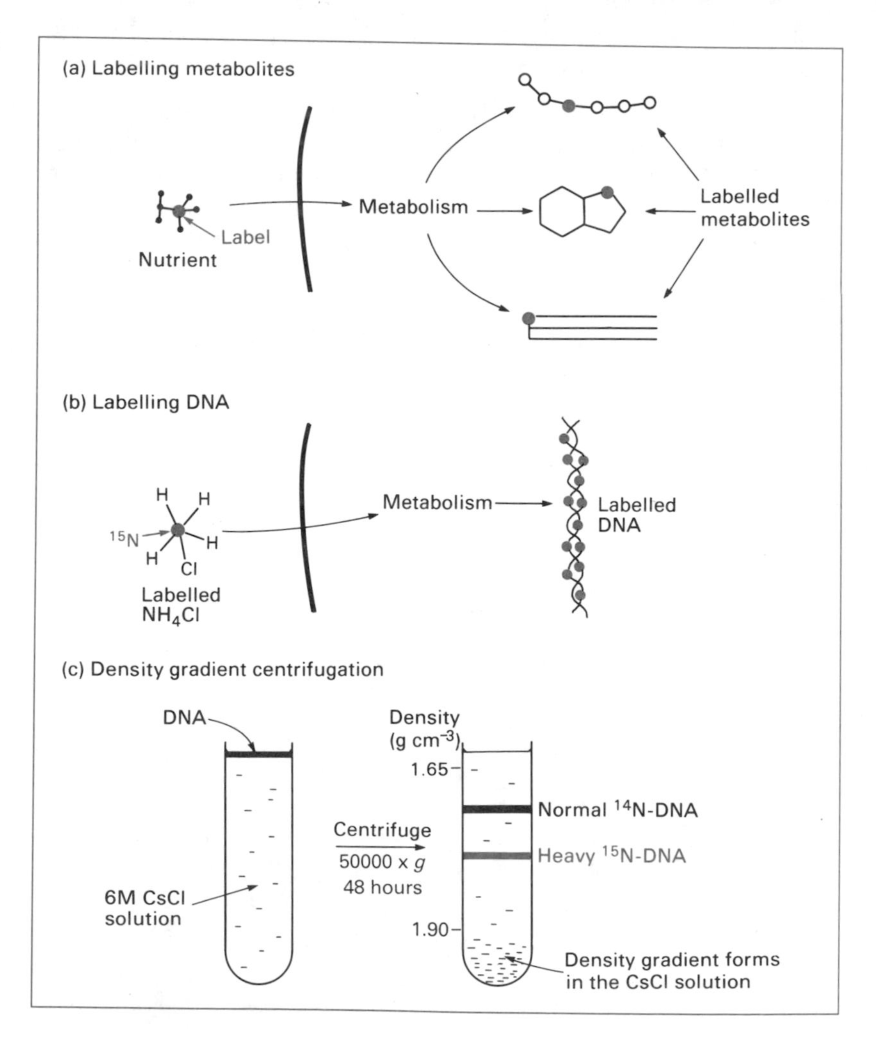

Figure 11.3 The use of biochemical labels. (a) A labelled nutrient produces labelled metabolites that can be traced. (b) Labelled $^{15}NH_4Cl$ gives rise to labelled DNA. (c) ^{14}N- and ^{15}N-DNA can be separated by density gradient centrifugation.

molecule (that is, the biochemical reactions that it undergoes in the cell) can be followed by identifying all the cellular metabolites that become labelled with the isotope (Fig. 11.3(a)). Of relevance to the Meselson–Stahl experiment is the fact that *E. coli* cells provided with heavy nitrogen in the form of $^{15}NH_4Cl$ will incorporate the labelled nitrogen into their DNA molecules (Fig. 11.3(b)).

Heavy nitrogen is a stable isotope so does not emit radiation. However, DNA molecules containing heavy nitrogen can be distinguished from DNA molecules containing the normal isotope by density gradient centrifugation (p. 19). This is because ^{15}N-DNA has a greater buoyant density than ^{14}N-DNA and will therefore form a band at a lower position in a CsCl density gradient (Fig. 11.3(c)).

Matthew Meselson b. 24 May 1930, Denver

The experiment that proved that DNA replication is semi-conservative was carried out by Matthew Meselson and Franklin Stahl at the California Institute of Technology in 1957. Meselson had just completed his PhD thesis at the Institute and stayed there a further 4 years before moving to Harvard. He was involved in the first identification of mRNA (with Sydney Brenner and François Jacob in 1961) and has been responsible for several of the advances in understanding the process of recombination between DNA molecules (Section 12.3). In 1968 with R. Yuan he achieved one of the first successful isolations of a restriction endonuclease, a type of enzyme that has become of central importance in recombinant DNA technology (see Chapter 20).

(a) The experiment. Meselson and Stahl utilized heavy nitrogen in the following way (Fig. 11.4). First a culture of *E. coli* was grown in the presence of $^{15}NH_4Cl$ so that the DNA molecules in the cells became labelled with heavy nitrogen. Then the culture was spun in a low-speed centrifuge, the heavy medium discarded, and the bacteria resuspended in medium containing only $^{14}NH_4Cl$. New polynucleotides synthesized after resuspension therefore contained only the normal isotope of nitrogen.

The bacteria were then allowed to undergo one round of cell division, which takes roughly 20 minutes for *E. coli*, during which time each DNA molecule replicates just once. Some cells were then taken from the culture, their DNA purified and a sample analysed by density gradient centrifugation. The result, shown in Fig. 11.4, was a single band of DNA at a position corresponding to a buoyant density intermediate between the values expected for ^{15}N-DNA and ^{14}N-DNA, showing that after one round of replication each DNA double helix contained roughly equal amounts of ^{15}N-polynucleotide and ^{14}N-polynucleotide. If we re-examine the three schemes for DNA replication (Fig. 11.2) we see that this result is compatible with both semi-conservative replication (each double helix comprises one ^{15}N-polynucleotide and one ^{14}N-polynucleotide) and dispersive replication (both strands are made up of a mixture of ^{15}N-polynucleotide and ^{14}N-polynucleotide). However, conservative replication can be ruled out at this stage, because this scheme would give two populations of DNA molecules, some entirely ^{15}N and some entirely ^{14}N: no hybrids would be seen.

To distinguish between semi-conservative and dispersive replication, Meselson and Stahl allowed the *E. coli* culture to undergo a second round of cell division (see Fig. 11.4). Again, cells were removed and their DNA analysed in a density gradient. Two bands appeared: one representing the same hybrid molecules as before, but with an additional band corresponding to wholly

Figure 11.4 The Meselson–Stahl experiment.

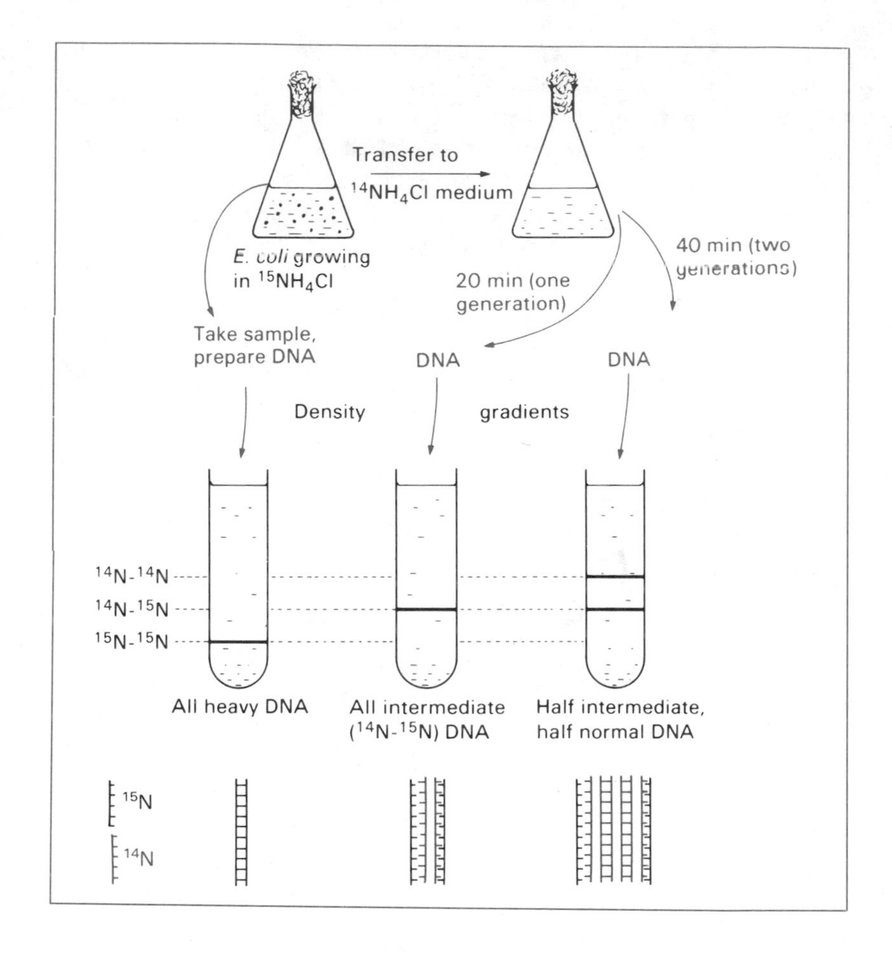

^{14}N-DNA. This result is entirely in agreement with the semi-conservative mode of replication, because according to this scheme there will now be some granddaughter molecules that are made up entirely of ^{14}N-polynucleotides. In contrast, the dispersive mode can be discounted because that method would still produce only hybrid molecules, and in fact would continue to do so for a very large number of cell generations.

The Meselson–Stahl experiment is rightly considered a classic example of scientific technique. The problem to be resolved was carefully delineated, the available techniques were marshalled, and the experiment was designed so that the results would be unequivocal. This masterpiece enabled the attention of biochemists and molecular biologists to move directly on to the second aspect of DNA replication, the question of which enzymes are involved and how the process actually occurs.

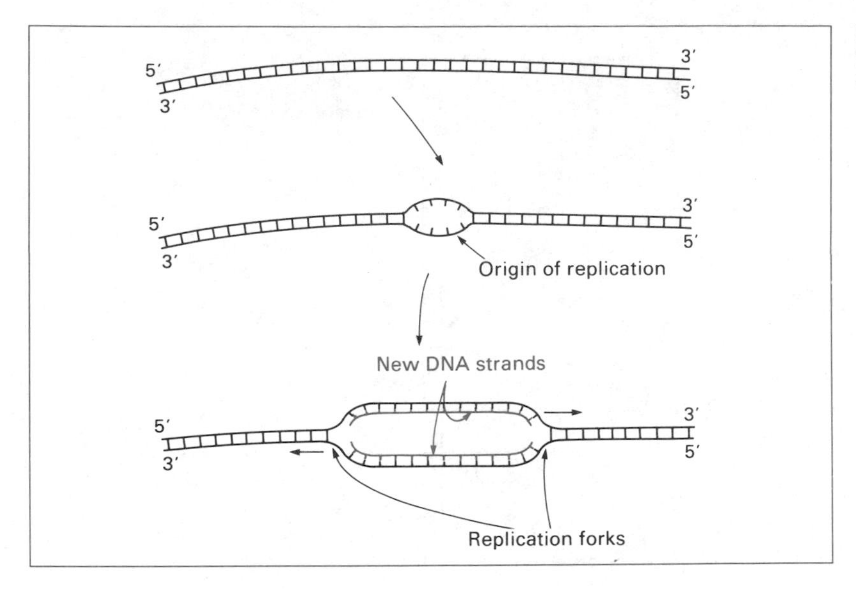

Figure 11.5 A replication origin and the movement of replication forks along a double helix.

11.2 The mechanism of DNA replication in *E. coli*

When a DNA molecule is being replicated only a limited region is ever in a non-base-paired form. The breakage in base pairing starts at a distinct position, called the **replication origin** (Fig. 11.5) and gradually progresses along the molecule, possibly in both directions, with synthesis of the new polynucleotides occurring as the double helix unzips. The important region at which the base pairs of the parent molecule are broken and the new polynucleotides are synthesized, is called a **replication fork**.

11.2.1 DNA polymerase

An enzyme able to synthesize a new DNA polynucleotide is called a **DNA polymerase**. The chemical reaction catalysed by a DNA polymerase is very similar to that of RNA polymerase (Section 5.3) except of course that the new polynucleotide that is assembled is built up of deoxyribonucleotide subunits rather than ribonucleotides (Fig. 11.6). In general terms the reactions are the same: the sequence of the new polynucleotide is dependent on the sequence of the template and is determined by complementary base pairing and, as with RNA synthesis, DNA polymerization can occur only in the 5′ → 3′ direction.

(a) DNA polymerase I and DNA polymerase III. In 1957 Arthur Kornberg and his colleagues isolated from *E. coli* an enzyme capable of DNA synthesis on a polynucleotide template. Naturally

Figure 11.6 The reaction for DNA synthesis, catalysed by a DNA polymerase enzyme during DNA replication. Compare this figure with the scheme for RNA synthesis during transcription (Fig. 5.3).

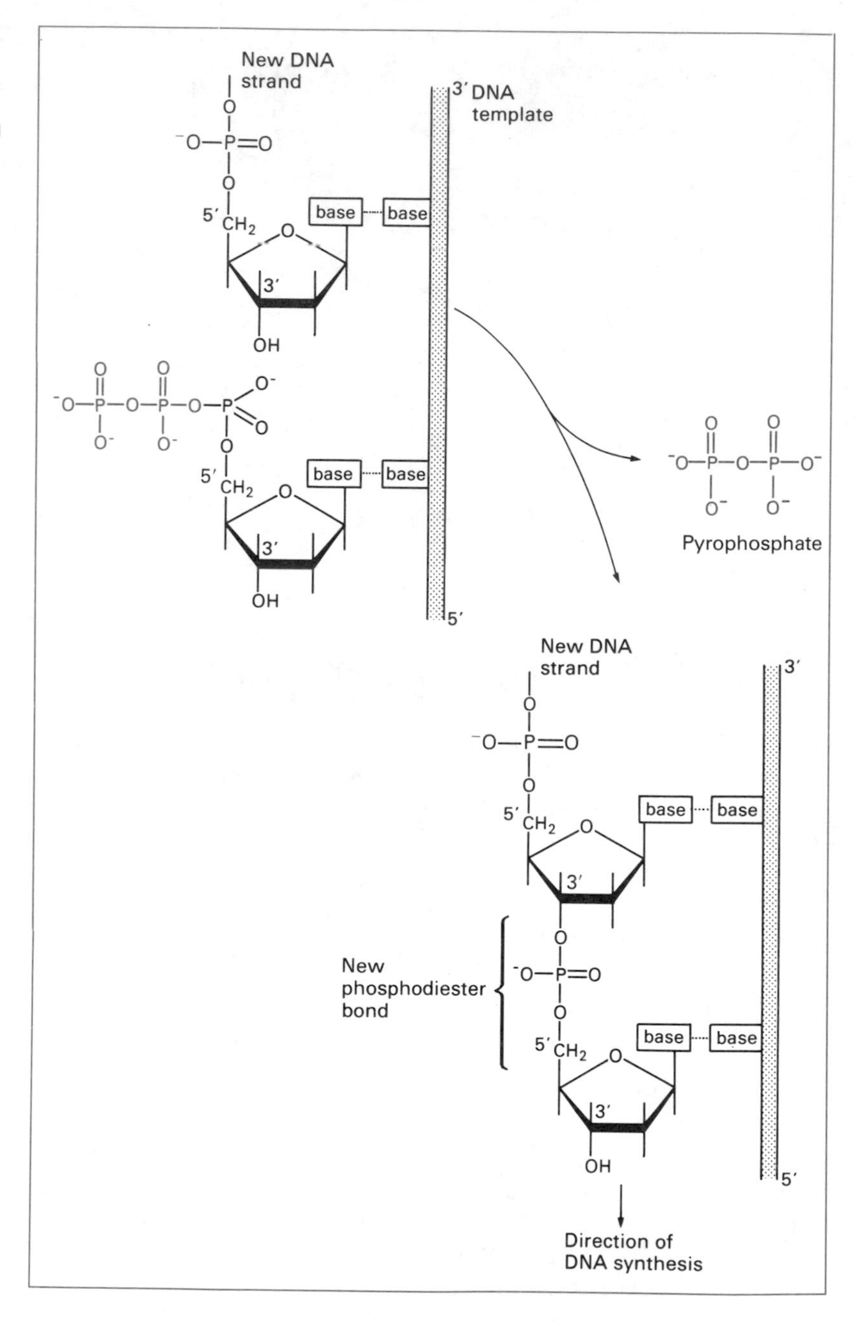

it was believed that this enzyme was the DNA polymerase responsible for DNA replication in the bacterial cell. Unfortunately, as the enzyme was studied in greater detail facts inconsistent with this role began to emerge. The most problematic of these arose with the discovery in 1969 of *E. coli* cells in which the gene coding for

Arthur Kornberg b. 3 March 1918, Brooklyn

'I have never met a dull enzyme'.

Kornberg is one of several biochemists whose work has been of immense value in understanding gene expression and DNA replication. Kornberg gained his MD from Rochester University in 1941 and then spent 10 years at the National Institutes of Health laboratories at Bethesda, Maryland. His research at this time centred on coenzymes, non-protein molecules that bind to and are essential for the activity of certain enzymes. The isolation of DNA polymerase I occurred later when Kornberg had moved to Washington University, which is neither in the city nor the State of Washington, but in fact in St Louis, Missouri. This work was recognized by the 1959 Nobel Prize, shared with Severo Ochoa. In the same year Kornberg became Professor of Biochemistry at Stanford University, where he has continued his studies on the mechanism and control of DNA replication. He has written a highly entertaining account of his life in science: *For the Love of Enzymes: The Odyssey of a Biochemist*, Harvard University Press, Boston, 1989.

The three DNA polymerases of *E. coli*

E. coli possesses three distinct DNA polymerases. Their details are as follows:

	DNA polymerase		
	I	II	III
Molecular mass (daltons)	109 000	120 000	>250 000
Approximate number of molecules per cell	400	?	10–20
Function	DNA replication, repair of mutated DNA	?	DNA replication

the enzyme, now called **DNA polymerase I**, was inactive: these mutant cells were still able to replicate their DNA. Eventually it was determined that although DNA polymerase I is involved in DNA replication it is not the main replicating enzyme. The enzyme now believed to be primarily responsible for DNA replication in *E. coli* was isolated in 1972 and is called **DNA polymerase III**. The complexity of its role in the cell is reflected by the fact that it is a very large enzyme consisting of an as yet undetermined number of subunits and with a molecular mass in excess of 250 000 daltons.

11.2.2 Events at the replication fork

Our knowledge of exactly how DNA is replicated has been built up over several years through the efforts of numerous molecular biologists and biochemists in laboratories all over the world. The

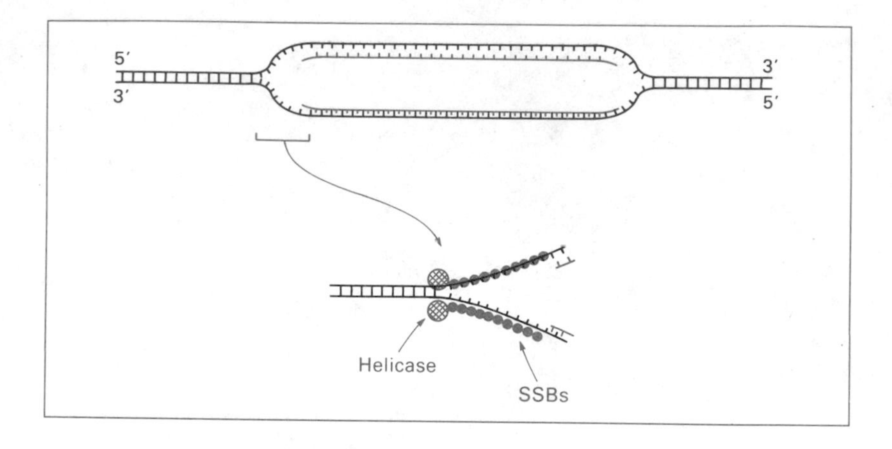

Figure 11.7 The roles of helicases and single-strand binding (SSB) proteins in breaking the double helix and preventing reannealing during DNA replication.

consensus that has been reached is of a complex process involving a number of enzymes and other proteins.

(a) Breakage of the parent double helix. During DNA replication there is a continual need for breakage of the base pairing between the two strands of the parent DNA molecule. This is carried out by a **helicase** enzyme, which works in conjunction with **single-strand binding proteins** (SSBs). The latter attach to the single-stranded DNA that results from helicase action and prevent the two strands from immediately reannealing (Fig. 11.7). The result is the replication fork that provides the templates on which DNA polymerase III can work.

(b) The leading and lagging strands. The main complication in DNA replication is that the two disassociated strands of the parent molecule cannot be treated in the same way. This is because DNA polymerase III can synthesize DNA only in the 5′ to 3′ direction, which means that the template strands must be read in the 3′ to 5′ direction (see Fig. 11.6). For one strand of the parent, called the **Leading strand**, this is no problem because the new polynucleotide can be synthesized continuously (Fig. 11.8(a)). However, the second or **lagging strand** cannot be copied in a continuous fashion because this would necessitate 3′ to 5′ DNA synthesis. Instead the lagging strand has to be replicated in sections: a portion of the parent helix is disassociated and a short stretch of the lagging strand replicated, a bit more of the helix is disassociated and another segment of the lagging strand replicated, and so on (Fig. 11.8(b)). At first this process was just a hypothesis but the isolation by Reiji Okazaki in 1968 of short fragments of DNA between 100 and 1000 nucleotides in length, associated with DNA replication, confirmed that the suggestion is correct.

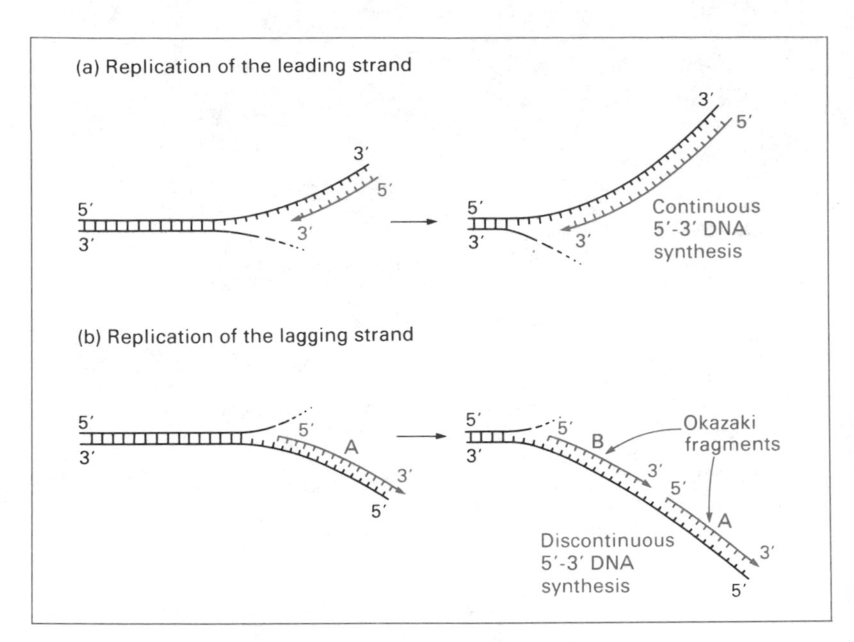

Figure 11.8 The distinction between the leading and lagging strand during DNA replication.

(c) The priming problem and joining up the Okazaki fragments. Unlike RNA polymerase, DNA polymerase III cannot initiate DNA synthesis unless there is already a short double-stranded region to act as a **primer** (Fig. 11.9(a)). How can this occur during DNA replication? The answer appears to be that the very first few nucleotides attached to either the leading or the lagging strand are not deoxyribonucleotides but ribonucleotides that are put in place by an RNA polymerase enzyme (Fig. 11.9(b)). Once a few ribonucleotides (between six and 30) have been polymerized on the template the RNA polymerase moves aside, and polymerization, now of DNA, is continued by DNA polymerase III. The RNA

The primosome

Much of the recent research on DNA replication has centred on the primosome. This structure consists of the primase enzyme together with at least six additional proteins. The primosome moves along the lagging strand in the 5′ to 3′ direction, following the replication fork as the double helix unzips. At appropriate places the primase component of the primosome synthesizes an RNA primer. During primer synthesis the template is read in the 3′ to 5′ direction, the opposite direction to that in which the primosome is moving. This implies that the primosome pauses in its passage along the template whenever it needs to make a primer, but this has not yet been proven.

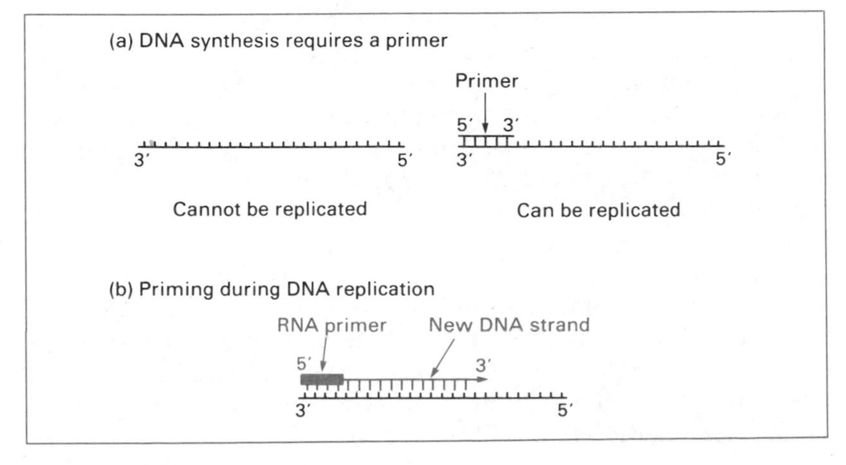

Figure 11.9 DNA synthesis must be primed.

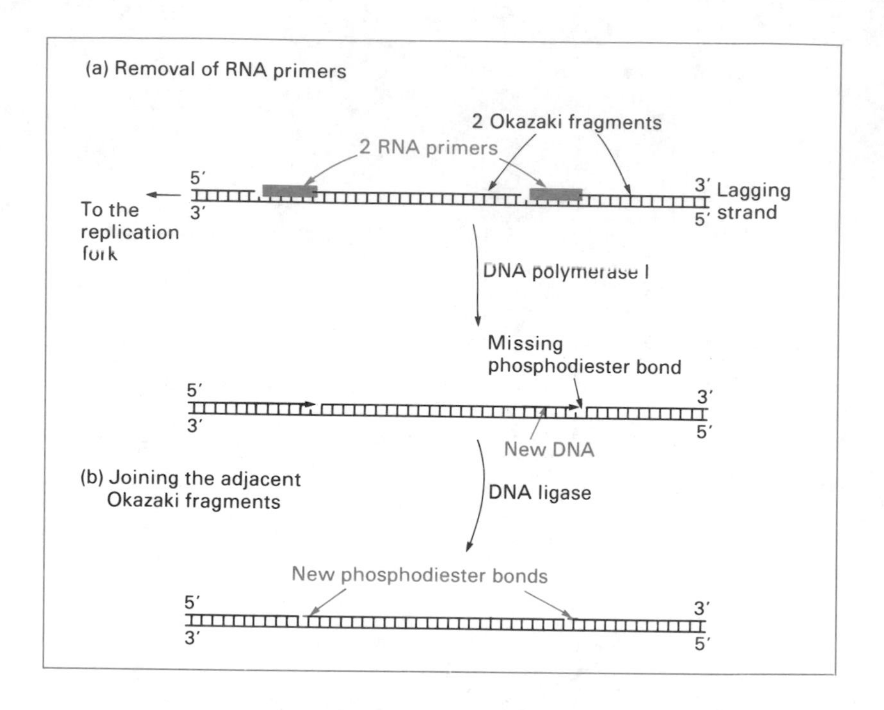

Figure 11.10 Completion of DNA synthesis on the lagging strand.

The exonuclease activities of DNA polymerases I and III

DNA polymerase III has to stop DNA synthesis when it reaches the start of the next Okazaki fragment as it does not have the 5′ to 3′ exonuclease activity that it would need to remove the ribonucleotides of the primer. DNA polymerase I does have such an activity, and so takes over from DNA polymerase III at this point.

Both enzymes possess exonuclease activities in the 3′ to 5′ direction, the **reverse** of the synthesis direction. This means that, if a nucleotide is inserted incorrectly during DNA synthesis, the enzyme is able to retrace its steps, removing the mismatch and replacing it with the correct nucleotide. This is called the **proofreading** function and helps to ensure that DNA replication is virtually error-free. It has been calculated that DNA polymerase III makes an uncorrected error just once every 5 000 000 000 nucleotides.

polymerase is called the **primase** and in *E. coli* is a single polypeptide with a molecule mass of about 60 000 daltons (and so is distinct from the transcribing RNA polymerase described in Chapter 5). It acts in conjunction with six or more additional proteins which make up a structure referred to as the **primosome.**

On the lagging strand, DNA polymerase III can synthesize DNA only for a certain distance before it reaches the RNA primer at the 5′ end of the next Okazaki fragment. Here DNA polymerase III stops and DNA polymerase I comes into action, continuing DNA synthesis and carrying out a further important function by removing the ribonucleotides of the primer of the adjacent Okazaki fragment and replacing them with deoxyribonucleotides (Fig. 11.10(a)). When all the ribonucleotides are replaced, DNA polymerase I either stops or possibly carries on a short distance into the DNA region of the Okazaki fragment, continuing to replace nucleotides before it dissociates from the new double helix.

The final reaction needed to complete replication of the lagging strand is to join up adjacent Okazaki fragments, which are now separated by just a single gap between neighbouring nucleotides. All that is needed is to synthesize a phosphodiester bond at this position, a reaction catalysed by another new enzyme, this one called **DNA ligase** (Fig. 11.10(b)).

11.2.3 The topological problem

The system just described brings about replication of the leading and lagging strands of the parent molecule and produces two daughter double helices. However, there is one outstanding problem to consider. The two polynucleotides of the parent helix are, of course, wound round one another; this means that progression of the replication fork along the parent molecule requires the double helix not just to be unzipped but also to be unwound. This is in fact a significant problem if you consider that the *E. coli* DNA molecule is 4000 kb, or 400 000 turns of the helix, in length and must be replicated in 20 minutes. The implication is that the double helix is rotating at a rate of 20 000 rpm!

This is so inconceivable that for many years molecular biologists sought solutions that avoided unwinding the helix. At one stage around 1979 things became so desperate that some scientists even suggested that the double helix was incorrect and that in fact the polynucleotides in a double-stranded DNA molecule are laid side by side and not wound round each other. Fortunately a group of enzymes that solve this topological problem was eventually discovered.

(a) DNA topoisomerases. The enzymes that unwind the double helix are called **DNA topoisomerases** and they fall into two classes, Type I and Type II. Both unwind a DNA molecule without actually rotating the double helix; they achieve this feat by causing transient breakages in the polynucleotide backbone.

Type I topoisomerases break just one of the polynucleotides and pass the other strand through the gap before reforming the backbone (Fig. 11.11). Type II enzymes, which include the well-

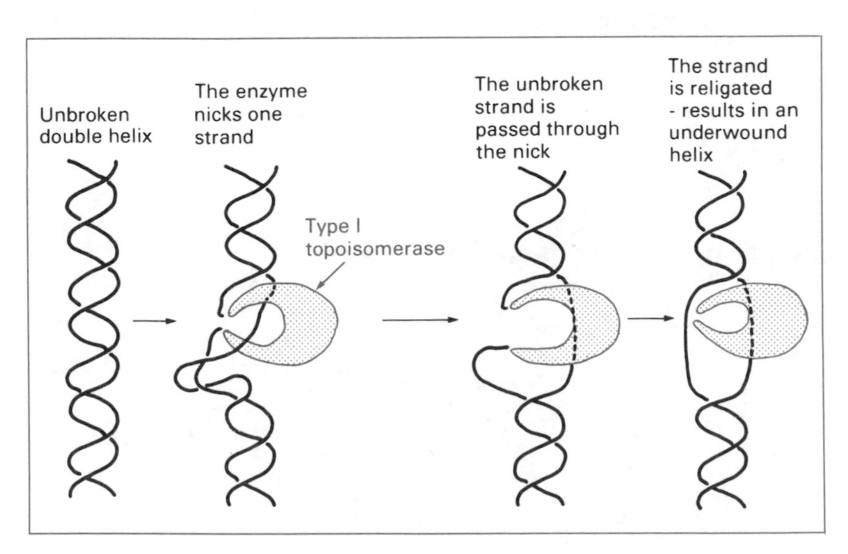

Figure 11.11 The mode of action of a Type I topoisomerase. Type II enzymes carry out a similar reaction but break both polynucleotides.

characterized **DNA gyrase** of *E. coli*, carry out the same sort of reaction but break both polynucleotides at adjacent positions whilst doing so. Both types of topoisomerase produce the same result: portions of the double helix just in advance of the replication fork are unwound allowing the fork to progress unhindered along the parent molecule. DNA topoisomerases can also carry out the reverse reaction and introduce extra turns into a DNA molecule: this results in **supercoiling** (p. 246) as opposed to unwinding.

11.3 DNA replication in eukaryotes

DNA replication occurs in a very similar manner in all organisms, and the scheme just described for events at the replication fork in *E. coli* also holds for eukaryotes. Mammalian cells have five different DNA polymerases, called α, β, γ, δ and ε, with DNA polymerase α and DNA polymerase δ being responsible for replication of nuclear DNA. Their precise roles are still not completely understood, but DNA polymerase δ probably replicates the leading strand, with the lagging strand being copied by DNA polymerase α. The latter enzyme has a tightly associated primase activity, which is thought to be responsible for synthesizing the RNA primer for each Okazaki fragment, implying that the primosome either does not exist in eukaryotic cells or has a drastically modified function. Another puzzle surrounds the fact that neither DNA polymerase α nor DNA polymerase δ have the 5′ to 3′ exonuclease activities that are needed to remove the RNA primers from the lagging strand. Proofreading the lagging strand is also a problem as DNA polymerase α also lacks the 3′ to 5′ exonuclease, and therefore cannot retrace its steps to replace an incorrect nucleotide.

Outstanding problems in replication

Although the events at the replication fork are now understood in some detail, there is still little information on how replication of a DNA molecule is initiated. Initiation takes place at one or more specific points, the origins of replication (see Fig. 11.5), which are recognized by proteins that open up the double helix and enable the DNA synthesis reactions to begin. In *E. coli* a **prepriming complex**, consisting of multiple copies of six proteins (including DNA gyrase and single-strand binding proteins) is known to be built at the origin of replication, but the difficulties in isolating origins have meant that equivalent work has not yet been carried out with eukaryotes. Even in *E. coli* the picture is far from complete, especially as different bacteriophages, whose DNA is replicated in *E. coli*, appear to use a variety of strategies for replication initiation.

The five DNA polymerases of mammalian cells

DNA polymerase	*Molecular mass (daltons)*	*Function*
α	>500 000	Replication of nuclear DNA
β	40 000	DNA repair
γ	140 000	Replication of mitochondrial DNA
δ	173 000	Replication of nuclear DNA
ε	215 000	DNA repair?

*Refers to the *Xenopus laevis* enzyme.

11.4 Replication of molecules

From the foregoing discussion of the events at the replication fork it might appear that all DNA molecules must be replicated in exactly the same way. This is not the case and a variety of strategies have evolved for DNA replication. These centre on how many replication forks there are (one or two, giving unidirectional or bidirectional replication) and whether the molecule being replicated is linear (as in a eukaryotic chromosome) or circular (as in *E. coli* and many bacteriophages and viruses). We will return to

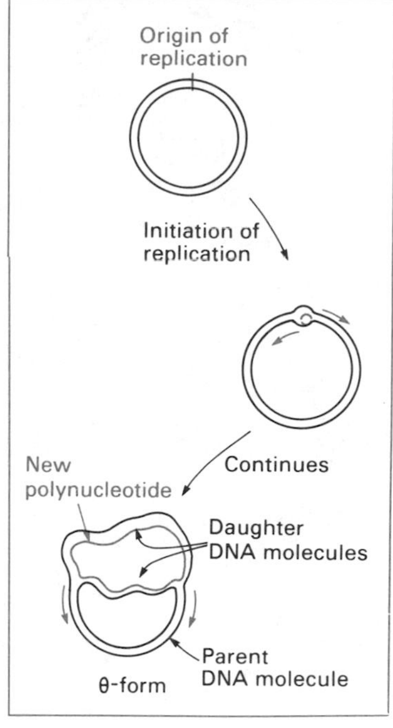

Figure 11.12 Replication of a circular DNA molecule via a θ-intermediate, so called because an electron microscopic image of the structure resembles the Greek letter 'θ'.

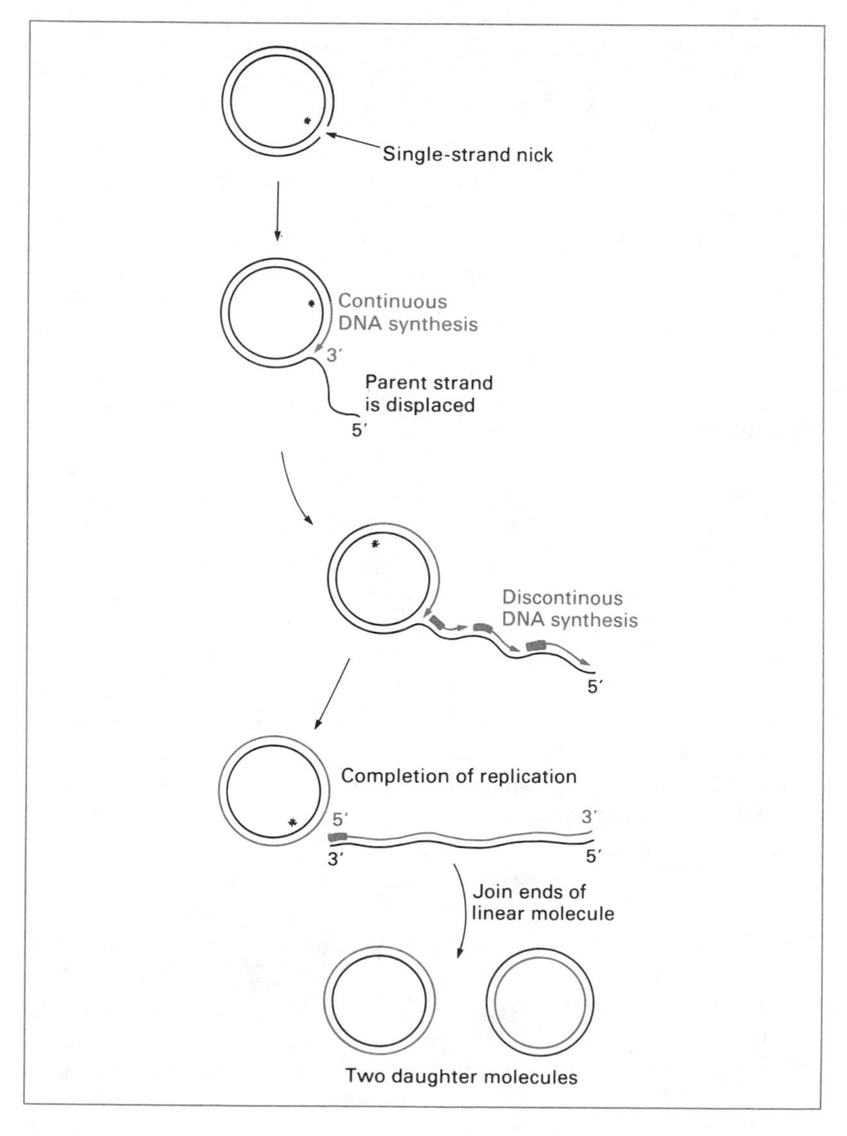

Figure 11.13 Rolling circle replication.

DNA replication strategies in Part Two when we look at the types of DNA molecule present in different organisms. Before leaving DNA replication we should, however, consider two of the ways in which a circular DNA molecule can be replicated.

The simplest way to replicate a circular DNA molecule is by having a single origin of replication, from which two replication forks progress, in opposite directions, around the molecule. This results in an intermediate ***θ*-form** (Fig. 11.12). Most bacterial DNA molecules and many bacteriophage and viral DNAs replicate in this way.

The second system, employed by several types of phage, is called **rolling circle replication** and begins with a nick in one strand of the parent molecule and extension of the free 3′-OH end by DNA polymerase (Fig. 11.13). The original parent strand is therefore displaced and 'rolled off' the molecule. DNA synthesis can stop after one revolution, resulting in two daughter molecules as shown in the figure, or can continue round the circle for several revolutions, very rapidly producing a series of **concatamers** (single-stranded copies of the genome linked end to end). In either case, discontinuous replication of the displaced strand will produce a long double-stranded molecule that, if necessary, can then be cut into individual genomes. Ligation of the ends of these will produce new circular molecules.

Reading

M. Meselson and F. Stahl (1958) The replication of DNA in *Escherichia coli. Proceedings of the National Academy of Sciences USA*, **44**, 671–82 – the famous experiment.

A. Kornberg (1960) Biologic synthesis of deoxyribonucleic acid. *Science*, **131**, 1503–8 – a description of DNA polymerase I.

B. Alberts and R. Sternglanz (1977) Recent excitement in the DNA replication problem. *Nature*, **269**, 655–61; A. Kornberg (1984) DNA replication. *Trends in Biochemical Sciences*, **9**, 122–4 – two descriptions of events at the replication fork.

T. Okazaki and R. Okazaki (1969) Mechanisms of DNA chain growth. *Proceedings of the National Academy of Sciences USA*, **64**, 1242–8 – the discovery of Okazaki fragments.

M. Radman and R. Wagner (1988) The high fidelity of DNA duplication. *Scientific American*, **259**(2), 24–30 – describes how errors are avoided and corrected during DNA replication.

J. C. Wang (1982) DNA topoisomerases. *Scientific American*, **247**(1), 84–95 – although now out of date, still a good source of information on these enzymes.

J. Blow (1989) Eukaryotic chromosome replication requires both α and δ polymerases. *Trends in Genetics*, **5**, 134–6 – ideas on the roles of these two DNA polymerases in eukaryotic DNA replication.

J. Cairns (1963) The bacterial chromosome and its manner of replication as seen by autoradiography. *Journal of Molecular Biology*, **6**, 208–13 – visualization of DNA replication in bacteria.

Problems

1. Define the following terms:

   semi-conservative replication
   replication origin
   replication fork
   leading strand
   lagging strand
   Okazaki fragment
   primer
   primosome
   concatamer

2. Describe the role of the following enzymes and proteins during DNA replication in *E. coli*:

   DNA polymerase I
   DNA polymerase III
   helicase
   single-strand binding protein
   primase
   DNA ligase
   DNA topoisomerases
   DNA gyrase

3. Distinguish between the three possible ways in which the double helix could be replicated and describe an experiment that determines which method is operative in *E. coli*.

4. Distinguish between the roles of DNA polymerase I and DNA polymerase III during DNA replication in *E. coli*.

5. Explain why the first few nucleotides laid down during DNA replication are in fact ribonucleotides.

6. Explain why replication of the lagging strand is discontinuous.

7. What is the solution to the topological problem concerning DNA replication?

8. Describe two ways by which circular DNA molecules are replicated.

9. *Discuss why the semi-conservative mode of DNA replication was favoured even before the Meselson–Stahl experiment was carried out.

10. Draw the density gradient banding patterns that Meselson and Stahl would have obtained if DNA replication had turned out to be (a) conservative, or (b) dispersive.

11. Draw the banding patterns that Meselson and Stahl would have obtained if they had followed DNA replication through more than two rounds of replication. Be sure to indicate the relative amounts of DNA in each band that you draw.

Alterations in the genetic material

12

Mutation – different types – reversing the effect of a mutation – mutagens • DNA repair • Recombination – generalized recombination – role of recombination in genetics

Although DNA replication is virtually error-free, mistakes do occasionally occur. An error in replication will result in a daughter molecule that contains a **mismatch**, a position at which a base pair will not form as the nucleotides opposite each other in the double helix are not complementary (Fig. 12.1(a)). When the mismatched daughter molecule is itself replicated the two granddaughter molecules will not be identical: one will have the correct nucleotide sequence, but the second will contain a **mutation**. A mutation is simply an alteration in the nucleotide sequence of a DNA molecule.

Mutations are also caused by chemical and physical **mutagens**. Several chemicals react with DNA molecules and change the structure of one or more nucleotides within the double helix. Heat, ultraviolet radiation and other physical mutagens have similar effects. Changing the structure of a nucleotide may affect its base pairing properties, resulting in a mutation when the DNA molecule is replicated (Fig. 12.1(b)).

As we shall see, mutations may be harmless, or they may have a serious, even lethal, effect on the organism. To minimize the number of mutations that become established, all organisms have **DNA repair** mechanisms, with which they attempt to correct mismatches and structural alterations in their DNA molecules. Despite these measures a few mutations slip through.

A mutation is a small-scale change in a DNA molecule. Larger-scale alterations occur by the process called **recombination**. Recombination can have any one of several effects, including the exchange of segments of polynucleotides between different DNA molecules, and the **transposition** of a piece of DNA from one position to another in a molecule (Fig. 12.2). Far from being harmful, recombination plays several important roles in the life of a cell.

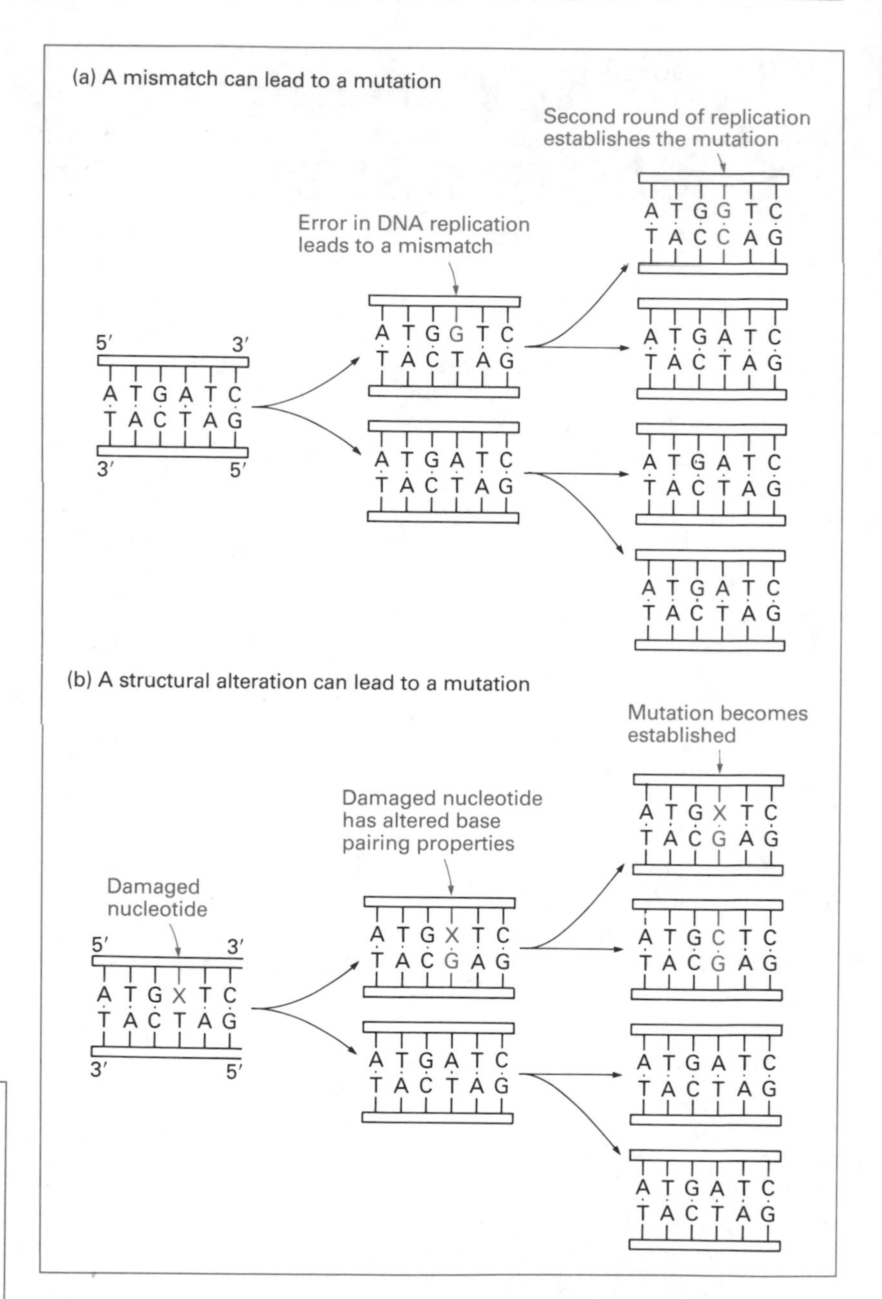

Figure 12.1 Two ways in which a mutation can arise.

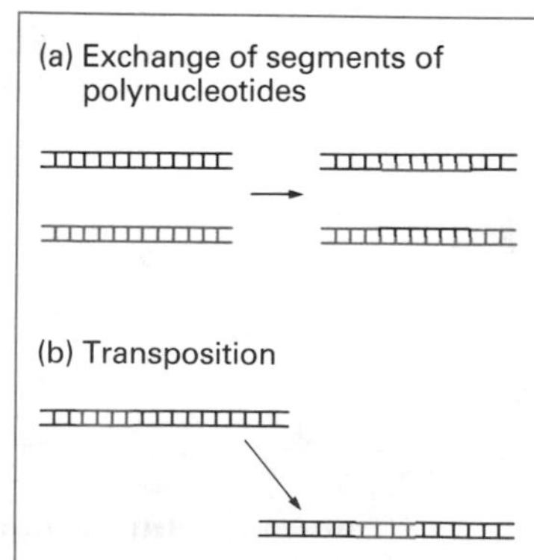

Figure 12.2 Two possible effects of recombination.

In this chapter we will look at mutations, mutagens, repair mechanisms and recombination. We will cover a lot of ground and come across a number of new jargon words. The jargon is tiresome but must be dealt with: mutation and recombination are central events in genetics.

12.1 Mutations

Before looking at mutations and their effects a few terms must be defined. As already stated, a mutation is a change in the nucleotide sequence of a DNA molecule. A mutation may occur within a gene or it may occur in the intergenic regions. If a mutation occurs in an intergenic region it will probably be **silent** and have no discernible effect on the cell. However, if a mutation occurs in a gene it may alter the gene product and generate an observable change in the organism: this is referred to as a change in **phenotype**. The phenotype of an organism is its set of observable characteristics; in contrast the **genotype** is its genetic constitution. An organism displaying the usual phenotype for that species is called a **wild-type**; an organism whose phenotype has been altered by mutation is called a **mutant**.

Genotype and phenotype

The distinction between genotype and phenotype can be confusing. To clarify the meaning of the two terms, consider an *E. coli* mutant that lacks β-galactosidase activity:

The **genotype** is described as *lacZ*$^-$, with the superscript 'minus' signifying that the gene is mutated. Remember that a gene designation is set in italic typeface when printed, or underlined when written (p. 158).

The **phenotype** is described as 'lacks β-galactosidase'. A geneticist would write this as LacZ$^-$. Note that here we use roman script and a capital 'L' so there is no confusion with the genotype designation.

12.1.1 Different types of mutation

Mutations can be looked at from three different angles. First, we can consider just the DNA sequence itself and list the different types of nucleotide sequence alterations that can occur. Second, there is the question of what effect the different types of mutation will have if they occur in a gene. And finally, there is the point of view of the organism and the different kinds of phenotypic change that can arise as a result of mutation. We will consider each of these three levels of mutation in turn.

(a) Mutations at the level of the DNA sequence. The various types of alteration in DNA sequence that can occur as a result of mutation are as follows (Fig. 12.3):

1. A **point mutation** is the replacement of one nucleotide by another. A point mutation is classified as a **transition** if it is a purine to purine (A ↔ G) or a pyrimidine to pyrimidine (T ↔ C) change, or a **transversion** if the alteration is purine to pyrimidine or vice versa (A or G ↔ T or C).
2. An **insertion** or **deletion** is the addition or removal of anything from one base pair up to quite extensive pieces of DNA.
3. An **inversion** is the excision of a portion of the double helix followed by its reinsertion at the same position but in the reverse orientation.

(b) Mutations at the level of the gene. Whether a point, insertion, deletion or inversion, a mutation will probably be silent if it occurs in an intergenic region. But what are the likely effects if a mutation occurs in the coding region of a gene? To answer this

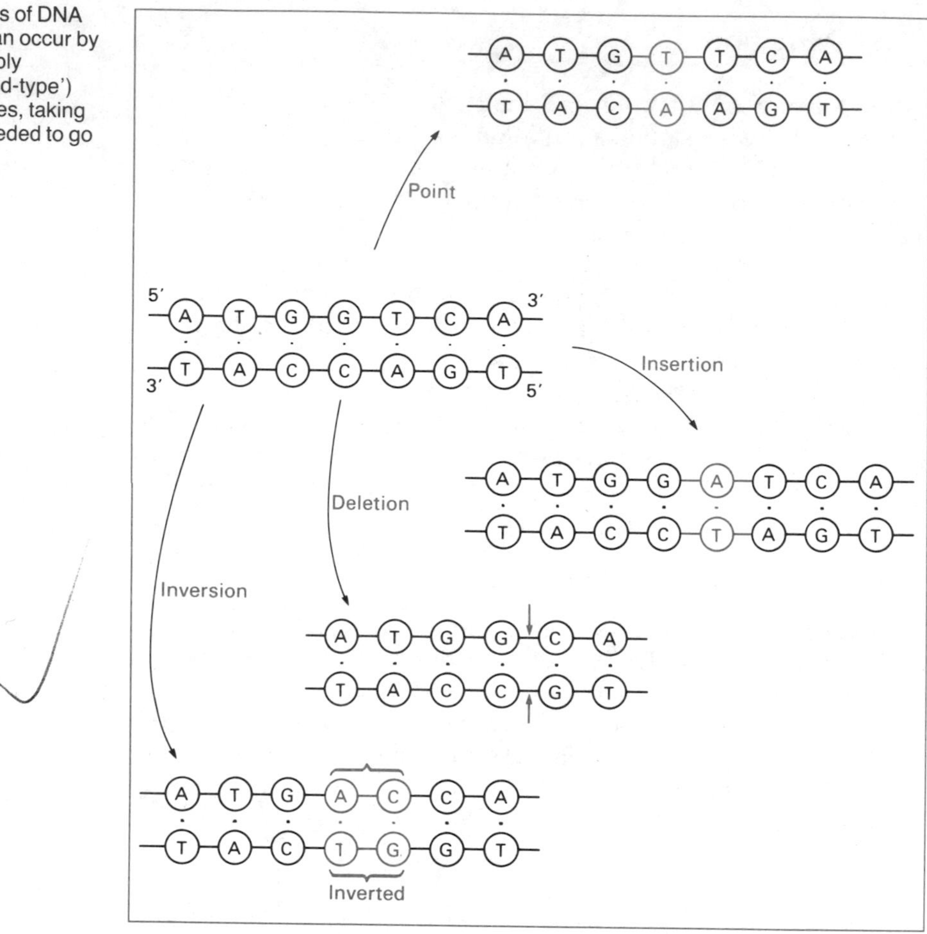

Figure 12.3 Different types of DNA sequence alteration that can occur by mutation. Here we are simply comparing the original ('wild-type') and mutated DNA molecules, taking no account of the steps needed to go from one to the other.

question we will invent a short imaginary gene (see Fig. 12.4) which, although coding for a peptide only six amino acids in length, will show us most of the possible consequences.

1. ***Silent mutation within a gene*** (Fig. 12.4). This will occur if a point change takes place at the third nucleotide position of a codon, and changes the codon but, owing to the degeneracy

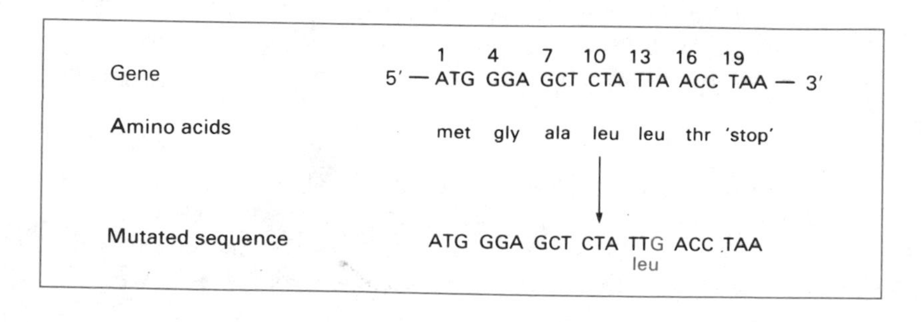

Figure 12.4 A silent mutation within a gene.

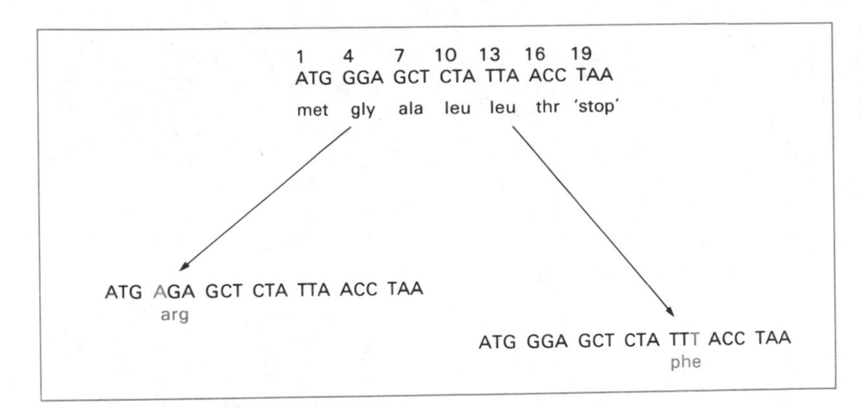

Figure 12.5 Two missense mutations.

of the genetic code, not the amino acid (e.g. nucleotide 15 changed from A to G). A silent mutation will have no effect on the amino acid sequence of the gene product and will not give rise to a mutant phenotype.

2. ***Missense mutation*** (Fig. 12.5). This is also a point change but in this case it does change the amino acid. Most point mutations at the first or second nucleotide positions of a codon will be missense, as will a few third position changes. For example, changing nucleotide 4 from G to A produces an arginine codon instead of a glycine codon; similarly, changing nucleotide 15 from A to T specifies phenylalanine rather than leucine. A missense mutation will give rise to a polypeptide with a single amino acid change; whether or not this will cause a mutant phenotype depends on the precise role of the mutated amino acid in the structure and/or function of the protein. Most proteins can tolerate some changes in their amino acid sequence, although a missense mutation that alters an amino acid essential for structure or function will inactivate the protein and lead to a mutant phenotype.
3. ***Nonsense mutation*** (Fig. 12.6). This is a point mutation that changes a codon specifying an amino acid into a termination codon (e.g. changing nucleotide 14 from T to G). The result will be a truncated gene which codes for a polypeptide that has lost

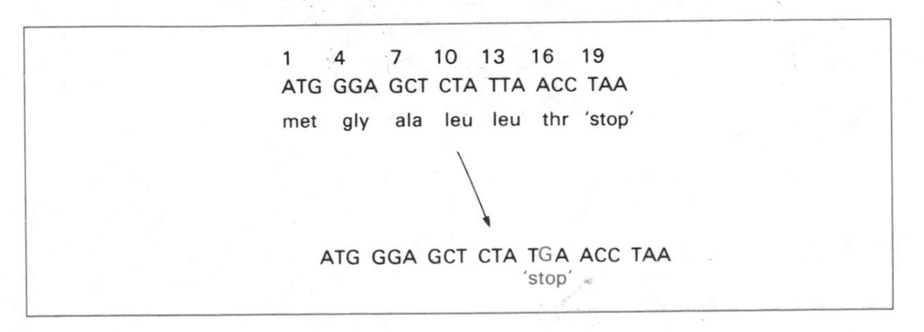

Figure 12.6 A nonsense mutation.

Figure 12.7 Frameshift mutations.

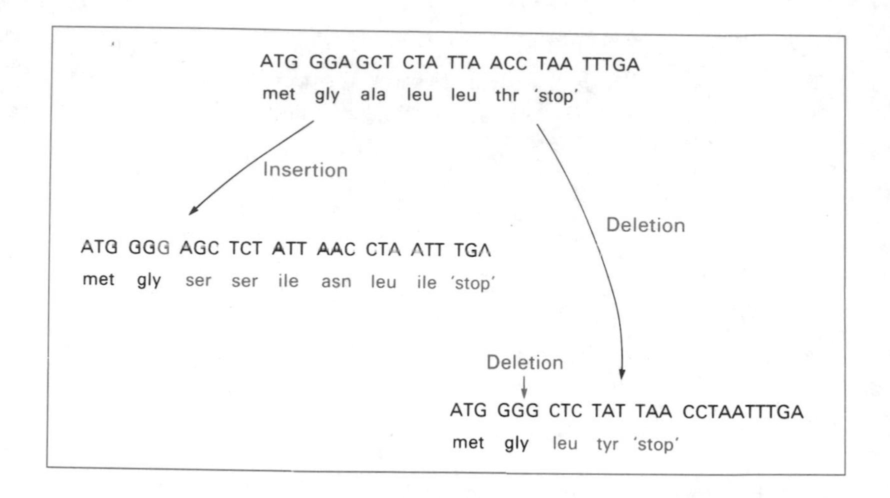

a segment at its carboxyl-terminus. In many cases, although not always, this segment will include amino acids essential for the protein's activity and a mutant phenotype will result.

4. ***Frameshift mutation*** (Fig. 12.7). This is the usual consequence of an insertion or deletion event because the addition or removal of any number of base pairs that is not a multiple of three will cause the ribosome to read a completely new set of codons downstream from the mutation. As with nonsense mutations, frameshifts usually produce mutant phenotypes.

(c) Mutations at the level of the organism. In order to produce a mutant phenotype the nucleotide sequence alteration must give rise to a mutated gene product that is unable to fulfil its function in the cell. In many cases the cell will be unable to tolerate the loss of the function and will die. Such mutations are therefore called **lethal mutations**.

However, not all mutations that produce a change in phenotype are so drastic in their effect. Some mutations will inactivate pro-

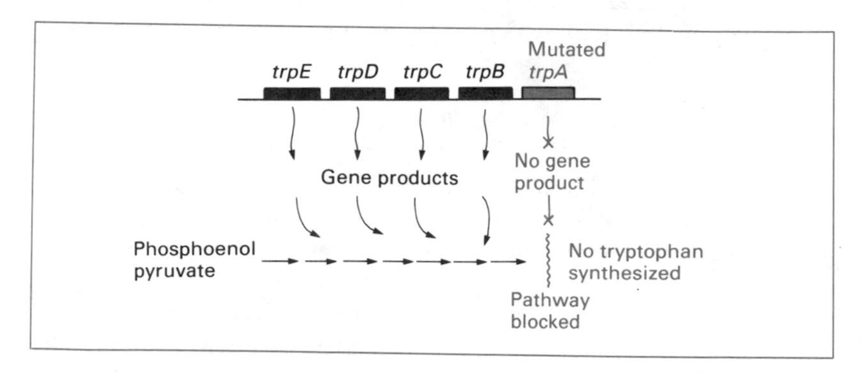

Figure 12.8 An example of an auxotrophic mutation. The top line shows the *trp* operon (p. 163), which consists of five genes that code for enzymes that convert phosphoenol pyruvate to tryptophan. A mutation in one of these genes, *trpA* for instance, prevents tryptophan from being synthesized.

teins that are not essential to the cell, and others will result in proteins with reduced or modified activities. This is true for both eukaryotes and prokaryotes but is most easily understood by looking at a few examples in *E. coli*.

1. ***Auxotrophic mutants***. Such a mutant lacks a gene product involved in synthesis of an essential metabolite such as an amino acid. However, these mutants can be kept alive if the metabolite is supplied as a nutrient in the culture medium. For example, a tryptophan auxotroph of *E. coli* is a mutant that lacks one of the enzymes involved in tryptophan biosynthesis (Fig. 12.8); the organism is unable to survive on a **minimal medium** that lacks tryptophan but can grow on a supplemented medium to which this amino acid has been added. The opposite of an auxotroph is a **prototroph**, a cell that has no nutritional requirements beyond those of the wild-type.
2. ***Conditional-lethal mutants***. This type of mutant can survive, but only if cultured under a particular set of conditions. The most common examples are **temperature-sensitive mutants**, which are able to survive at one temperature range (less than 30°C, say) but die if the temperature is raised above this permissive threshold. The mutation carried by a temperature-sensitive organism is often one that affects an amino acid essential for maintaining the tertiary structure of the protein product. The mutated protein is able to retain its correct structure at a low temperature, but is easily denatured and inactivated by heat (Fig. 12.9).
3. ***Antibiotic resistant mutants***. Antibiotics kill wild-type bacteria but have no effect on resistant mutants. Resistance to an antibiotic can arise in several ways but the commonest is where the molecule that is the target for the antibiotic becomes altered. Streptomycin, for example, interferes with protein synthesis by binding to ribosomal protein S12, one of the components of the small subunit in *E. coli* (p. 144). When streptomycin is bound to the small subunit, the initiator tRNA cannot entre the P-site, so no mRNAs are translated and the bacterium dies. In a streptomycin-resistant mutant the gene for the ribosomal protein S12 is mutated. This leads to an altered S12 protein which is still able to fulfil its role as a ribosomal protein, but can no longer bind to streptomycin. The antibiotic therefore has no effect on the mutated bacterium.
4. ***Regulatory mutants***. Regulatory mutants have lost the ability to control expression of a gene or operon normally subject to regulation. For instance, it is possible to obtain *E. coli* mutants that express the genes of the *lac* operon even in the absence of lactose (p. 159). These are called **constitutive mutants** and usually arise through a mutation in the gene for the *lac*

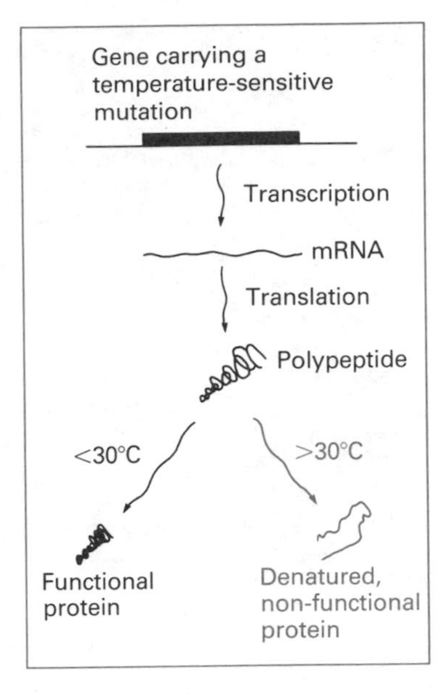

Figure 12.9 The possible effect of a temperature-sensitive mutation. In this example the permissive temperature is 30°C, above which the mutated polypeptide denatures.

Leaky mutants

A leaky mutant is intermediate between wild-type and fully mutant. For instance, a leaky $trpA^-$ auxotroph will be able to synthesize some tryptophan, but not as much as the wild-type. The mutant will therefore grow slowly on minimal medium, distinguishing it from a full $trpA^-$ mutant, which will not grow at all in the absence of tryptophan. A leaky mutant is usually one in which the mutated gene product is not totally inactivated and so can function with reduced efficiency. Leaky-lethal, -conditional and -resistant mutants are also possible.

The effects of mutations in multicellular organisms

In a multicellular organism the effect of a mutation will depend both on the nature of the mutation and the cell that is involved. A lethal mutation in a single cell will generally go unnoticed as cells in tissues and organs are continually being replaced by new cells. A non-lethal mutation may have an effect if the mutated cell gives rise to a lineage of progeny cells that also carry the mutation. An example is provided by the variegated colours that appear on the leaves of some ornamental plants. In some cases these arise because a gene involved in synthesis of the pigment becomes mutated in a cell whose progeny will eventually make up a sector of the leaf:

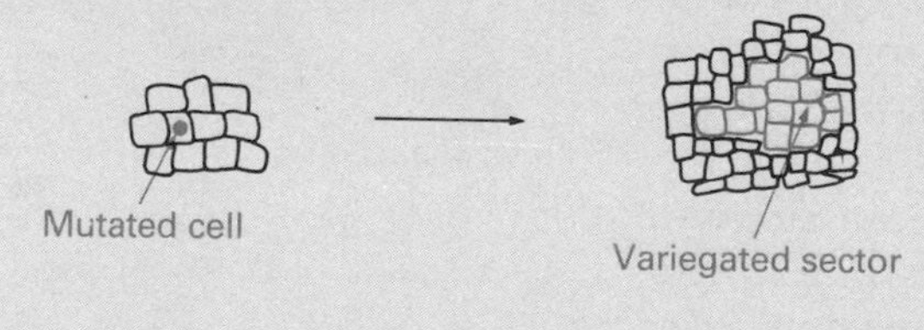

In higher eukaryotes such as man the important distinction is between the **somatic cells**, which make up the bulk of the organism but are not passed on during reproduction, and the **sex cells**, which give rise to female eggs and male sperm. A mutation in a somatic cell will only be important if it confers on the cell some dangerous property such as the ability to divide in an uncontrolled fashion, which may lead to cancer. On the other hand, a mutation in a sex cell will be passed on to the offspring via the egg or sperm so that all the somatic cells of the offspring will be mutated. Queen Victoria almost certainly suffered a mutation in her sex cells as several of her children and grandchildren suffered from haemophilia, which is caused by a mutation in the gene for a blood-clotting protein. You could argue that this mutation caused World War I and goodness knows what else.

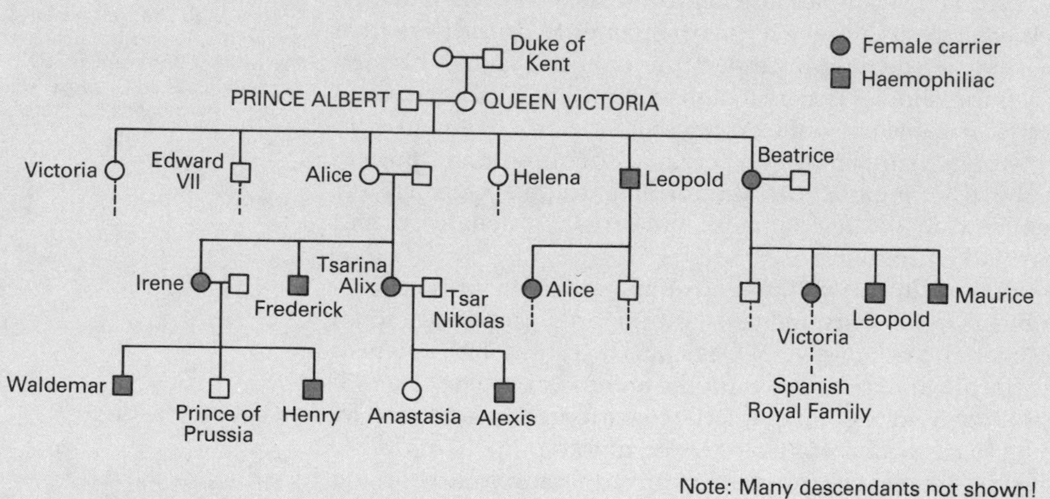

repressor, so that the repressor is either not produced, or has an altered structure and can no longer bind the operator. A mutant with an identical phenotype could also arise from a mutation within the operator sequence itself.

12.1.2 Reversing the effect of a mutation

Mutant organisms can sometimes return to the wild-type phenotype by virtue of a second mutation. There are several

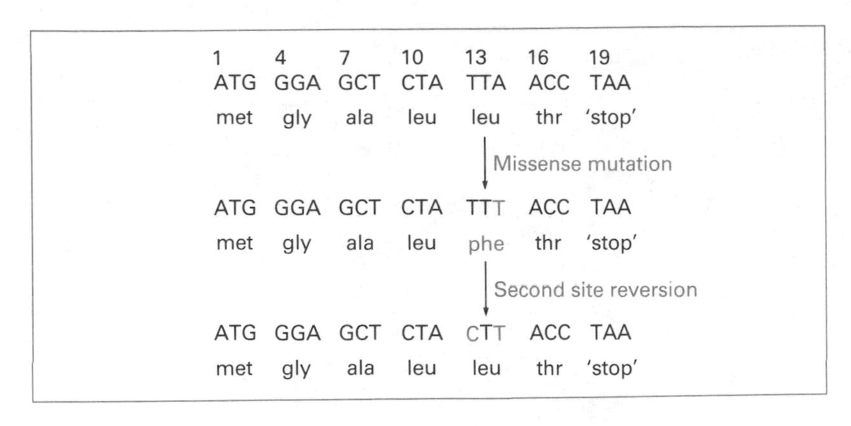

Figure 12.10 Second site reversion.

mechanisms by which this can occur, the simplest being when a second mutation restores the original nucleotide sequence of the DNA molecule. A point mutation can be reversed by a second point mutation, an insertion event by a subsequent deletion, and so on. These events are called **back mutations** and are not very likely unless the site at which the original mutation occurred has some natural predisposition towards mutation.

Many mutations can also be corrected by **second site reversions**, a second mutation that restores the original phenotype but does not return the DNA sequence to its precise unmutated form. To illustrate one of the ways in which this might occur we will return to our example gene. One of the missense mutations shown in Fig. 12.5 involves changing nucleotide 15 from A to T, changing a leucine codon (TTA) into a codon for phenylalanine (TTT). If we now introduce a second mutation, changing nucleotide 13 from T to C, the codon becomes CTT, which codes for leucine once more (Fig. 12.10). The missense mutation has been corrected and the gene product returned to normal even though the nucleotide sequence remains different from the original.

A second site reversion occurs within the same gene as the original mutation. The final way of reversing the effect of a mutation is by **suppression**, in which the second mutation occurs in a different gene. There are several examples of this phenomenon but the most common is the reversal of a nonsense mutation by a suppressive mutation in the anticodon of a tRNA. For instance, the anticodon for $tRNA^{trp}$ is 5′-CCA-3′, to decode the tryptophan codon 5′-UGG-3′. If the first nucleotide of the anticodon in the tRNA is changed by mutation from C to U, then the anticodon will become 5′-UCA-3′ and will now decode the termination codon 5′-UGA-3′ (Fig. 12.11(a)). This termination codon is therefore 'read through' during translation and the shortened polypeptide produced by the original nonsense mutation restored to full length (Fig. 12.11(b)). Note though that the amino acid inserted into the

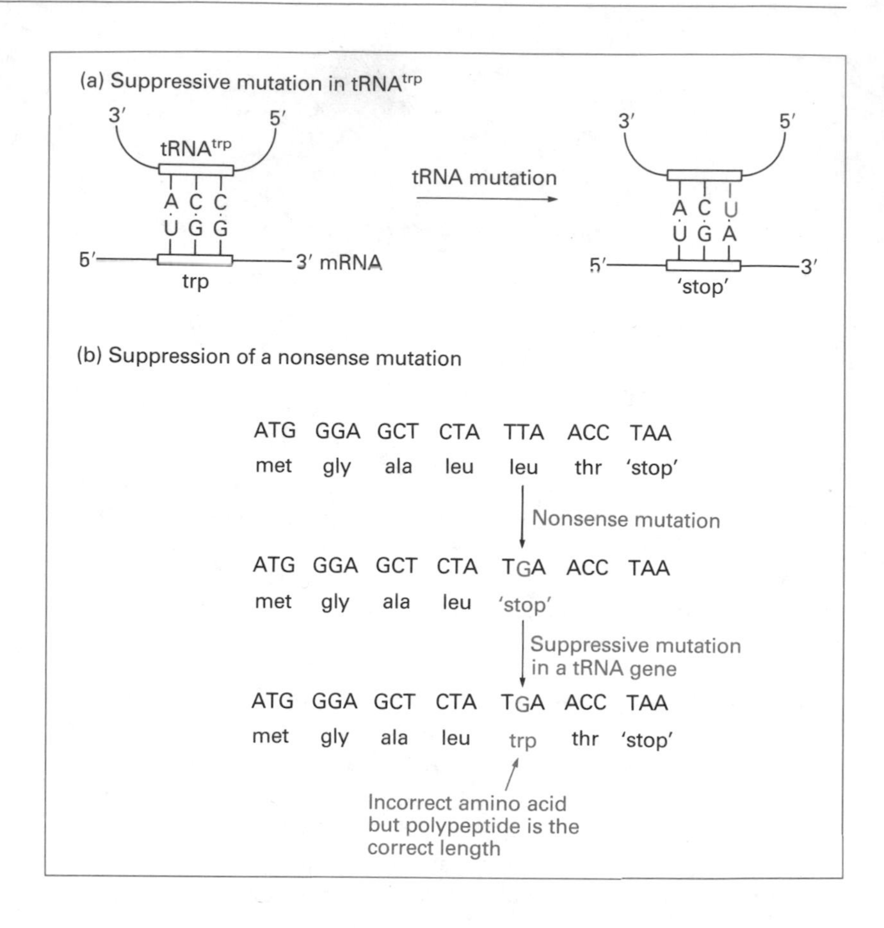

Figure 12.11 Suppression.

peptide when the suppressed termination codon is read is not the one found at the corresponding position in the unmutated version. However, as with a missense mutation, such a small alteration in the amino acid sequence will probably be tolerated by the protein, so the activity is restored.

12.1.3 Mutagens

We have seen that various kinds of mutation can occur in DNA molecules. Now we must look at how these alterations arise.

It has been calculated that, in nature, a mutation occurs in a bacterial cell once every 10^8 cell divisions, and in higher organisms such as man this **spontaneous mutation rate** is probably higher. A large proportion of these mutations occur at random, simply through errors in DNA replication that escape the proofreading functions of the replicating enzymes (p. 184). However, the mutation rate will be increased if the organism is exposed either in nature or in the laboratory to any one of several mutagens. These may be chemicals or possibly physical agents such as ultraviolet

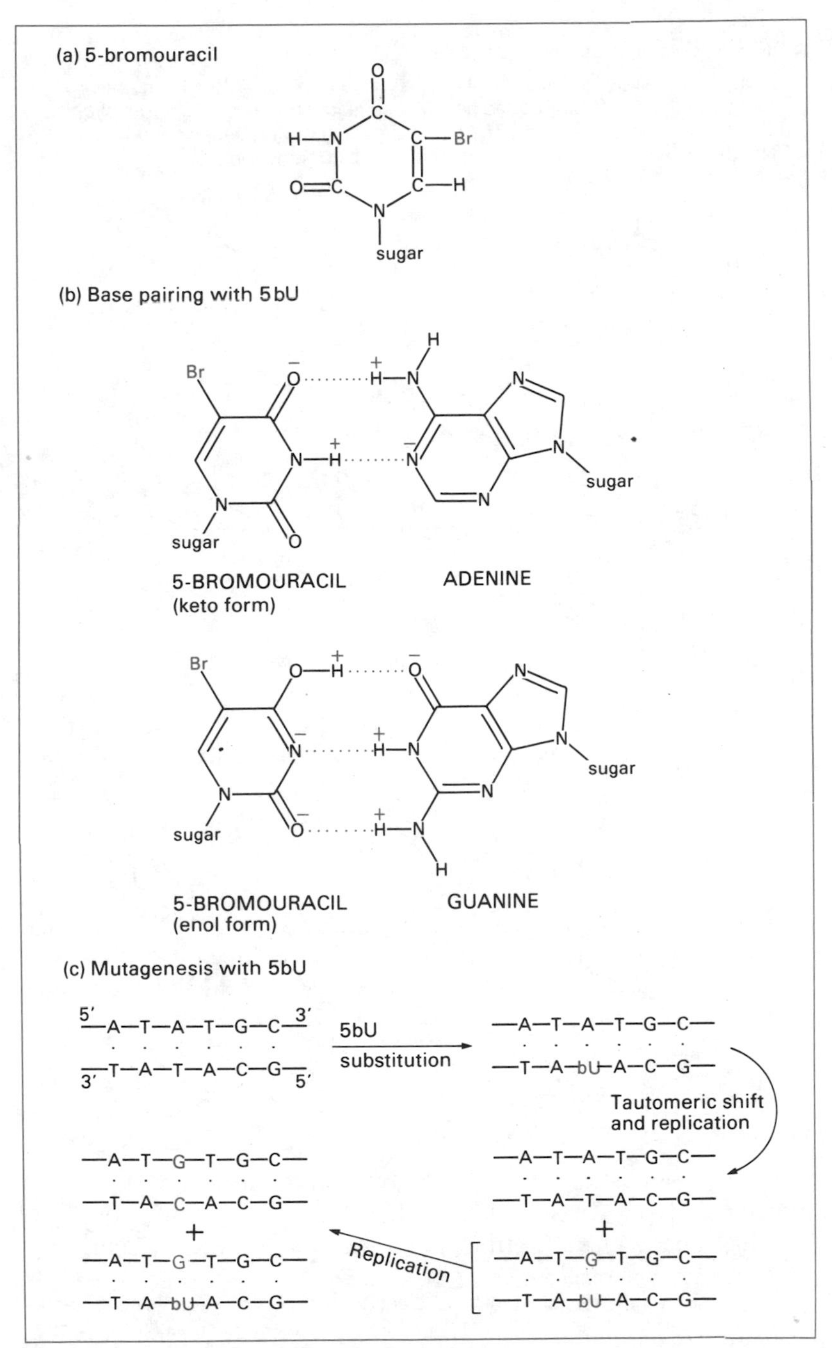

Figure 12.12 5-Bromouracil (5bU) and its mutagenic effect.

radiation. These agents interact directly with DNA molecules by a variety of means, resulting in structural changes that may lead to errors being introduced during replication.

There are several types of mutagen, each acting in a different

Tautomers

Tautomers are structural isomers that exist in equilibrium with one another. Tautomers of a single compound have the same chemical formulae but slightly different molecular structures. The pyrimidine and purine bases occurring in DNA and RNA can each exist as either of two tautomeric forms:

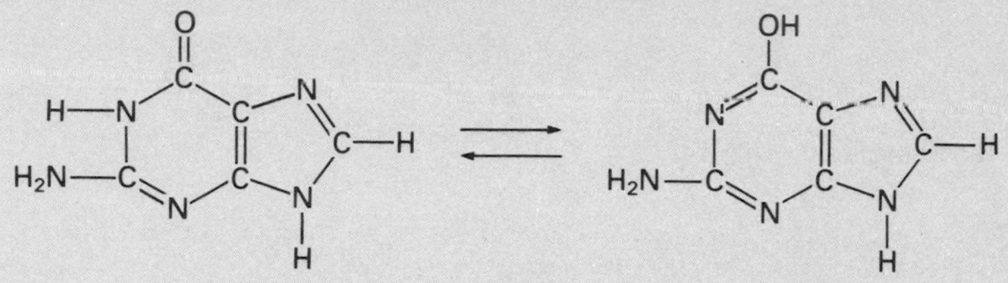

Keto GUANINE Enol

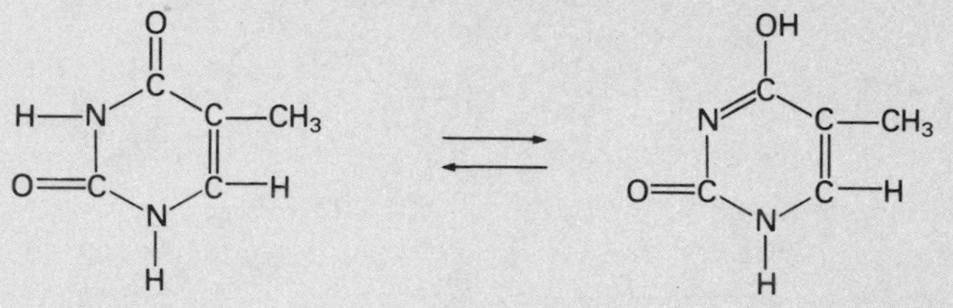

Keto THYMINE Enol

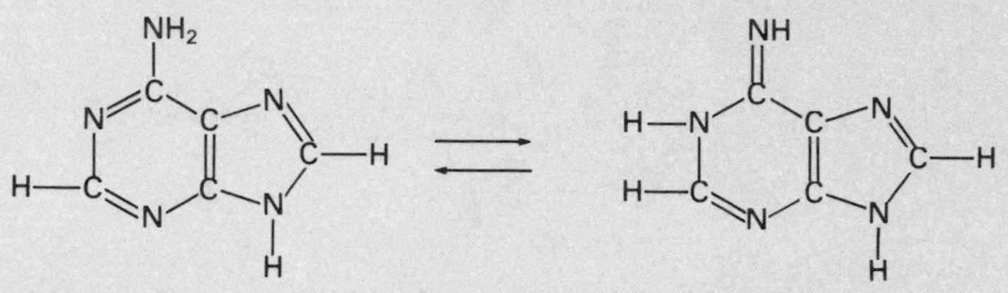

Amino ADENINE Imino

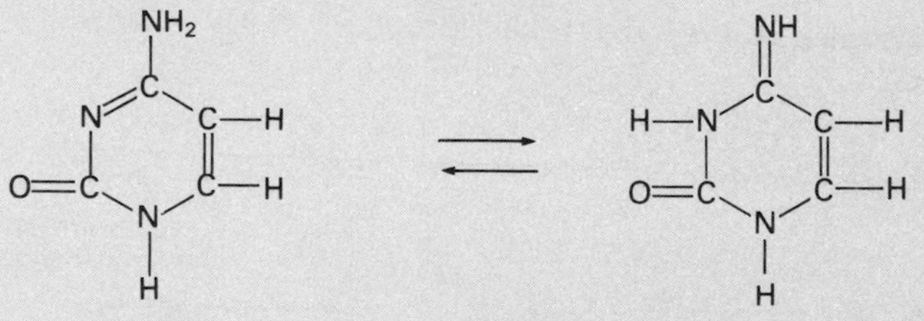

Amino CYTOSINE Imino

Under physiological conditions the equilibrium is biased almost completely towards the keto and amino forms.

A tautomeric shift (from one tautomer to another) in a DNA molecule could lead to a mutation as the change will alter the positions on the molecule of the groups able to form hydrogen bonds, leading to different base pairing properties. The rare enol form of T, for instance, base pairs with G, not A. Similarly, the rare imino form of A base pairs with C. Tautomeric shifts may therefore cause spontaneous mutations. In addition, some base analogues, such as 5bU, have a greater predisposition towards the rarer tautomer, so will increase the mutation rate if incorporated into DNA.

way, but there is no need to look in detail at each one. We will cover just four: the **base analogues**, the **intercalating agents**, heat and ultraviolet radiation.

(a) Base analogues can substitute during replication. 5-Bromouracil (Fig. 12.12(a)) is derived from thymine by replacement of the methyl (—CH_3) group with bromine (—Br). It is sufficiently similar to thymine to be incorporated into a polynucleotide chain in place of the normal nucleotide, as may occur during DNA replication in the presence of 5bU.

The important fact with regard to 5bU is that as well as base pairing with A, as would be expected of an analogue of T, 5bU can also undergo a slight change in its chemical structure, called a **tautomeric shift**, after which it base pairs with G (Fig. 12.12(b)). If this happens during DNA replication then one of the daughter molecules will have a 5bU–G base pair instead of the original A–T (Fig. 12.12(c)); note, however, that the other daughter double helix is correct and not mutated. A further round of replication of the mutant molecule will produce a G–C pair in one double helix, in which the mutation is now established, and a 5bU–G or possibly 5bU–A in the other daughter helix. This is an example of how a point mutation can be brought about.

(b) Intercalating agents cause insertions. An important group of mutagens used in the laboratory are the intercalating agents such as the acridine dyes, of which ethidium bromide (EtBr; Fig. 12.13(a)) is an example. EtBr is a four-ringed molecule whose dimensions are similar to those of a purine–pyrimidine base pair, so much so that the compound can intercalate into a double helix (Fig. 12.13(b)), moving adjacent base pairs slightly apart. Exactly how this disrupts DNA replication is not known but the effect is clear: an insertion of a single nucleotide is likely to occur at the intercalation position and cause a frameshift mutation if the position lies within a gene. Because frameshifts generally give rise to phenotypic changes, intercalating agents are popular mutagens for generating mutants to be used in research.

(c) Heat and ultraviolet radiation are mutagenic. Heat is probably the most important environmental mutagen. Its effect on a DNA molecule is to cleave the bonds between purine bases and their sugars, resulting in apurinic sites in the polynucleotides (Fig. 12.14). Up to 10 000 apurinic sites are created every day in every human cell. Those that evade repair can cause point or deletion mutations when the DNA is replicated.

In addition, several types of radiation are mutagenic. Ultraviolet radiation of about 260 nm wavelength is absorbed by purines and pyrimidines and can cause structural changes, in particular the

Figure 12.13 Ethidium bromide (EtBr) and its mutagenic effect.

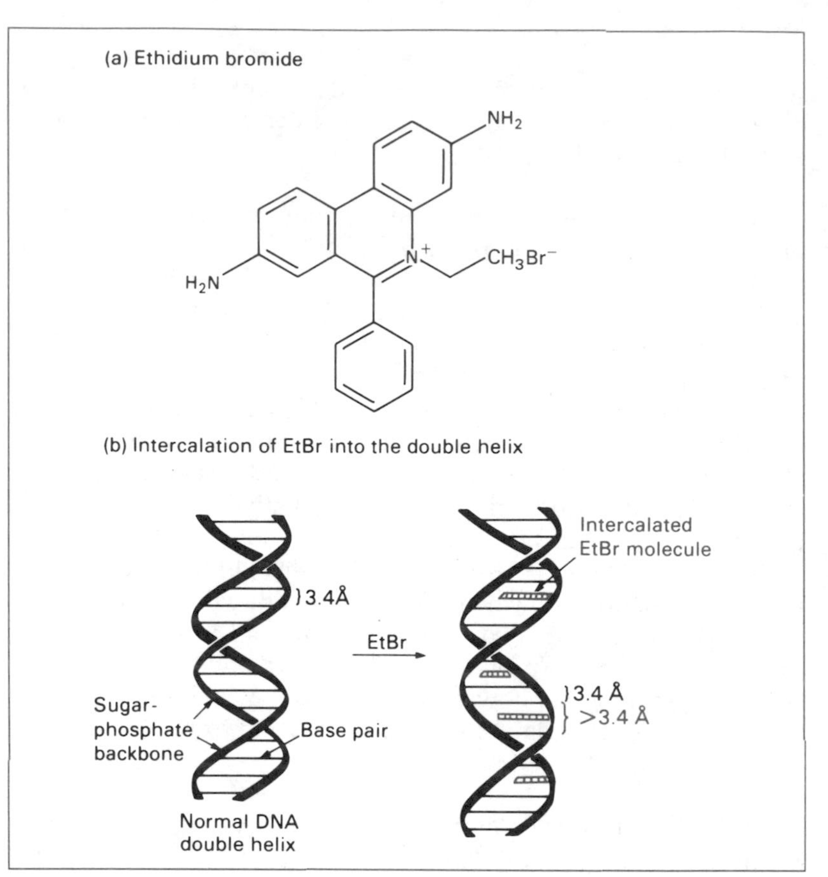

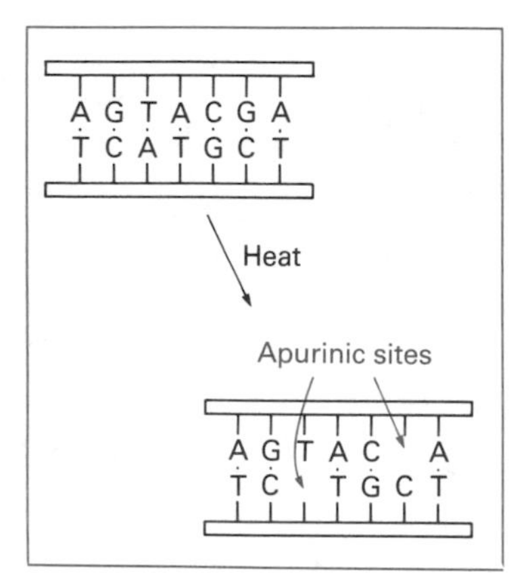

Figure 12.14 The mutagenic effect of heat.

Isolation of mutants

Mutants play important roles in genetical research (p. 159 and Chapters 17–19). Methods are therefore needed for the isolation of organisms carrying mutations in genes of interest.

With microorganisms such as bacteria and yeast, mutant isolation is quite straightforward. First, a culture of the microorganism is mutagenized by treatment with a chemical mutagen such as ethidium bromide (added directly into the growth medium) or by exposure to ultraviolet radiation. The result will be a large number of mutated cells, including a few whose mutations lie in the desired genes. Now a method for identifying those mutants of interest is needed. Numerous strategies can be employed, depending on the nature of the mutated gene. For example, to isolate a tryptophan auxotroph of *E. coli*, the mutagenized cells would first be spread onto a minimal medium supplemented with tryptophan. All of the cells, including the tryptophan auxotrophs, will give rise to colonies. Once the colonies have grown, they are transferred *en masse* onto a minimal medium (lacking tryptophan). The tryptophan auxotrophs will not grow on this medium, and so can be identified by comparing the positions of the colonies on the two plates.

Grow mutagenized bacteria on minimal medium plus tryptophan

All cells produce colonies

Transfer colonies to minimal medium – no tryptophan

Colonies grow = *trp*$^+$

Colony does not grow = *trp*$^-$ (recover sample from the first plate)

With higher organisms the procedure is less easy. Mutations can be induced in organisms such as *Drosophila* by exposure to ultraviolet radiation, but selection methods specific for a mutation in a particular gene are often impossible. Individual mutants must therefore be examined for morphological changes or for biochemical abnormalities that may indicate a mutation in a gene of interest.

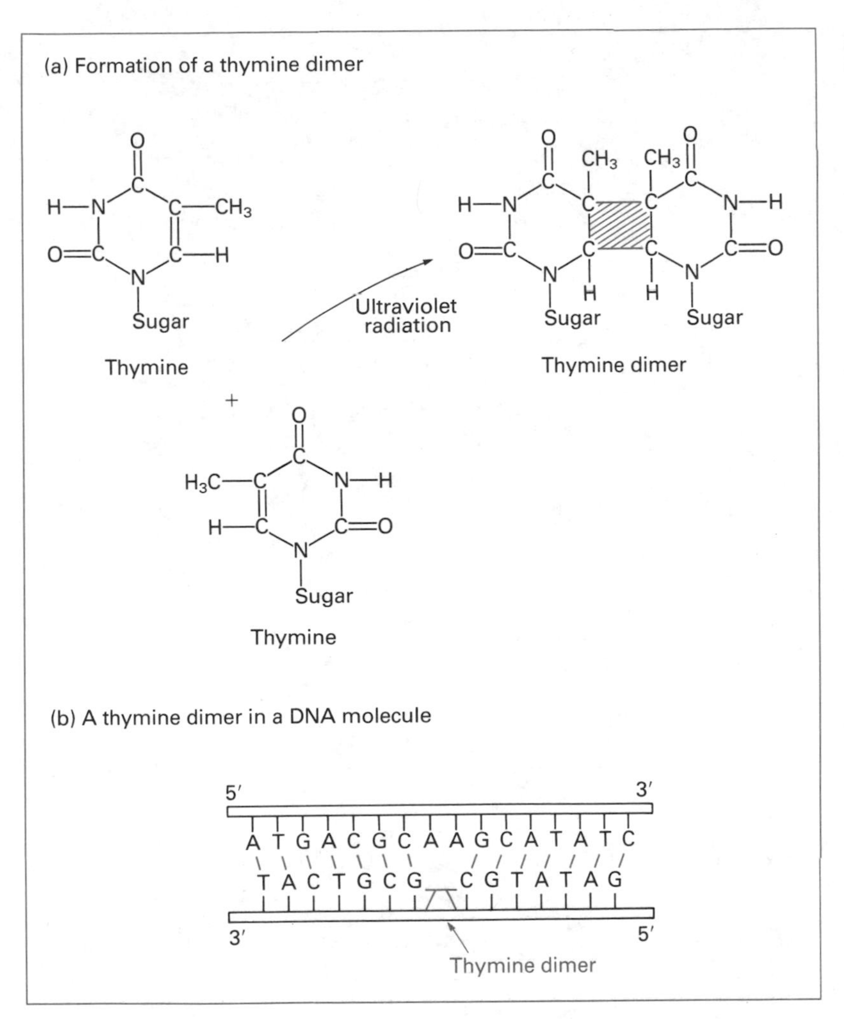

Figure 12.15 Thymine dimers, caused by ultraviolet radiation.

formation of **cyclobutyl dimers** between adjacent pyrimidines in a polynucleotide, most often between thymines (Fig. 12.15). Dimerization causes the bases to stack closer together and can give rise to deletions during DNA replication.

12.2 DNA repair

In view of the number of chemical and physical agents that can cause structural alterations in DNA molecules, it is reassuring that all organisms, not least man, possess extensive mechanisms for repairing damaged DNA. DNA repair is a complicated subject, with at least 24 different proteins in *E. coli* being involved in

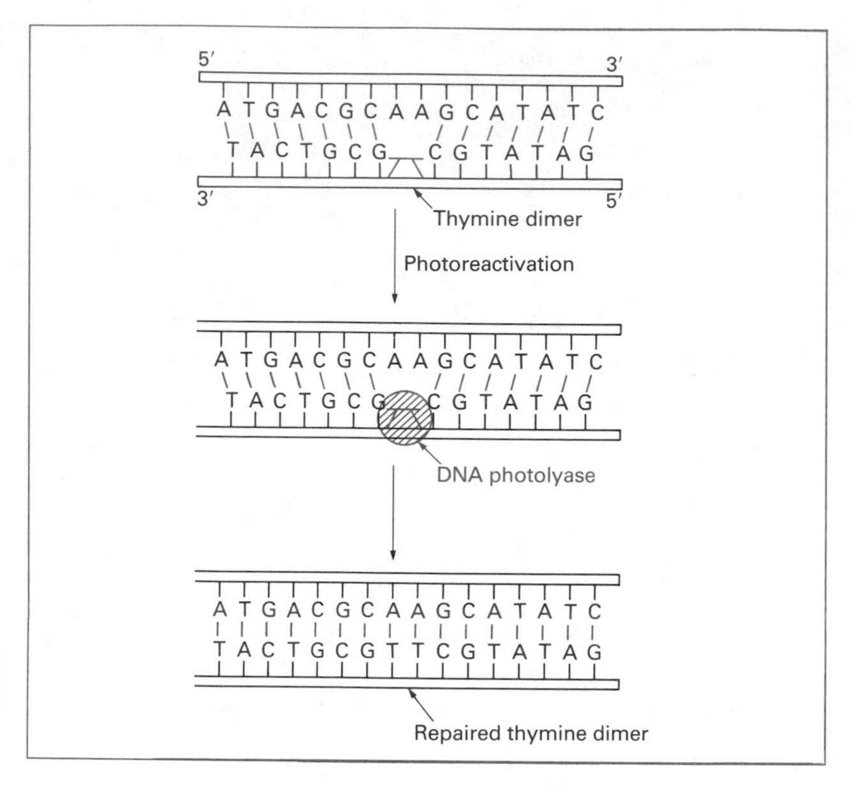

Figure 12.16 Photoreactivation repair of a thymine dimer.

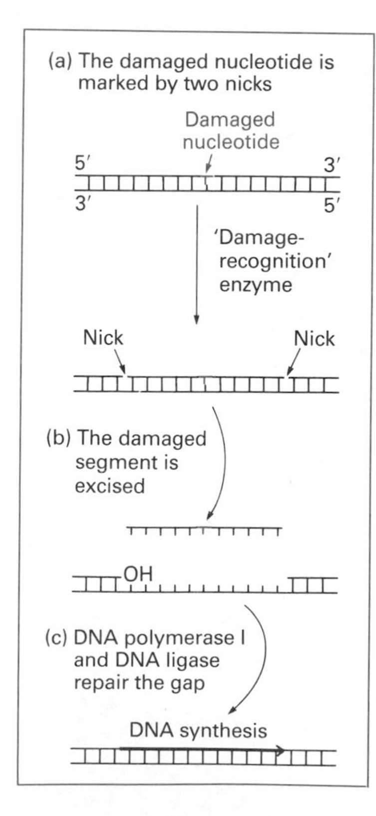

Figure 12.17 One way in which a damaged nucleotide can be corrected by excision repair.

correcting damage of one kind or another. The three main types of repair mechanism are as follows:

1. ***Direct repair*** involves the simple reversal of a structural alteration. It is not a common means of DNA repair but there is one important example, called **photoreactivation**. Photoreactivation is carried out by enzymes called DNA photolyases which, when activated by visible light, repair thymine dimers by cleaving the links responsible for the dimerization (Fig. 12.16). DNA photolyases are known in bacteria, microbial eukaryotes and plants.
2. ***Excision repair*** is more complicated but is probably the main form of DNA repair in most organisms. It is initiated by any one of several enzymes that recognize and 'mark' a nucleotide that is damaged. The marker may be in the form of one or two single-stranded nicks adjacent to the damaged nucleotide, or may involve cleavage of the altered base to produce an apurinic or apyrimidinic site (Fig. 12.17(a)). Many types of damage can be recognized, including thymine dimers. In the second stage of the repair process a nuclease excises the damaged nucleotide along with a few of its neighbours, leaving a gap (Fig. 12.17(b)). The gap is then filled in by a DNA polymerase

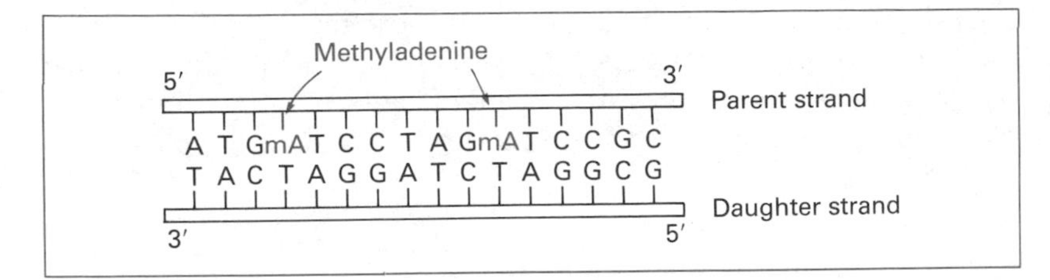

Figure 12.18 In *E. coli* the parental strand of a DNA molecule is tagged with methyl groups attached to adenines within sequences such as 5′-GATC-3′.

(DNA polymerase I in *E. coli*), with DNA ligase sealing up the polynucleotide (Fig. 12.17(c)).

3. ***Mismatch repair*** is responsible for correcting errors introduced during DNA replication. Again, there are enzymes that can recognize that something is wrong and either mark the mismatch or repair it directly. In order to do this the repair system must be able to distinguish the parental polynucleotide from the newly-synthesized daughter strand. This is because the incorrect nucleotide will be in the daughter strand, so it is this strand that must be repaired, not the parent. In *E. coli* the distinction is possible because the parent strand is tagged with methyl groups, attached to adenine nucleotides that occur within specific sequences (e.g. within the sequence 5′-GATC-3′). These modified adenines act as labels that say 'parent strand' and enable the repair enzymes to recognize which polynucleotide should be repaired at mismatch positions (Fig. 12.18). The daughter strand will also become methylated, but not until some time after synthesis.

12.3 Recombination

Before moving on to the second major type of event that leads to alteration in the genetic material we should pause to refresh our minds and place the preceding information on mutation to one side. Recombination and mutation are not related except that both cause alterations to the genetic material.

Recombination results in a rearrangement of the genetic material of the cell. It comes in three slightly different forms:

1. **Generalized recombination**, also called **homologous recombination**, in which the two molecules that recombine have extensive sequence homology.
2. **Site-specific recombination**, which can occur between two molecules that share just a short stretch of homologous nucleotide sequence.
3. **Illegitimate recombination**, or **non-homologous recombination**, in which there is no homology between the two molecules that recombine.

Understanding recombination is one of the great intellectual challenges of molecular biology. We are fairly confident that we know, at least in outline, how generalized recombination takes place, but there are still problems in understanding the other two versions.

12.3.1 Generalized recombination

Generalized recombination requires two double-stranded DNA molecules (or two separate parts of the same molecule) that have a region of homology where the nucleotide sequences are the same or at least very similar (Fig. 12.19(a)). The molecules line up adjacent to one another and interact by exchanging portions of polynucleotide. First, single-stranded cleavages occur at equivalent

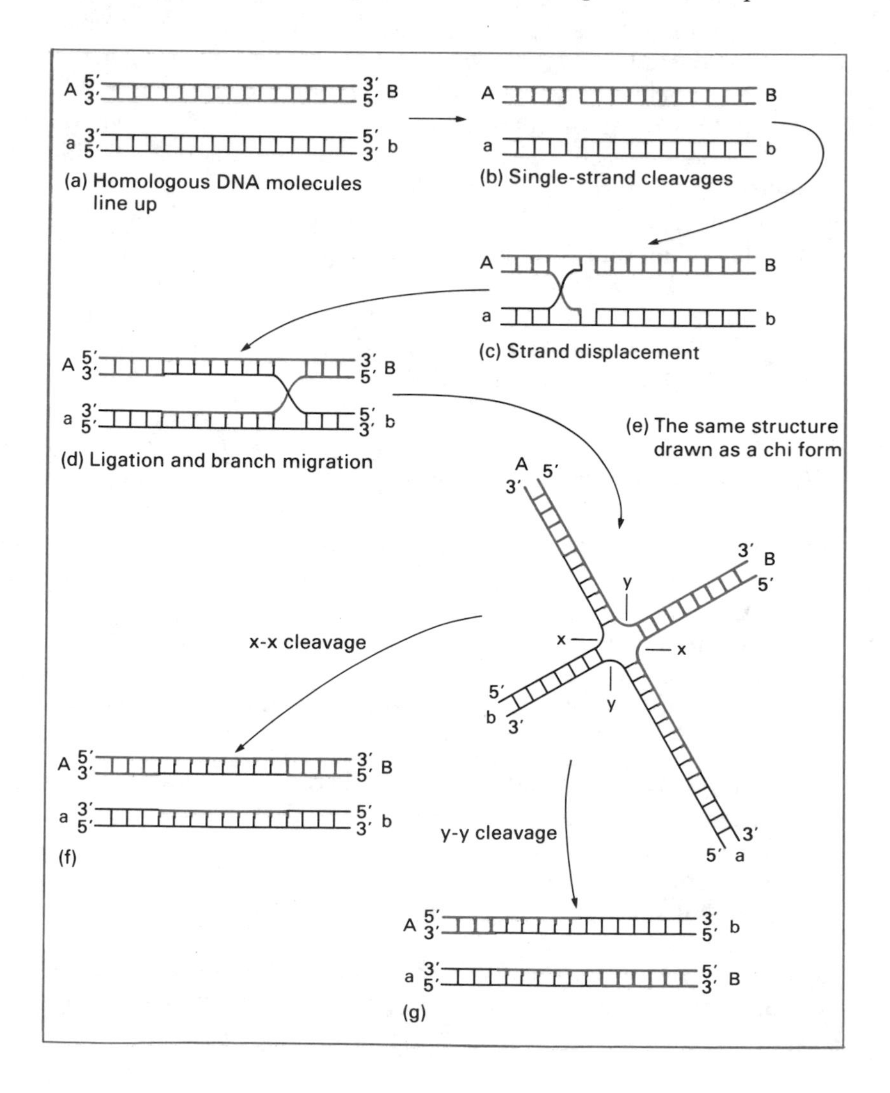

Figure 12.19 Recombination between two homologous, linear, double-stranded DNA molecules. See the text for details.

points of the identical polynucleotides by endonuclease action (Fig. 12.19(b)) and strand displacement takes place so that a crossover branch is formed, giving a **heteroduplex** in which the nicked polynucleotides are shared between the two double helices (Fig. 12.19(c)). The nicks are sealed by DNA ligase to stabilize the heteroduplex and strands are exchanged by branch migration, with the crossover point migrating along the two molecules (Fig. 12.19(d)). Remember that as the molecules are homologous the polynucleotides that are exchanged can base pair with the complementary strands of either DNA molecule. The sealed heteroduplex is called the **Holliday structure**, after Robin Holliday who first proposed its existence.

The main complication arises when the heteroduplex becomes cleaved to produce two separate molecules. The two-dimensional representation used so far in Fig. 12.19 is rather unhelpful at this stage because it obscures the true topological structure of the heteroduplex and makes the cleavage events difficult to follow. The heteroduplex shown in Fig. 12.19(d) is more correctly drawn in the **chi form** (Fig. 12.19(e)), an intermediate that has been confirmed by electron microscopic observation of recombining DNA molecules. Two different cleavage events can resolve the chi form back into separate double-stranded DNA molecules. The first of these, shown in Fig. 12.19(f), is equivalent to cleavage directly across the crossover shown in Fig. 12.19(d), so that after ligation of the cut strands two molecules that have exchanged a portion of polynucleotide result. As the exchanged strands are homologous the effect on the genetic constitution of each molecule will be relatively minor. More important is the second way of resolving the chi structure (Fig. 12.19(g)), the method that is not readily apparent from the representation shown in Fig. 12.19(d). Now two radically different DNA molecules emerge, with the end of one molecule having been exchanged for the end of the other molecule, clearly a much more dramatic rearrangement.

(a) The Meselson–Radding modification. One problem with the model just described is that it requires two single-stranded nicks to occur at adjacent positions of homologous DNA molecules. It is unlikely that these will occur at exactly the right positions by chance, making it difficult to see how the recombination process could be initiated. To deal with this problem, Meselson and Radding have suggested that a nick in one molecule may be sufficient to start the process, as the free polynucleotide that is produced might be able to 'invade' the second, double-stranded molecule at the appropriate position (Fig. 12.20). Possibly this could stimulate the cleavage needed in the invaded molecule, enabling the heteroduplex to be set up. At subsequent stages the events shown in Fig. 12.19 would be followed.

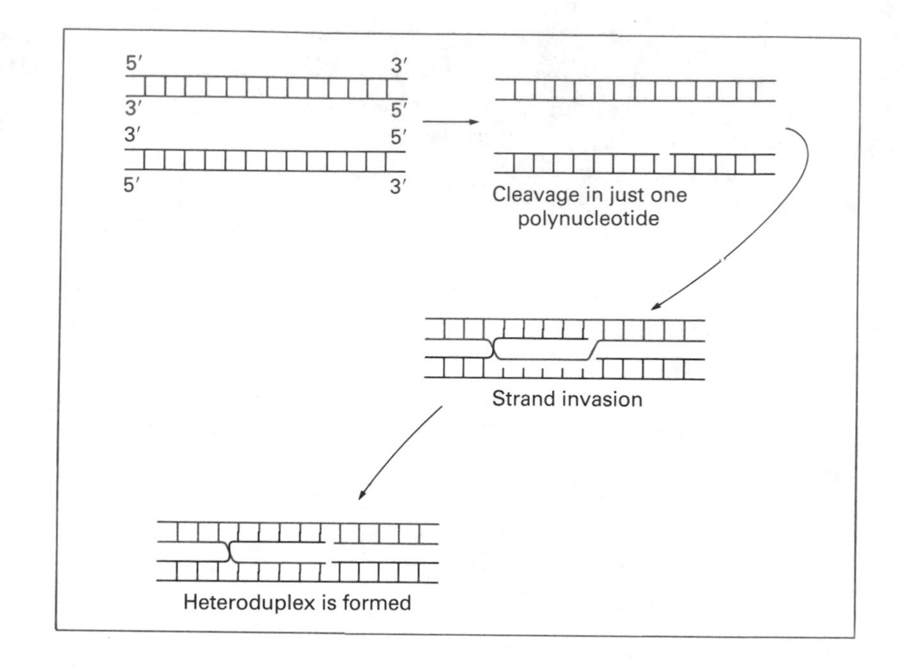

Figure 12.20 The Meselson–Radding model for the initiation of recombination.

12.3.2 The role of recombination in genetics

In the following chapters we will encounter a number of important events that result from recombination. Three should be mentioned before we progress any further.

(a) Generalized recombination results in crossing over between chromosomes. Generalized recombination can result in the exchange of genetic material between individual linear, double-stranded DNA molecules, as shown in Fig. 12.19. This type of recombination occurs between chromosomal DNA molecules in eukaryotic cells, and results in the merging of parental characteristics during sexual reproduction. We will return to it when we deal with eukaryotic cell division in Chapter 15, and when we examine gene analysis techniques in Part Three.

(b) Site-specific recombination between circular DNA molecules. Consider the effect of recombination if the two DNA molecules are not linear but circular. The result of resolution of the chi structure could now be insertion of one circle into the other (Fig. 12.21(a)). The two molecules need only have very limited regions of homology, as site-specific recombination requires just a short stretch of homologous nucleotides to set up the heteroduplex. After recombination the homologous sequences will flank the inserted element. Intramolecular recombination between these two flanking sequences could then bring about excision of the

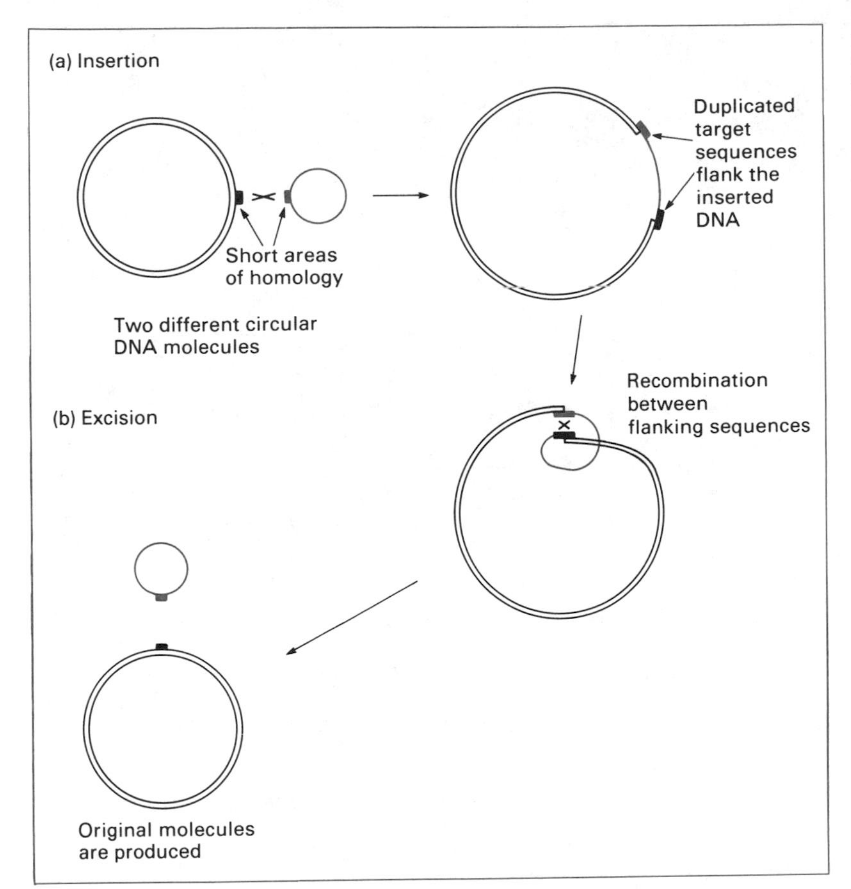

Figure 12.21 Recombination between two circular, double-stranded DNA molecules can insert one molecule into the other. A second recombination event can then produce the two original molecules again.

inserted region and recreation of the two original molecules (Fig. 12.21(b)). These events are very important in bacterial genetics and we will meet them again in Chapter 13 when we look at the life cycles of bacteriophages, and in Chapter 14 when we tackle plasmids and bacterial sex.

(c) Transposition. This is a special type of recombination event which enables a piece of DNA (usually between 750 bp and 10 kb) to move from one position to another on the same DNA molecule, or between two different DNA molecules (Fig. 12.22(a)). When in place, the **transposable element** is flanked by short repeated nucleotide sequences, and transposition is thought to involve recombination between these sequences, possibly with formation of an intermediate circular molecule. Transposition often involves illegitimate recombination, as the transposable element usually does not have any sequence homology with the site that it moves into.

Transposable elements are important in bacteria as they often

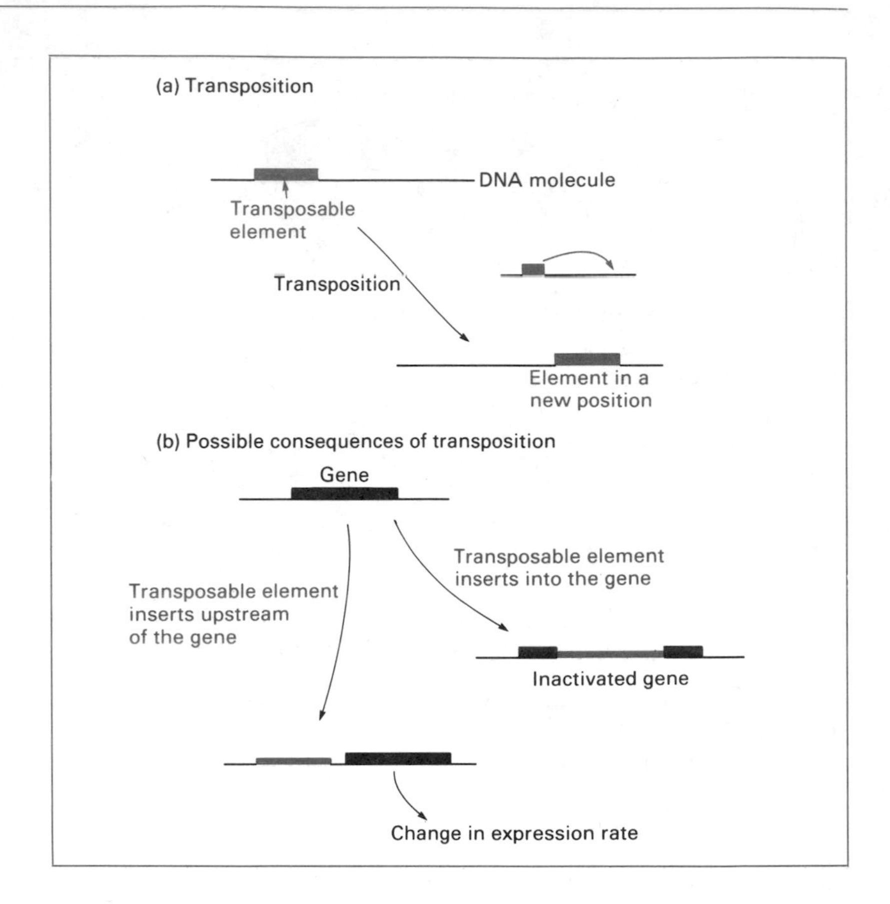

Figure 12.22 Transposition and its possible consequences.

carry genes for antibiotic resistance and may mediate spread of resistance in natural populations (see Chapter 14). In higher organisms the retroviruses (see Chapter 13), which include some viruses implicated in cancer as well as the AIDS virus, resemble transposable elements. In the laboratory, transposable elements are used as mutagens because they can inactivate genes that they transpose into, or possibly modulate the expression of a gene downstream from an insertion site (Fig. 12.22(b)), in both cases causing a phenotypic change indistinguishable from a mutation.

Reading

J. W. Drake, B. W. Glickman and L. S. Ripley (1983) Updating the theory of mutation. *American Scientist*, **71**, 621–30 – a general review of mutation.

P. Howard-Flanders (1981) Inducible repair of DNA. *Scientific American*, **245**(5), 56–64 – mechanisms for correcting mutations.

V. A. Bohr and K. Wassermann (1988) DNA repair at the level of the gene. *Trends in Biochemical Sciences*, **13**, 429–33 – describes recent research concerning repair in mammalian cells.

F. Stahl (1987) Genetic recombination. *Scientific American*, **256**(2), 53–63 – a review that makes this difficult subject a bit easier to understand.

R. Holliday (1964) A mechanism for gene conversion in fungi. *Genetical Research*, **5**, 282–304 – the central paper in recombination.

M. Meselson and J. J. Weigle (1961) Chromosome breakage accompanying genetic reconstruction in bacteriophage. *Proceedings of the National Academy of Sciences USA*, **47**, 857–68; N. Sigal and B. Alberts (1972) Genetic recombination: the nature of a crossed strand-exchange between two homologous DNA molecules. *Journal of Molecular Biology*, **71**, 789–93; M. Meselson and C. M. Radding (1975) A general model for genetic recombination. *Proceedings of the National Academy of Sciences USA*, **72**, 358–61 – three papers propounding different aspects of recombination models.

D. Leach (1987) Making Holliday junctions. *Nature*, **329**, 290–1 – a short summary of the role of recombination in integration of a bacteriophage genome into the *E. coli* chromosome.

A. I. Bukhari (1981) Models of DNA transposition. *Trends in Biochemical Sciences*, **6**, 56–60.

Problems

1. Define the following terms:

mismatch
mutation
mutagen
DNA repair
recombination
transposition
phenotype
genotype
wild-type
mutant
transition
transversion
silent mutation
lethal mutation
auxotroph
prototroph
minimal medium
temperature-sensitive mutant
constitutive mutation
back mutation
second site reversion
suppression
spontaneous mutation rate
base analogue
intercalating agent
tautomeric shift
cyclobutyl dimer
photoreactivation
heteroduplex
chi form
transposable element

2. Distinguish between the following terms: (a) point mutation, (b) insertion mutation, (c) deletion mutation, (d) inversion mutation.

3. Explain why nonsense and frameshift mutations are more likely to result in a mutant phenotype than are missense mutations.

4. Which of the following types of mutation could be caused by a point change in a nucleotide sequence: (a) missense, (b) nonsense, (c) frameshift, (d) back, (e) second site reversion, (f) suppression. Which could be caused by an insertion?

5. Why are only some mutations lethal?

6. Distinguish between the terms 'auxotrophic mutant' and 'conditional-lethal mutant'.

7. Describe three ways in which the phenotypic effect of a mutation could be reversed.

8. Distinguish between the mutagenic activities of 5-bromouracil, ethidium bromide, heat and ultraviolet radiation.

9. Describe two ways in which a thymine dimer can be corrected by DNA repair mechanisms in *E. coli*.

10. How does the mismatch repair system distinguish between the parental and daughter polynucleotides?

11. Outline the differences between generalized, site-specific and illegitimate recombination.

12. Draw the series of events involved in recombination between two linear, double-stranded DNA molecules.

13. Describe what happens during recombination between two circular, double-stranded DNA molecules.

14. *Explain why a purine to purine or pyrimidine to pyrimidine point mutation is called a transition, whereas a purine to pyrimidine (or vice versa) change is called a transversion.

15. *The example of suppression described in the text ought to result in all UGA termination codons being read through. Clearly this would result in a lot of elongated polypeptides. Would you expect this to be lethal? If not, why not? If so, then what might be the precise nature of this type of suppressive mutation?

16. *Propose a scheme for a suppressive mutation that does not involve changing a tRNA gene.

17. Here is the nucleotide sequence of a short gene:

    5′-ATGGGTCGTACGACCGGTAGTTACTGGTTCAGTTAA-3′

    Write out the amino acid sequence of the polypeptide coded by this gene. Now introduce the following mutations and describe their effects on the polypeptide: (a) a silent mutation, (b) a missense mutation, (c) a nonsense mutation, (d) a frameshift mutation.

18. Explain how point mutations could result in each of the following codon changes:

(a) a glycine codon to an alanine codon
(b) a tryptophan codon to a termination codon
(c) a cysteine codon to an arginine codon
(d) a serine codon to an isoleucine codon
(e) a serine codon (either AGU or AGC) to a threonine codon
(f) the previous mutation, back to serine but without recreating the original nucleotide sequence.

19. Describe how you would isolate a streptomycin-resistant mutant of *E. coli*.

Part Two
Genomes

In Part One we examined the structure of genes and the way in which they are expressed and replicated. Occasionally we thought about interactions between different genes, in gene regulation for example, but on the whole our emphasis has been on genes as independent physical units. Now we must correct the balance by looking at the **genomes** that exist within the different types of organism on planet Earth.

The word 'genome' is used to describe the full complement of DNA molecules possessed by an organism. It does not mean simply 'all the genes' but instead refers to 'all the genes and all the intergenic regions as well'. In the next three chapters we will look at the genomes of viruses, bacteria and eukaryotes. The emphasis will be slightly different in each chapter, reflecting the different questions that geneticists have attempted to answer with each type of organism. There will, however, be one recurring theme: the method used by the organism to replicate its genome and ensure that a complete copy is passed to each of its progeny.

Having dealt with the genomes of viruses, bacteria and eukaryotes we will turn our attention to man. Man is of course just another eukaryote but the human genome is so important to us that it warrants a chapter of its own. The key issues that we will examine are the way in which the genes and intergenic regions are organized within the human genome, and how nucleotide sequences can be used to trace the origins and evolution of the human race.

13 Viruses – the simplest forms of life

Bacteriophages – genomes – life cycles – gene organization and expression • Viruses of eukaryotes – structures and infection cycles – replication strategies – the AIDS virus – retroviruses and cancer

The viruses are the first and simplest form of life that we will investigate. In fact viruses are so simple in biological terms that the question has frequently been asked as to whether they can actually be considered to be living organisms. Doubts arise partly because viruses are constructed along lines different from all other forms of life – viruses are not cells – and partly because of the nature of the virus life cycle. Viruses are obligate parasites of the most extreme kind; they can reproduce only within a host cell and in order to express and replicate their genes viruses must subvert at least part of the host's genetic machinery to their own ends. Some viruses

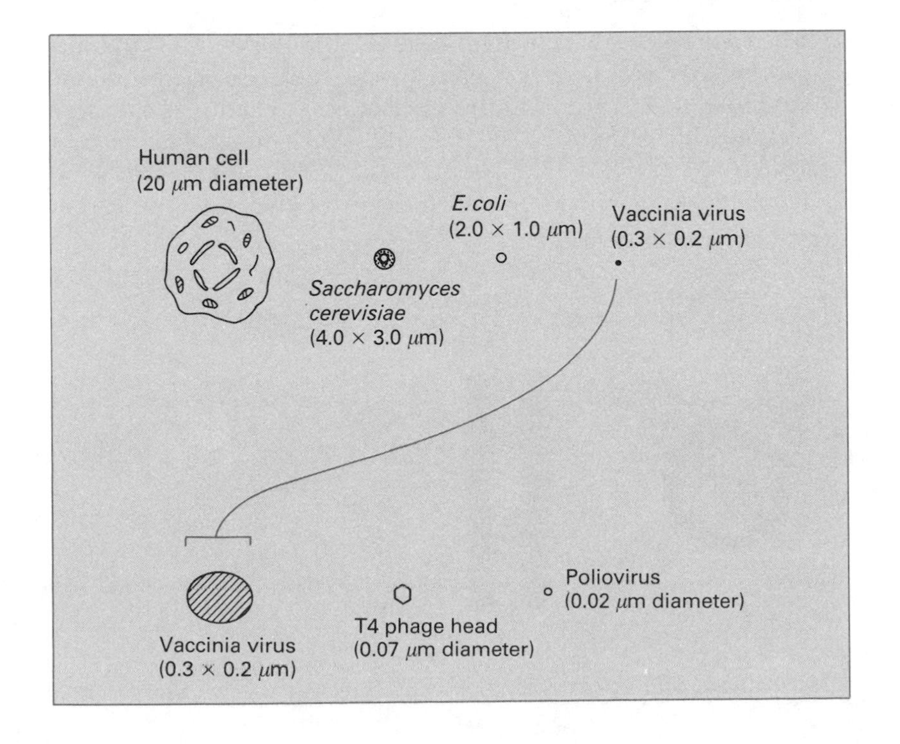

possess genes coding for their own RNA polymerase and DNA polymerase enzymes but many depend on the host enzymes for transcription and DNA replication. All viruses make use of the host's ribosomes and translation apparatus for synthesis of the polypeptides that will make up the protein coats of their progeny. This means that virus genes must be matched to the host genetic system. Viruses are therefore quite specific for particular organisms, and individual types cannot infect a broad spectrum of species.

There are a multitude of different types of virus but the ones that have received most attention are those that infect bacteria. These are called bacteriophages and have been studied in great detail since the 1930s, when the early molecular biologists, notably Max Delbrück (Section 1.4), chose phages as convenient model organisms with which to study genes. We will follow the lead taken by Delbrück and use bacteriophages as the starting point for our own investigation of viral genetics.

13.1 Bacteriophages

Bacteriophages are constructed from two basic components: protein and nucleic acid. The protein forms a coat or **capsid** within which the nucleic acid genome is contained. There are three basic capsid structures (Fig. 13.1):

1. ***Icosahedral***, in which the individual polypeptide subunits (**protomers**) are arranged into a three-dimensional geometric structure that surrounds the nucleic acid. Examples are MS2 phage, which infects *E. coli*, and PM2, which infects *Pseudomonas aeruginosa*.
2. ***Filamentous*** or ***helical***, in which the protomers are arranged in

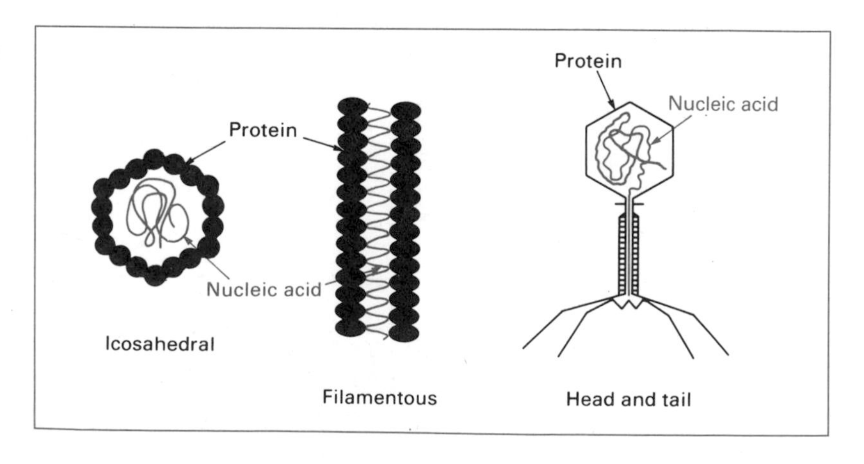

Figure 13.1 The three types of capsid structure commonly displayed by bacteriophages. The structures are shown in cross-section; in each case the DNA is enclosed within the protein capsid.

a helix producing a rod-shaped structure. The *E. coli* phage called M13 is an example.

3. ***Head-and-tail***, a combination of an icosahedral head, containing the nucleic acid, attached to a filamentous tail plus possibly additional structures that facilitate entry of the nucleic acid into the host cell. This is a common structure possessed by, for example, the *E. coli* phages T4 and λ, and phage SPO1 of *Bacillus subtilis*.

13.1.1 Bacteriophage genomes can be DNA or RNA

The term 'nucleic acid' has been used when referring to phage genomes because in some cases these molecules are made of RNA. Viruses are the one form of 'life' that contradict the conclusion of Avery and his colleagues and of Hershey and Chase that the genetic material is DNA (see Chapter 2). Viruses, including phages, also break another rule: their genomes, whether of DNA or RNA, can be single-stranded as well as double-stranded. A whole range of different structures are possible, as summarized in Table 13.1. With most types of phage there is a single DNA or RNA molecule that comprises the entire genome. However, this is not always the case and a few RNA phages have **segmented genomes**, meaning that their genes are carried by a number

Table 13.1 Features of some typical bacteriophages

Phage	*Host*	*Shape*	*Genome type**	*Genome size (kb)*	*Number of genes†*
MS2	*E. coli*	Icosahedral	ss linear RNA	3.6	3
ϕX174	*E. coli*	Icosahedral	ss circular DNA	5.4	11
M13	*E. coli*	Filamentous	ss circular DNA	6.4	10
ϕ6	*Pseudomonas phaseolica*	Icosahedral	ds linear RNA	3.3, 4.6, 7.5‡	?
PM2	*Pseudomonas aeruginosa*	Icosahedral	ds linear DNA	13.5	?
T7	*E. coli*	Head + tail	ds linear DNA	39.9	55+
λ	*E. coli*	Head + tail	ds linear DNA	49.5	48
SPO1	*Bacillus subtilis*	Head + tail	ds linear DNA	150	?
T2, T4, T6	*E. coli*	Head + tail	ds linear DNA	166	150+

* The form given is that in the phage capsid; some genomes exist in different forms within the host cell. Abbreviations; ss, single-stranded; ds, double-stranded.
† The numbers refer to genes with identified functions. Additional unassigned genes may also be present. λ, for example, has a total of 63 genes, but the functions of 15 of these are unknown.
‡ ϕ6 has a segmented genome; the sizes are for the three components.

of different RNA molecules. The sizes of phage genomes vary enormously (Table 13.1), from about 1.6 kb for the smallest phages to over 150 kb for large ones such as T2, T4 and T6. The number of genes also varies, from just three to over 200.

Visualizing bacteriophage infection

In research, infection of bacteria with bacteriophages is often studied on a solid agar medium. A concentrated sample of bacteria is mixed with a few bacteriophages and poured on to the agar. The plate is incubated for 18 hours or so, time for the bacteria to multiply and produce a confluent lawn of cells on the agar plate. However, the lawn will not be absolutely unbroken, as the phages in the sample will cause zones of clearing, called **plaques**. Each plaque arises from a single infected bacterium, which bursts, releasing new phages that infect the surrounding cells, which themselves burst, and so on:

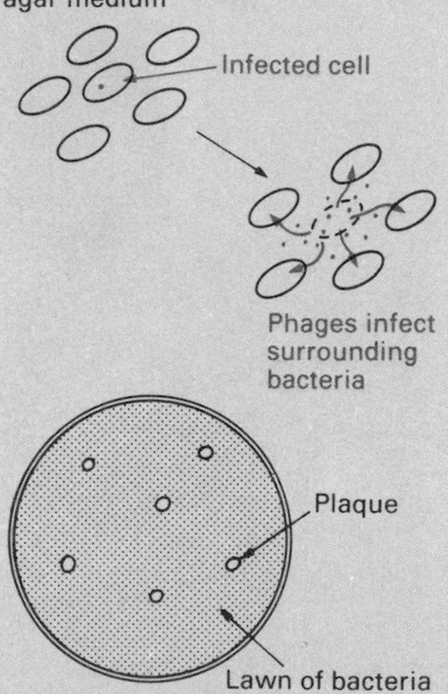

The number of plaques on the agar plate provides an estimate of the number of phages in the original sample. To the expert, the size, shape and general appearance of a plaque enables the type of bacteriophage to be identified, and the effects of mutations in its genes to be assessed.

13.1.2 Bacteriophage life cycles

Bacteriophages are classified into two groups according to their life cycle: **lytic** and **lysogenic**. The fundamental difference between these groups is that a lytic phage kills its host bacterium very soon after the initial infection, whereas a lysogenic phage can remain quiescent within its host for a substantial period of time, even throughout numerous generations of the host cell. These two life cycles are typified by two *E. coli* phages: the lytic (or **virulent**) T4 and the lysogenic (or **temperate**) λ.

(a) The lytic infection cycle of T4. The T series of *E. coli* phages (T1, T2, T3, etc.) were isolated between 1939 and 1941 by Mstislav Demerec, a Yugoslavian geneticist working in the USA. These were the first phages to become available to molecular geneticists and have been the subject of much study since. T2 for instance was used by Hershey and Chase to prove that genes are made of DNA. T4 is one of the best characterized.

The initial event in infection of *E. coli* by T4 is attachment of the phage particle to a receptor protein on the exterior of the bacterium (Fig. 13.2(a)). Different types of phage have different receptors: for T4 the receptor is a protein called OmpC ('Omp' stands for 'outer membrane protein'), which is a type of porin, the proteins that form channels through the outer cell membrane and facilitate the uptake of nutrients. After attachment, the phage injects its DNA genome into the cell through its tail structure (Fig. 13.2(b)).

The next 22 minutes are called the **latent period** when, to the outside observer, nothing very much seems to happen. Inside the cell, however, all is activity directed at synthesis of new phage particles. Immediately after entry of the phage DNA, the synthesis of host DNA, RNA and protein stops and transcription of the phage genes begins (Fig. 13.2(c)). Within 5 minutes the bacterial DNA molecule has depolymerized and the resulting nucleotides are being utilized in replication of the T4 genome (Fig. 13.2(d)). After 12 minutes new phage capsid proteins start to appear and the first complete phage particles are assembled (Fig. 13.2(e)). Finally, at the end of the latent period, the cell bursts and the new phage particles are released. A typical infection cycle produces 200 to 300 T4 phages per cell, all of which can go on to infect other bacteria.

(b) The lysogenic life cycle of λ. λ and most other temperate phages can follow a lytic infection cycle but they are more usually

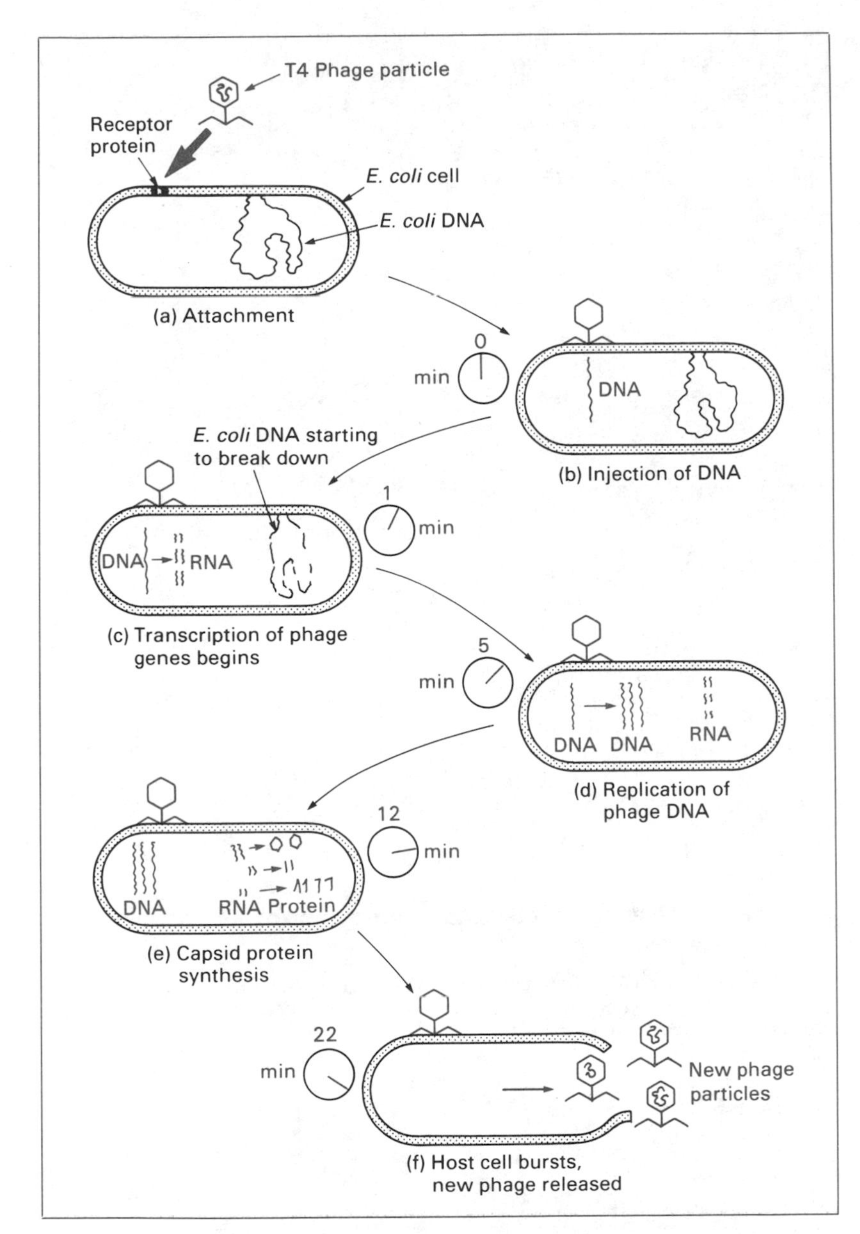

Figure 13.2 The lytic infection cycle of bacteriophage T4.

associated with the alternative lysogenic cycle. The distinction is that during a lysogenic cycle the phage genome becomes integrated into the host DNA. This occurs immediately after entry of the phage DNA into the cell, and results in a quiescent form of the bacteriophage, called the **prophage** (Fig. 13.3(a)). Integration is an example of one of the guises of recombination (Section 12.3). The λ DNA molecule and the *E. coli* genome share a short region of nucleotide homology, only 15 bp in length, but sufficient for the

The one-step growth curve

The T4 infection cycle was first studied by Emory Ellis and Max Delbrück in 1939. They added phages to a culture of *E. coli*, waited 3 minutes for the phages to attach to the bacteria, and then measured the number of infected cells over a period of 60 minutes. The results were as follows:

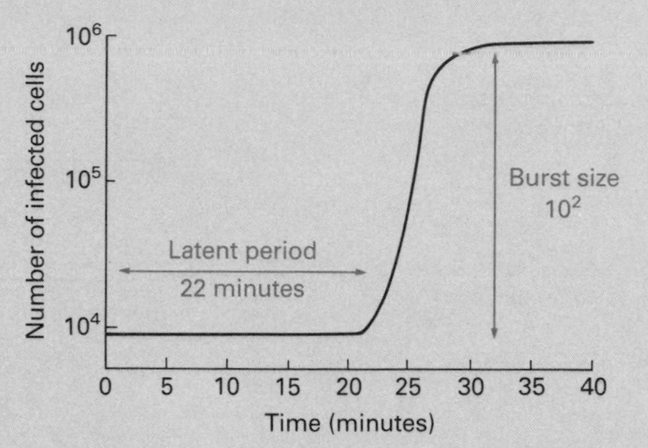

The graph shows that there was no change in the number of infected cells during the first 22 minutes of the experiment. Ellis and Delbrück interpreted this as the time needed for the phages to reproduce within their hosts. After 22 minutes the number of infected cells started to increase, showing that lysis of the original hosts had occurred, and that the new phages that had been produced were now infecting other cells in the culture.

The one-step growth curve provides a measure of the latent period and of the burst size, the latter being the number of phage produced per infection. In the experiment shown here the latent period is 22 minutes, and the burst size about 100 phages per cell.

Site-specific recombination leads to integration of the λ genome into *E. coli* DNA

The site-specific recombination event that leads to integration of the λ genome into the host DNA molecule proceeds as shown. The 15 bp sequence that sets up the recombination is labelled 'O'.

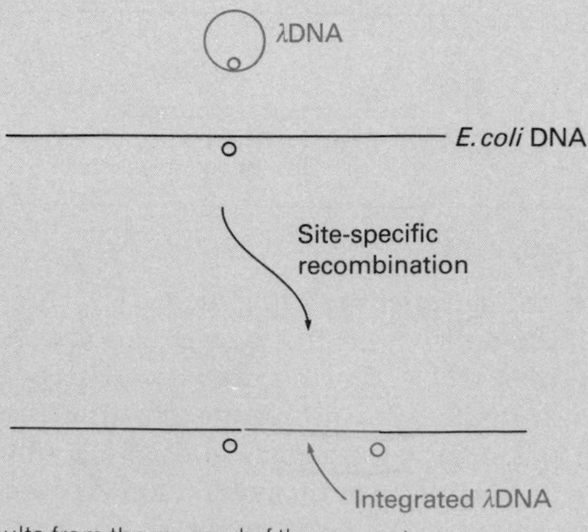

Excision results from the reversal of these events.

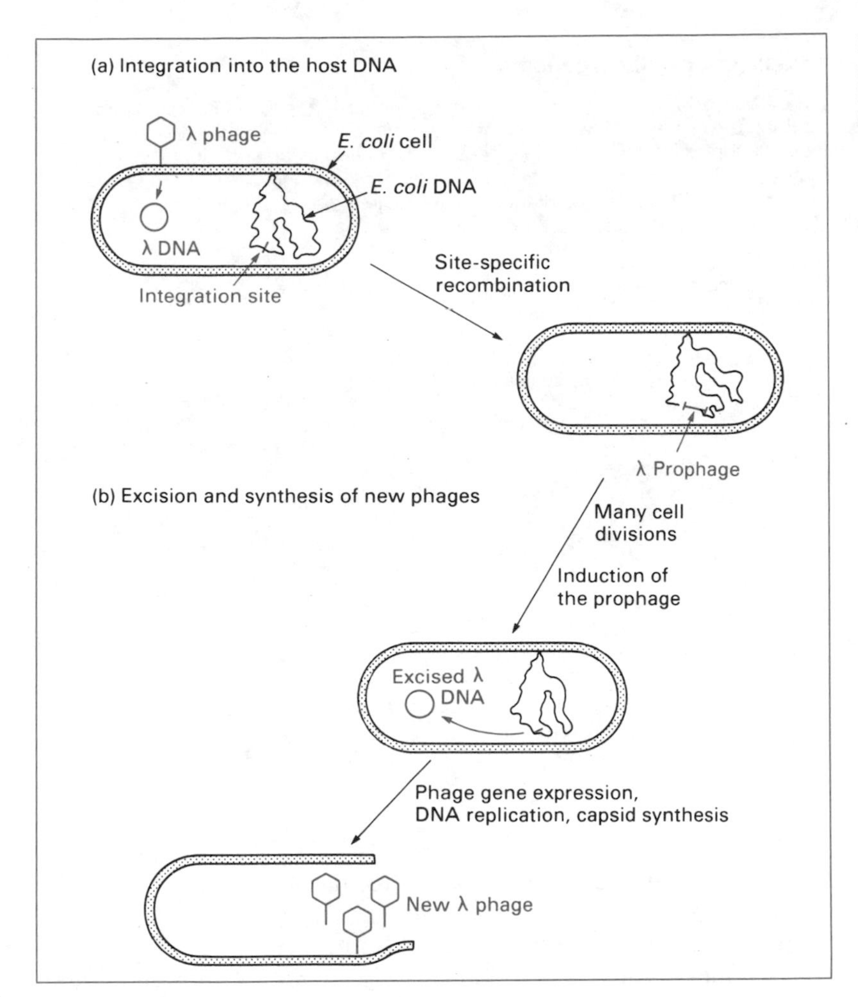

Figure 13.3 The lysogenic infection cycle of phage λ. After induction the infection cycle is similar to the lytic mode shown in Fig. 13.2 (c–f).

recombination event to be initiated. Note that this means that the λ genome always integrates at the same position on the *E. coli* DNA molecule.

The integrated prophage can be retained in the host DNA molecule for many cell generations, being replicated along with the bacterial genome and passed with it to the daughter cells. However, the switch to the lytic mode of infection will occur if the prophage is **induced** by any one of several chemical or physical stimuli. Each of these appears to be linked to DNA damage and possibly therefore signals the imminent death of the host by natural causes. In response to these stimuli, a second recombination event excises the phage genome from the host DNA, phage DNA replication begins, and phage coat proteins are synthesized (Fig. 13.3(b)). Eventually, the cell bursts and new λ phages are released. Lysogeny adds an additional level of complexity to the

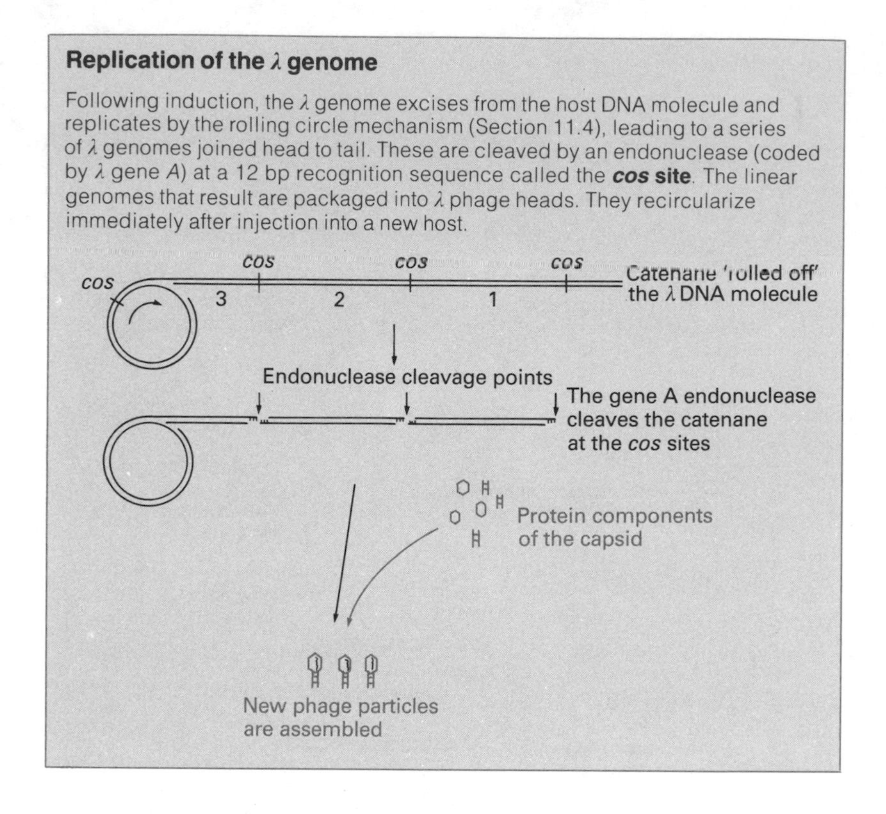

Replication of the λ genome

Following induction, the λ genome excises from the host DNA molecule and replicates by the rolling circle mechanism (Section 11.4), leading to a series of λ genomes joined head to tail. These are cleaved by an endonuclease (coded by λ gene *A*) at a 12 bp recognition sequence called the ***cos* site**. The linear genomes that result are packaged into λ phage heads. They recircularize immediately after injection into a new host.

phage life cycle, and ensures that the phage is able to adopt the particular infection strategy best suited to the prevailing conditions.

(c) Unusual phage life cycles. Although lysis and lysogeny are the two most typical phage life cycles they are not the only ones. One or two other bacteriophages display unusual infection cycles that are neither truly lytic nor truly lysogenic. An example is provided by a third *E. coli* phage, M13, which has assumed importance in recent years in recombinant DNA technology as a source of single-stranded DNA versions of cloned genes. The M13 phage particle contains a single-stranded circular genome which, after injection into the bacterium, is replicated by synthesis of the complementary strand, producing a double-stranded, circular DNA molecule (Fig. 13.4). This molecule then undergoes further replication until there are over 100 copies of it in the cell. At this stage the infection cycle takes on characteristics of both lytic and lysogenic phages. As with lytic phages, M13 coat proteins are synthesized, and new phage particles are assembled and released from the cell. However, as with lysogenic phages, cell bursting does not occur and the infected bacteria continue to grow and divide. Copies of the phage genome are passed to daughter bacteria during cell division and M13 assembly and release con-

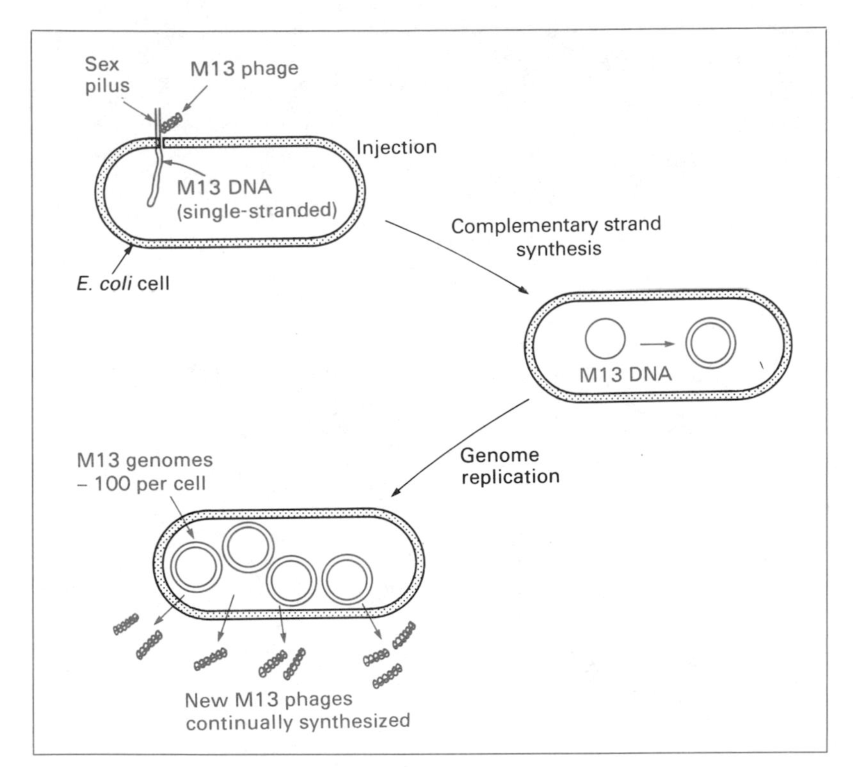

Figure 13.4 The unusual life cycle of phage M13. This bacteriophage injects its DNA into the cell by way of the structure called the sex pilus (Section 14.3.2). New phages are continually produced by infected cells.

tinues. The M13 infection cycle is therefore partly lytic and partly lysogenic.

13.1.3 Organization and expression of bacteriophage genes

We have established that there are two distinct types of bacteriophage infection cycle: lytic and lysogenic. The next question to ask is how expression of the phage genes is regulated to ensure that the lytic or lysogenic cycle proceeds according to plan. Before addressing this question we must first examine how genes are arranged in phage genomes.

(a) Organization of bacteriophage genes. Our knowledge of bacteriophage genes has benefited enormously from the development during the mid-1970s of rapid and efficient DNA sequencing methods. The complete nucleotide sequences of several phage genomes have now been worked out, showing in detail how the genes are organized. The smaller phage genomes of course contain relatively few genes. For instance, ten have been identified on the 6407 nucleotide genome of M13; these code for the phage coat proteins and for enzymes involved in replication of the phage DNA (Fig. 13.5(a)). However, a small genome may be organized in

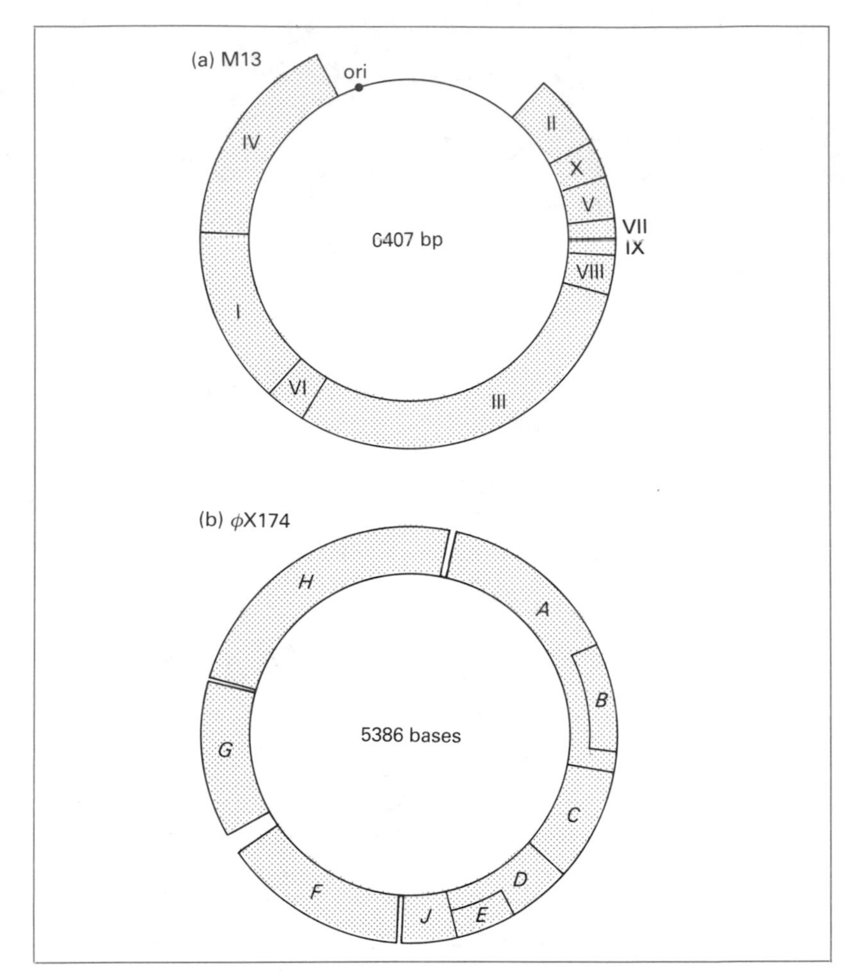

Figure 13.5 The organization of genes on the genomes of: (a) M13 and (b) ϕX174. The function of the genes are as follows. In M13: II, V, replication; III, VI, VIII, capsid proteins. In ϕX174: *A*, genome replication; *B*, *C*, *D*, capsid assembly; *E*, lysis of host; *F*, *G*, *H*, capsid proteins; *J*, injection of DNA into host. The ϕX174 map omits the positions of two additional genes (*A** and *K*) whose functions are not known.

a very complex manner. Phage ϕX174 (5386 nucleotides) manages to pack into its genome 'extra' biological information as several of its genes overlap (Fig. 13.5(b)). The **overlapping genes** share nucleotide sequences (gene *B*, for example, is contained entirely

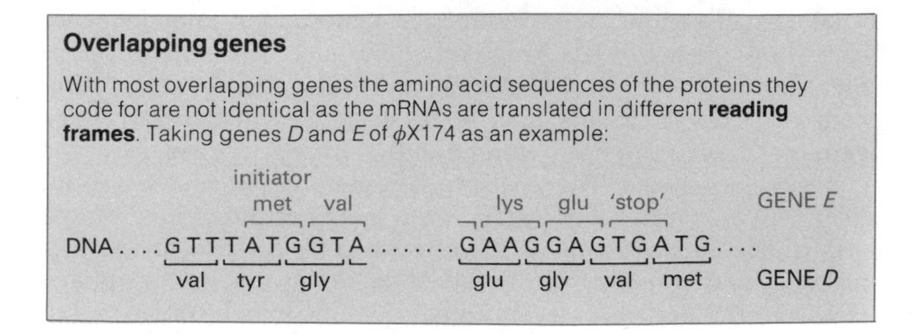

Overlapping genes

With most overlapping genes the amino acid sequences of the proteins they code for are not identical as the mRNAs are translated in different **reading frames**. Taking genes *D* and *E* of ϕX174 as an example:

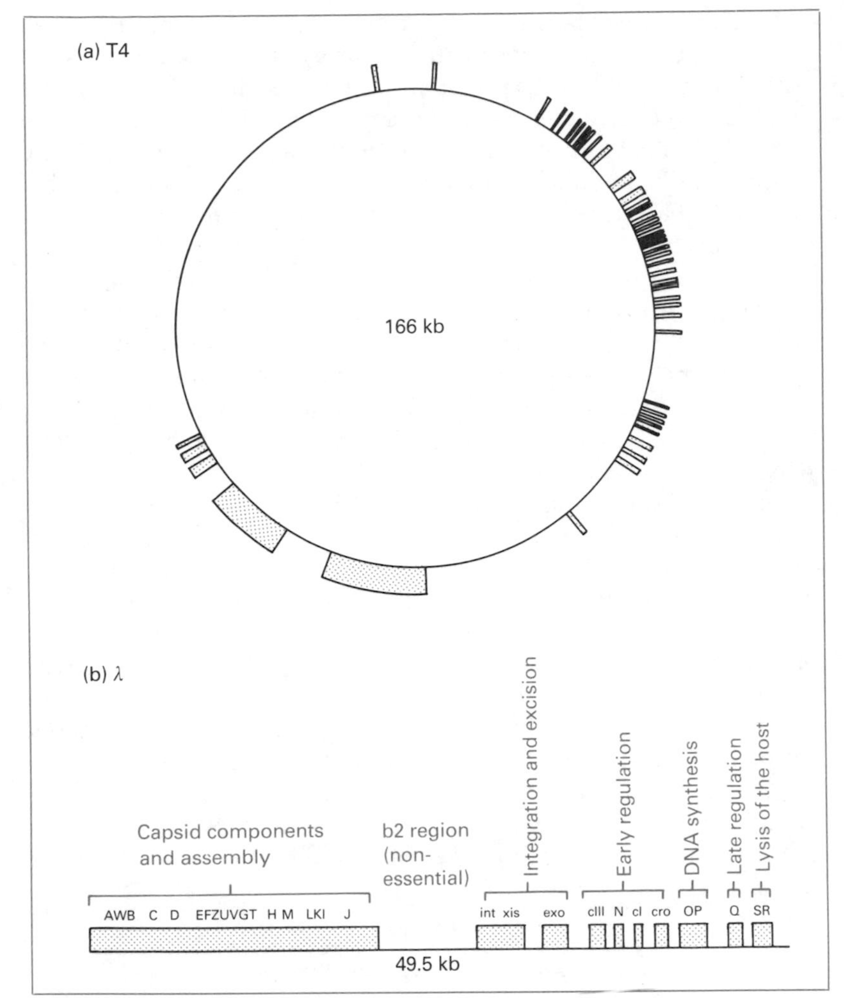

Figure 13.6 The organization of genes on the genomes of (a) T4, and (b) λ. On the T4 map only those genes coding for components of the phage capsid are shown; about 100 additional genes involved in other aspects of the phage life cycle are not marked. The λ map shows just the important gene clusters, with only a few representative genes marked.

within gene *A*), but code for different gene products as the transcripts are translated from different start positions. Overlapping genes are not uncommon in viruses.

The larger phage genomes contain more genes, reflecting the more complex capsid structures of these phages and a dependence on a greater number of phage-encoded enzymes during the infection cycle. T4 (166 kb, about 150 genes) has some 50 genes involved solely in construction of the phage capsid (Fig. 13.6(a)), and λ (49.5 kb, 48 genes with identified functions) has genes specifically for integration of the phage genome into the host DNA (Fig. 13.6(b)). Despite their complexity, even these large phages still require at least some host encoded proteins and RNAs in order to carry through their infection cycles.

(b) Gene expression during the lytic infection cycle. With all phages, but especially those with the larger genomes, the question arises as to how gene expression is regulated in order to ensure that the correct activities occur at the right time during the infection cycle. With most phages, genome replication precedes synthesis of capsid proteins. Similarly, synthesis of **lysozyme**, the enzyme that causes the bacterium to burst, must be delayed until the very end of the infection cycle.

With some phages, gene regulation is fairly straightforward. ϕX174 exerts no control over transcription of its genes: all 11 genes are transcribed by the host RNA polymerase as soon as the phage DNA enters the cell. Genome replication and capsid synthesis occur more or less at the same time, but lysozyme synthesis is delayed as the mRNA for this enzyme is translated slowly. Exactly why is not known.

With most other phages there are distinct phases of gene expression. Traditionally, two sets of genes are recognized: **early genes**, whose products are needed during the early stages of infection, and **late genes**, which remain inactive until towards the end of the cycle. A number of strategies are employed by bacteriophages to ensure that the early and late genes are expressed in the correct order. Many of these schemes utilize a **cascade system**, meaning that the appearance in the cell of the translation products from one set of genes switches on transcription of the second set of genes (Fig. 13.7). With T4, for example, the very first genes to be expressed are transcribed by the *E. coli* RNA polymerase from a few standard *E. coli* promoter sequences (Section 5.4.1) present on

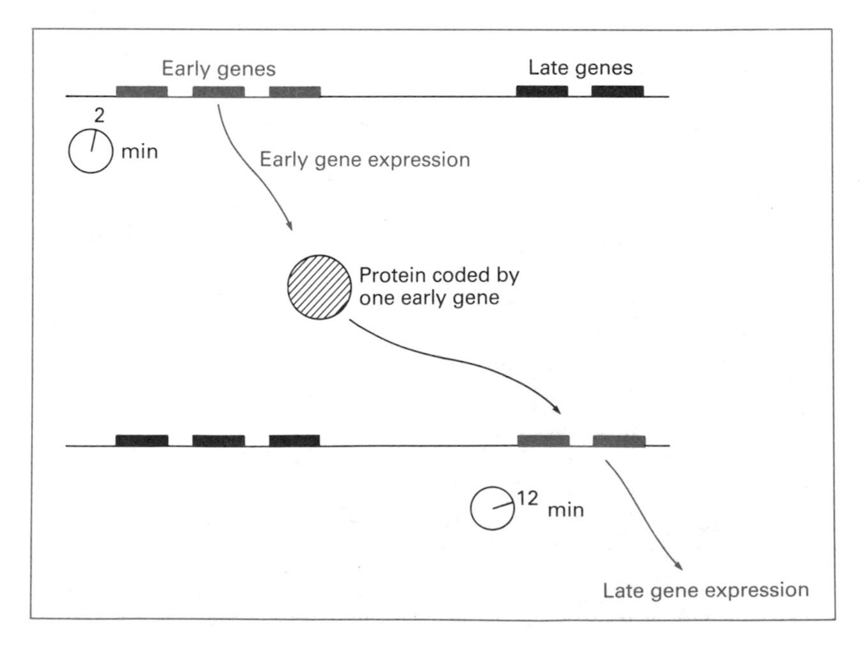

Figure 13.7 A cascade system to ensure the correct timing of early and late gene expression.

the phage genome. The very early gene products include proteins that modify the host RNA polymerase so it no longer recognizes *E. coli* promoters (thereby switching off host gene expression) but now specifically transcribes a second set of phage genes. The products of this second set of genes include proteins that modify the RNA polymerase a second time, so it now transcribes another new set of phage genes. The individual gene sets are therefore expressed in the correct order, the products of one set switching on expression of the next set.

(c) Gene expression during the lysogenic infection cycle. The ability of bacteriophages such as λ to follow a lysogenic infection cycle raises three questions: how does the phage 'decide' whether to follow the lytic or lysogenic cycle – how is lysogeny maintained – and how is the prophage induced to break lysogeny? A considerable amount is known about gene expression during λ infection, so much so that very detailed and complex answers can be given for these three questions. We will attempt to distinguish the wood from the trees.

1. **Lysogeny is maintained by the cI repressor.** Question number 2 must be dealt with before numbers 1 and 3. The first step in the lytic infection cycle is expression of the early λ genes. These are transcribed from two promoters, P_L and P_R, located either side of a regulatory gene called *c*I (Fig. 13.8). During lysogeny, P_L and P_R are switched off because the *c*I gene product, which is a repressor protein (Section 10.3.1), is bound to the operators O_L and O_R. As a result, the early genes are not expressed and the phage cannot enter the lytic cycle. Lysogeny is maintained for numerous cell divisions because the *c*I gene is continuously expressed, albeit at a low level, so that the amount of cI repressor present in the cell is always enough to keep P_L and P_R

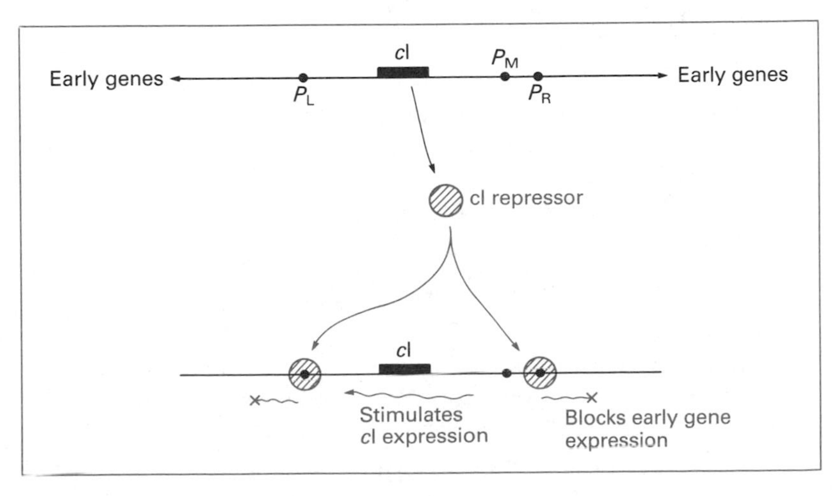

Figure 13.8 The cI repressor blocks transcription from P_L and P_R (early gene promoters) but stimulates transcription from P_M (promoter for cI).

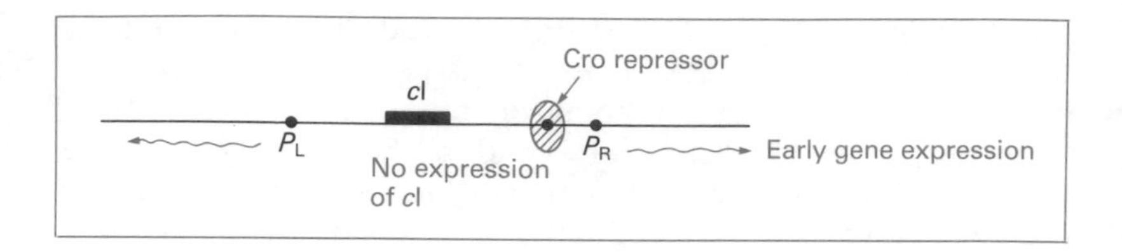

Figure 13.9 The Cro repressor blocks transcription of *c*I.

switched off. This continued expression of *c*I occurs because the cI repressor, when bound to O_R, not only blocks transcription from P_R, but also *stimulates* transcription from the promoter for the *c*I gene. The dual role of the cI repressor is therefore the key to lysogeny.

2. **The race between *c*I and *cro* decides between lysis and lysogeny.** When a λ DNA molecule enters an *E. coli* cell, the host's RNA polymerase enzymes attach to the various promoters on the molecule and start transcribing the λ genes. Once *c*I is expressed the cI repressor binds to O_L and O_R and blocks expression of the early genes, preventing entry into the lytic cycle. But λ does not always enter the lytic cycle. This is because a second gene, *cro*, also codes for a repressor, but in this case one that prevents transcription of *c*I (Fig. 13.9). Both *c*I and *cro* are expressed immediately after the λ DNA molecule enters the cell. If the cI repressor is synthesized more quickly than the Cro repressor, then early gene expression will be blocked and lysogeny will follow. However, if *cro* wins the race, then the Cro repressor will block *c*I expression before enough cI repressor has been synthesized to switch the early genes off. As a result, the phage will enter the lytic infection cycle. The decision between lysis and lysogeny therefore depends on which of the two gene products, cI and Cro, accumulates the quickest. The decision appears to be completely random, depending on chance events that lead to either the cI or the Cro repressor predominating in the cell.
3. **Lysogeny is broken by inactivation of the cI repressor.** Lysogeny will be maintained as long as the cI repressor is bound to the operators O_L and O_R. The prophage will therefore be induced if the levels of active cI repressor decline below a certain point. This may happen by chance, leading to spontaneous induction, or may occur in response to physical or chemical stimuli, as mentioned in Section 13.1.2. These stimuli activate a general protective mechanism in *E. coli*, the **SOS response**. Part of the SOS response is expression of an *E. coli* gene, *recA*, whose product inactivates the cI repressor by cleaving it in half (Fig. 13.10). This switches on expression of the early genes, enabling the phage to enter the lytic cycle. Inactivation of the cI repressor also means that transcription of *c*I is no longer stimulated,

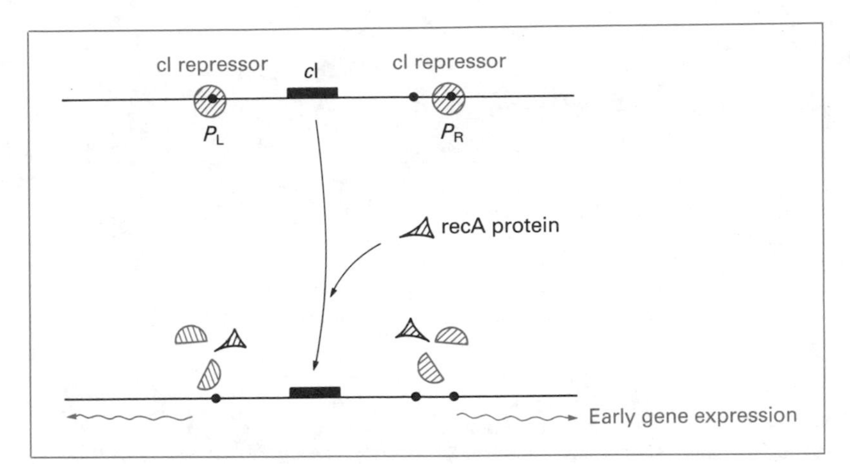

Figure 13.10 The recA protein cleaves the cI repressor, enabling the early genes to be expressed.

avoiding the possibility of lysogeny being re-established through the synthesis of more cI repressor. Inactivation of the cI repressor therefore leads to induction of the prophage.

13.2 Viruses of eukaryotes

All eukaryotes act as hosts for viruses of one kind or another. Indeed most eukaryotes are susceptible to infection by a broad range of virus types: think of the number of viral diseases that humans can catch. Because of the medical relevance, the viruses of man and animals have received most research attention, with plant viruses, capable of destroying crops, a rather distant second.

In many respects eukaryotic viruses resemble bacteriophages, but the fact that their hosts are eukaryotes rather than bacteria forces some differences upon them. Their genes have to be expressed within eukaryotic cells and so must resemble eukaryotic genes; they need the complex upstream sequences required to activate transcription by RNA polymerase II (Section 10.4.1), and they may contain introns (Section 7.4). In fact much of what we know about eukaryotic gene expression has been based on research with eukaryotic viruses.

We will not spend too much time on the basic features of eukaryotic viruses, limiting ourselves to those aspects that are novel in comparison with bacteriophages. Our emphasis will be on the roles of viruses in human diseases such as AIDS and cancer.

13.2.1 Structures and infection cycles of eukaryotic viruses

The capsids of eukaryotic viruses are either icosahedral or filamentous: the head-and-tail structure is unique to bacteriophages. One

Table 13.2 Features of some typical eukaryotic viruses

Virus	*Host*	*Type of genome**	*Size of genome (kb)*	*Number of genes*
Parvovirus	Mammals	ss linear DNA	1.6	5
Hepatitis B	Mammals	partly ds circular DNA	3.2	4
Tobacco mosaic virus	Plants	ss linear RNA	6.4	6
SV40	Monkeys	ds circular DNA	5.0	5
Influenza virus	Mammals	ss linear segmented RNA	22.0	12
Retroviruses	Mammals, birds	ss linear RNA	6.0–9.0	3
Poliovirus	Mammals	ss linear RNA	7.6	8
Reovirus	Mammals	ds linear segmented RNA	22.5	22
Adenovirus	Mammals	ds linear DNA	36.0	30
Vaccinia virus	Mammals	ds circular DNA	240	240

* The form given is that of the virus particle; several genomes exist in different forms within the host cell. ss, single-stranded; ds, double-stranded.

distinct feature of eukaryotic viruses, especially those with animal hosts, is that the capsid may be surrounded by a lipid membrane, forming an additional component to the virus structure (Fig. 13.11). This membrane is derived from the host when the new virus particle leaves the cell, and may subsequently be modified by insertion of virus-specific proteins.

Virus genomes display a great variety of structures. They may be DNA or RNA, single- or double-stranded (or partly double-stranded with single-stranded regions), linear or circular, segmented or non-segmented. For reasons that no one has ever explained, the vast majority of plant viruses have RNA genomes. Genome sizes cover approximately the same range as seen with phages, although the largest viral genomes (e.g. vaccinia virus at 240 kb) are rather larger than the largest phage genomes. Eukaryotic virus structure is summarized in Table 13.2.

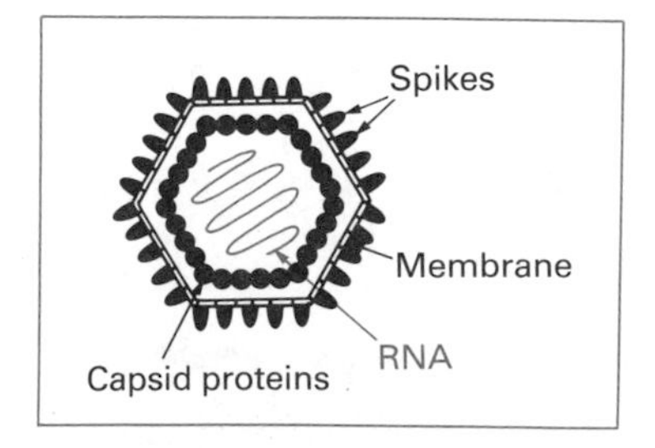

Figure 13.11 The structure of a retrovirus. The capsid is surrounded by a membrane envelope to which additional virus proteins are attached.

Most eukaryotic viruses follow only the lytic infection cycle, but few take over the host cell's genetic machinery in the way that a bacteriophage does. Many viruses coexist with their host cells for long periods, possibly years, with the host cell functions ceasing only towards the end of the infection cycle, when the virus progeny that have been stored in the cell are released (Fig. 13.12(a)). Other viruses have infection cycles similar to M13 in *E. coli*, continually synthesizing new virus particles that are extruded from the cell (Fig. 13.12(b)). These long-term infections can occur even if the viral genome does not integrate into the host DNA.

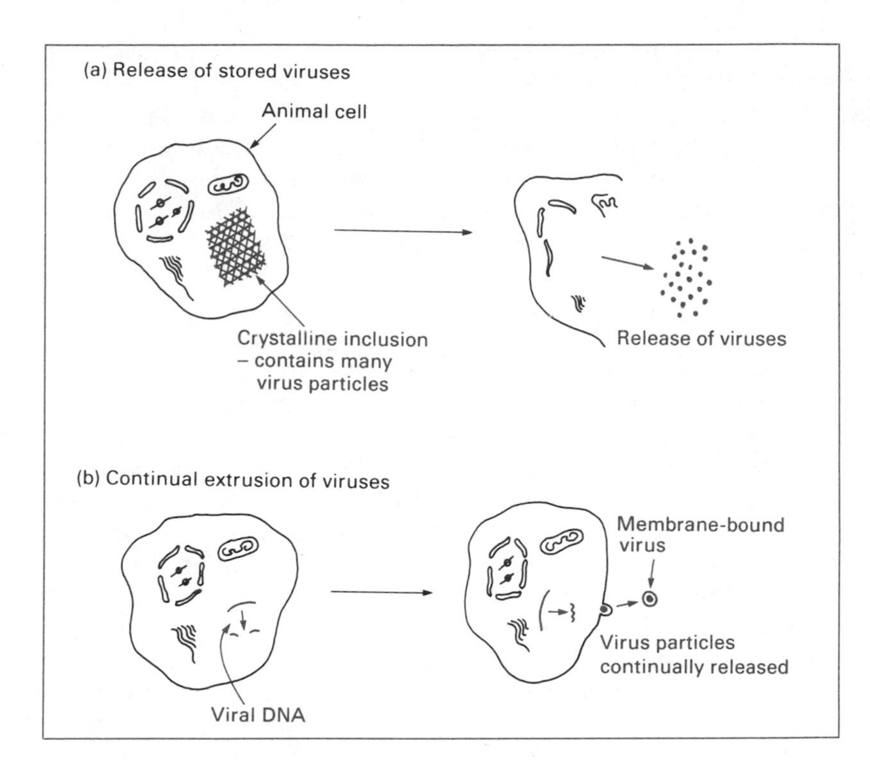

Figure 13.12 The progeny of a eukaryotic virus may be stored in the host cell (a) or released continually from the infected cell (b).

This does not mean that there are no eukaryotic equivalents to lysogenic bacteriophages. A number of DNA and RNA viruses are able to integrate into the genomes of their hosts, sometimes with drastic effects on the host cell. Renato Dulbecco in 1960 made the conceptual link between the lysogeny of *E. coli* by λ, and the ability of certain animal viruses to cause **cell transformation**. This phenomenon (not to be confused with genctic transformation, as studied by Griffith, Section 2.2.1) involves changes in cell

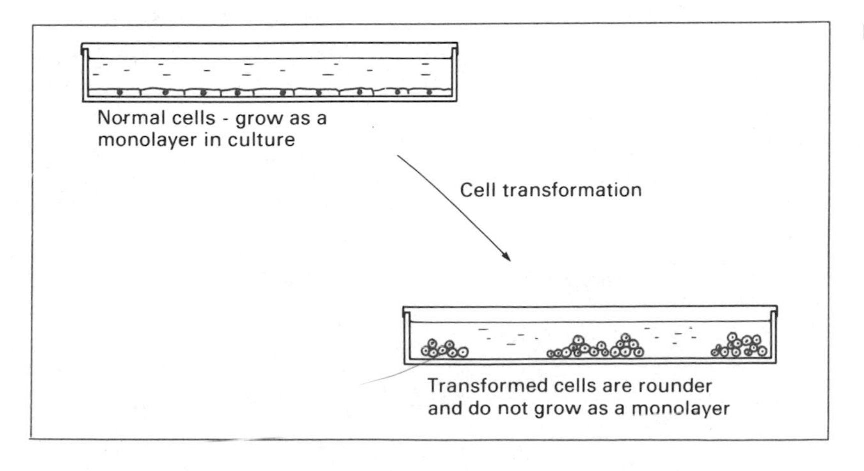

Figure 13.13 Cell transformation.

morphology and physiology. In cell cultures, transformation results in a loss of control over growth, so that transformed cells grow as a disorganized mass, rather than as a monolayer (Fig. 13.13). In whole animals, cell transformation is thought to underlie the development of tumours. This suggests that cancer may be caused by integration of eukaryotic viral genomes into host DNA. We will pursue this possibility further in Section 13.2.3.

13.2.2 Replication strategies for eukaryotic viruses

Many eukaryotic viruses, adenovirus for example, have DNA genomes that are replicated in a straightforward fashion by a DNA polymerase enzyme. Others, such as poliovirus, have RNA genomes that are replicated by an RNA polymerase. There is nothing particularly exciting about these two replication strategies.

Rather different replication strategies are used by the eukaryotic viruses called **viral retroelements**. Their replication pathways include a novel step in which an RNA version of the genome is converted into DNA. There are two kinds of viral retroelement: the **retroviruses**, whose capsids contain the RNA version of the genome, and the **pararetroviruses**, whose encapsidated genome is made of DNA. The ability to convert RNA into DNA contravenes Crick's Central Dogma (Section 4.3.1), which states that information flows from DNA to RNA to protein. With viral retroelements it is possible for information to flow from RNA to DNA (Fig. 13.14). This fact was confirmed independently in 1970 by Howard Temin of the University of Wisconsin and David Baltimore of the Massachusetts Institute of Technology. Working with cells infected with retroviruses, both Temin and Baltimore isolated an enzyme, now called **reverse transcriptase**, capable of making a DNA copy of an RNA template.

(a) The retroviral infection cycle. The typical retroviral genome is a single-stranded RNA molecule, between 6000 and 9000 nucleotides in length. After entry into the cell the genome is copied into double-stranded DNA by a few molecules of reverse transcriptase

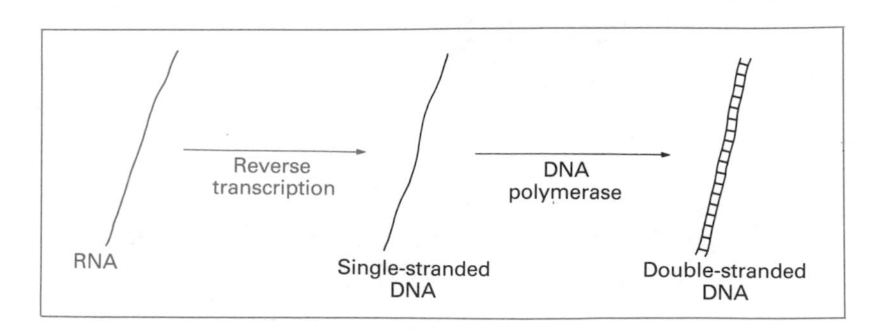

Figure 13.14 Reverse transcription of an RNA template into DNA.

Howard Temin

David Baltimore

Howard Temin b. 10 December 1934, Philadelphia, and **David Baltimore** b. 7 March 1938, New York

Howard Temin and David Baltimore, who independently discovered reverse transcriptase in 1970, are both protégés of Renato Dulbecco, one of the major influences in animal virology. Temin worked with Dulbecco at the California Institute of Technology in the late 1950s, Baltimore with him at the Salk Institute, San Diego, a few years later. This was the period when the possibility that cell transformation may be linked to integration of a viral genome into chromosomal DNA was first being talked about. Temin and Baltimore both attended the same undergraduate college, Swarthmore in Pennsylvania, Temin graduating in 1955 and Baltimore in 1960. Temin has spent most of his career at the University of Wisconsin at Madison; Baltimore has spent his at the Massachusetts Institute of Technology, leaving in 1990 to spend a year as President of the Rockefeller University. Temin, Baltimore and Dulbecco shared the 1975 Nobel Prize.

'Starting as a biochemist and later integrating genetics and cell biology into my work, I have come to realize the great power provided by joining these three streams of modern biological research.' (Baltimore)

'Scientific research is fascinating because of never-ending surprises. Top quality work on any subject can suddenly lead to fundamentally important knowledge.' (Temin)

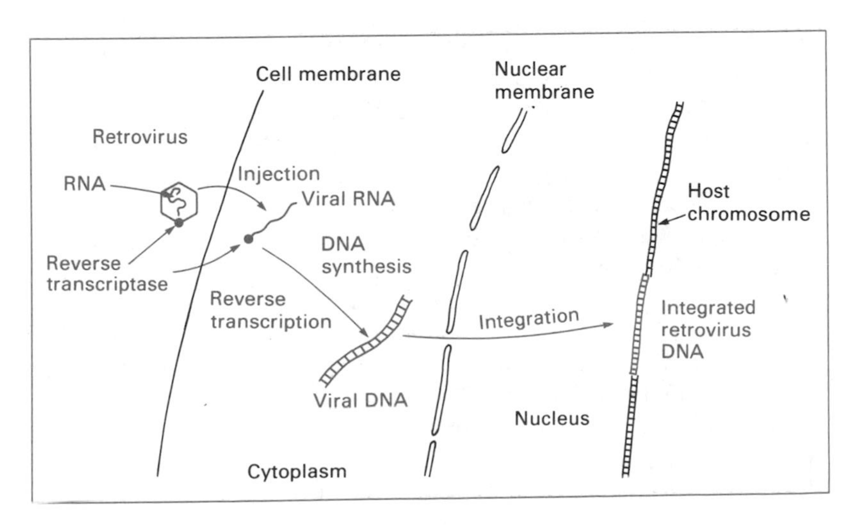

Figure 13.15 The infection cycle of a retrovirus. After injection of the viral RNA, reverse transcriptase synthesizes a DNA copy which integrates into the host genome.

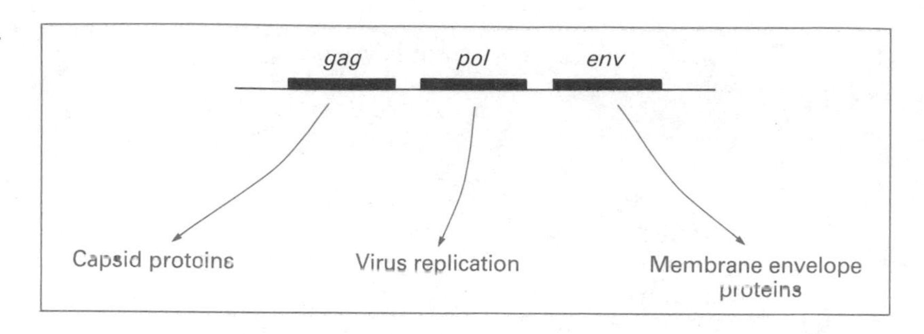

Figure 13.16 The retroviral genome.

that the virus carries in its capsid. The double-stranded version of the genome then integrates into the host DNA (Fig. 13.15). Unlike λ, retrovirus integration appears to be an example of illegitimate recombination (Section 12.3), as the viral genome has no sequence homology with its insertion site in the host DNA.

Integration of the viral genome into the host DNA is a prerequisite for expression of the retrovirus genes. There are three of these, called *gag*, *pol* and *env* (Fig. 13.16). Each codes for a **polyprotein** that is cleaved after translation into two or more functional gene products. These products include the viral coat proteins (from *env*) and the reverse transcriptase (from *pol*). The protein products combine with full-length RNA transcripts of the retroviral genome to produce new virus particles.

(b) The AIDS viruses. The causative agents of AIDS (acquired immune deficiency syndrome) were shown to be retroviruses in 1983–84. The first AIDS virus was isolated independently by two groups, led by Luc Montagnier at the Pasteur Institute, Paris, and Robert Gallo of the National Cancer Institute, Maryland. This virus is called human immunodeficiency virus, or HIV-1, and is responsible for the most prevalent and pathogenic form of AIDS. A related virus, HIV-2, discovered by Montagnier in 1985, is less widespread and causes a milder form of the disease. The HIV viruses attack certain types of lymphocyte in the bloodstream, thereby depressing the immune response of the host. These lymphocytes carry on their surfaces multiple copies of a protein called CD4, which acts as a receptor for the virus. A virus particle binds to a CD4 protein and then enters the lymphocyte after fusion between its lipid envelope and the cell membrane.

In most respects HIV-1 and HIV-2 are typical retroviruses, but they do display some unusual features. In particular the HIV genome contains a set of short genes which are not known in other viruses of this type (Fig. 13.17). The exact functions of these extra genes have not yet been delineated, but it is clear that they are involved in regulating expression of *gag*, *pol* and *env*. For instance, in the absence of the *rev* gene product the mRNAs transcribed from

Luc Montagnier b. 18 August 1932, Chabris, France

Luc Montagnier has made many important contributions to the understanding of how HIV viruses cause AIDS. He studied at the University of Paris, obtaining his MD in 1960, and in 1972 was invited by Jacques Monod to set up a Viral Oncology Unit at the Pasteur Institute. His work on mammalian retroviruses led in 1983 to isolation of a human retrovirus originally named lympadenopathy-associated virus (LAV). Between May 1983 and December 1984 Montagnier and his colleagues gradually accumulated evidence that LAV (which we now call HIV-1) is the causative agent of AIDS. The conclusion was based on many intertwined observations, including the presence of antibodies against LAV in the blood of AIDS patients, and the fact that LAV attacks T lymphocytes, the malfunction of which is one of the central features of AIDS. In 1985 Montagnier's group discovered the second AIDS virus, HIV-2. Montagnier then moved on to study the interaction between HIV-1 and T lymphocytes, and demonstrated that the virus attaches to the cells by recognizing a protein, called CD4, that is found on the surface of T lymphocytes. Montagnier is also tackling the important question of why HIV-1 infection only infrequently leads to the immediate onset of AIDS. Cofactors must be required to activate the virus, and Montagnier's most recent research suggests that small bacteria called mycoplasmas may be involved.

Robert Gallo b. 23 March 1937, Waterbury, Connecticut

Robert Gallo has devoted much of his research career to the study of retroviruses and has been responsible for many of the important discoveries concerning HIV and AIDS. He was a graduate of Providence College in 1959 and obtained his MD from Jefferson Medical College in 1963. Since 1972 he has been Chief of the Laboratory of Tumor Cell Biology at the National Cancer Institute in Maryland. In 1980 Gallo and his co-workers isolated the first known human retrovirus, which they called human T-cell leukaemia virus Type I or HLTV-I, a virus initially shown to cause some human leukaemia and later also shown to cause a fatal neurological disease. Two years later they discovered a second, related virus called HLTV-II. This was the time when AIDS was first being recognized as a new disease and Gallo deduced that the virus responsible for AIDS was probably another retrovirus related to HLTV-I and -II. He and his co-workers subsequently isolated a third human retrovirus and in a remarkable series of four papers published in *Science* in May 1984 demonstrated that this virus (which we now call HIV-1) is the causative agent of AIDS. The virus was independently isolated from a patient with lymphadenopathy by scientists at the Pasteur Institute in 1983. Gallo's group also overcame the difficult problem of how to grow the AIDS virus in cultured cells, enabling them to obtain enough virus particles to be able to produce antibodies against HIV-1. This led to blood tests that allow the HIV status of an individual to determined reliably.

> 'Medical research is rewarding. There are trying and difficult times, but in my experience from these low times some good also comes forward. All in all it is often much fun.' (Robert G. Gallo, November 1991)

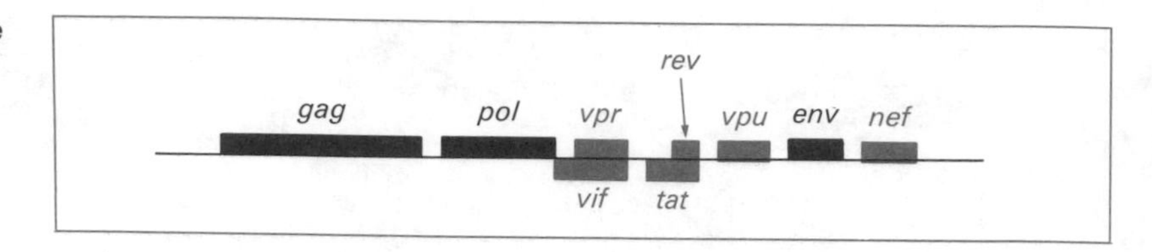

Figure 13.17 The extra genes in the HIV-1 genome.

gag, *pol* and *env* are not spliced and so the proteins coded by these genes are not synthesized. Uncovering the functions of these extra genes is a major area of current research and could lead to ways of preventing or delaying the onset of AIDS in HIV-positive individuals.

13.2.3 Retroviruses and cancer

The human immunodeficiency viruses are not the only retroviruses capable of causing diseases. Several retroviruses can also induce cell transformation, possibly leading to cancer. There appear to be two distinct ways in which this can happen. With some retroviruses, such as the leukaemia viruses, cell transformation is a natural consequence of infection, although it may be induced only after a long latent period during which the integrated provirus lies quiescent within the host genome.

Other retroviruses cause cell transformation because of abnormalities in their genome structures. These **acute transforming viruses** carry cellular genes that they have captured by some undefined process. With at least one transforming retrovirus (Rous sarcoma virus) this cellular gene is in addition to the standard retroviral genes (Fig. 13.18(a)); with others the cellular gene replaces part of the retroviral gene complement (Fig. 13.18(b)). In the latter case the retrovirus may be **defective**, meaning that it is unable to replicate and produce new viruses, as it has lost genes coding for vital replication enzymes and/or capsid proteins. These defective retroviruses are not always inactive as they can make

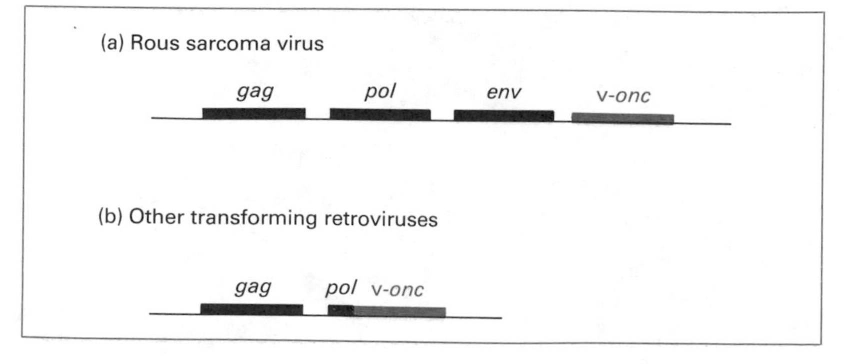

Figure 13.18 Schematic representation of the genomes of (a) Rous sarcoma virus, and (b) other acute transforming viruses. In (b) the v-*onc* gene has replaced part of *pol* and all of *env*.

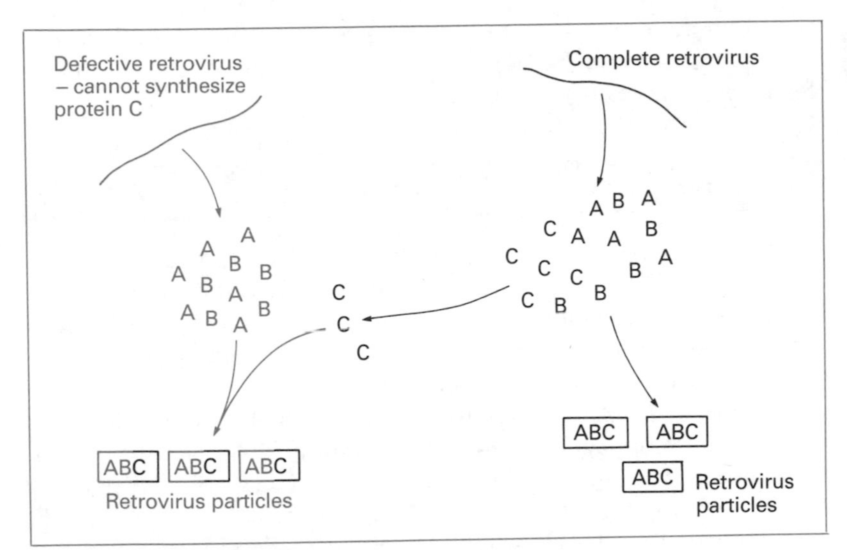

Figure 13.19 A defective retrovirus may be able to give rise to infective virus particles if it shares the cell with a non-defective retrovirus. The latter acts as a 'helper', providing the proteins that the defective virus is unable to synthesize.

use of proteins provided by other retroviruses in the same cell (Fig. 13.19).

The ability of an acute transforming virus to cause cell transformation lies with the nature of the cellular gene that has been captured. Often this captured gene (called a v-*onc*, with '*onc*' standing for **oncogene**) codes for a protein involved in cell proliferation (Table 13.3). The normal cellular version of the gene (the c-*onc*) will be subject to strict regulation, and would be expressed only in limited quantities when needed. It is thought that expression of the v-*onc* follows a different, less controlled pattern due either to changes in the gene structure or to the influence of

Table 13.3 Functions of cellular oncogenes

Oncogene	*Virus and host*	*Size of the c-*onc* (codons)*	*Function of the c-*onc
src	Rous sarcoma (hen)	533	Protein kinase – a type of cell surface receptor
sis	Simian sarcoma (monkey)	220	Growth factor
erbB	Avian erythroblastosis (hen)	1210	Cell surface receptor for epidermal growth factor
fms	Feline sarcoma (cat)	>1200	Cell surface receptor involved in phagocyte development

Prions

Prions are infectious agents that are responsible for scrapie in sheep and goats. Their transmission to cattle has led to the new disease called BSE – bovine spongiform encephalopathy. At first prions were thought to be viruses but it is now clear that they are rather different. In fact prions are quite remarkable in that the infectious particles appear not to contain any nucleic acid: they are pure protein. Genes that code for the prion protein have been identified in mammals, but the protein synthesized from these genes has important differences from infectious prion particles. For example, the cellular protein is degraded by proteinases whereas prion particles are not. Exactly where prions come from, how they cause disease, and how they reproduce, are all questions for which we currently do not have answers.

promoters within the retrovirus. One result of this altered expression pattern could be a loss of control over cell development, leading to the transformed state.

Reading

W. B. Wood and R. S. Edgar (1967) Building a bacterial virus. *Scientific American*, **217**(1), 60–74; P. J. G. Butler and A. Klug (1978) The assembly of a virus. *Scientific American*, **239**(5), 52–9 – cover various topics including the basic features of virus structure.

F. Sanger, G. M. Air, B. G. Barrell *et al.* (1977) Nucleotide sequence of bacteriophage ϕX174 DNA. *Nature*, **265**, 687–95; V. B. Reddy, B. Thimmappaya, R. Dhar *et al.* (1978) The genome of simian virus 40. *Science*, **200**, 494–502; F. Sanger, A. R. Coulson, G. F. Hong *et al.* (1982) Nucleotide sequence of bacteriophage λ DNA. *Journal of Molecular Biology*, **162**, 729–73; J. J. Dunn and F. W. Studier (1983) Complete nucleotide sequence of bacteriophage T7 DNA and the locations of T7 genetic elements. *Journal of Molecular Biology*, **166**, 477–535 – the complete sequences and gene organizations of four viral genomes.

E. L. Ellis and M. Delbrück (1939) The growth of bacteriophage. *Journal of General Physiology*, **22**, 365–83; M. Delbrück (1940) The growth of bacteriophage and lysis of the host. *Journal of General Physiology*, **23**, 643–60; A. H. Doermann (1952) The intracellular growth of bacteriophage. *Journal of General Physiology*, **35**, 645–56 – some of the early papers that delineated the phage infection cycle.

A. Lwoff (1953) Lysogeny. *Bacteriological Reviews*, **17**, 269–337 – a masterful review.

A. M. Campbell (1976) How viruses insert their DNA into the DNA of the host cell. *Scientific American*, **235**(6), 102–13 – a review on genome integration during virus infection.

P. Tiollais and M.-A. Buendia (1991) Hepatitis B virus. *Scientific American*, **264**(4), 48–54 – deals with a number of additional aspects of virology.

H. M. Temin (1972) RNA-directed DNA synthesis. *Scientific American*, **226**(1), 24–31; H. Varmus (1987) Reverse transcription. *Scientific American*, **257**(3), 48–54 – Temin's article was written when reverse transcriptase was first discovered; Varmus's gives a more up-to-date account.

H. E. Varmus (1982) Form and function of retroviral proviruses. *Science*, **216**, 812–20 – a useful review.

L. Montagnier (1991) AIDS pathogenesis. *The Biochemist*, **13**(4), 3–7 – this is an excellent review that is worth searching for. Ask a

member of the Biochemical Society for a copy.
T. Hunter (1985) Oncogenes and growth control. *Trends in Biochemical Sciences*, **10**, 275–80 – a little out of date but still a good account.

Problems

1. Define the following terms:

capsid	cascade system
protomer	SOS response
segmented genome	cell transformation
lytic phage	viral retroelement
lysogenic phage	reverse transcriptase
latent period	polyprotein
prophage	acute transforming virus
overlapping genes	defective virus
lysozyme	oncogene

2. Describe the three different capsid structures seen with bacteriophages.
3. Describe the different types of bacteriophage genome.
4. Distinguish between the infection cycles of lytic and lysogenic bacteriophages.
5. Explain in what respects the M13 infection cycle is lytic and in what respects it is lysogenic.
6. Describe what is meant by the terms 'early genes' and 'late genes'. Outline a typical strategy by which a phage ensures that its genes are expressed at the correct time.
7. Explain how λ makes the decision to enter lysogeny and how the state is maintained.
8. Outline the link between the *E. coli* SOS response and induction of a λ prophage.
9. Explain in what ways the structures of eukaryotic viruses are: (a) similar to, and (b) distinct from the structures of bacteriophages.
10. Distinguish between a retrovirus and a pararetrovirus.
11. Explain how retroviruses contravene the Central Dogma.
12. Describe the unusual features of the HIV viruses.
13. What is the distinctive feature of an acute transforming virus and how does this relate to its ability to cause cell transformation?

14. *To what extent can viruses be considered a form of life?
15. *Bacteriophages with small genomes (for example, ϕX174) are able to replicate very successfully in their hosts. Why then should other phages, such as T4, have large and complicated genomes?
16. *Propose a hypothesis to account for the ability of phages such as T4 to modify the host RNA polymerase so that it no longer recognizes *E. coli* promoters, but transcribes phage genes instead.

Prokaryotic genomes 14

The bacterial nucleoid • Prokaryotic genes – eubacteria – archaea • Plasmids – types – bacterial sex – copy number and incompatibility • Bacterial transposons

The distinction between prokaryotes and eukaryotes is based on the internal structure of the cell. As described in Section 5.1, eukaryotic cells have a complex internal architecture, with membranous organelles and a separate nucleus to house the genetic material. Prokaryotic cells lack this internal architecture and have no distinct nuclear compartment (see Fig. 5.1).

Until 1977 it was thought that all prokaryotes were more or less the same. It was recognized that the eubacteria (or 'true' bacteria, including *E. coli*) are different in some respects from the cyanobacteria (or blue-green algae), and that groups such as the myxobacteria and spirochaetes are slightly different again, but these were thought to be variations on a common theme. The assumption was shattered when Carl Woese and George Fox of the University of Illinois showed that the ribosomal RNA molecules of the unusual group of organisms called the **archaebacteria** are very dissimilar from the rRNAs of *E. coli*. Since 1977 more and more differences between archaebacteria and other prokaryotes have been found, so much so that microbiologists now favour the term **archaea**, to emphasize that these organisms are distinct from bacteria. In fact, at the fundamental biochemical and genetic levels, archaea are as different from other prokaryotes as they are from eukaryotes.

Despite the diversity among the prokaryotes, this chapter is, of necessity, weighted towards *E. coli*. This is because we know much more about *E. coli* and other eubacteria than we do about the other groups of prokaryotes. We will refer to the archaea again when we look at prokaryotic genes, but elsewhere in the chapter our attention will be on *E. coli*.

14.1 The bacterial nucleoid

Although prokaryotic cells are not divided into membrane-bound compartments, their internal structure is not entirely featureless. The electron microscope reveals two distinct regions within an *E. coli* cell: a central area called the **nucleoid**, taking up about one-

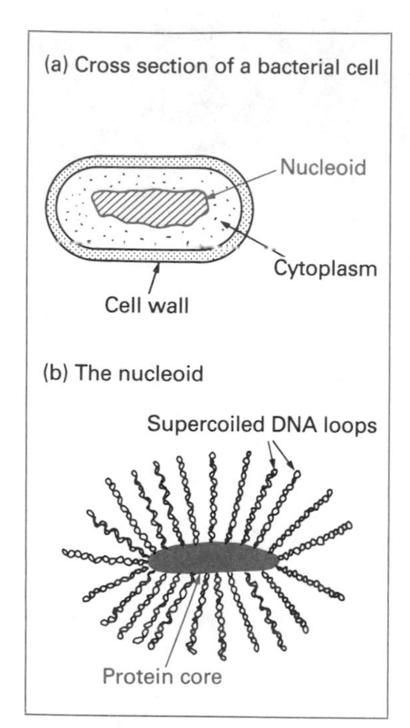

Figure 14.1 The bacterial nucleoid.

third of the volume of the cell, surrounded by a peripheral region that is usually referred to as the **cytoplasm** (Fig. 14.1(a)).

The nucleoid is made up of DNA and protein. The DNA is a single, circular molecule that carries between 99 and 100% of the bacterium's genes. In *E. coli* this DNA molecule has a contour length (i.e. circumference) of approximately 1 mm, and has to be squeezed into a cell that is about 1 μm by 2 μm. This is not impossible as a DNA molecule is very thin and so does not take up much space when it is folded up tightly by **supercoiling**. The problem is that the DNA molecule has to be accessible, as the genes it carries must be transcribed, and it must also be possible to replicate the DNA and separate the daughter molecules without the whole thing getting tangled up. This means that the DNA molecule must be folded up in a very precise way, probably with the protein component of the nucleoid helping to build and maintain the ordered structure.

The exact organization of the nucleoid is not known although it is clear that there is a central protein core from which up to 100 supercoiled loops of DNA radiate (Fig. 14.1(b)). Several DNA-binding proteins, thought to be involved in packaging the DNA

Supercoiling

Supercoiling is a common means by which the space taken up by a long double-helical molecule is compacted as much as possible. To become supercoiled, a DNA molecule must be circular or, if linear, it must be attached to proteins or other structures that prevent the free rotation of the ends of the molecule. Enzymes such as DNA topoisomerases (Section 11.2.3) can then introduce additional turns into the double helix (positive supercoiling) or remove existing turns (negative supercoiling). In each case the torsional stress that is introduced causes the molecule to wind around itself.

Breakage of one of the polynucleotides allows the two strands to rotate freely, so the torsional stress is released and the helix returns to its normal configuration.

Supercoiling is important in packaging bacterial DNA molecules into nucleoids and also underlies DNA packaging into eukaryotic chromosomes (see Chapter 15). Many plasmids (Section 14.3) are supercoiled in the cell.

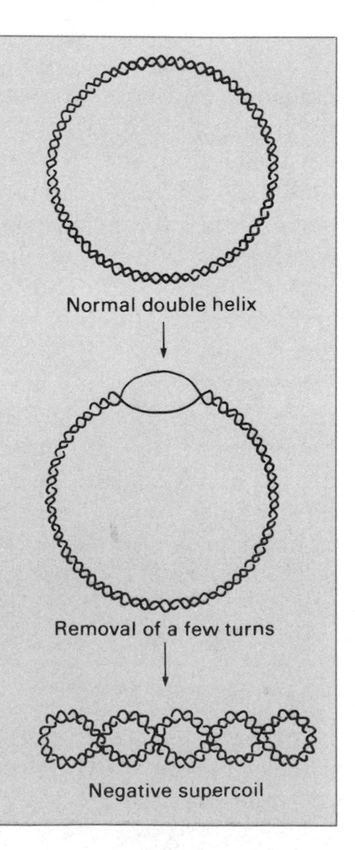

Table 14.1 Histone-like proteins of *E. coli*

Protein	*Molecular mass (daltons)*	*Copies per cell*
HU	9000	60000
HLP1	17000	20000–100000
H	28000	30000
H1	16000	20000

molecule, have been isolated from *E. coli* nucleoids but their precise roles have not yet been assigned. They may be components of the protein core or they may be responsible for maintaining the supercoiling within the DNA loops. Four of these proteins (Table 14.1) are particularly interesting as in some respects they resemble the histone proteins that are responsible for DNA packaging in eukaryotic chromosomes (Section 15.2.1).

With such a great length of DNA in such a small volume, replicating the molecule and partitioning the daughters during cell division is clearly going to be a difficult task. Very little is known

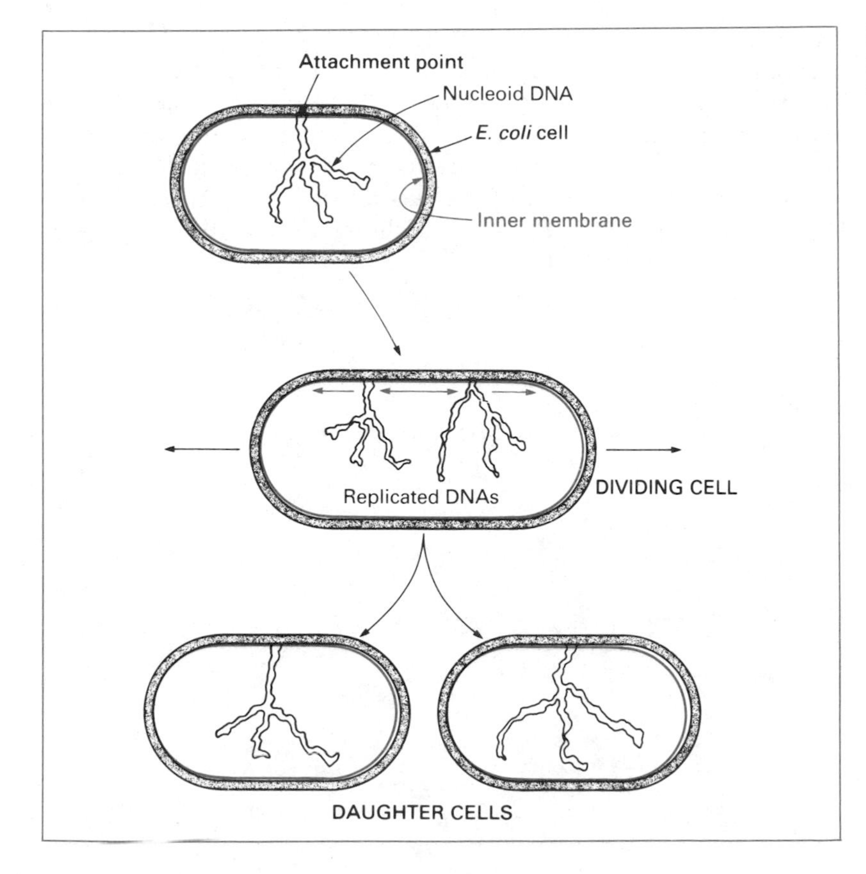

Figure 14.2 Membrane attachment points ensure nucleoid partitioning during bacterial cell division.

about how the process works. The best suggestion remains the one made in 1963 by François Jacob and Sydney Brenner, who proposed that each daughter molecule has an attachment point on the cell membrane, and that partitioning at cell division occurs by movement of the attachment points away from each other (Fig. 14.2). Unfortunately, this hypothesis throws no light on the important question of how the daughter molecules avoid getting tangled up.

Bacterial 'chromosomes'

The large, circular DNA molecule that usually carries 99–100% of a bacterium's genes is sometimes called the bacterial 'chromosome'. This is acceptable shorthand although not entirely accurate. The term 'chromosome' (meaning 'coloured body') was first used by Waldeyer in 1888 to describe the densely-staining structures seen in the nuclei of eukaryotic cells. A eukaryotic chromosome consists of DNA and protein, in roughly equal amounts, and is equivalent to the bacterial nucleoid, although quite different in structure. It is therefore inaccurate to use the word 'chromosome' to refer either to the bacterial nucleoid, or to the bacterial DNA molecule. It takes longer to use the right words but makes for better understanding in the end.

14.2 Prokaryotic genes

We have already mentioned that 99–100% of a bacterium's genes are carried on the DNA molecule present in the nucleoid. How are these genes arranged?

14.2.1 Eubacterial genes

The *E. coli* DNA molecule is approximately 4000 kb in length and carries some 2800 genes. The relative positions of just over 1000 of these have been determined by a combination of gene mapping (Chapter 19) and recombinant DNA techniques such as DNA sequencing (Chapters 20 and 21). Of the mapped genes, 260 are organized into 75 different operons, with the remaining 740 genes scattered around the rest of the molecule apparently at random. Most of the genes are present as just a single copy each, the main exception being the rRNA gene cluster, of which there are seven copies (Section 6.1.2). Figure 14.3 shows the positions on the *E.*

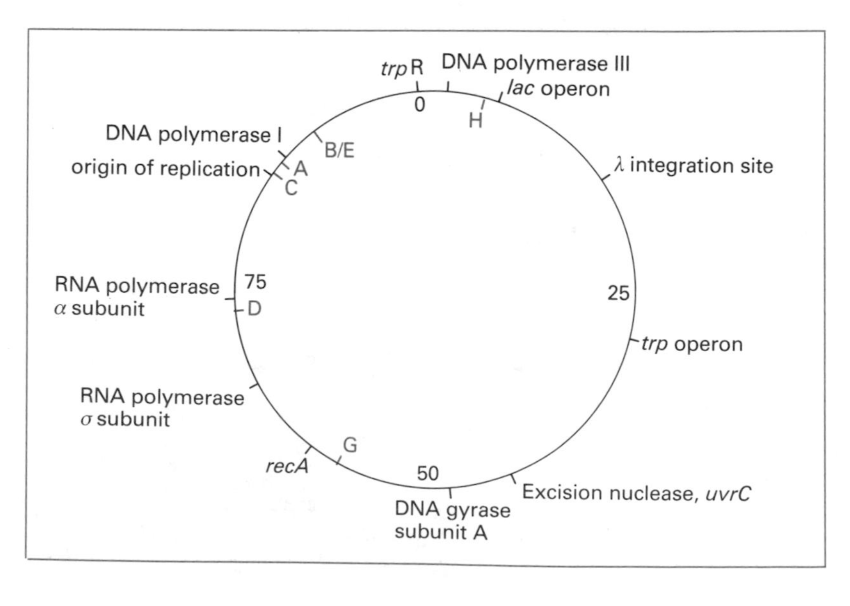

Figure 14.3 The positions of a few genes on the *E. coli* DNA molecule. The molecule is split into 100 units, as labelled on the inside of the circle. The positions of the seven rRNA operons (A, B, C, D, E, G and H) are marked in colour.

coli DNA molecule of some of the genes we have encountered in previous chapters.

The 2800 *E. coli* genes are estimated to take up approximately 75% of the DNA molecule, with the remaining 1000 kb being made up of intergenic regions between the genes. The term 'spacer' is sometimes used instead of 'intergenic region', but this is misleading as at least some intergenic regions have important functions beyond just keeping the genes apart. For example, one region contains approximately 1 kb of nucleotide sequence that acts as the single origin of replication for the *E. coli* DNA molecule. Other intergenic regions may contain sequences that interact with DNA-binding proteins to package the DNA molecule into the nucleoid.

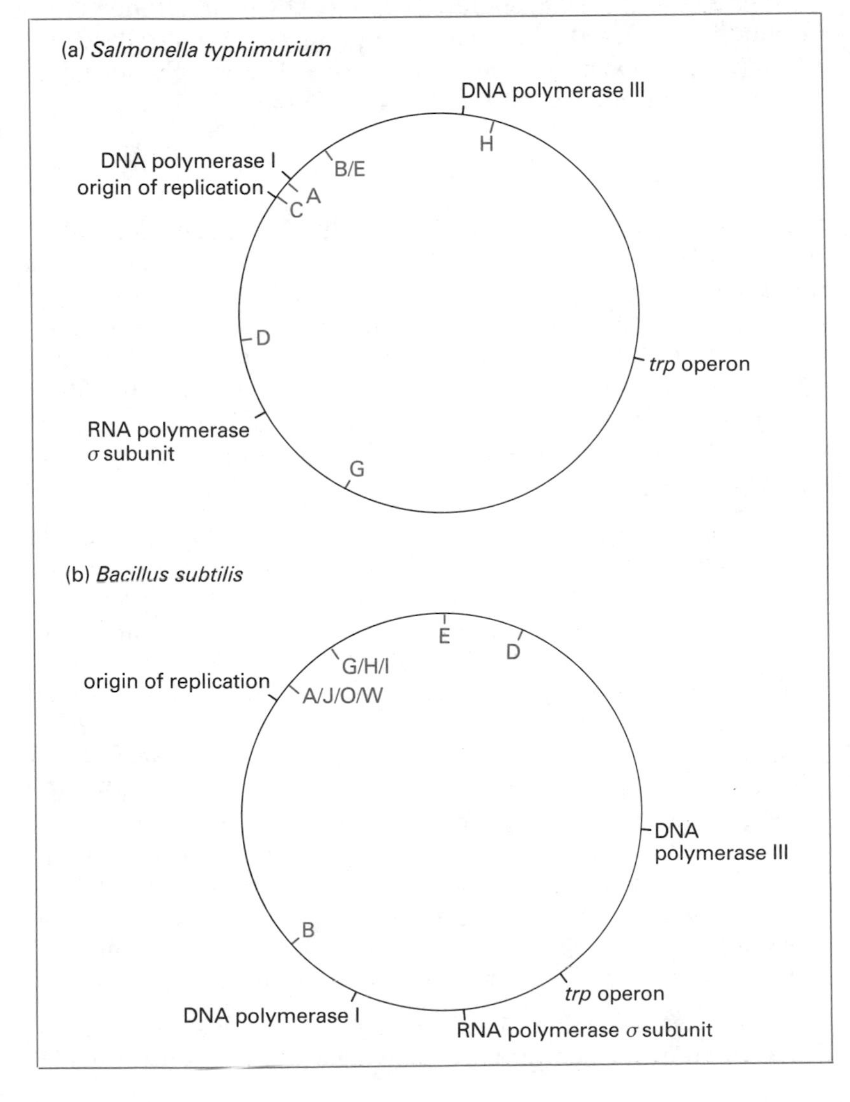

Figure 14.4 Comparisons between the gene maps for *Salmonella typhimurium* and *Bacillus subtilis*. The two maps are orientated so that the replication origins of the two molecules are in the same relative positions. The *S. typhimurium* map is virtually the same as that for *E. coli* (see Fig. 14.3). In contrast, the gene positions on the *B. subtilis* DNA molecule are quite different. The positions of the rRNA operons are shown on the inside of the circle. Note that *B. subtilis* has 10 rRNA operons, three more than either *S. typhimurium* or *E. coli*.

How typical is *E. coli* of the eubacteria as a whole? Gene mapping techniques are not as advanced with other species but some comparisons can be made. The basic features of gene organization, with numerous operons but few repeated genes, appear to be the same in all eubacteria. As might be expected, bacteria related to *E. coli* (such as *Salmonella typhimurium*) have their genes arranged in a similar order, whereas the map for a more distant species (for example *Bacillus subtilis*) is more different (Fig. 14.4). There are also variations in the sizes of the DNA molecules, from about 1200 kb for *Haemophilus influenzae* (near the bottom of the range) to over 30 000 kb for *Bacillus megaterium* (near the top). To a certain extent these size differences result from differences in the number of genes, with *B. megaterium* having extra genes that enable it to synthesize heat-resistant spores. The extra DNA cannot, however, be entirely accounted for in this way and the larger molecules must also have more extensive intergenic regions. This is a phenomenon that we will meet again with eukaryotic genomes.

14.2.2 Archaeal genes

The first indications that the archaea are very different from other prokaryotes came when their rRNA genes were compared with those of eubacteria. It was shown that the archaeal rRNAs take up a base-paired secondary structure that is significantly different from that of *E. coli* (Section 6.1.1). There is still relatively little information on archaeal genes and DNA molecules, but it is already clear that the unusual rRNAs are accompanied by a number of other distinctive features. The most important of these are as follows:

1. Some archaeal rRNA genes contain introns. Introns are absent from eubacteria, although a very limited number are known in bacteriophage genes. The archaeal introns are unrelated to eukaryotic introns (Section 7.4) and appear to fall into two separate classes.
2. The archaeal RNA polymerase is unlike the *E. coli* enzyme, and in fact resembles the eukaryotic RNA polymerases in some respects (Chapter 5). As yet the structures of promoters and transcription terminators in the archaea are not known, although stem-loops do not appear to be involved in termination.
3. Some archaeal species have only one copy of the rRNA transcription unit, and the most that has been found is four. This contrasts with the seven copies in *E. coli*. In at least one archaeal species, *Thermoplasma acidophilum*, the rRNA genes are not organized into a cluster and are transcribed individually.

There are also a number of biochemical distinctions between the archaea and eubacteria, in particular regarding the structure and

The archaea

The archaea consist of three related groups of organisms:

1. ***The methanogens***, which live in sediments at the bottom of lakes and other bodies of water, and produce methane by anaerobic respiration. Example: *Methanobacterium vannielii*.
2. ***The extreme thermophiles***, which live in hot springs and other environments where the temperatures generally exceed 60°C. Examples: *Thermofilium pendens*, *Desulphurococcus mobilis*.
3. ***The extreme halophiles***, which have an absolute requirement for sodium chloride, growing optimally at 25% NaCl. They live in brine pools and other high-salt environments such as the Dead Sea. Example: *Halobacterium halobium*.

chemical composition of the cell wall. The fact that most archaea live in extreme environments means that it is difficult to isolate and culture them, but progress is gradually being made in understanding their unique features.

14.3 Plasmids

The nucleoid DNA molecule is not the only repository for genes in the bacterial cell. Bacteria often also harbour small independent DNA molecules called **plasmids**, which carry genes not found on the main DNA molecule. It is to these plasmids that we must now turn our attention.

Plasmids are circular DNA molecules that lead an independent existence in the bacterial cell (Fig. 14.5). Plasmids almost always carry one or more genes, and often these genes are responsible for a useful characteristic. For example, the ability to survive in normally toxic concentrations of antibiotics such as chloramphenicol or ampicillin is often due to the presence in the bacterium of a plasmid that carries antibiotic-resistance genes.

All plasmids possess at least one DNA sequence that can act as an origin of replication, so they are able to multiply in the cell independently of the main DNA molecule (Fig. 14.6(a)). The

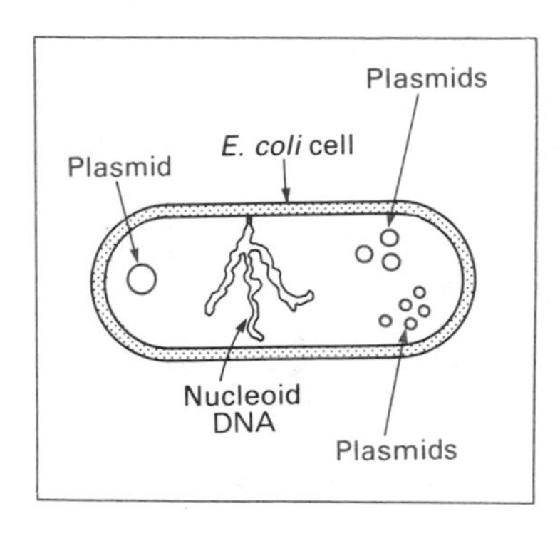

Figure 14.5 Plasmids.

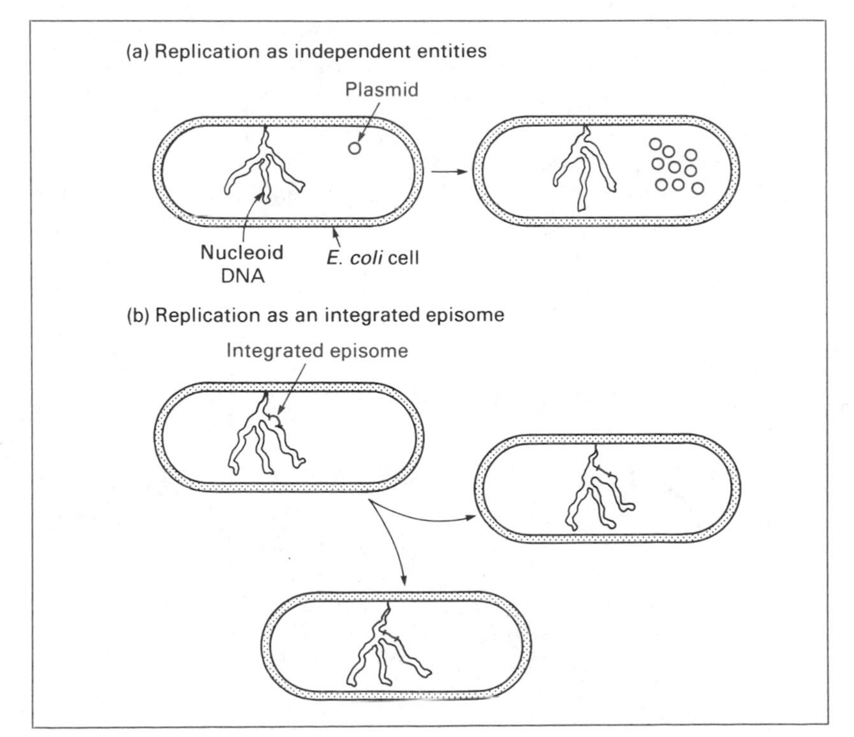

Figure 14.6 Two strategies for plasmid replication.

smaller plasmids make use of the cell's own DNA replicative enzymes to make copies of themselves, whereas some of the larger ones carry genes that code for special enzymes that are specific for plasmid replication. A few types of plasmid are also able to replicate by inserting themselves into the main DNA molecule (Fig. 14.6(b)). These integrative plasmids or **episomes** may be stably maintained in this form through numerous cell divisions, but will at some stage exist as independent elements.

14.3.1 Different types of plasmid

Virtually all species of bacteria harbour plasmids and a large number of different types are known. These are most usefully classified according to the genes that they carry and the characteristics that these genes confer on the host bacterium. According to this classification there are five main types of plasmid.

1. ***Fertility (F) plasmids*** are able to direct **conjugation** (Section 14.3.2) between different bacteria. An example is the F plasmid of *E. coli*.
2. ***Resistance (R) plasmids*** carry genes conferring on the host bacterium resistance to one or more antibacterial agents, such as chloramphenicol, ampicillin or mercury. R plasmids are very important in clinical microbiology because their spread through natural populations can have profound consequences for the treatment of bacterial infections. An example is RP4, commonly found in *Pseudomonas* but also occurring in other bacteria.
3. ***Col plasmids*** carry genes coding for colicins, proteins that kill other bacteria. An example is ColE1 of *E. coli*.
4. ***Degradative plasmids*** allow the host bacterium to metabolize unusual molecules such as toluene and salicylic acid. Several examples occur in the *Pseudomonas* genus of bacteria, for instance TOL of *Pseudomonas putida*.

Table 14.2 Properties of various plasmids

Plasmid	*Type*	*Size (kb)*	*Copy number*	*Bacterium*
F	Fertility	95	1–2	*E. coli*
Rbk	Resistance	37	12	*E. coli* + others
ColE1	Colicin	6.36	10–15	*E. coli*
TOL	Degradative	117	1–2	*Pseudomonas putida*
Ti	Virulence	213	1–2	*Agrobacterium tumefaciens*
pBR322	Cloning vector*	4.362	15	*E. coli*

* See Chapter 20.

5. ***Virulence plasmids*** confer pathogenicity on the host bacterium. The best-known example is the Ti plasmid of *Agrobacterium tumefaciens*, which induces crown gall disease in dicotyledonous plants.

The sizes of plasmids vary from approximately 1 kb for the smallest to over 250 kb (6.25% of the size of the main DNA molecule in *E. coli*) for the larger ones (Table 14.2). Some plasmids are restricted to just a few related species and are found in no other bacteria; others, such as RP4, have a broad host range and can exist in numerous species, although still displaying a preference for a few kinds of bacteria in which they will be more commonly found.

14.3.2 Plasmids and bacterial sex

Sex is always important to geneticists because it is the basis of analytical techniques that allow the relative positions of genes on DNA molecules to be determined (see Chapters 18 and 19). The discovery that bacteria have distinct sexes was therefore a major advance because it opened up a new avenue for genetic studies of prokaryotes. We will look more closely at this topic in Chapter 19

Joshua Lederberg b. 23 May 1925, Montclair, New Jersey

Joshua Lederberg's personality has maintained a towering influence over bacterial genetics since his entry into the field in 1946. His own account of his early life and the discovery of bacterial sex (*Annual Review of Genetics*, **21**, 23–46, 1987) captures his enthusiasm for research and the positive atmosphere of the post-War USA that allowed scientific ingenuity to bloom. In 1944, Lederberg was a pre-medical undergraduate at Columbia University, New York, when he heard of Avery's demonstration at the nearby Rockefeller Institute that the transforming principle is DNA. Lederberg realized that the significance of Avery's work would be lost if it could not be demonstrated that bacteria possess genes. So in 1945 he set up experiments to search for proof of bacterial genes, moving the following year to Yale, in order to work more directly with Edward Tatum, who had just completed a similar, successful analysis of fungal genes with George Beadle. The discovery of gene transfer in *E. coli* was announced at a Cold Spring Harbor meeting in July 1946, stimulating a discussion that Lederberg describes as 'lively'. The next years were spent characterizing the conjugation system and then in 1952, with Norton Zinder, Lederberg discovered a second type of gene transfer, called bacteriophage transduction (Section 19.3.1). Six years later Lederberg shared the 1958 Nobel Prize with Beadle and Tatum. With Crick, Lederberg has been responsible for stimulating, directing and interpreting research in molecular genetics for the past 45 years.

> 'Our incoming students at the Rockefeller University will continue in their own life the struggle for the advancement of new knowledge, and against disease. At the same time, they come at an age when they are still untrammelled by too many of the received wisdoms of a prior generation. They ask many new kinds of questions, and so often the most creative and the most revolutionary findings are going to be made by very young people.'

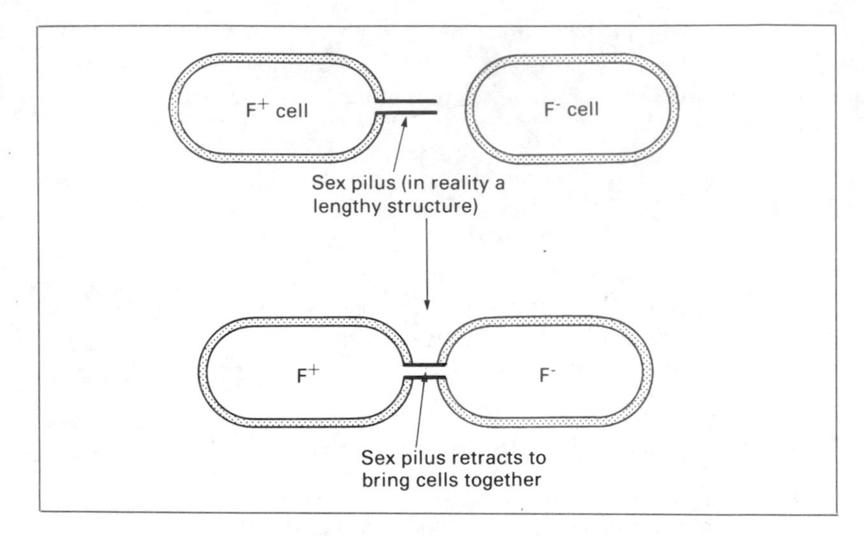

Figure 14.7 Bacterial conjugation. The sex pilus is shown forming a cytoplasmic connection between the two cells. Although this has not been confirmed, we will assume in subsequent figures that the connection exists and that DNA transfer occurs through the sex pilus.

but must at this stage examine the role of plasmids in sexual differentiation of bacteria.

(a) The discovery of bacterial conjugation. The breakthrough was announced at the Cold Spring Harbor Symposium of 1946 by Joshua Lederberg, a 21-year-old graduate student supervised by Edward Tatum at the University of Yale. They had discovered that individual cells of *E. coli* can exchange genetic material by a process that was subsequently called conjugation, and which has become the basis of genetic analysis in bacteria (Section 19.2).

This quite unexpected discovery stimulated several geneticists and microbiologists around the world to examine conjugation. During the next 10 years or so Joshua Lederberg and his wife Esther with, amongst others, William Hayes of the Hammersmith Hospital, London, the Italian Luca Cavalli-Sforza, and Elie Wollman and François Jacob of the Pasteur Institute, Paris, demonstrated that transfer of genes is unidirectional from donor cells referred to as $\mathbf{F^+}$ (F for fertility) to the recipients $\mathbf{F^-}$, and that transfer involves a tube-like structure, called the **sex pilus**, which F^+ cells are able to construct (Fig. 14.7). The exact role of the sex pilus is controversial. An F^+ cell forms a connection with an F^- cell via its sex pilus, and as the pilus is hollow it has been assumed that DNA transfer occurs through it. This has never been demonstrated, and it may be that the pilus acts only to draw the F^+ and F^- cells into close contact.

(b) The F plasmid is responsible for bacterial sex. The first indication that a plasmid is involved in bacterial sex came during the 1950s when several biologists demonstrated that under certain conditions F^+ cells could lose their ability to set up conjugation,

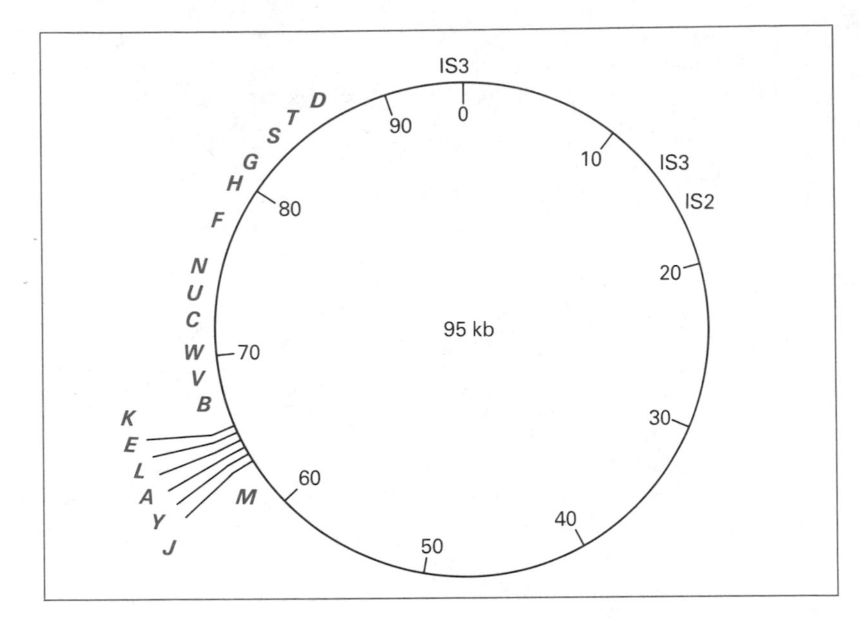

Figure 14.8 The F plasmid. The numbers inside the circle indicate the distance (kb) in a clockwise direction from the replication origin at the top. The positions of the *tra* genes, whose products are involved in DNA transfer, are marked. The F plasmid also contains three insertion sequences (IS2 and two of IS3) whose positions are also marked. Insertion sequences will be described later in this chapter.

becoming indistinguishable from F^- cells. This led to the idea that fertility is controlled by a factor distinct from the main bacterial DNA molecule. Eventually the F factor was shown to be a plasmid, the **F plasmid**, which in *E. coli* is 95 kb in size (Fig. 14.8). This plasmid carries approximately 30 genes (although additional genes may still remain to be discovered) including a large operon that contains the *tra* (for transfer) genes. These code for proteins involved in synthesis and assembly of the pilus and in the DNA transfer process itself. You will now realize that the explanation of F^+ and F^- cells is that F^+ cells contain the F plasmid, whereas F^- cells do not.

(c) The mechanism of gene transfer during conjugation. During conjugation a copy of the F plasmid passes into the F^- cell, presumably through the sex pilus, converting the F^- cell into F^+ (Fig. 14.9). A copy of the plasmid also remains in the donor bacterium so this cell remains F^+ also. However, bacterial conjugation is important to geneticists not because the F plasmid is

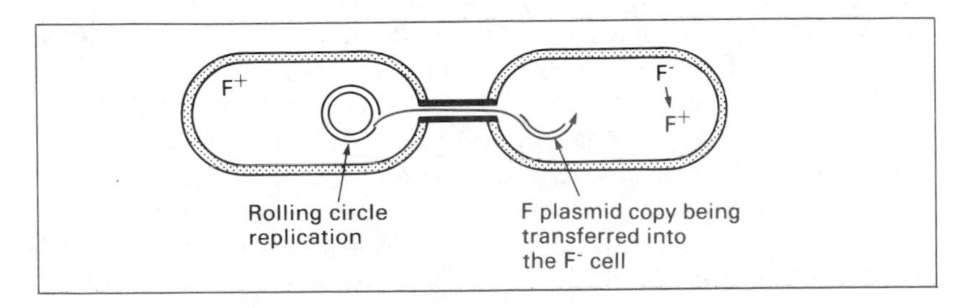

Figure 14.9 Transfer of the F plasmid during conjugation between an F^+ and an F^- cell.

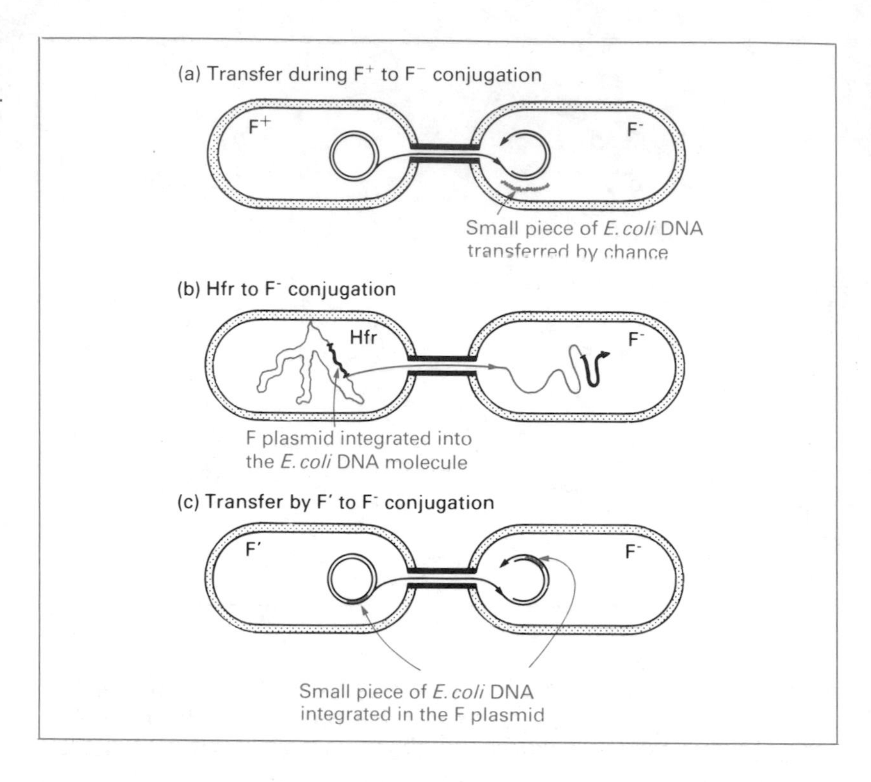

Figure 14.10 Three ways in which *E. coli* DNA can be transferred from donor to recipient during conjugation.

transferred, but because bacterial genes can also be passed from donor cell to recipient. How does this come about? There are several possible ways, the first being simply that a small random piece of the donor cell's DNA molecule is transferred along with the F plasmid (Fig. 14.10(a)). This is thought to account for the gene transfer phenomenon originally observed by Lederberg and Tatum, but probably occurs fairly infrequently. More important in genetics are the gene transfer properties of two additional types of donor cell, called **Hfr** and **F′** ('F-prime'). In Hfr cells, the F plasmid has become integrated into the *E. coli* DNA molecule. The integrated form of the F plasmid can still direct conjugal transfer, but in this case as well as transferring itself it will also carry into the F^- cell at least a portion of the *E. coli* DNA molecule to which it is attached (Fig. 14.10(b)). Consequently a mating involving an Hfr cell will virtually always bring about transfer of some *E. coli* genes. As we will see in Chapter 19, gene transfer with Hfr cells is the basis to one of the techniques for mapping genes in *E. coli*.

F′ cells, the second type of donor cell regularly associated with transfer of bacterial genes, occasionally arise from Hfr cells when the integrated F plasmid excises from the main DNA molecule. Normally this event results in an F^+ cell, but sometimes excision of the F plasmid is not entirely accurate, and a small segment of the adjacent bacterial DNA is also snipped out. This will lead to an F′

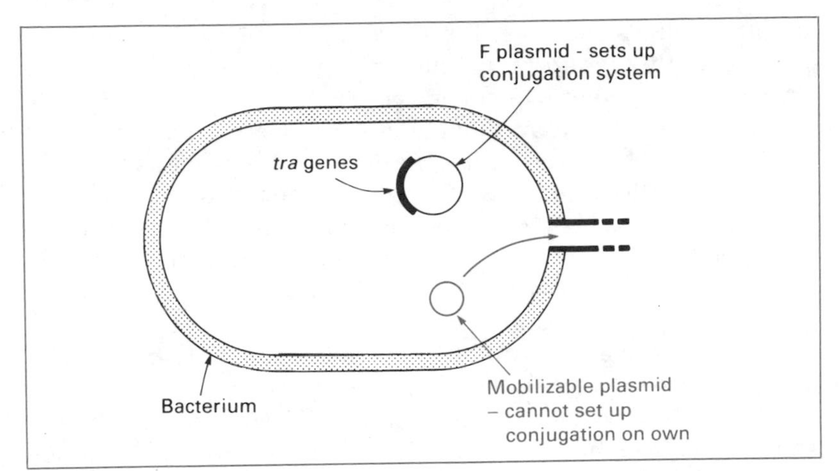

Figure 14.11 Some plasmids cannot set up conjugation, but can be mobilized if they share a cell with a conjugative plasmid.

plasmid that carries a small piece of bacterial DNA, possibly including a few genes (Fig. 14.10(c)). Conjugation involving the F′ cell will always result in transfer of these plasmid-borne bacterial genes.

(d) Conjugative transfer in other species. A large number of bacterial species are able to conjugate in the same way as *E. coli* and in all of these the process is controlled by a plasmid analogous to the F plasmid. Fertility plasmids of this type are called self-transmissible, which means that they can set up conjugation and mobilize themselves into the recipient cell. Conjugation (i.e. setting up the initial contact) and mobilization (passage of plasmid DNA from donor to recipient) are two distinct characteristics and not all plasmids are able to direct both. A few can set up only the conjugation contact between cells and not mobilize on their own; others can mobilize, but only if they are coexisting in the cell with a second plasmid that can set up the initial contact (Fig. 14.11). Other plasmids are totally non-fertile and can neither conjugate nor mobilize.

14.3.3 Copy number and incompatibility

Two further features of plasmid biology must be mentioned. First, **copy number**, which refers to the number of molecules of a plasmid that are present in a single cell and is a characteristic value for each plasmid type. Some plasmids, called **stringent** plasmids, have a low copy number of perhaps just one or two per cell; others, called **relaxed** plasmids, are present in multiple copies of 10 or more per cell (see Table 14.2). Second, **incompatibility**, which refers to the fact that certain plasmid types cannot coexist in the same cell. An *E. coli* cell can contain up to seven or more types of plasmid at the same time, but these must all belong to different **incompatibility groups**; two different plasmids of the same group cannot exist together in a single cell.

Copy number and incompatibility appear on the surface to be quite straightforward characteristics, but explanations of the mechanisms that underlie them have remained elusive. One theory holds that copy number is determined by an inhibitor molecule that prevents further replication of the plasmid once the characteristic value is reached. It has also been suggested that replication of incompatible plasmids is controlled by the same inhibitor molecule. A second theory suggests that plasmid replication requires attachment to a specific binding site on the cell membrane (recall that replication of the nucleoid involves attachment to the membrane) and that incompatible plasmids compete for the same attachment sites. The divergence between these two theories indicates that our current understanding of the topic is limited.

14.4 Bacterial transposons

The next aspect of bacterial genes that we must examine concerns the genetic elements called **transposons**. During the 1960s geneticists were puzzled by a set of unusual mutations occasionally found in operons of *E. coli* and other bacteria. These are called **polar mutations** because they affect not only the gene in which the mutation occurs but also other genes downstream of the mutation (Fig. 14.12). Several types of polar mutation are known and often they are ones that introduce a transcription termination signal into the operon. However, the particular set of mutations that caused the problems in the 1960s are unusual in that unlike most other

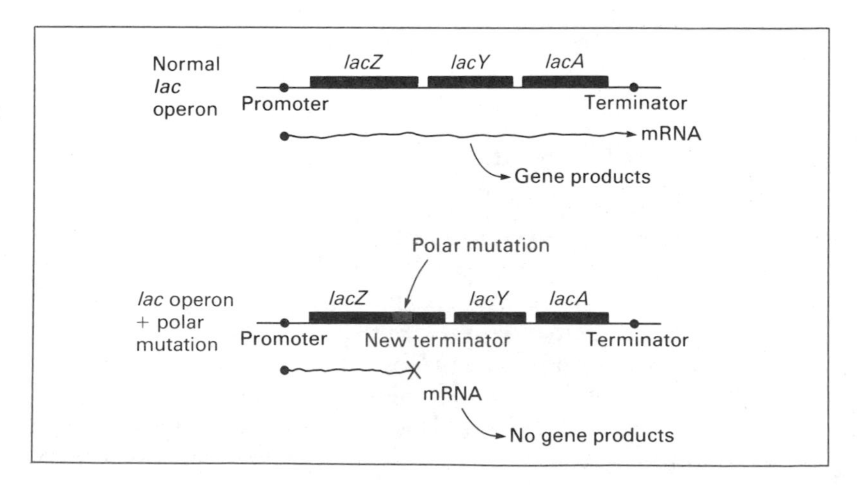

Figure 14.12 The effect of a polar mutation in the *lac* operon. The polar mutation in *lacZ* prevents expression of *lacY* and *lacA*, even though the last two genes are not themselves altered.

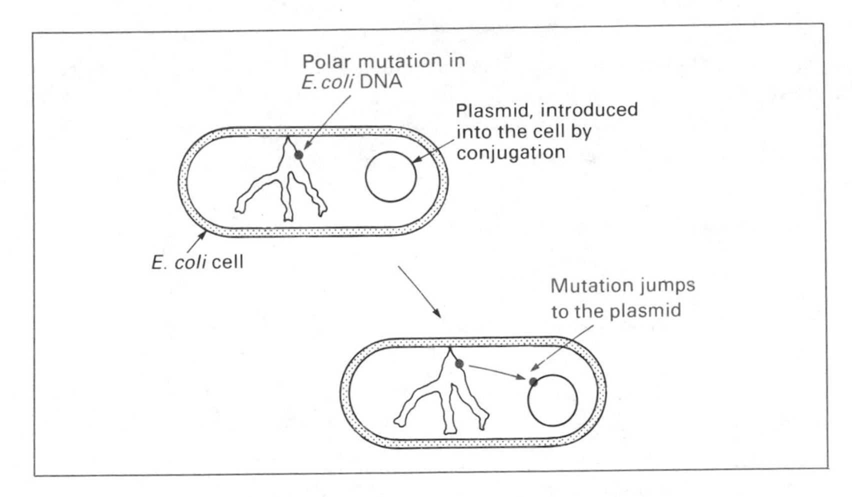

Figure 14.13 The unusual type of polar mutation that is able to jump from one position to another on *E. coli* DNA molecules.

mutations they are almost always irreversible. A second very unexpected feature becomes apparent if a plasmid is introduced into a bacterium that carries a polar mutation of this type. Similar, irreversible polar mutations suddenly appear in genes carried by the plasmid (Fig. 14.13), almost as though the mutation is able to jump from one DNA molecule to another!

(a) Insertion sequences can move around the genome. The first inkling of an explanation for this strange phenomenon appeared when the plasmids that had become mutated were studied. It was found that these plasmids had increased in size, generally by a kilobase or more. When the DNA sequences of the mutated plasmids were examined it was discovered that the mutations were caused by the insertion into the target genes of new pieces of DNA, which were subsequently called **insertion sequences** or ISs (Fig. 14.14(a)). These insertion sequences possess the remarkable capability of being able to move around the genome from site to site by a process dependent on recombination and called transposition (Section 12.3.2). When an IS transposes to a new site it leaves behind a short duplication of the target site, usually between 5 and 15 bp, which means that the gene which carried the IS probably remains mutated even after the element has departed.

Several different IS types are known in *E. coli* (Table 14.3). An individual isolate of the bacterium may contain a number of different IS types, each with a copy number of anything up to approximately 20. Transposition does not occur that frequently, and a single cell may undergo a thousand cell divisions before one of its insertion sequences transposes. Insertion sequences can transfer between bacteria during conjugation, and can also transfer between related species. The bulk of each IS is taken up by a single gene (except IS1 which has two) coding for a **transposase** enzyme

Figure 14.14 Insertion sequences.

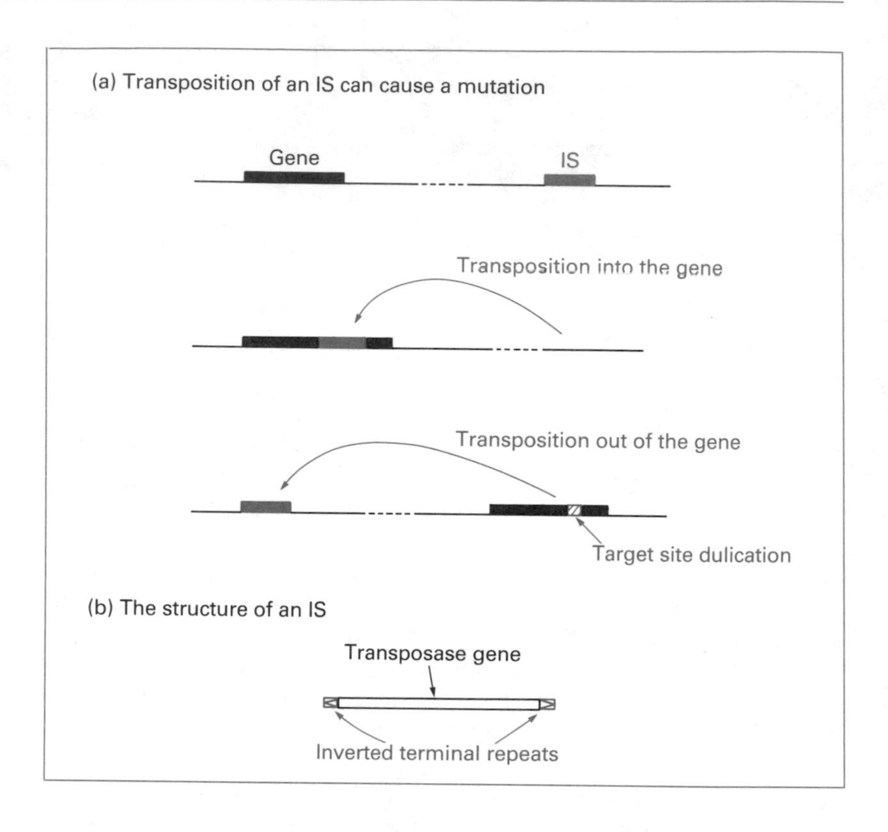

Table 14.3 Insertion sequences of *E. coli*

Insertion sequence	*Length (kb)*	*Flanking repeat (bp)*	*Target site duplication (bp)*
IS1	768	23	9
IS2	1327	41	5
IS4	1428	18	11
IS5	1195	16	4
IS10	1329	22	9
IS50	1531	9	9

Inverted repeats

Consider the sequence:
5′-AGGCAT-3′
3′-TCCGTA-5′

An inverted repeat is:

5′-AGGCATATGCCT-3′
3′-TCCGTATACGGA-5′

Repeats may occur adjacent to one another, as shown here, or may flank a different DNA sequence, as with the repeats at the ends of insertion sequences and Tn3-type transposons.

that catalyses the transposition event. The element has at its ends a pair of **inverted repeats** between 9 and 41 bp in length depending on the IS type (Fig. 14.14(b)).

(b) Insertion sequences are just one type of transposon. Insertion sequences were unique when they were first discovered in 1961. No similar type of genetic element was known at that time. Now we appreciate that IS elements are just one class of transposable element and that others, similar in many respects to IS elements, also exist:

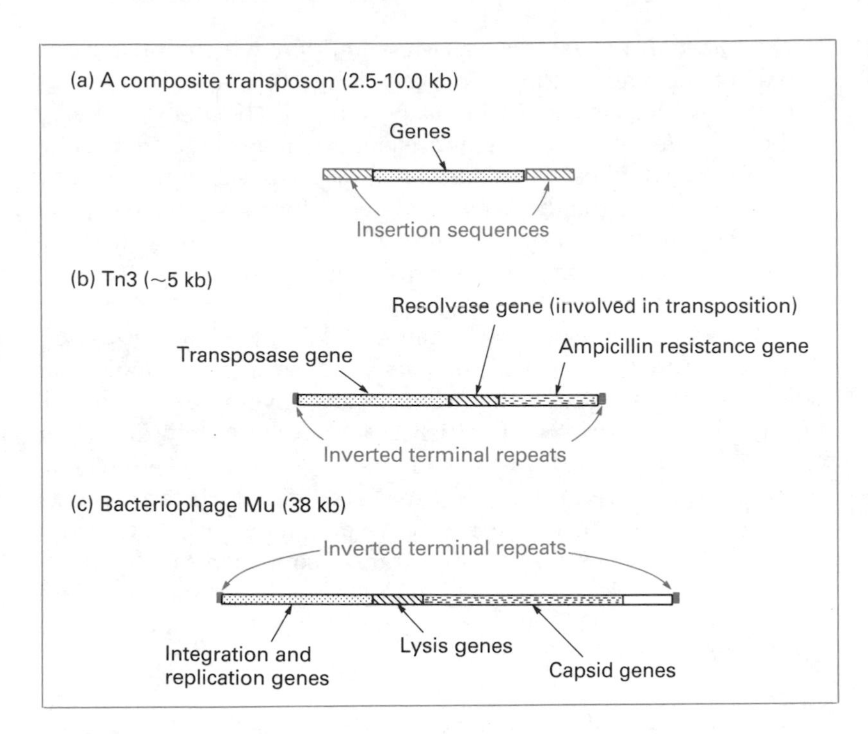

Figure 14.15 Three types of bacterial transposable element.

Table 14.4 Composite and Tn3-type transposons

Transposon	Size (bp)	Flanking elements	Gene carried
(a) Composite transposons			
Tn5	5700	IS50	Kanamycin resistance
Tn9	2500	IS1	Chloramphenicol resistance
Tn10	9300	IS10	Tetracycline resistance
(b) Tn3-type transposons			
Tn3	~5000	38 bp inverted repeats	Ampicillin resistance
Tn501	~5000	38 bp inverted repeats	Mercury resistance

1. ***Composite transposons*** consist of two insertion sequences flanking an internal region which usually includes at least one gene, often coding for resistance to an antibiotic (Fig. 14.15(a)). Transposition occurs because of the transposases coded by the IS elements. There are several different composite transposons (Table 14.4(a)), some over 10 kb in size. Often these types of element are carried by plasmids.
2. ***Tn3-type transposons*** (Table 14.4(b)), which are not flanked by IS elements but are still able to transpose because they carry a different transposase (Fig. 14.15(b)).

3. ***Transposable phages***, such as Mu and D108, which transpose as part of their normal infection cycle (Fig. 14.15(c)). This means that, as well as inserting themselves into the host DNA (like λ), the genomes of these phages are able to move about from place to place in the host DNA (unlike λ). Transposable phages may be transposons that have evolved genes for protective proteins, enabling them to exist outside of the cell, or they may be ordinary phages that have learned to transpose.

(c) Why have transposable elements? We do not as yet understand the function of transposons. Indeed several molecular biologists have suggested that transposons have no function at all and in fact are examples of **selfish DNA**, existing only to further their own existence and contributing nothing to the cell as a whole. However, it has been established that bacteria which carry a large number of transposons have a selective advantage over those with few transposons, possibly because transposons increase the mutation rate of the cell, perhaps improving the ability of the bacterium to adapt to changing environmental conditions. These speculations are clearly tenuous.

Reading

D. S. Goodsell (1991) Inside a living cell. *Trends in Biochemical Sciences*, **16**, 203–6 – fascinating drawings that illustrate the relative sizes of the molecules and structures within an *E. coli* cell.

J. A. Hobot, W. Villiger, J. Escaig *et al.* (1985) Shape and fine structure of nucleoids observed on sections of ultrarapidly frozen and cryosubstituted bacteria. *Journal of Bacteriology*, **162**, 960–71 – the appearance of bacterial cells under the electron microscope.

M. B. Schmid (1988) Structure and function of the bacterial chromosome. *Trends in Biochemical Sciences*, **13**, 131–5 – details of the *E. coli* DNA molecule and how it is thought to be packaged into the nucleoid.

R. A. Garrett (1985) The uniqueness of Archaebacteria. *Nature*, **318**, 233–5; W. F. Doolittle (1985) Archaebacteria coming of age. *Trends in Genetics*, **1**, 268–9; R. A. Garrett, J. Dalgaard, N. Larsen *et al.* (1991) Archaeal rRNA operons. *Trends in Biochemical Sciences*, **16**, 22–6 – reviews describing what is known about the archaea.

R. C. Clowes (1973) The molecule of infectious drug resistance. *Scientific American*, **228**(4), 18–27; R. P. Novick (1980) Plasmids. *Scientific American*, **243**(6), 76–90 – two general reviews.

J. Lederberg (1986) Forty years of genetic recombination in bacteria. *Nature*, **324**, 627–8; J. Lederberg (1987) Genetic recombination in bacteria: a discovery account. *Annual Review of Genetics*, **21**, 23–46 – personal reminiscences.

P. Starlinger (1980) IS elements and transposons. *Plasmid*, **3**, 241–59; S. N. Cohen and J. A. Shapiro (1980) Transposable genetic elements. *Scientific American*, **242**(2), 36–45 – two useful reviews.

Problems

1. Define the following terms:

   archaebacteria
   archaea
   nucleoid
   cytoplasm
   supercoiling
   plasmid
   episome
   conjugation
   sex pilus
   F plasmid
   tra genes
   copy number
   stringent plasmid
   relaxed plasmid
   incompatibility group
   transposon
   polar mutation
   insertion sequence
   transposase
   inverted repeat
   selfish DNA

2. To what extent can prokaryotes be considered a single group of organisms?
3. Describe the structure of the *E. coli* nucleoid.
4. Outline how *E. coli*'s genes are arranged on its DNA molecule.
5. Why are the archaea considered to be different from the eubacteria?
6. Explain what a plasmid is and provide examples of the different types of plasmid that are known.
7. What is the link between plasmids and bacterial sex?
8. Distinguish between F^+, F^-, Hfr and F′ cells.
9. Describe three ways in which transfer of bacterial genes can occur during conjugation.
10. Distinguish between insertion sequences and other types of bacterial transposable element.
11. Describe how insertion sequences were first discovered.
12. *Is there any justification for using the adjective 'primitive' to describe the archaebacteria?

13. *Discuss possible ways in which plasmid copy number could be controlled. Could your proposed mechanisms also account for copy number control of insertion sequences and other transposons?

14. *Genetic elements that reproduce within or along with a host genome, but confer no benefit on the host, are called 'selfish'. Discuss this concept, in particular as it applies to transposons.

15. Plasmid R100 carries a gene for resistance to streptomycin. You have an F^+ strain of *E. coli* that contains R100. Design an experiment to test if R100 is mobilizable.

Eukaryotic genomes 15

The nuclear genome • DNA packaging • Chromosome morphology – centromeres – telomeres • The cell cycle – events – control • Extrachromosomal genes – organellar genetic systems – organellar genomes

The next two chapters concern eukaryotic genomes. In this chapter we will make a broad survey of genes and genomes in all eukaryotes, concentrating on topics that are of general importance. We will look at the sizes of eukaryotic genomes, the way in which the genome is packaged into chromosomes, how chromosomes replicate, and the special features of the DNA molecules present in mitochondria and chloroplasts. Then, in Chapter 16, we will make a more detailed examination of the human genome, paying particular attention to those topics that are currently being researched most actively.

A eukaryotic cell may have as many as three distinct genetic systems. When Sutton proposed in 1903 that the genes of higher organisms reside on structures called chromosomes he was of course correct, although not completely so as all eukaryotes also have an additional complement of **extrachromosomal genes**. These are located within the energy-producing organelles called mitochondria and, in the case of plants and algae, in the chloroplasts also (Fig. 15.1). We therefore talk about the nuclear, mitochondrial and chloroplast genomes. The three genomes have little in common, both in terms of gene organization and the way the genes are expressed. In fact they can be looked on as separate and independent genetic systems. We will study the nuclear genome first, moving on to the organellar systems later in the chapter.

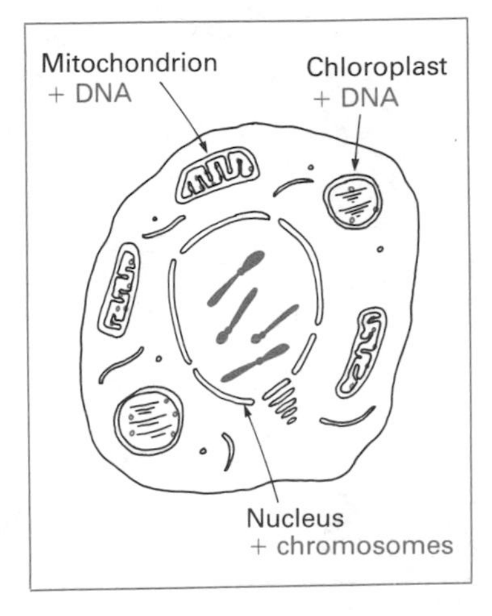

Figure 15.1 Stylized representation of a eukaryotic cell showing the three distinct genetic systems.

15.1 The nuclear genome

In the last chapter we saw that the genes in a prokaryotic cell, with the exception of the few carried by plasmids, are organized into a single circular DNA molecule. In the eukaryotic nucleus the situation is rather more complex (Fig. 15.2):

1. The nuclear genome is split into a number of individual DNA molecules, each of which is contained in a different chromosome.

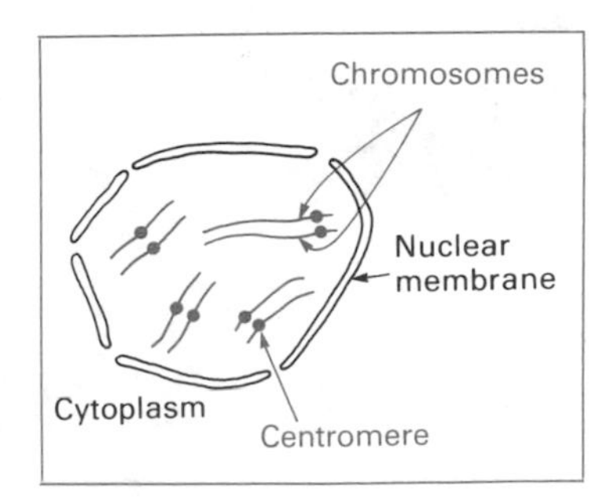

Figure 15.2 The nuclear genetic system.

Table 15.1 Sizes of prokaryotic and eukaryotic genomes

Organism	*Genome size (kb)*	*Number of chromosomes*	*Average amount of DNA per chromosome (kb)*
Prokaryotes			
E. coli	4 000	1	4 000
Eukaryotes			
S. cerevisiae	20 000	16	1 250
Fruit fly	165 000	4	41 250
Man	3 000 000	23	130 000
Maize	15 000 000	10	1 500 000
Salamander	90 000 000	12	7 500 000

For eukaryotes the data refer to the haploid complement. The values for genome sizes are approximate. Note that each chromosome in an organism does not contain the same amount of DNA: the DNA content of human chromosomes ranges from 55 000 kb (chromosome 21) to 250 000 kb (chromosome 1).

2. These DNA molecules are linear not circular.
3. In most cells of most eukaryotes there are two copies of each chromosome, and hence two copies of each gene. This is called the **diploid** complement: the word **haploid** refers to the situation usually found only in reproductive cells, where the nucleus contains just a single copy of each chromosome.

Part of the reason for having more than one chromosome is probably because eukaryotic genomes are substantially larger than

Genome sizes

The size ranges for genomes in different types of organism are as follows:

Organism	*Smallest genome size (kb)*	*Largest genome size (kb)*
Fungi	9 400	175 000
Nematodes	75 000	620 000
Coelenterates	280 000	685 000
Molluscs	375 000	5 100 000
Insects	47 000	12 000 000
Crustaceans	660 000	21 250 000
Fish	2 650 000	6 950 000
Amphibians	950 000	10 150 000
Reptiles	1 600 000	5 100 000
Birds	1 125 000	1 975 000
Mammals	2 350 000	5 550 000
Ferns	600 000	4 050 000
Flowering plants	95 000	120 000 000

Data from T. A. Brown (1991) *Molecular Biology Labfax*, BIOS, Oxford.

their prokaryotic counterparts (Table 15.1). However, there is no direct correlation between genome size and chromosome number. The yeast *Saccharomyces cerevisiae* has a relatively small genome of 20 000 kb, split into 16 chromosomes (the haploid number is 16; diploid would be 32). In contrast, the cells of human beings have about 150 times more DNA but a haploid complement of only 23 chromosomes. It is worth noting that as the size of the yeast genome is only about five times that of *E. coli*, each yeast chromosome contains a DNA molecule smaller than that present in the *E. coli* nucleoid.

These considerations illustrate the tremendous variation in genome size as we progress up the evolutionary tree from bacteria through yeast to man. But does it actually take more genes to make a human being that it does to make a yeast cell? The answer is yes, but this is not the only factor that influences genome size. Human genomes are not the largest in existence: certain amphibians and flowering plants have ten to a hundred times more DNA than man. What is the extra DNA for?

15.1.1 The organization of nuclear genomes

It has been calculated that only 2–3% of the human nuclear genome is actually made up of coding DNA, by which we mean the exon regions of genes. Even if we take into account introns and gene relics such as pseudogenes (Section 4.2.2) we are still left with about 70–80% of the genome apparently unused. Similar calculations offer the same conclusions for other vertebrates; only when we look at lower eukaryotes such as yeast and fungi do we find a closer correspondence between the 'expected' and actual genome sizes.

Part of the answer to this paradox was provided by biochemical studies of purified eukaryotic DNA in the 1960s and 1970s. A substantial part of the eukaryotic nuclear genome is made up of **repetitive DNA**: individual sequence elements that are repeated many times over, either in tandem arrays or interspersed throughout the genome (Fig. 15.3). Repetitive DNA is conventionally thought of as falling into two classes:

1. ***Highly repetitive DNA***, made up of sequences that are repeated several hundred to several million times in the genome.

Calculating the amount of coding DNA in the human genome

The human genome is thought to contain between 50 000 and 100 000 genes. We can estimate the average size of a gene by working back from what is known about the proteins in human cells. The average molecular mass of a human polypeptide is about 40 000 daltons, which corresponds to a gene 1 kb in length (the average molecular mass of a single amino acid is 120, so the average polypeptide is approximately 340 amino acids in length). The calculation therefore tells us that 50 000–100 000 kb of the human genome consists of coding DNA. As the human genome is 3 000 000 kb, this represents about 2–3% of the total.

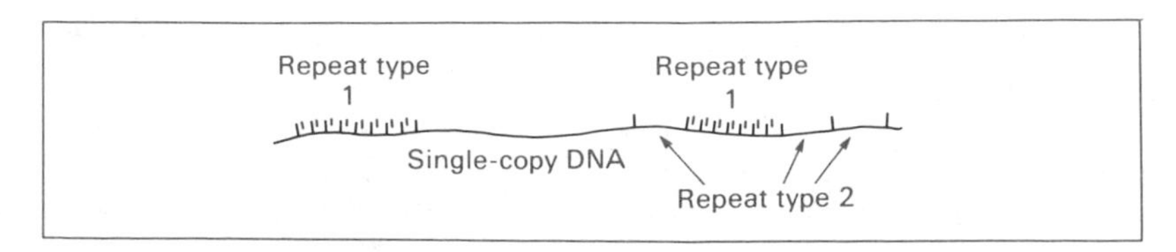

Figure 15.3 Repetitive DNA in a eukaryotic genome. Repeat 1 is highly repetitive, repeat 2 is moderately repetitive.

2. ***Moderately repetitive DNA***, comprising sequences repeated less frequently, between ten and several hundred times.

In fact this distinction is rather artificial and it probably places misleading emphasis on the two extremes of what is more probably a continuous spectrum of repetitive DNA subclasses.

The amount of repetitive DNA is probably the main factor that determines the size of an organism's genome. In man, repetitive DNA makes up about 20% of the genome, but in amphibians and plants the figure is generally larger: the toad *Bufo bufo*, for example, has a genome of 6 600 000 kb, but over 80% of this is repetitive DNA. When we examine the human genome in Chapter 16 we will see that it possesses a number of different repeated sequence elements, the unit lengths anything from a few base pairs to several kb. Functions have been assigned to a few of these sequences (we will encounter two later in this chapter), but the bulk of the repetitive fraction apparently has no role in the genome. Repetitive DNA may therefore be a second example of 'selfish DNA' (Section 14.4).

15.2 Packaging DNA into chromosomes

The amount of repetitive DNA in the eukaryotic genome has a direct bearing on the next topic that we must consider. If the human genome were a single DNA molecule its total length would be about one metre. As this molecule is shared between 23 chromosomes the average length of DNA in a single chromosome is a bit less than 5 cm. Yet the sizes of individual chromosomes are measured in micrometres – thousandths of a millimetre. There must be a highly organized packaging system to fit such a lengthy DNA molecule into such a small structure. We are now beginning to understand this packaging system, thanks to a clever combination of molecular biology, biochemistry and electron microscopy.

15.2.1 Proteins, DNA and chromatin

Cytologists towards the end of the nineteenth century coined the term **chromatin** for the component of chromosomes that stains most strongly with chromosome-specific dyes. Chromatin is now looked on as the complex association between the DNA of the chromosomes and the proteins to which it binds, the latter being responsible for packaging the DNA in a regular fashion within the chromosome.

Chromosomes are about half protein and half DNA. Of the protein component a portion is made up of a heterogeneous mixture of molecules including DNA and RNA polymerases and

Table 15.2 Histones

Histone	*Molecular mass (daltons)*	*Basic amino acid content (% total amino acids)**
H1[†]	23 000	30
H2A	13 960	20
H2B	13 774	22
H3	15 342	23
H4	11 282	25

*That is, the amount of lysine, histidine and arginine in each protein.
[†] H1 is in fact a family of closely related proteins. The molecular mass that is given is the average for the members of the family.

regulatory proteins. The remainder, associated most intimately with the DNA component of the chromosomes, consists of a group of proteins called **histones**, five of which are known at the present time (Table 15.2). Histones are very basic proteins (they contain a high proportion of basic amino acids), and are remarkable in that they are highly conserved between different species. If for example the H2A proteins of pea and cow are compared we find that only two of the 102 amino acids in the polypeptides are different. This is a much greater degree of similarity than we expect for proteins, even ones with equivalent functions, in such widely divergent organisms. This conservation indicates that histones were among the first proteins to evolve (before the common ancestor of peas and cows lived) and that their structures have not changed for billions of years.

(a) Histones are constituents of the nucleosome. In 1973 several investigators demonstrated that when purified chromatin is digested with an endonuclease (an enzyme that cuts DNA molecules at internal phosphodiester bonds) and the DNA extracted, the resulting fragments are not of random sizes as might be expected, but instead are all multimers of about 200 bp in length (Fig. 15.4(a)). This is an example of a DNase protection experiment (p. 64), one of the standard methods for analysing DNA–protein interactions. It shows that the protein in chromatin is associated with the DNA in a regular fashion and that the protein complexes protect regions of the DNA molecule from attack by the endonuclease (Fig. 15.4(b)).

These biochemical results were complemented the following year by the first electron microscopic observations of chromatin that had been subjected to a novel preparative method. These preparations showed linear arrays of spherical structures (**beads-**

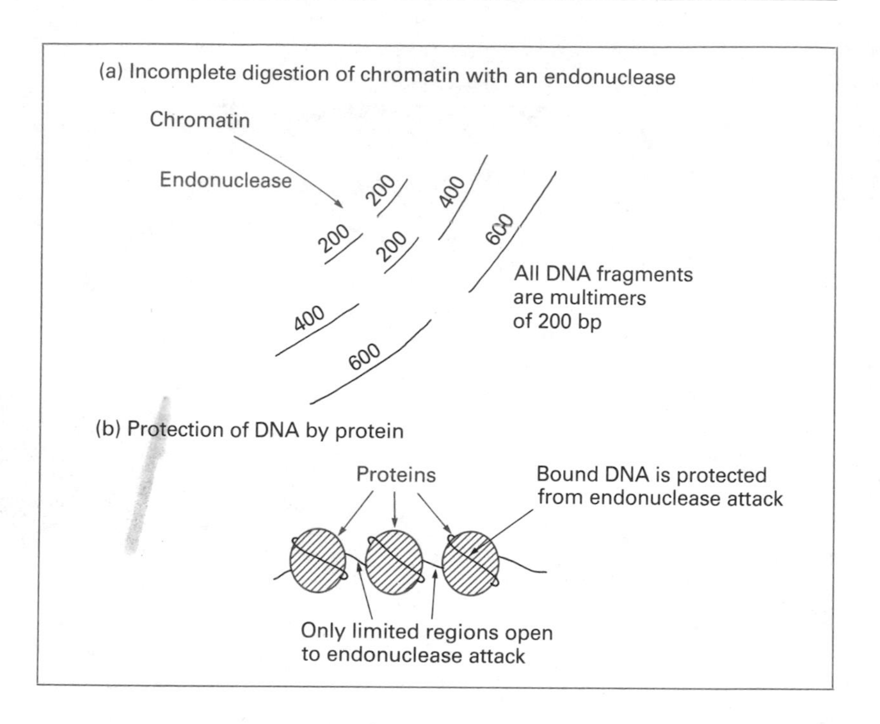

Figure 15.4 Digestion with an endonuclease indicates that chromatin has a regular structure.

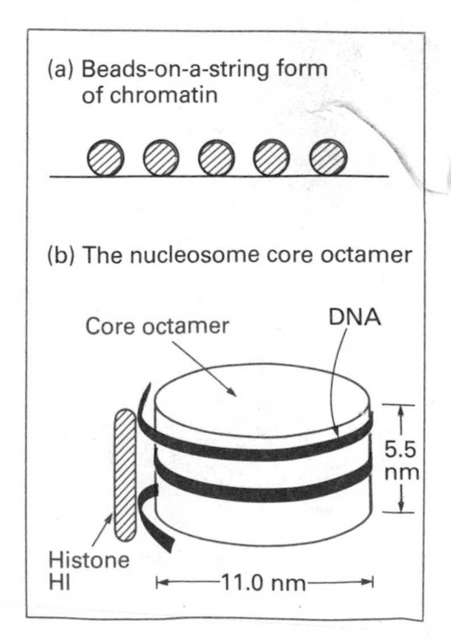

Figure 15.5 Nucleosomes.

on-a-string; Fig. 15.5(a)), generally believed to be the direct visualization of the protein complexes attached to a DNA molecule. The spherical particles are called **nucleosomes** and were shown in 1975 by Robert Kornberg to be made up of equal amounts of each histone except H1. It subsequently emerged that the histones form a barrel-shaped **core octamer** consisting of two molecules each of H2A, H2B, H3 and H4, with the DNA molecule wound twice round each nucleosome (Fig. 15.5(b)). Histone H1 is thought to bind to the outer surface of the nucleosome, possibly keeping the DNA in place in the same way that you would use your finger to stop a knot coming undone before tying the bow. Exactly 146 bp of DNA is associated with the nucleosome and hence protected from endonuclease digestion. The only parts of the DNA molecule available to the endonuclease are the 50–70 bp stretches of **linker DNA** that join the individual nucleosomes to one another. A single cleavage in each linker region will therefore give rise to 200 bp fragments of DNA.

Packaging of DNA round nucleosomes reduces its length by a factor of about six (so a 6 cm length of DNA, a bit above average for a single chromosome, would now take up 1 cm). This is still much more than the few micrometres taken up by DNA in chromosomes; we must therefore look for additional, higher levels of chromatin structure.

(b) Nucleosomes associate to form the 30 nm chromatin fibre. The second level of chromatin structure was characterized between 1977 and 1980 from electron microscopic observations carried out by Aaron Klug, which revealed a chromatin fibre, less dispersed than the beads-on-a-string form, about 30 nm in diameter. This fibre is formed by packing together individual nucleosomes, although the precise arrangement is not known. Klug's original 'solenoid' model still looks the most attractive (Fig. 15.6).

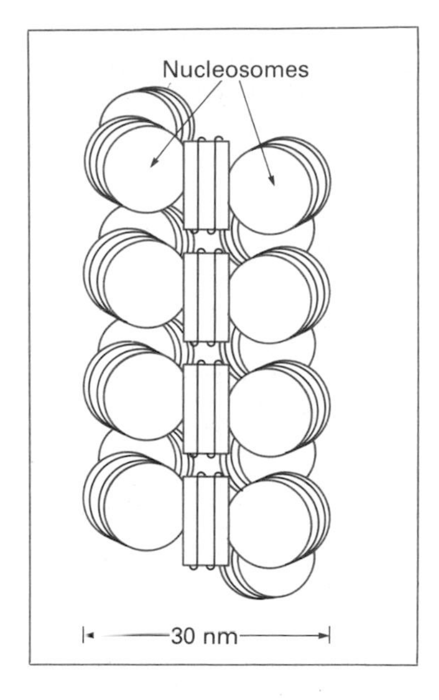

Figure 15.6 The solenoid model for the 30 nm chromatin fibre.

The **30 nm chromatin fibre** would reduce the length of the beads-on-string structure by about seven times, so our original 6 cm piece of DNA will now be about 1.4 mm in length. In cells that are not actively dividing the 30 nm fibre is probably the most organized state of chromatin in the nucleus. Individual chromosomes are not directly visible in these cells and it has long been recognized that the higher levels of packaging come into play only during cell division when **metaphase chromosomes** (the most highly organized structures) become visible. To achieve the additional size reduction needed to produce a metaphase chromosome, the 30 nm fibre takes up a conformation reminiscent of the bacterial nucleoid (Section 14.1), with supercoiled loops radiating from a core comprising non-histone proteins (Fig. 15.7). About 85 kb of DNA (in the form of the 30 nm fibre) is contained in each loop, rather less than in a loop of *E. coli* nucleoid DNA. The width of the structure is about 0.75 μm, in good agreement with the width of a metaphase chromosome, suggesting that this is the highest level of structural organization displayed by the chromosomal DNA.

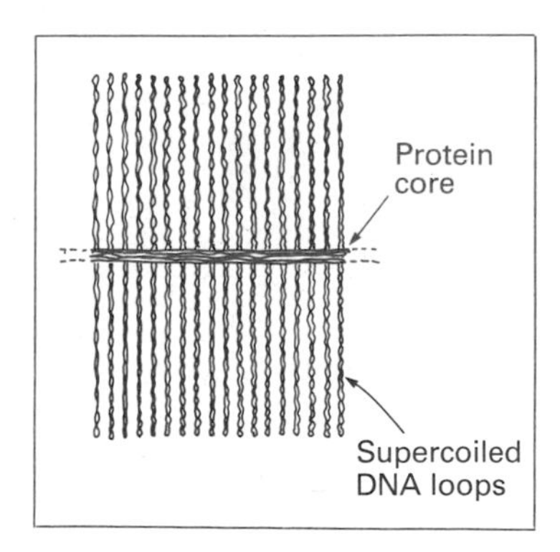

Figure 15.7 The supercoiled loops structure thought to exist in metaphase chromosomes.

15.3 Chromosome morphology

The typical appearance of a metaphase chromosome is shown in Fig. 15.8(a).This is in reality two chromosomes joined together because by the stage of the cell cycle when chromosomes condense and become visible by light microscopy, DNA replication has already occurred. Metaphase chromosomes therefore contain the two daughter DNA molecules linked together by a structure called the **centromere**. The position of the centromere is characteristic for a particular chromosome and is one of the features used to distinguish individual members of the **karyotype**, the entire complement of chromosomes in the nucleus (Fig. 15.8(b)).

The arms of the chromosome are called **chromatids** and their appearance provides the second distinguishing feature of different chromosomes. This is because chromatids do not take up chromosome-specific stains in a uniform manner: instead, a pattern of light and dark bands emerges, with the banding pattern characteristic for an individual chromosome (Fig. 15.8(c)). Five different staining techniques can be used to produce chromosome

Nucleosomes and gene expression

If the DNA in the nucleus is organized into the 30 nm chromatin fibre, with nucleosomes attached every 200 bp or so, then how can RNA polymerase gain access to the genes in order to transcribe them? Part of the answer is that RNA polymerase is able to strip off the nucleosomes as it progresses along the DNA, the nucleosomes re-attaching once the enzyme has passed. This explains what happens once transcription is underway, but is not the full answer as RNA polymerase does not appear to be able to attach to a fully condensed 30 nm chromatin fibre. This implies that DNA in the region of an active gene is packaged in a slightly unorthodox manner, and recent research has confirmed that this is the case. Nucleosomes are still attached but the 30 nm fibre is not completely condensed, enabling RNA polymerase to attach to the DNA at the appropriate point. Histone H1 may play a central role in determining which structure the chromatin fibre takes up.

One intriguing observation is that not all genes are in the active conformation in any one cell. Those genes that are expressed in the cell are accessible to RNA polymerase, but other genes (for example, genes expressed only in other tissues or cell types) are closed off in 30 nm fibres. Chromatin structure may therefore be responsible for the tissue-specific regulation of gene expression.

bands, but with only one (replication staining, which shows which parts of the chromosome contain DNA that is in the process of replicating) is the exact reason for the differential staining known. One of the most useful techniques, G-banding, is thought to highlight regions of chromosomes that are rich in A and T nucleotides. This would be particularly interesting as AT-rich regions are probably deficient in genes, so a region that does not produce a G-band may be one that contains a cluster of genes. This is still conjecture though.

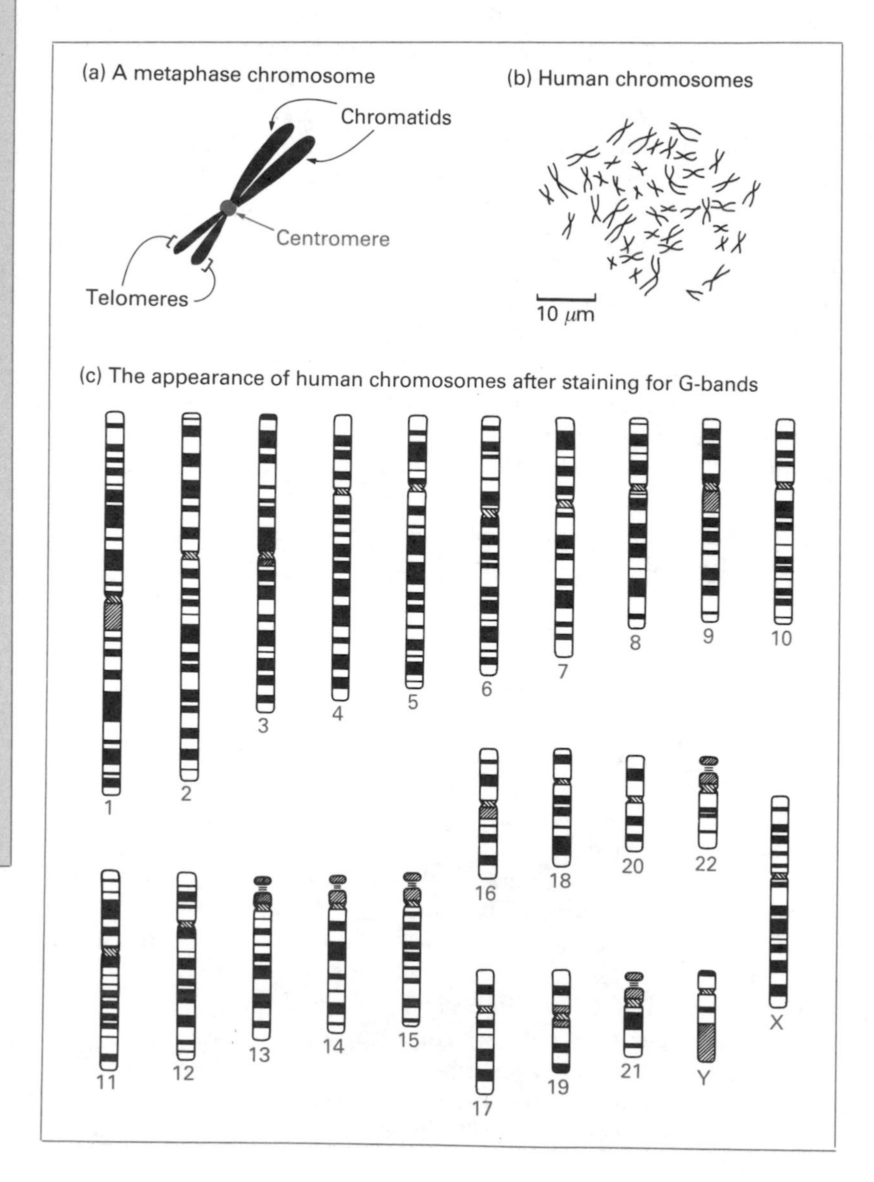

Figure 15.8 Chromosome morphology.

15.3.1 Centromeres

The centromere is the point on the chromosome where the two daughter chromatids are joined together. It is also the position at which the chromosome attaches to the microtubules that draw the daughters into their respective nuclei during cell division (see Figs 15.16 and 15.17). Because of these two functions the centromeric region of the chromosome contains special proteins and is characterized by specific centromeric DNA sequences. This is most clearly understood with the yeast *Saccharomyces cerevisiae*, whose relatively small chromosomes are more amenable to research than the much larger ones of other eukaryotes. The essential component of a yeast centromere is a 125 bp segment of DNA that has a related sequence in each of the 11 yeast centromeres that have been looked at so far (Fig. 15.9(a)). This 125 bp sequence is the nucleation point for a series of proteins that combine to form the **kinetochore**, the specialized structure to which a single microtubule attaches (Fig. 15.9(b)).

Unfortunately, *S. cerevisiae* chromosomes display a number of unusual features and are not good models for chromosome structure in other eukaryotes. In particular, yeast kinetochores appear to be permanent structures to which microtubules are attached at all times. This contrasts with most other organisms, whose mature kinetochores develop only when the chromosomes condense into the metaphase forms. In higher eukaryotes, such as man, a kinetochore is a relatively complex structure that appears to be built onto the surface of the metaphase chromosome (Fig. 15.10(a)). Whether the kinetochore is made just of protein, or includes DNA from the chromosome as well, is not known. At this

Centromere positions

The exact position of the centromere is characteristic of a particular chromosome.

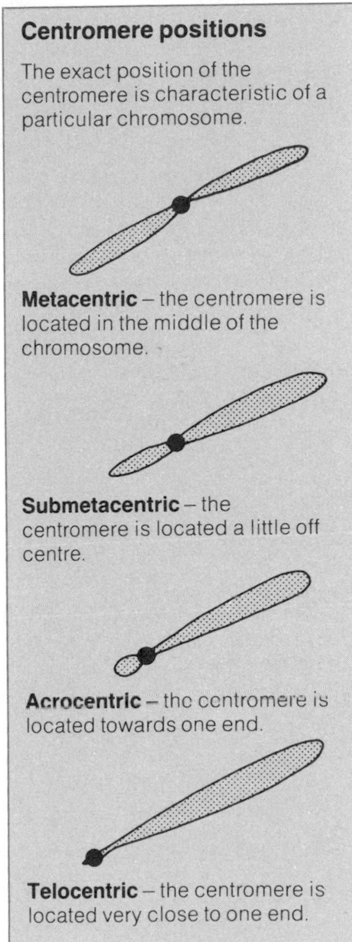

Metacentric – the centromere is located in the middle of the chromosome.

Submetacentric – the centromere is located a little off centre.

Acrocentric – the centromere is located towards one end.

Telocentric – the centromere is located very close to one end.

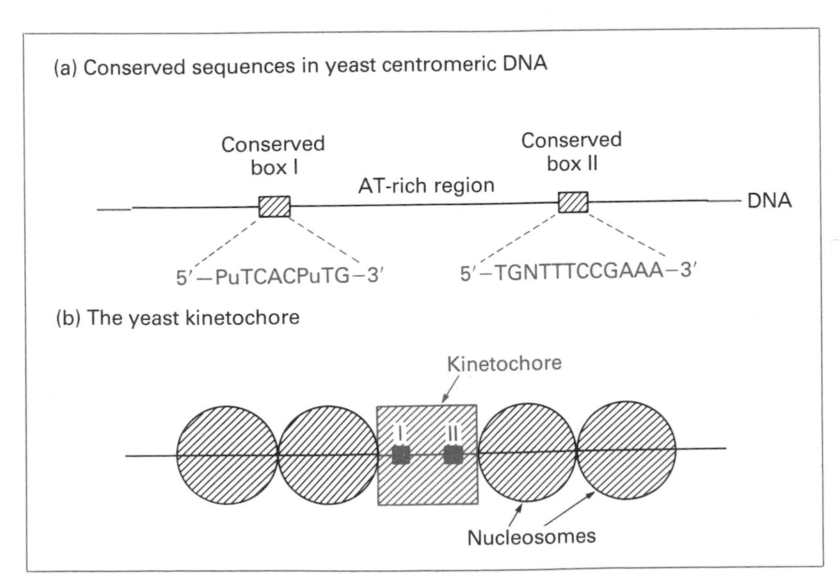

Figure 15.9 Yeast centromeres. The kinetochore is associated with 220 bp of DNA, of which the central 125 bp, comprising the two conserved boxes, the AT-rich region, and some nucleotides to either side, is very similar in different centromeres. Abbreviations: Pu, purine nucleotide; N, any nucleotide.

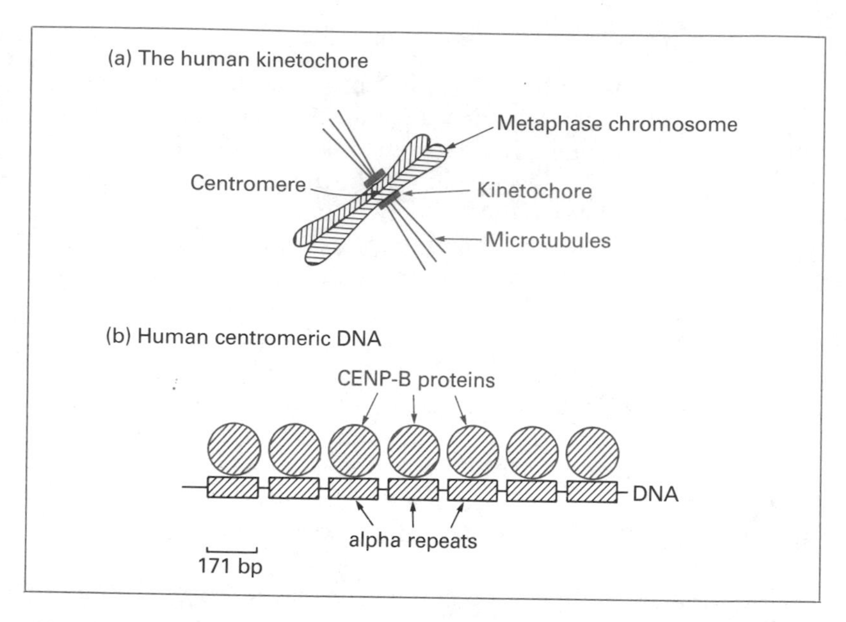

Figure 15.10 Human centromeres.

Proteins associated with mammalian centromeres and kinetochores

Protein	*Molecular mass (daltons)*	*Proposed function*
CENP-A	18 000	Centromere-specific histone
CENP-B	80 000	Binds to centromeric DNA
CENP-C	140 000	Component of kinetochore
CENP-D	50 000	Component of kinetochore
INCENP-A	135 000	Attachment between sister chromatids
INCENP-B	155 000	Attachment between sister chromatids
CLIPs	?	Attachment between sister chromatids

substructural level the only information available at present is that the DNA in the centromeric region has a highly repetitive sequence, and interacts in some way with special proteins. In humans, for instance, the centromeric DNA is composed of multiple copies of a 171 bp sequence, called **alpha DNA**, each repeat acting as a binding site for a single molecule of the centromere-specific protein CENP-B (Fig. 15.10(b)). At least five other proteins are associated with the centromere–kinetochore, but how they interact with each other, with the DNA, or with nucleosomes (if these are present) is still a mystery.

15.3.2 Telomeres

The ends of a chromosome are called the **telomeres**. Although the end of a chromosome may seem rather mundane, telomeres are in fact specialized structures that have to play three important roles:

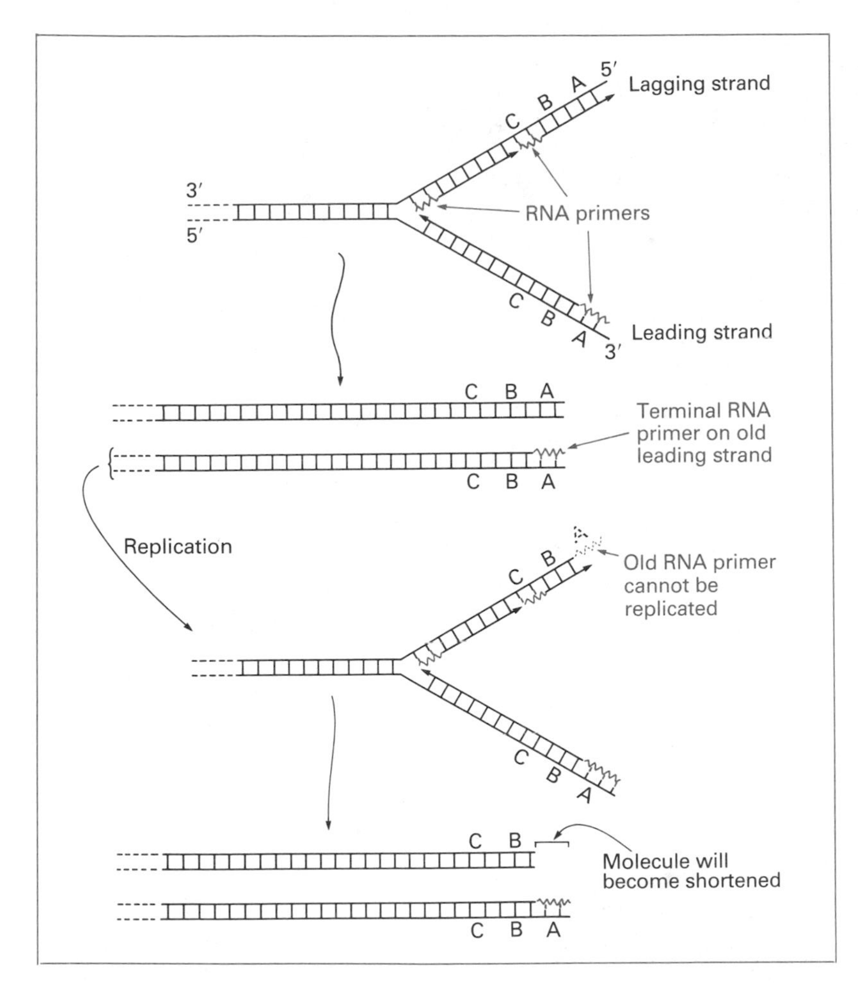

Figure 15.11 The problem at the ends of a linear DNA molecule. DNA synthesis requires an RNA primer (Section 11.2.2), which means that the extreme 5′ end of the leading strand will not be converted into DNA: it will remain as an RNA primer. During a second round of replication this primer will not be able to act as a template for DNA replication, so the resulting molecule will be shortened. If this happens enough times the molecule will eventually disappear.

1. The telomeres must protect the ends of the chromosomes from attack by nuclease enzymes.
2. They must prevent chromosomes from joining together: broken chromosomes attach to one another very efficiently via their internal breakpoints, implying that the ends of normal chromosomes have special structures that prevent joining.
3. They must overcome the problem posed by DNA replication, which will apparently lead to the progressive shortening of the linear DNA molecules in chromosomes. Recall that DNA synthesis occurs exclusively in the 5′ to 3′ direction and that initiation requires a short RNA primer (see Chapter 11). This means that the extreme 5′-terminus of a linear DNA molecule will comprise an RNA primer that in normal circumstances will not be replaced with DNA (Fig. 15.11). The result ought to be that chromosomes gradually decrease in length, something that does not happen.

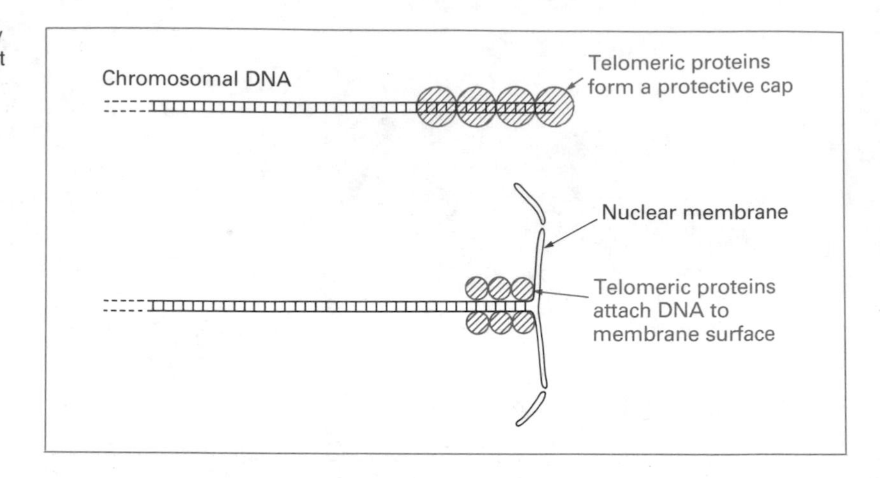

Figure 15.12 Two possible ways by which telomeric proteins may protect the ends of chromosomes.

Telomeres, like centromeres, contain special DNA sequences that appear to underlie their functional specialization. Telomeric DNA is made up of multiple copies of a short sequence, 5′-AGGGTT-3′ in man, which is repeated possibly a thousand times or more at the extreme ends of each chromosomal DNA molecule. The actual structure is different for each telomere, but resembles:

5′ . . . AGGGT T AGGGT T AGGGT T AGGGT T AGGGTTAGGGTT-3′
3′ . . . T CCCAAT CCCAAT CCCAAT CCCAAT CC-5′

There are two points here. The first is that one strand is rich in G nucleotides, with the other rich in Cs. This is true for all eukaryotes, even though the actual repeat sequence varies, and the G-rich strand is always the one that contributes the 3′ end of the DNA molecule. The second point is more obvious: the G-rich strand extends for 12–16 nucleotides beyond the terminus of the C-rich strand.

Nucleosomes are not present in the telomeric regions of chromosomes, being replaced with other protein structures. These telomere-specific proteins have been isolated for a few species (not man) and are presumed to act as a cap, preventing the ends of the chromosomes being degraded or fusing with other chromosomes. Alternatively, they may protect the chromosome ends by mediating attachment of the telomeres to other structures within the nucleus, such as the inner surface of the nuclear membrane (Fig. 15.12).

(a) Telomerase resynthesizes ends lost during replication. The third function of telomeres, prevention of gradual shortening of the chromosomes during DNA replication, has proven the most difficult to understand. A number of hypotheses were proposed

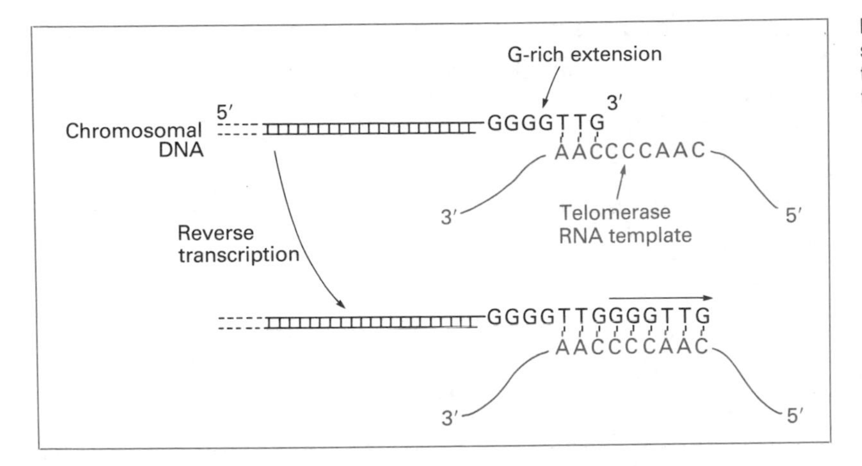

Figure 15.13 Extension of the G-rich strand at a telomere by reverse transcription catalysed by the enzyme telomerase.

during the early 1980s to explain why chromosomal DNA molecules do not eventually disappear, but progress was stalled until 1985 when Elizabeth Blackburn and Carol Greider of the University of California discovered a novel enzyme associated with the telomeres of the protozoan *Tetrahymena*. This enzyme is called **telomerase** and consists of a protein subunit and an RNA molecule, the latter 159 nucleotides in length. The key to how telomerase works is provided by the sequence of this RNA molecule. In *Tetrahymena*, the telomerase RNA contains at one position the sequence 3′-AACCCCAAC-5′, which compares with the *Tetrahymena* telomeric repeat of:

5′-GGGGTT-3′
3′-CCCCAA-5′

What this means is that the telomerase RNA can act as a template for synthesis of the repeat sequence of the G-rich strand of *Tetrahymena* telomeres (Fig. 15.13). As the template is RNA, and the product DNA, the reaction is a second example of reverse transcription (Section 3.2.2). By repeatedly extending the G-rich strand the shortening effect of DNA replication can be counterbalanced.

Telomerase has been isolated from other organisms, including man, but only in *Tetrahymena* and one other protozoan has the RNA component been sequenced. The model for telomerase action is therefore still tentative, but additional work, especially with the human enzyme, will soon confirm or disprove it.

15.4 The cell cycle

Human cells growing in culture divide about once every 24 hours. In the body the time taken to pass through a single **cell cycle** is usually longer but the events that take place are the same in all cases. These events are coordinated and controlled to ensure that the chromosomal DNA molecules are replicated at the appropriate time, and that cell division, when it occurs, results in the orderly partitioning of the replicated chromosomes into the daughter cells.

15.4.1 Events during the cell cycle

The cell cycle is conveniently divided into four phases (Fig. 15.14):

1. ***G1 phase.*** This is generally the longest phase of the cell cycle and for some cells may last for weeks or even months. 'G' stands for 'gap', implying that this is an unimportant period, which is misleading as it is the time when the cell is growing, metabolizing and, within a multicellular organism, performing its specialized function. During G1 the cell has its 'normal' complement of chromosomes, 46 (23 pairs) for a diploid human cell.
2. ***S phase.*** During S phase the chromosomal DNA molecules are replicated. At the start of S phase the replication origins are activated and bidirectional DNA synthesis progresses along the linear molecules until replication is complete. In eukaryotes, DNA replication generally takes place at a rate of between 2000 and 4000 bp per minute, much slower than in *E. coli*. This means that it would take an impossibly long time to replicate an entire DNA molecule if there was just one origin. Eukaryotic DNA molecules therefore have multiple origins (Fig.

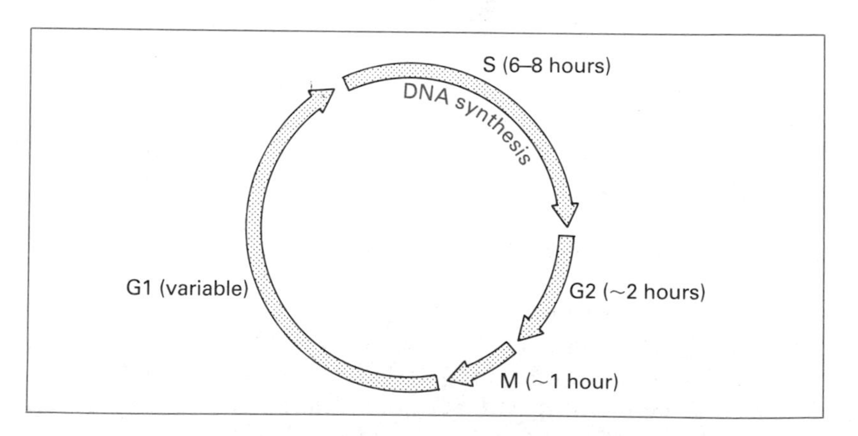

Figure 15.14 The cell cycle.

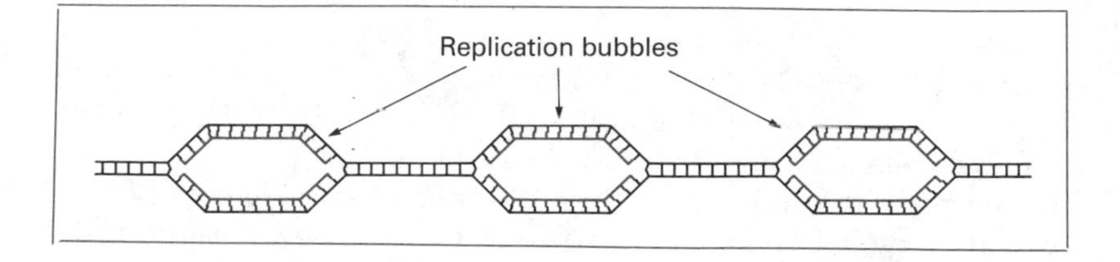

Figure 15.15 Eukaryotic DNA molecules have multiple replication origins, giving rise to a series of replication 'bubbles' on a molecule that is in the process of being replicated.

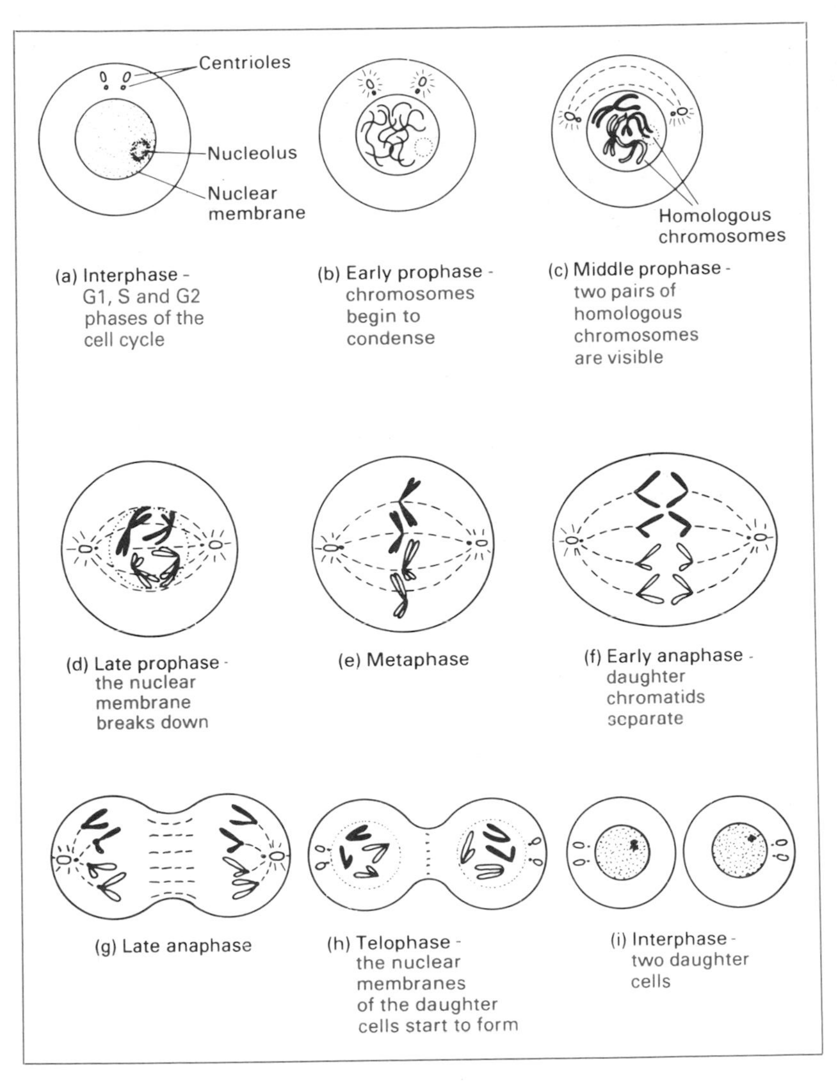

Figure 15.16 Mitosis of a diploid nucleus. At the start of mitosis each chromosome has already undergone DNA replication, and is seen as two daughter chromatids, attached to each other at their centromeres (as described on p. 271). During mitosis the daughter chromatids separate and then segregate into the daughter nuclei in such a way that each daughter nucleus receives a full complement of chromosomes. Each nucleus produced by mitosis therefore has the same number of chromosomes as the parent nucleus.

15.15), about one every 40 kb for yeast and one every 150 kb (over 25 000 altogether) for mammals. Despite these multiple origins, S phase can still take 6–8 hours to complete.

3. ***G2 phase.*** G2 is a second 'gap' phase, although the cell is now committed to division and G2 rarely lasts longer than 3–4 hours. During G2 the cell is effectively tetraploid, as each chromosome (which is present in two copies) has been replicated and so contains two DNA molecules. At the end of G2 the chromosomes condense to form the metaphase structures (see Fig. 15.8).
4. ***M phase.*** This is when nuclear and cell division take place. Two distinct types of nuclear division can occur: **mitosis** and

Figure 15.17 Meiosis of a diploid nucleus containing two pairs of homologous chromosomes. Meiosis results in four daughter nuclei, each containing the haploid complement of chromosomes. Each daughter nucleus therefore has a chromosome number that is half that of the parent nucleus. Meiosis is two cycles of nuclear division. At the outset each chromosome has already undergone DNA replication and is seen as two daughter chromatids attached to each other at their centromeres. During the first cycle of meiosis, homologous chromosomes line up and crossing-over (see Section 12.3.2) may occur. The chromosomes then segregate into the first two daughter nuclei. During the second stage of meiosis, each chromosome splits into its daughter chromatids, which then segregate into the second set of daughter nuclei.

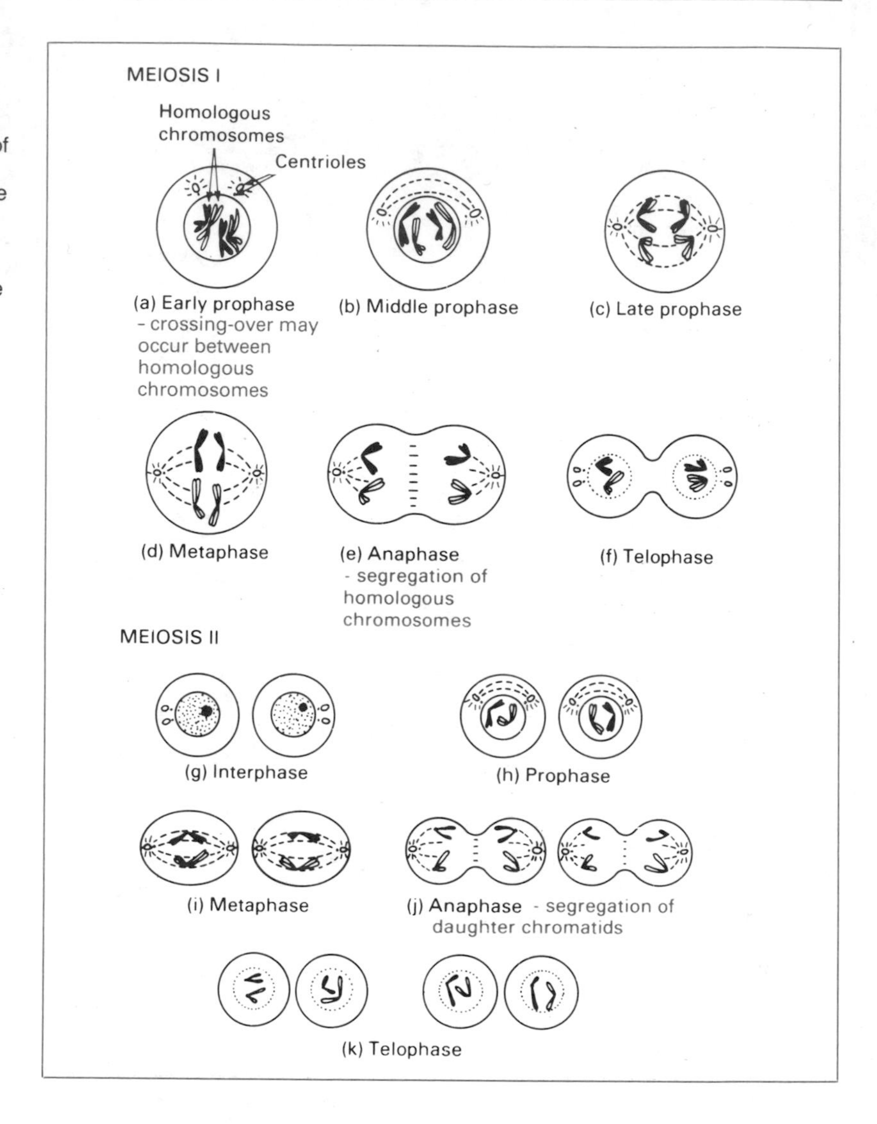

meiosis, each usually taking less than 1 hour to complete. Mitosis (Fig. 15.16) gives rise to two daughter nuclei, each containing the same chromosome complement as the parent, and is the standard type of cell division for **vegetative** (non-reproductive) cells. Meiosis (Fig. 15.17) results in a reduction of the chromosome number by half, so meiosis of a cell that was diploid in G1 (tetraploid in G2) will give rise to four haploid daughter cells. Usually these haploid daughters are either the male or female reproductive cells (sperm and egg cells respectively). Fusion with the complementary partner from another, or the same, individual results in a fertilized, diploid egg cell. During meiosis, **homologous chromosomes** (the two identical members of each pair in a diploid nucleus) can line up next to

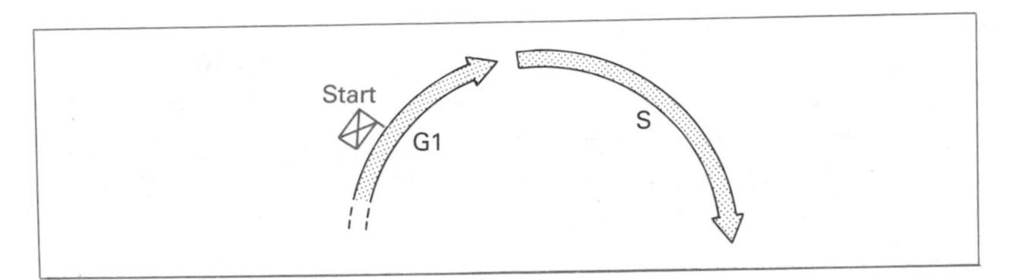

Figure 15.18 Start.

each other and exchange DNA by recombination ('crossing-over' – Section 12.3.2). This provides a basis for analysis of gene organization in eukaryotic chromosomes, as described in Chapter 18.

15.4.2 Control of the cell cycle

Completion of M phase returns the cell to G1. It will now grow and metabolize until it again enters S phase. However, G1 can last for a few hours or many days. How is the length of the cell cycle controlled?

There appear to be a number of critical transitions during the cell cycle, but the most important regulatory point is probably the phase called 'start', which occurs towards the end of G1 (Fig. 15.18). Once the cell passes through start it cannot turn back: it must replicate its DNA and undergo division, progressing through the cell cycle until it reaches G1 again. Whether the cell passes start depends on its size, the nutrients that are available, and, in multicellular organisms, the presence or absence of external signals telling it to divide. Research with two species of yeast, *S. cerevisiae* and its distant cousin *Schizosaccharomyces pombe*, has identified a number of genes that are involved in the cell cycle, one of which, *cdc2* of *S. pombe*, could play a central role in determining whether the cell passes start. The product of *cdc2* is a protein, called **maturation promoting factor** (MPF), which activates other proteins, but not in an entirely uniform manner: its properties change during the cell cycle. In contrast to the problems with chromosome structure, yeasts are good model organisms for the cell cycle in higher eukaryotes, so the current work on MPF could have a direct bearing on mammalian cell division, including regulatory problems that may lead to cancer.

15.5 Extrachromosomal genes

Genetic and biochemical studies during the 1950s led to the suspicion that eukaryotes might possess extrachromosomal genes. One or two genes with unusual inheritance patterns were discovered, so unusual that it seemed impossible for the genes involved to be carried by nuclear chromosomes. At the same time it was established that DNA is present in chloroplasts of algae such as *Chlamydomonas reinhardtii* and in mitochondria of fungi such as *Neurospora crassa*. It was then realized that the unusual genes must be carried by the DNA molecules present in these organelles.

15.5.1 Organellar genetic systems

The mitochondria of all organisms, as well as the chloroplasts of all photosynthetic cells, contain DNA molecules that carry a limited number of genes. These code for RNAs and proteins required in the organelle.

Mitochondrial and chloroplast DNA molecules are generally circular and double-stranded, although a few linear mitochondrial genomes are known. Copy numbers are not well understood: each human mitochondrion appears to contain about 10 identical molecules, which means that there are about 8000 per cell. In yeast the figure is probably much smaller. The sizes of the organellar genomes depend on species, with mitochondrial DNA molecules displaying a 200-fold variation, from 16 569 bp in humans to over 2000 kb in some higher (that is, flowering) plants (Table 15.3). Chloroplast genomes are less variable: those of all higher plants are between 120 and 180 kb and in other photosynthetic eukaryotes, with the exception of one or two algae, the sizes are not greatly different (Table 15.4).

Most of the RNA molecules, enzymes and proteins involved in transcription and translation of organellar genes are distinct from the equivalent molecules in the nucleus and cytoplasm of the cell. A case in point is provided by the transcribing enzyme RNA polymerase. In the eukaryotic nucleus there are three different RNA polymerases, each responsible for transcribing a particular set of genes (Section 5.3.1). Neither of these enzymes is the same as either the mitochondrial RNA polymerase, which is typically a

Table 15.3 Mitochondrial genome sizes

Species	*Type of organism*	*Genome size (kb)*
Ascaris suum	Metazoan	14.5
Homo sapiens	Metazoan	16.569
Drosophila melanogaster	Metazoan	19
Schizosaccharomyces pombe	Yeast	19
Aspergillus nidulans	Fungus	33.3
Neurospora crassa	Fungus	60
Saccharomyces cerevisiae	Yeast	68–78
Brassica campestris	Plant	218
Zea mays	Plant	570
Cucumis melo	Plant	2 500

Table 15.4 Chloroplast genome sizes

Species	*Type of organism*	*Chloroplast genome size (kb)*
Pisum sativum	Higher plant	120
Marchantia polymorpha	Liverwort	121
Euglena gracilis	Alga	130–152
Nicotiana tabacum	Higher plant	156
Spirodela sp.	Higher plant	180
Chlamydomonas reinhardtii	Alga	195

single polypeptide with a molecular mass of about 65 000 daltons, or the chloroplast RNA polymerase, which is similar to the *E. coli* enzyme in complexity. The same picture holds true for organellar ribosomes: mitochondrial and chloroplast ribosomes are not the same as each other, and both are different from their cytoplasmic counterparts, being smaller and having different rRNA molecules and different ribosomal proteins. In fact mitochondrial and chloroplast ribosomes more closely resemble prokaryotic ribosomes than the versions in the eukaryotic cytoplasm.

15.5.2 Organellar genomes

Organellar genomes were among the first DNA molecules to be studied when rapid nucleotide sequencing techniques were developed in the late 1970s. In particular the group led by Frederick Sanger at Cambridge, who had been responsible for many of the breakthroughs in DNA sequencing technology, set out to sequence the entire human mitochondrial genome. The result was published in April 1980.

(a) The human mitochondrial genome is packed full of genes. The human mitochondrial genome (Fig. 15.19) is 16 569 bp and carries genes for 13 proteins, 2 rRNAs and 22 tRNAs. These genes take up 15 368 bp of the molecule, leaving only 1201 bp of non-coding sequences, part of which comprises the specific nucleotide sequence of the replication origin. Very little of the DNA molecule, only 87 bp or so, is genetically unimportant. This is much less than would be the case for a similarly sized portion of a chromosomal DNA molecule.

In fact the human mitochondrial genome seems to have gone to great lengths to become as compact as possible. The following features illustrate this point:

1. The rRNA genes are among the smallest known, coding for a large subunit rRNA with a sedimentation coefficient of only 16S

Frederick Sanger b. 13 August 1918, Rendcombe, England

'If the planned experiment doesn't work, don't worry, start planning the next experiment.'

Frederick Sanger has been responsible for two of the most important technical advances of the past 40 years. First, in the early 1950s, he developed a method for determining the amino acid sequence of a polypeptide, and put it to use by working out the sequence of amino acids in insulin. This work, completed in 1955, provided the first real proof that the amino acids in a protein are present as a constant, genetically determined sequence. This gave molecular biologists the confidence to tackle other problems such as the genetic code. Then, in 1977, Sanger perfected a rapid method for sequencing DNA, opening the door to precise examination of gene structure and organization. Both advances have been invaluable and both were recognized by Nobel Prizes, in 1958 and 1980. Sanger has spent his career at the University Department of Biochemistry and the Medical Research Council Laboratory for Molecular Biology at Cambridge, UK.

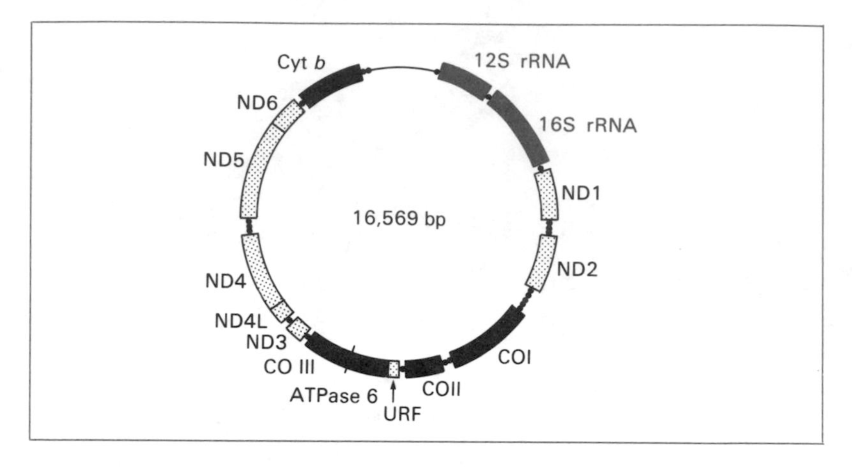

Figure 15.19 The genetic map of the human mitochondrial genome, as revealed by DNA sequencing. Abbreviations: Cyt *b*, apocytochrome *b*; ND, NADH dehydrogenase subunit; CO, cytochrome *c* oxidase subunit; ATPase, ATPase subunit; URF, unidentified reading frame. The positions of tRNA genes are indicated by dots.

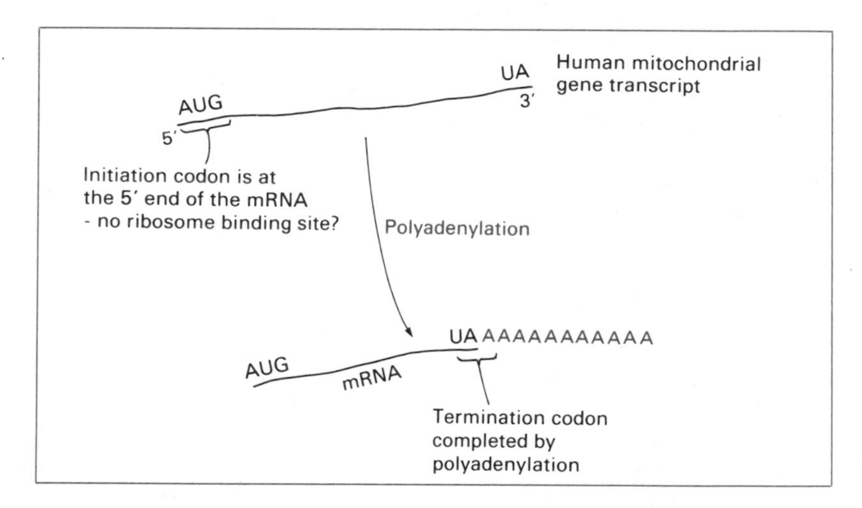

Figure 15.20 Problems with human mitochondrial mRNAs.

and a small subunit RNA of 12S (compare these figures with the typical values given in Table 6.2).

2. There are no intergenic regions, so the mRNAs are copies of just the gene: they have no leader or trailer sequences. This means that at the 5′ ends of the mRNAs there are no obvious ribosome binding sites, so how translation is initiated is a mystery (Fig. 15.20).
3. Several mRNAs are so truncated that the termination codons are incomplete: five mRNAs end with just a U or UA. The rest of the termination codon is provided by polyadenylation after transcription (see Fig. 15.20), a form of RNA editing (Section 7.5).
4. Genes are present for only 22 tRNAs, the absolute minimum needed to decode the mRNAs. In fact the standard codon–anticodon pairing rules (Section 9.1.2) have to be modified by

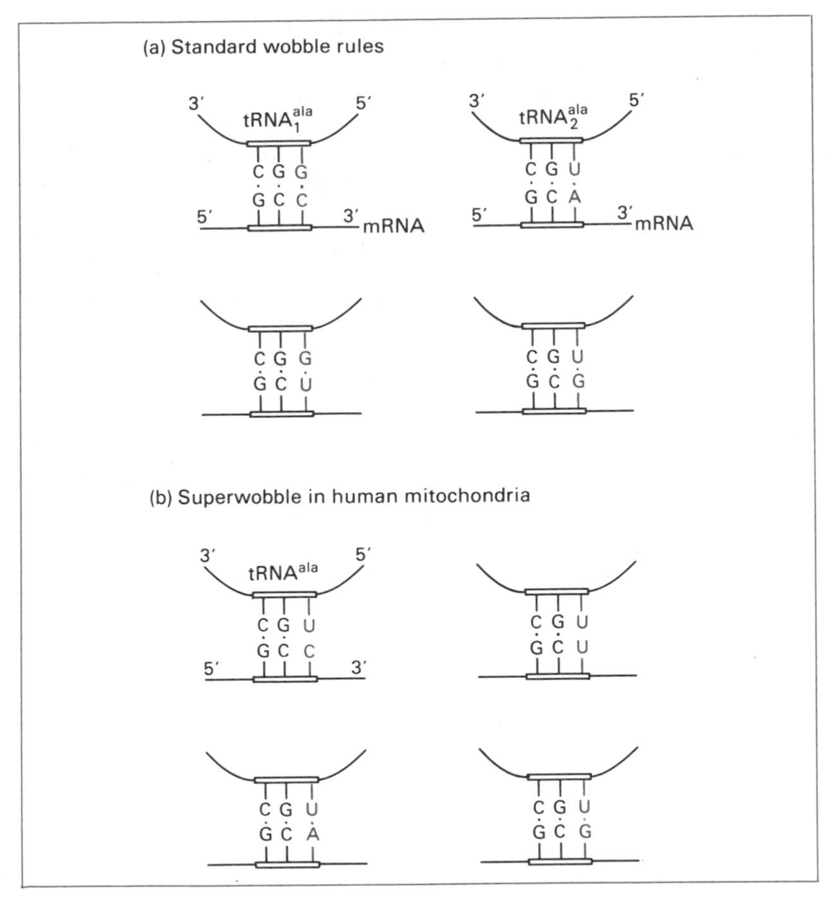

Figure 15.21 'Superwobble' in human mitochondria means that a single tRNA can decode each of the four alanine codons. In effect the third nucleotide of the codon is ignored. In the cytoplasm, the stricter wobble rules mean that two $tRNA^{ala}$ molecules, with different anticodons, are needed. Mitochondrial tRNAs for other amino acids also display superwobble.

'superwobble', which allows some tRNAs to recognize all four codons of a single family (Fig. 15.21).

This last point has important implications. In order to minimize the number of tRNAs needed, the human genome has taken the extreme measure of changing the assignations of the genetic code. In fact there are a total of four changes in codon assignations and two additions to the number of triplets that can act as initiation codons. We now know that non-standard codes are general features of mitochondrial genomes, but that the differences are not the same in all organisms (Section 8.2.3).

(b) Fungal and plant mitochondrial genomes are less compact. DNA sequencing has now provided information about mitochondrial genomes from a variety of organisms. All metazoan mitochondrial DNAs are organized along the same lines as the human genomes, but different features appear when fungi and plants are examined.

Mitochondrial gene products

The human mitochondrial genome contains genes for the two rRNAs of the mitochondrial ribosome and all the tRNAs needed to decode the transcripts of the protein coding genes. This means that there is no need to import rRNAs or tRNAs into the mitochondrion from the cytoplasm: the mitochondrion can make a full complement of these molecules. In contrast there are only 13 protein coding genes, each specifying one of the components of the electron transport chain. Many more than 13 different proteins are present in a mitochondrion, the vast majority of these being coded by nuclear genes, synthesized on cytoplasmic ribosomes, and then transported into the mitochondrion. This is also the case with mitochondria of other organisms, and for chloroplasts as well. None of these organellar genomes code for more than a small number of the proteins needed in the functioning organelle.

Why not have all the necessary protein coding genes in the organelle? Or, to put it another way, why have organellar genomes at all?

Mitochondrial genomes in fungi and yeast are up to ten times larger than the human genome (see Table 15.3) but contain no more than one or two additional genes. The extra DNA means that the genomes are less compact, with longer intergenic regions, and introns (Group I or Group II) are frequently present. Plant mitochondrial DNAs are much larger again, the smallest ones being the same size as the largest fungal genomes.

With plant mitochondrial DNA there is less information from DNA sequencing, although it is not thought that many additional genes are present. The only indication of extra sequences comes from the discovery that the mitochondrial genomes of at least a few plant species contain DNA segments obtained from the chloroplast. These sequences, called **promiscuous DNA**, may be several kb in length and may contain complete copies of some chloroplast genes. As yet we have no idea whether these genes are functional in the mitochondrion, or how the promiscuous DNA is transferred from one organelle to another.

(c) Chloroplast genomes. The relatively large sizes of chloroplast genomes meant that initially they were less accessible to analysis by DNA sequencing. Thanks to the remarkable industry of two Japanese groups, one led by Kazou Shinozaki at Nagoya University, the other by Kanji Ohyama at Kyoto, the complete

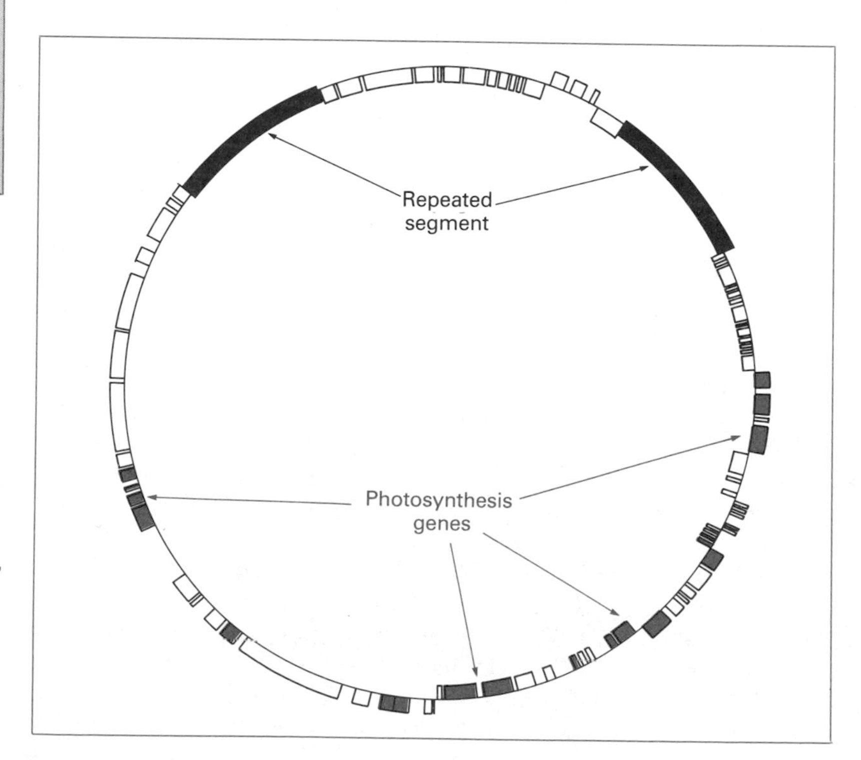

Figure 15.22 Map of the chloroplast genome of the liverwort *Marchantia polymorpha*. The repeated segments, containing the rRNA genes and a few tRNA genes, are indicated, as are those protein coding genes involved in photosynthesis. The functions of many of the other genes are not known. Redrawn from K. Ohyama *et al.* (1986) *Nature*, **322**, 572–4.

sequences of two chloroplast genomes – those of tobacco and the liverwort *Marchantia polymorpha* – were completed in 1986 (Fig. 15.22). These show that chloroplast genomes contain a substantially larger number of genes than their mitochondrial counterparts. The genomes are fairly compact but introns (mainly Group II) are present in a few genes. Both the tobacco and liverwort chloroplast genomes have a large repeated segment, containing the genes for the chloroplast rRNAs plus a few tRNAs. This repeat is a characteristic feature of many, although not all, chloroplast genomes.

Tobacco and liverwort are not closely related plants. It was therefore interesting to discover that equivalent genes in the two genomes have a relatively high amount of nucleotide sequence homology. This indicates that evolutionary change is acting relatively slowly on these molecules.

The origin of extrachromosomal genes

Ever since organellar genomes were first discovered, biologists have debated the reason for having extrachromosomal genes and the implications for theories on the evolution of cells. Most of these ideas centre around the fact that, in some respects, the transcription and translation processes occurring in organelles resemble the equivalent processes in prokaryotes, rather than events in the nucleus and cytoplasm of the eukaryotic cell. This leads directly to the proposal that mitochondria and chloroplasts were once free-living prokaryotes that formed symbiotic relationships with the proto-eukaryotic cell at the earliest stages of cellular evolution. Two competing hypotheses have been proposed, both assuming that the newly acquired organelles originally had a full complement of genes, most of these eventually being transferred to the nucleus.

It is doubtful whether definitive evidence for either of these hypotheses will ever be available. However, at least with chloroplasts, the comparison with possible prokaryotic relatives (in this case the photosynthetic cyanobacteria) provides a number of striking similarities, not only at the genetic level but also in terms of their biochemistries and morphologies.

Reading

D. E. Olins and A. L. Olins (1978) Nucleosomes: the structural quantum in chromosomes. *American Scientist*, **66**, 704–11; R. D. Kornberg and A. Klug (1981) The nucleosome. *Scientific American*, **244**(2), 48–60 – two reviews.

T. J. Richmond, J. T. Finch, B. Rushton *et al.* (1984) Structure of the nucleosome core particles at 7 Å resolution. *Nature*, **311**, 532–7; R. W. Burlinghame, W. E. Love, B. C. Wang *et al.* (1985) Crystallographic structure of the octameric histone core of the nucleosome at a resolution of 3.3 Å. *Science*, **228**, 546–53 – probing the fine structure of the nucleosome.

G. Felsenfeld and J. D. McGlee (1986) Structure of the 30 nm chromatin fibre. *Cell*, **44**, 375–7.

S. M. Dilworth and C. Dingwall (1988) Chromatin assembly *in vitro* and *in vivo*. *Bioessays*, **9**, 44–9 – a useful review of this topic.

F. Thoma (1991) Structural changes in nucleosomes during transcription: strip, split or flip? *Trends in Genetics*, **7**, 175–7 – speculations on how RNA polymerase gains access to DNA that is packaged into nucleosomes.

L. Clarke (1990) Centromeres of budding and fission yeasts. *Trends in Genetics*, **6**, 151–4; H. F. Willard (1990) Centromeres of mammalian chromosomes. *Trends in Genetics*, **6**, 410–15 – two interesting reviews on centromeres in yeast and mammals.

E. M. Blackburn (1991) Telomeres. *Trends in Biochemical Sciences*, **16**, 378–81 – recent news on telomerase.

R. K. Moyzis (1991) The human telomere. *Scientific American*, **265**(2), 34–41 – a blow-by-blow account of how human telomeres were first isolated.

D. Mazia (1961) How cells divide. *Scientific American*, **205**(3), 100–20;

D. Mazia (1974) The cell cycle. *Scientific American*, **230**(1), 54–64 – reviews about mitosis and meiosis.
L. A. Grivell (1983) Mitochondrial DNA. *Scientific American*, **248**(3), 60–73.
S. Anderson *et al.* (1981) Sequence and organization of the human mitochondrial genome. *Nature*, **290**, 457–65; M. J. Bibb *et al.* (1981) Sequence and gene organization of mouse mitochondrial DNA. *Cell*, **26**, 167–80; S. Anderson *et al.* (1982) Complete sequence of bovine mitochondrial DNA. *Journal of Molecular Biology*, **156**, 683–717 – the complete sequences of three mammalian mitochondrial genomes.
K. Umesono and H. Ozeki (1987) Chloroplast gene organization in plants. *Trends in Genetics*, **3**, 281–7 – a review of the information to come out of the tobacco and liverwort sequences.

Problems

1. Define the following terms:

extrachromosomal gene	chromatid
diploid	kinetochore
haploid	alpha DNA
repetitive DNA	telomere
chromatin	telomerase
histone	cell cycle
nucleosome	mitosis
core octamer	meiosis
linker DNA	vegetative cell
30 nm chromatin fibre	homologous chromosomes
metaphase chromosome	maturation promoting factor
centromere	promiscuous DNA
karyotype	

2. What comprises the 80% of the human genome that is apparently unused?
3. What is meant by repetitive DNA?
4. Describe the role of histones in packaging DNA into eukaryotic chromosomes.
5. Distinguish between the different levels of chromosome structure from naked DNA to the metaphase chromosome.
6. Describe the main morphological features of a human metaphase chromosome.
7. Compare the structures of centromeres in *Saccharomyces cerevisiae* and man.

8. Describe the problems that arise in replicating the ends of a linear DNA molecule. How may these problems be solved?

9. Explain what is meant by the 'cell cycle' and distinguish between the different stages of the cell cycle.

10. Draw the processes of mitosis and meiosis and point out the distinctive features of each.

11. Explain why meiosis reduces the chromosome number by half whereas mitosis does not.

12. Describe how the genetic systems within mitochondria and chloroplasts differ from the genetic system of the cell as a whole.

13. Describe the important features of the human mitochondrial genome, especially those that contribute to its compactness.

14. To what extent do the size variations displayed by mitochondrial genomes of different organisms reflect different gene compositions?

15. *Several major research projects are aimed at working out the complete nucleotide sequences of all the DNA molecules in a particular organism. Is this a worthwhile goal?

16. *What might be the basis to the banding patterns seen when chromosomes are stained?

17. *The proteins coded by mitochondrial genes are all important components of the oxidative phosphorylation system. For organisms other than yeast, most mutations in mitochondrial genes are lethal. Describe the types of non-lethal mutation that might be obtained in order to study mitochondrial genes in these organisms.

18. *What reasons might there be for having some proteins coded by genes in the mitochondrion and others coded in the nucleus?

16 The human genome

Structure of the human genome – genes and gene families – extragenic DNA – the Human Genome Project • Tracing mankind's origins – polymorphisms – studying population groups – the mitochondrial Eve – ancient DNA

An entire book could be devoted to the human genome. We have room for just a single chapter and so will inevitably have to omit many things, but the exercise will be worthwhile if we are careful in the choice of subjects with which we deal. By examining the human genome in more detail we will be able to understand the specific importance of some of the topics that we have covered in general terms in previous chapters, and we will also be able to familiarize ourselves with one or two of the major areas of current research.

16.1 The structure of the human genome

The figures are now at our fingertips: 50 000–100 000 genes in a genome of some 3 000 000 kb, split into 23 chromosomes, each containing a single, linear, double-stranded DNA molecule between 55 000 and 250 000 kb in length (Table 16.1). About a quarter of the genome consists of genes and gene-related sequences, the remainder being extragenic and having no known function (Fig. 16.1). We will look at the genes and gene-related sequences first.

16.1.1 Genes and gene families

Human genes cover a spectrum of sizes from less than 100 bp to over 2000 kb. The tRNA and U-RNA genes are the smallest, in some cases as short as 65–75 bp. The smallest protein coding genes are slightly longer, for example 406 bp for the histone H4 gene. At the other end of the size spectrum are genes whose bulk is made up not of coding sequence but of introns. The longest human gene so far discovered is the dystrophin gene, which contains over 100 introns in its 2300 kb length. Even though the dystrophin protein is relatively large (it has a molecular mass of over 400 000 daltons), its coding sequence (i.e. all of the exons added together) still makes up less than 1% of the length of the gene. At the moment the

Genes and transcription units

When we are thinking about the lengths of genes we must be clear in our minds about the distinction between a 'gene' and a 'transcription unit'. The gene is the DNA sequence that corresponds to the mature RNA or unprocessed protein product, plus any introns that may lie within that sequence. The transcription unit is the region of DNA that is transcribed, and so includes the gene and its surrounding leader and trailer regions. Although a tRNA gene may be just 75 bp, the tRNA transcription units are much longer (Section 6.2.2).

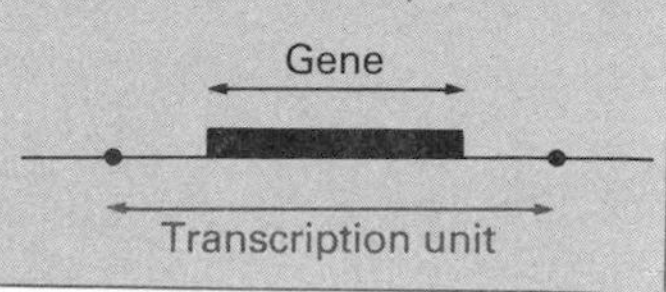

Table 16.1 Sizes of DNA molecules in human chromosomes

Chromosome number	*DNA content (kb)*
1	250 000
2	240 000
3	190 000
4	180 000
5	175 000
6	165 000
7	155 000
8	135 000
9	130 000
10	130 000
11	130 000
12	120 000
13	110 000
14	105 000
15	100 000
16	85 000
17	80 000
18	75 000
19	70 000
20	65 000
21	55 000
22	60 000
X	140 000
Y	60 000

Data taken from T. Strachan (1992) *The Human Genome*, BIOS, Oxford.

dystrophin gene is in a class of its own as it is ten times longer than the next largest human gene. However, it is difficult to identify long genes, so this may not be a true picture. As our techniques improve we may well discover other genes of this size, or even larger.

(a) Multigene families. The human genome illustrates each of the three ways in which a multigene family can be organized (Fig. 16.2):

1. With some families the individual genes are clustered together at a single position in the genome. Examples are the growth hormone gene family, whose five members are clustered on chromosome 17, and the 5S rRNA gene family, comprising 2000 genes in a tandem array on the long arm of chromosome 1.
2. With other families the genes are dispersed around the genome. For example, the five members of the aldolase gene family are located on chromosomes 3, 9, 10, 16 and 17.

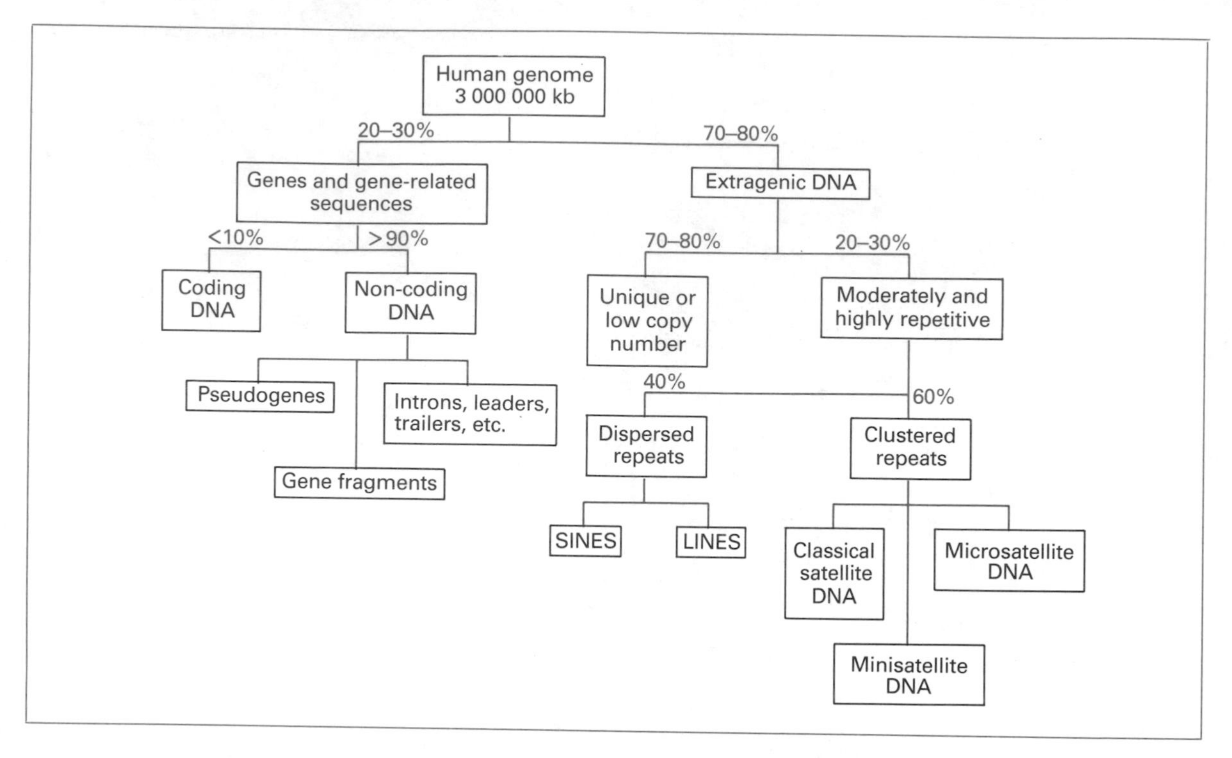

Figure 16.1 The human genome. Based on T. Strachan (see Reading).

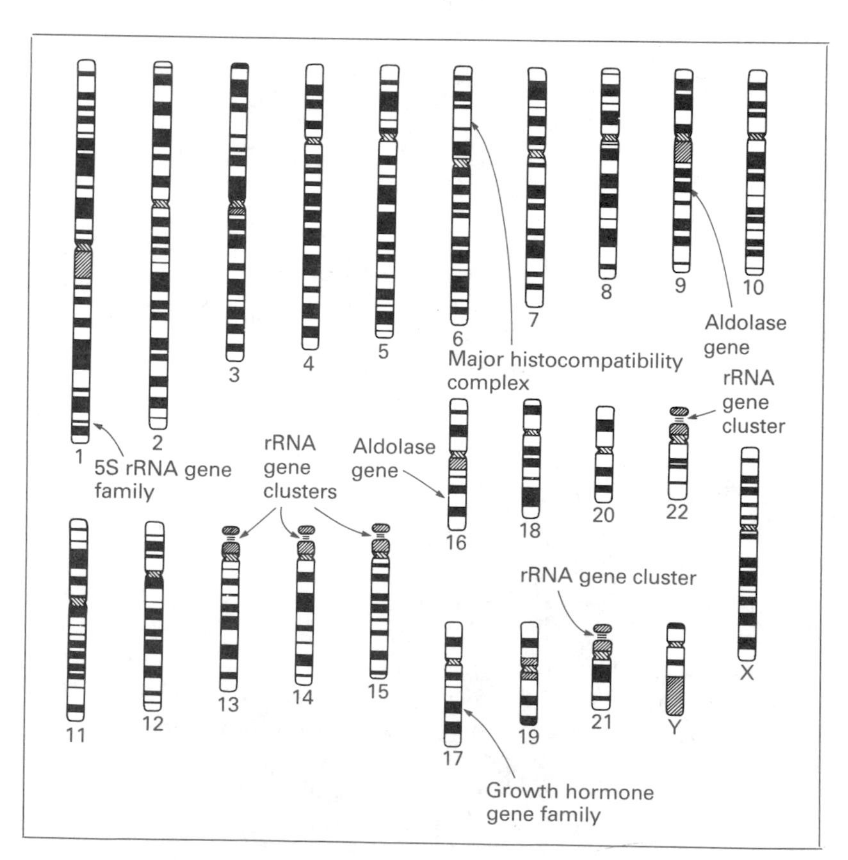

Figure 16.2 Chromosomal locations for various human genes and multigene families. The human chromosomes are drawn as they appear after staining for G-bands. Note that the aldolase genes on chromosomes 3, 10 and 17 are not shown as their precise positions are not known.

3. Some large gene families are both clustered and dispersed at the same time. There are about 280 copies of the rRNA transcription unit (Section 6.1.2), grouped into five clusters of 50–70 units each, the clusters located on the short arms of chromosomes 13, 14, 15, 21 and 22.

With some multigene families all the genes are identical. These families usually code for products that are needed in large quantities, such as rRNAs and histone proteins. By having multiple copies the cell is able to synthesize the products more rapidly than if it had to depend on a single gene. On the other hand, the members of some families display nucleotide sequence divergence, meaning that the individual genes are not all the same. The sequence differences may be so great that the gene products, although related to one another, have their own distinctive individual properties. The classic examples are provided by the α- and β-globin gene families (Section 4.2.1), whose members are expressed at different stages in development from embryo to adult.

(b) Gene relics. When we examine a multigene family we often find one or more members that have diverged to such an extent that they have lost their function altogether. Usually these decayed genes contain one or more inactivating mutations, such as a nonsense mutation that introduces a premature termination codon (Section 2.1.1). A gene that has lost its function in this way is called a **conventional pseudogene** (Fig. 16.3(a)).

A conventional pseudogene is distinct from a second type of gene relic, the **processed pseudogene**. A processed pseudogene is not simply a mutated version of its parent gene, but instead is a DNA copy of the mRNA derived from that gene. A processed pseudogene therefore lacks introns, which would be removed from the mRNA, and also lacks a promoter, which would lie upstream of the sequence transcribed into mRNA. The absence of a promoter means that a processed pseudogene is not transcribed. Nobody is really sure how a processed pseudogene arises, but it is presumed to involve reverse transcription of an mRNA molecule into double-stranded DNA, followed by insertion of the double-stranded DNA into the genome (Fig. 16.3(b)).

As well as pseudogenes, the human genome also contains truncated gene fragments, lacking either the 5′ or 3′ regions of the complete gene. A gene fragment is thought to result either from a deletion mutation that removes a segment of DNA from the parent gene, or from a recombination event that splits the parent gene in two (Fig. 16.4).

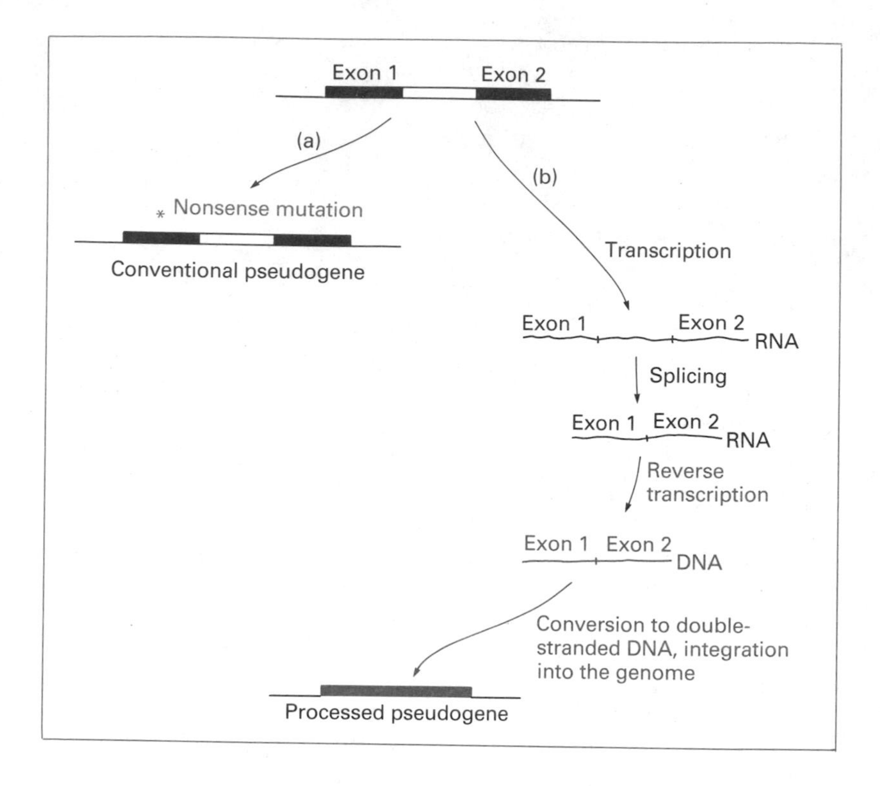

Figure 16.3 Conventional and processed pseudogenes.

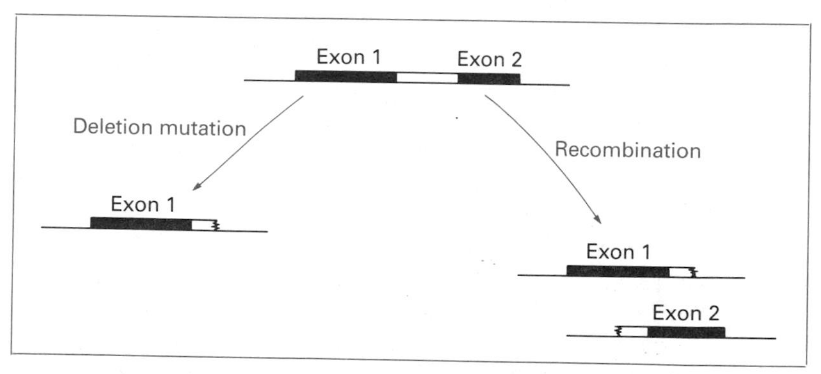

Figure 16.4 Two ways in which a truncated gene fragment can arise.

16.1.2 Extragenic DNA

By **extragenic DNA** we mean DNA that is not contained within a gene (as exons or introns), is not associated with a gene (leaders, trailers, promoters, upstream regulatory sites), and is not recognizably derived from a gene (pseudogenes, truncated gene fragments). Any DNA that does not fall into one of these three categories is extragenic. Extragenic DNA accounts for 70–80% of the human genome, possibly as much as 2 400 000 kb (see Fig. 16.1).

Most of the extragenic DNA consists of unique or low copy

number sequences. The remainder, some 700 000 kb of the entire genome, is made up of moderately and highly repetitive DNA, the repeat units either dispersed throughout the genome, or clustered into long tandem arrays.

(a) Dispersed repetitive DNA. Our current classification divides the dispersed repetitive DNA sequences in the human genome into two categories: **SINEs** and **LINEs**, standing for short and long interspersed nuclear elements.

The **Alu family** is the best known example of a human SINE. Not all Alu elements are identical, but they are similar enough to be considered members of one family. Their average length is about 250 bp and they are repeated some 700 000 times in the genome, occurring all over the place, even in introns. The Alu family is thought to have originated as one or more processed pseudogenes derived from the 7SL RNA, an RNA involved in secretion of proteins from the cell. One theory is that these pseudogenes possessed, or at a later stage acquired, transposable activity, enabling them to replicate and move to new sites in the genome. Recent results suggest that this propagation is still occurring.

LINEs are also thought to have propagated by transposition. **LINE-1** (copy number 60 000) is a type of **non-viral retroelement**, a transposon that can replicate and move around the genome by a process involving reverse transcription (Fig. 16.5(a)). Most LINE-1 elements are truncated, having lost their 5′ ends, but there are also several thousand full-length copies. These full-length elements have all the components needed for active transposition, including a gene that could code for a reverse transcriptase (Fig. 16.5(b)). Proving that they do actually move around in the genome is very difficult though.

Retroelements

A retroelement is a sequence that can propagate itself by a process involving reverse transcription. There are several different types of retroelement, the best known being the retroviruses (Section 13.2.2). The non-viral retroelements, such as LINE-1, are similar to retroviruses in that they carry genes for a reverse transcriptase enzyme and for proteins needed to integrate the double-stranded DNA product into the host genome. The difference is that the non-viral retroelements are unable to synthesize proteins for the viral capsid, and so cannot escape from the cell. Non-viral retroelements appear to be ubiquitous in most, if not all, eukaryotic genomes.

(b) Clustered repetitive DNA. The human genome contains extensive tracts made up of repeat sequences arranged into long tandem arrays. This type of repetitive DNA is called **satellite DNA** and is generally divided into three categories depending on the lengths of the clusters:

1. ***Classical satellite DNA***. This was the first type of satellite DNA to be identified in the human genome and consists of clusters between 100 and 5000 kb in length. The centromeric clusters of alpha repeats (Section 15.3.1) are typical examples of classical satellite DNA.
2. ***Minisatellite DNA***. Minisatellite DNA forms shorter clusters, between 100 bp and 20 kb in length. The telomeric repeat clusters (10–15 kb; Section 15.3.2) are classed as minisatellite DNA.

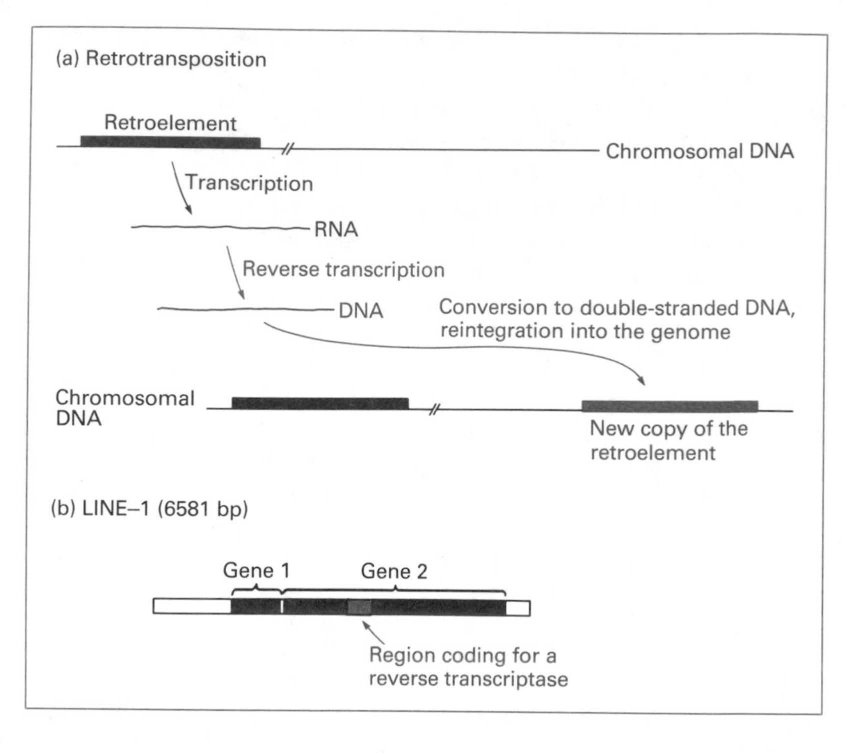

Figure 16.5 Retrotransposition and the structure of LINE-1.

3. ***Microsatellite DNA.*** There is overlap between the lengths of minisatellite and microsatellite clusters, but the latter are distinguished by the short length of the repeat units, which are rarely more than 4 bp. Dinucleotide repeats, such as CA/GT runs:

   5′-CACACACACACACACACACACACA-3′
   3′-GT GT GT GT GT GT GT GT GT GT GT GT -5′

   are very common in the human genome, contributing some 0.5% of the total. Mononucleotide A/T repeats:

   5′-AAAAAAAAAAAAAAAAAAAAAA-3′
   3′-TTTTTTTTTTTTTTTTTTTTTT-5′

   make up another 0.3%.

16.1.3 The Human Genome Project

Our current knowledge of the human genome has been built up through a combination of biochemical, genetical and molecular biological research. Gradually we have begun to understand how the genic and extragenic DNA is arranged in the genome, where individual genes are located, and how expression of these genes is regulated. Remarkable progress has been made over the past decade but our picture of the genome remains far from com-

Why does the human genome contain extragenic DNA?

The large amount of extragenic, apparently useless, DNA in the human genome is puzzling. Bacterial genomes are small and compact, with little wasted space, so why should the human genome be so large? There is no fully accepted answer but there are several possibilities. One is that the extragenic DNA is retained because there is no selective pressure to get rid of it. Genomes of bacteria are compact because these organisms must divide quickly to survive. Their genomes must be replicated as rapidly as possible, so extragenic DNA is a disadvantage. Human cells do not need to divide rapidly, so there is no selective pressure on the size of the genome. In the absence of a selective pressure the extragenic DNA is tolerated. The assumption here is that the extragenic DNA is purely neutral in its effect: it confers neither an advantage nor a disadvantage on the organism.

Satellite DNA

The term 'satellite DNA' is derived from the way in which repetitive DNA is prepared as a pure fraction separate from the rest of the DNA in a human cell.

Recall that in a density gradient (p. 19) a DNA molecule will migrate to the position where the density of the solute (e.g. caesium sulphate) is equal to its own buoyant density. The buoyant density of a DNA molecule depends on its GC content (Section 3.2), with human DNA (GC content 40.3%) having a buoyant density of 1.701 g cm^{-3}. Human DNA will therefore form a band at the 1.701 g cm^{-3} position in a density gradient. This will not, however, be the only band that appears: with human DNA there will be three 'satellite" bands, at 1.687, 1.693 and 1.697 g cm^{-3}:

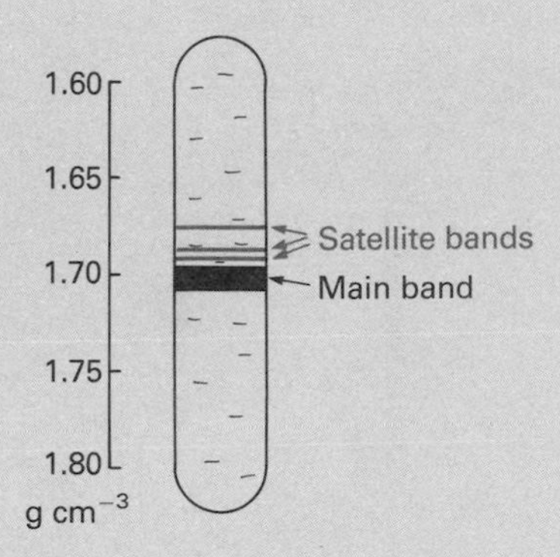

These additional bands contain repetitive DNA. They form because the long chromosomal molecules become cleaved into fragments, 50 kb or so in length, when the cell is broken open. Fragments containing predominantly single copy DNA will have GC contents close to the 'standard' human value of 40.3%, and so will move to the 1.701 g cm^{-3} position in the density gradient. However, fragments containing large amounts of repetitive DNA will behave differently. For instance, a fragment made up entirely of ATTAC repeats will have a GC content of just 20%, lower than the standard value, and so will have a buoyant density somewhat less than 1.701 g cm^{-3}. These repetitive DNA fragments will therefore migrate to a satellite position, above the main band in a density gradient.

plete. We have located only 2000 genes and obtained nucleotide sequences for just 1600 of these. Remember there are 50 000 to 100 000 genes altogether: still some way to go!

During the mid-1980s a number of biologists proposed that the human genome be studied in a more coordinated fashion. In particular, it was argued that research into many areas of human biology would benefit from a programme aimed at determining the entire nucleotide sequence of the genome. Genetic diseases provide a case in point. Many of the most serious incurable diseases of the present day result not from infections with bacteria or viruses but from defects in human genes. Finding cures for these diseases depends on first identifying the relevant genes and understanding how they are regulated. This would clearly be a much easier task if the genome sequence was known.

The Human Genome Project was set up in response to the

Sex

A diploid human cell, whether male or female, contains 23 pairs of chromosomes. In a female cell the two members of each pair are identical: two each of chromosomes 1–22 (the **autosomes**) and two X chromosomes (the **sex chromosomes**). Males are the same except that the two sex chromosomes are not an identical pair: there is one X and one Y. It does not take too much ingenuity to realise that the Y chromosome confers maleness. But the DNA molecule in the Y chromosome is 60 000 kb in length. Is all of it required for maleness, or are there one or more specific 'maleness' genes?

By 1966 it was established that maleness is specified by DNA sequences within the short arm of the Y chromosome: the long arm is unimportant in this respect. Little more was achieved during the next 20 years but since 1986 the region conferring maleness has been narrowed down step by step. In 1989 a 60 kb region containing the relevant genes was delineated, and in 1990 this was decreased to 35 kb. This segment contains just one gene, called *SRY* for 'sex determining region of the Y chromosome'.

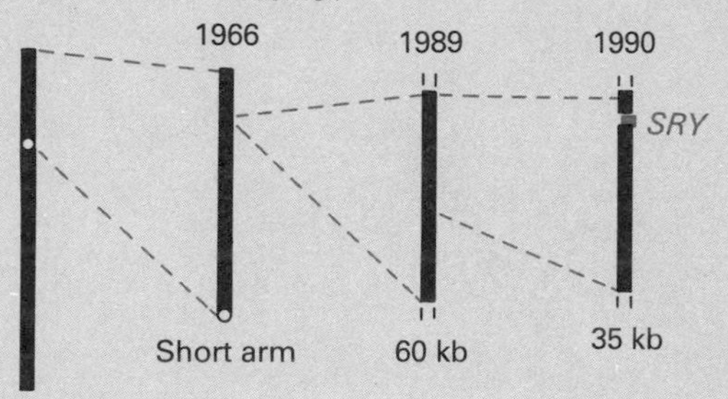

(Based on A. McLaren (1991) *Nature*, **351**, 96.)

Is this the only gene needed for maleness? The latest results suggest that it is. 'Female' mice turn out to be male even if the only component of the Y chromosome that they possess is a 14 kb fragment containing the mouse equivalent of *SRY*.

promptings from the biological research community. The Project is a coordinated programme involving laboratories in the USA, Europe, Japan and elsewhere, and has as its objective the complete genome sequence by the year 2005. In the initial phase of the work the technical and procedural problems posed by this massive undertaking are being tackled. DNA sequencing is now routine but a single person cannot obtain more than about 2 kb of sequence per day. This would be much too slow to make any significant inroad into the human genome, even if hundreds of people were sequencing different regions of the genome at the same time. Machines that sequence DNA automatically are therefore being developed, the most successful (from Japan) capable of 100 kb per day. A second problem is that it will be necessary to break the genome into small pieces, each 50–100 kb or so in length, to provide the molecules that will go into the sequencing machines. The relative positions of these molecules in the genome must be known, so that the sequences that are obtained can be fitted together in the correct order. Techniques for doing this are already available, but it will take some time to obtain and identify all the molecules that will be needed. Clearly the Human Genome Project represents a tremendous amount of work, but in the end the information that it will provide on human biology will make the effort well worthwhile.

16.2 The human genome as a tool to trace mankind's origins

The human genome can tell us many things. Not only does it contain the information needed to construct a functioning human being, it also carries a record of that person's ancestry and origins. To understand how ancestry can be read in a person's genes it will be necessary to deal with one or two concepts that in the natural course of events we would leave until Part Three of the book. This does not mean that the next few sections will be difficult to follow, but you may wish to postpone them until after you have covered Chapters 17 and 18.

16.2.1 Polymorphisms in the human genome

One important feature of the human genome is that it is variable: no two people have identical genomes. This is because a number of parts of the genome are **polymorphic**, meaning that they can exist in any of two or more forms. Extragenic regions in particular tend to be highly polymorphic: for instance, the number of repeat units

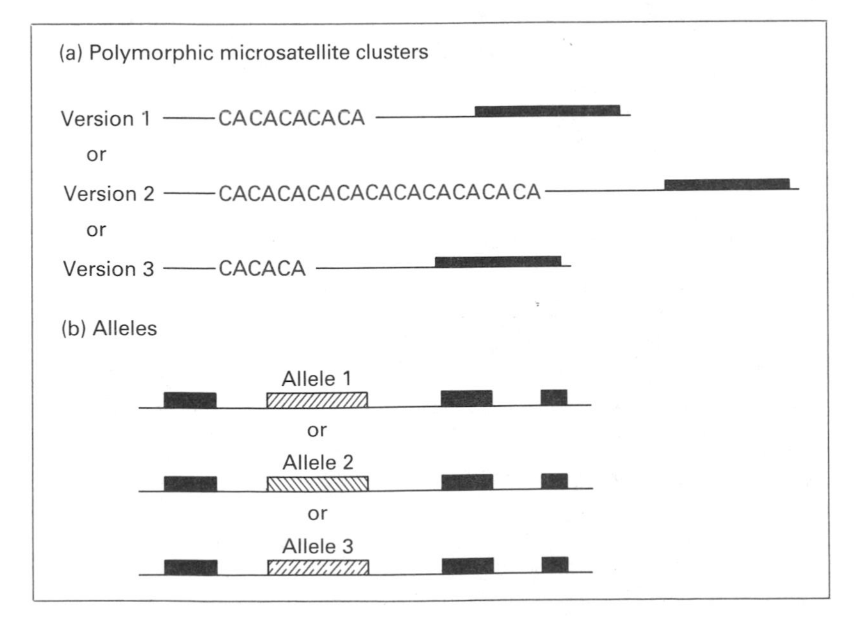

Figure 16.6 Polymorphisms.

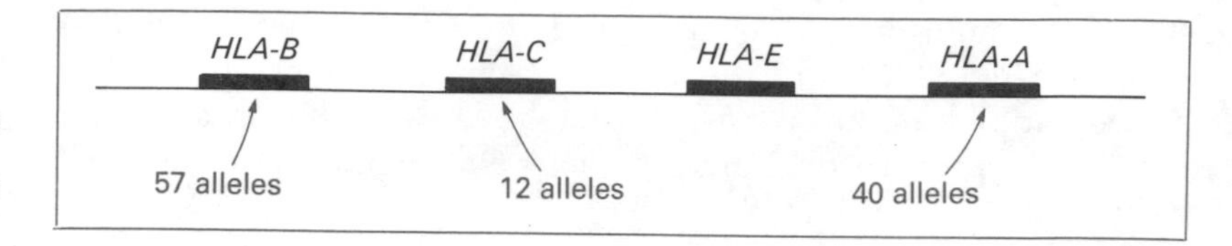

Figure 16.7 A part of the major histocompatibility locus showing the positions of four *HLA* genes, three of which are polymorphic. The numbers of alleles for the three genes are probably greater than shown. as new alleles are still being discovered.

in a microsatellite DNA cluster varies from person to person (Fig. 16.6(a)). Genes can also be polymorphic, the different versions being called **alleles** (Fig. 16.6(b)). Alleles of a gene are very similar in nucleotide sequence, but each codes for a slightly different protein product. This is not the same as a multigene family, in which multiple copies of a gene are found in the same genome; alleles refer to a single gene that can exist in more than one form.

Polymorphic genes rarely have more than a limited number of alleles, four or five at most, but a few exceptions are known. An example is provided by the **major histocompatibility complex**, which is located on chromosome 6 (see Fig. 16.2) and consists of a series of genes coding for proteins involved in the immune response. Several of these genes are highly polymorphic, one of them, *HLA-B*, having at least 57 different alleles (Fig. 16.7). An individual carries just two of these alleles, one on each of his or her two versions of chromosome 6. One allele will have been inherited from the father, one from the mother.

(a) Population groups can be distinguished by determining allele distributions. An individual inherits alleles from his or her parents. It therefore follows that two or more brothers and sisters will have related alleles. Even if you have a dozen brothers and sisters, you will still only have a maximum of four *HLA-B* alleles between you (Fig. 16.8). If we extend the group to include your cousins, aunts, uncles and grandparents then we will still find only a limited number of *HLA-B* alleles.

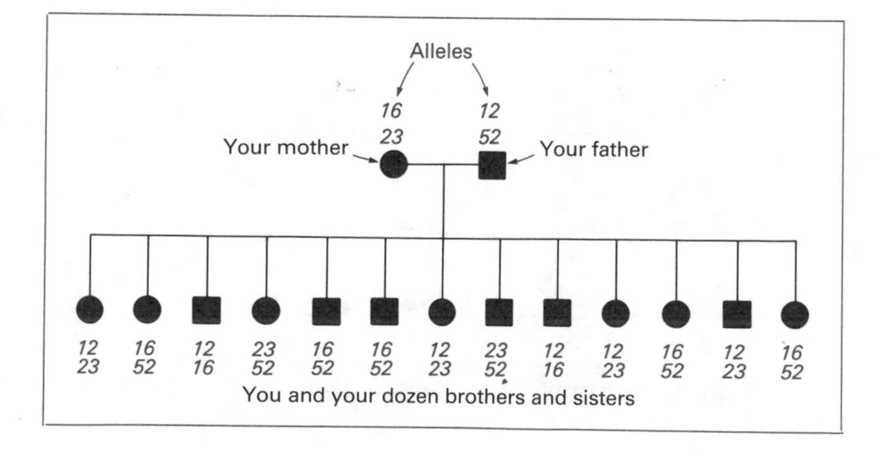

Figure 16.8 A family group possesses a maximum of just four different *HLA-B* alleles. In this example the mother possesses alleles 16 and 23, and the father has alleles 12 and 52. The children possess combinations of these alleles.

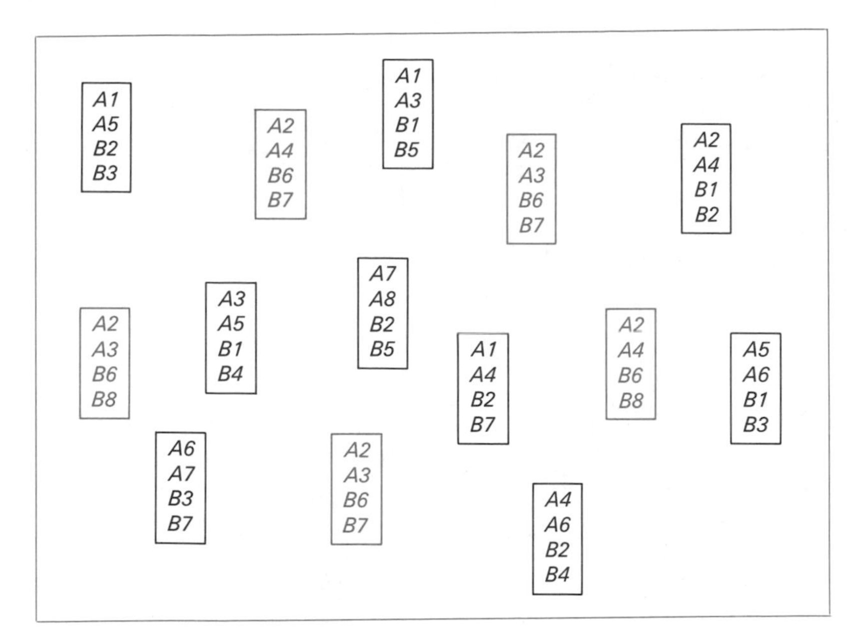

Figure 16.9 If we examine a number of individuals then we may identify a few with related alleles; these individuals may share a common ancestor. Each box represents an individual person and shows their allele pairs for genes *A* and *B*. The five individuals that are highlighted have very similar allele sets and so may be descended from a common ancestor.

Let us now reverse the argument and see what would happen if we look at the alleles possessed by individuals who are not closely related. We might be able to assign individuals to a series of population groups based on the alleles that they carry. To do this we would have to examine the alleles of a number of different genes, and then identify individuals who carry similar sets of alleles, and who are therefore possibly descended from the same ancestors (Fig. 16.9). The potential of this approach has been illustrated by a recent study involving 19 different genes, and a total of 63 alleles, in individuals at over 3000 places in Europe. The results have to be interpreted with care, but the distribution patterns of the alleles enable 33 'genetic boundaries' to be identified (Fig. 16.10). On either side of a genetic boundary the allele combinations are sufficiently different to suggest that relatively little mixing of the populations has occurred. When we examine the locations of the genetic boundaries we see that several represent geographical barriers, for example the North Sea, over which population mixing is unlikely to occur. Other genetic boundaries coincide with known cultural or linguistic divisions, such as between the Basque region and the rest of Spain, and again we can appreciate the reasons for the genetic distinctness. Still other genetic boundaries mark more subtle divisions between populations. Iceland, for example, is divided by a genetic boundary that delineates the west of the country, originally settled by people from the British Isles, from the east, where the population traces its roots back to the Vikings. These settlements were established a

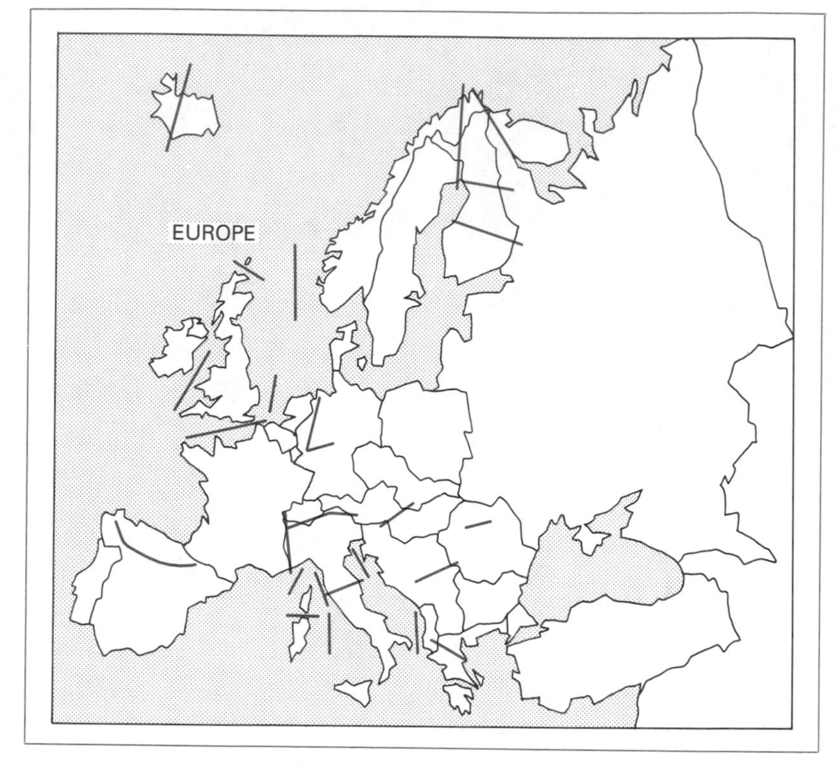

Figure 16.10 The positions of some of the 33 genetic boundaries identified by analysing allele distributions in Europe. Redrawn from G. Barbujani and R. R. Sokal (1990) *Proceedings of the National Academy of Sciences USA*, **87**, 1816–19.

thousand years ago, but the differences between the original populations can still be traced in the genes of the modern Icelanders.

(b) Further back to our earliest ancestors. How far back can human ancestry be traced in the genes of people living today? One answer to this question has been provided in dramatic fashion by Rebecca Cann, Mark Stoneking and Allan Wilson of the University of California at Berkeley. They compared polymorphic regions of the mitochondrial DNAs of 147 individuals and deduced that all modern humans are descended from a single woman who lived in Africa about 200 000 years ago.

This remarkable conclusion is based on the assumption that mutations accumulate at a gradual and constant rate in mitochondrial DNA molecules. If this is the case then the degree of difference between two mitochondrial DNAs will provide a measure of how closely related two individuals are. Not only can we say that two individuals with similar mitochondrial DNAs are 'closely related', we can also use the degree of difference between the mitochondrial DNAs to calculate when the common ancestor of the two individuals lived. Once this point has been grasped it is quite easy to follow the deductions that lead to our common maternal ancestor. The 147 mitochondrial DNAs that were compared by

Allan Wilson b. 18 October 1934, Ngaruawahia, New Zealand; d. 21 July 1991, Seattle

Allan Wilson studied in New Zealand and then in the USA, obtaining his Master's degree from Washington State University, Pullman and his PhD in 1961 from the University of California at Berkeley. He continued his research career at Berkeley and was one of the first proponents of the idea that the rate at which mutations accumulate in the genome can be used as a molecular clock to follow evolution. Most biologists accept that comparisons between the nucleotide sequences of the same genes in different species can be used to determine which organisms are closely related and which less so. Wilson's proposal – that polymorphisms can be used to follow lineages within a single species – is more controversial. Not surprisingly, the mitochondrial Eve hypothesis has stimulated much discussion among biologists and anthropologists, but if the reality of the molecular clock and the calibration of the rate of mitochondrial DNA evolution are accepted, then Wilson's conclusions cannot be doubted. Wilson's groups were also among the first to isolate ancient DNA, obtaining nucleotide sequences from a dried museum skin of a quagga, a type of zebra that died out at the end of the last century.

Allan Wilson's group were collected from all parts of the world and represent all modern populations. Comparisons between these mitochondrial DNAs enable a giant 'family tree' to be constructed (Fig. 16.11). The tree leads back to a common ancestral mitochondrial DNA that, according to the degrees of difference between the modern mitochondrial DNAs, must have existed between 140 000 and 290 000 years ago. The distribution of the mitochondrial DNA variants in modern populations suggests that this ancestral molecule was located in Africa. And it must have been in a woman as mitochondrial DNA is inherited solely from the mother: the father's mitochondrial DNA is not passed to the offspring. Eve has been identified!

We must not get carried away. The hypothesis does not state that the mitochondrial Eve was the only woman on the planet at that time, only that she carried the ancestral mitochondrial DNA from which all existing mitochondrial DNAs are descended. In fact, equivalent analyses with nuclear DNA suggest that the total human population has not dropped below 10 000 since *Homo sapiens* evolved from *Homo erectus* some 400 000 years ago. The mitochondrial Eve therefore had many female contemporaries. The distinction is that the mitochondrial lineages derived from all these other women have now died out, whereas Eve's has persisted.

The molecular clock

Allan Wilson calculated that the accumulation of mutations in mitochondrial DNA results in a divergence of 2–4% over a period of one million years. In other words, two mitochondrial DNAs whose common ancestor existed one million years ago should now differ at two to four nucleotide positions out of every 100. The rate at which mutations accumulate therefore provides a **molecular clock** that can be used to calibrate evolution, as demonstrated by the mitochondrial Eve hypothesis. Among the 147 modern mitochondrial DNAs studied by Wilson's group, the greatest divergence between any two pairs was 0.57%. According to the molecular clock this corresponds to a common ancestor 140 000 to 290 000 years ago.

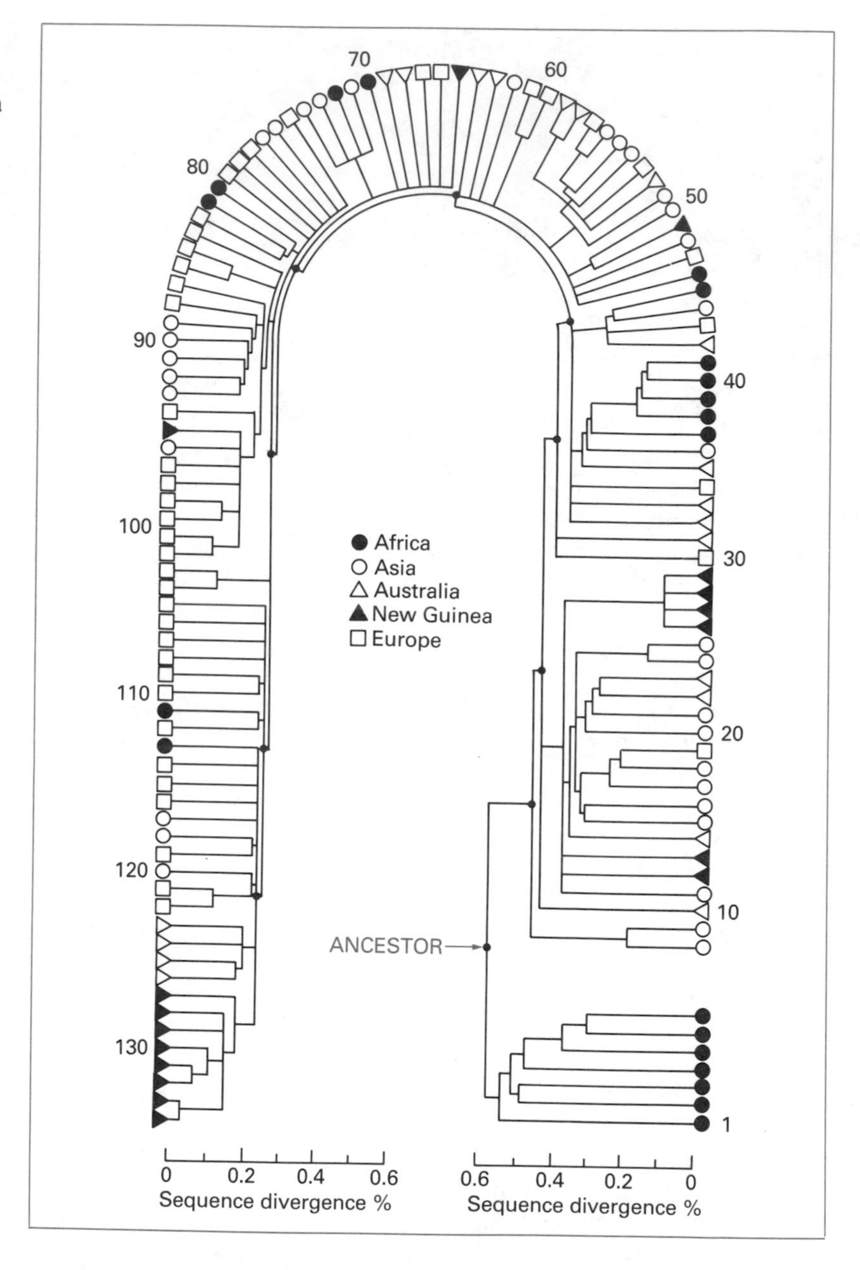

Figure 16.11 The mitochondrial DNA family tree that leads back to a common maternal ancestor. Each circle, square or triangle represents a single modern mitochondrial DNA, the shape of the symbol indicating its geographical origin. The branch points in the tree are positioned according to the degree of difference (i.e. sequence divergence) between that they link. Reprinted by permission from *Nature* vol. 325, pp. 31–36; Copyright © 1987 Macmillan Magazines Limited.

16.2.2 The ancient DNA time machine

Our ideas on past populations and the mitochondrial Eve hypothesis are based on studies of modern DNA sequences, with the critical deductions being made by working backwards towards ancestral molecules. Is this the only way to study ancient genomes?

Remarkably a time machine exists that enables us to travel back and examine the ancestral molecules directly. This time machine

...GTCTCCTGTAGTAGATCTTGGGGTCGTTCCATCAATTATAT...

Figure 16.12 A segment of the nucleotide sequence of an *HLA* gene from a 1-year-old boy who died approximately 2430 years ago. Taken from S. Pääbo (1985) *Nature*, **314**, 644–5.

was discovered independently in 1984 by Russell Higuchi of Berkeley and Svante Pääbo of the University of Uppsala. Both managed to extract small quantities of DNA from ancient biological material, in Pääbo's case from Egyptian mummies over 2400 years old. The DNA was very degraded but still sufficiently intact for nucleotide sequences to be obtained (Fig. 16.12).

Since this initial breakthrough **ancient DNA** has been extracted from a variety of sources, including human bones dating back 9000 years. Ancient nucleotide sequences are being used to study the movements of prehistoric populations in Neolithic Europe, and to follow the ancestors of the Amerindians as they crossed the Bering Strait and migrated throughout the New World. The time limit for ancient DNA has not yet been reached and nucleotide sequences are being obtained from increasingly older remains. We may not be able to find Eve herself, but by examining some of her long-dead descendants we may discover exactly who she was.

Reading

T. Strachan (1992) *The Human Genome*, BIOS, Oxford – the best description of the human genome currently available.

E. Ullu and C. Tschudi (1984) Alu sequences are processed 7SL RNA genes. *Nature*, **312**, 171–2; T. G. Fanning and M. F. Singer (1988) LINE-1: a mammalian transposable element. *Biochimica et Biophysica Acta*, **910**, 203–12 – two important research papers about SINEs and LINEs.

A. McLaren (1990) What makes a man a man? *Nature*, **346**, 216–17 – a review of the genetic basis to gender.

G. Barbujani and R. R. Sokal (1990) Zones of sharp genetic change in Europe are also linguistic boundaries. *Proceedings of the National Academy of Sciences USA*, **87**, 1816–19; N. H. Barton and J. S. Jones (1990) The language of genes. *Nature*, **346**, 415–6 – a research paper and a review that describe how European population groups can be distinguished by examining the distribution of alleles.

R. L. Cann, M. Stoneking and A. C. Wilson (1987) Mitochondrial DNA and human evolution. *Nature*, **325**, 31–6 – the mitochondrial Eve; this paper is well-written and is worth tackling.

J. Klein, J. Gutknecht and N. Fischer (1990) The major histocompatibility complex and human evolution. *Trends in*

Genetics, **6**, 7–11 – a quite difficult review that explains how *HLA* alleles can be used to study past populations.

S. Pääbo (1985) Molecular cloning of Ancient Egyptian mummy DNA. *Nature*, **314**, 644–5 – the first discovery of ancient DNA in archaeological remains.

Problems

1. Define the following terms:

   pseudogene
   extragenic DNA
   SINE
   LINE
   Alu family
   LINE-1
   non-viral retroelement
   satellite DNA
   polymorphism
   allele
   major histocompatibility complex
   ancient DNA

2. Provide examples of the different kinds of organization displayed by human multigene families.
3. Distinguish between a conventional and processed pseudogene.
4. Explain what is meant by the term 'extragenic DNA'.
5. Describe one example of a SINE and one of a LINE. What are the differences and similarities between these two types of element?
6. What is 'satellite DNA'? Distinguish between the three categories of satellite DNA in the human genome.
7. Outline the immediate and long-term objectives of the Human Genome Project.
8. Explain why a series of alleles is not the same thing as a multigene family.
9. How can allele distributions be used to study modern populations?
10. Who was the mitochondrial Eve?
11. Describe what ancient DNA is and how it is being used to study past populations.
12. *Discuss possible functions for the extragenic component of the human genome.
13. *The Human Genome Project is expected to cost $3 billion. Will this be money well spent?
14. *Can the mitochondrial Eve hypothesis be trusted?

Part Three
Studying Genes

In Part Three we will cover in some detail the major experimental techniques that have been responsible for the discoveries of the past and which will enable advances to be made in the future. We will start with Mendel, whose experiments laid the foundation to modern genetics, and then in Chapters 18 and 19 examine how the Mendelian approach can be used to study genes in eukaryotes and prokaryotes. Finally, in Chapters 20 and 21, we will look at the new techniques, centred on gene cloning, that have been responsible for the remarkable advances in molecular genetics that have taken place over the past 20 years.

17 What Mendel discovered

The scientific examination of inheritance – the monohybrid crosses – crosses involving two pairs of characteristics • How molecular genetics relates to Mendel – pairs of alleles – dominance and recessiveness – molecular explanations for Mendel's Laws

Mendel's contribution to biology cannot be overemphasized and it is not because of romanticism or nostalgia that his name is still revered by molecular geneticists. From the results of a series of straightforward although painstaking experiments, Mendel was able to deduce two fundamental Laws of Genetics, Laws that we still recognize today as encapsulating the underlying principles of inheritance and gene activity. The rediscovery of Mendel's work, and the acceptance of his results, led to the establishment of genetics as a central component of twentieth-century biology. In addition, Mendel's experimental approach forms the basis for more sophisticated techniques, developed by Thomas Hunt Morgan and his colleagues, that enable the positions of genes to be mapped on eukaryotic chromosomes. Our examination of the methodology for studying genes must therefore start with Gregor Mendel and his work, the first real experiments in genetics.

17.1 The scientific examination of inheritance

Mendel chose the garden pea, *Pisum sativum*, as his experimental organism and set out to search for a pattern to the way in which inherited characteristics are passed from parents to offspring. In the mid-nineteenth century the fact that a child displays characteristics of both mother and father was ascribed to vague events such as 'cytoplasmic mixing' and the accepted wisdom was that inheritance could not be studied in a scientific or statistical manner. Mendel was not so sure.

(a) Pairs of contrasting characteristics. Two considerations were taken into account by Mendel when choosing the particular characteristics to study with his garden peas. The first and overriding one was that there should be alternative forms to each

Table 17.1 The pairs of contrasting characteristics studied by Mendel

Feature	*Contrasting characteristics*
Stem height	tall : short
Pod location	axial : terminal
Pod morphology	full : constricted
Pod colour	green : yellow
Pea morphology	round : wrinkled
Pea colour	yellow : green
Flower colour	violet : white

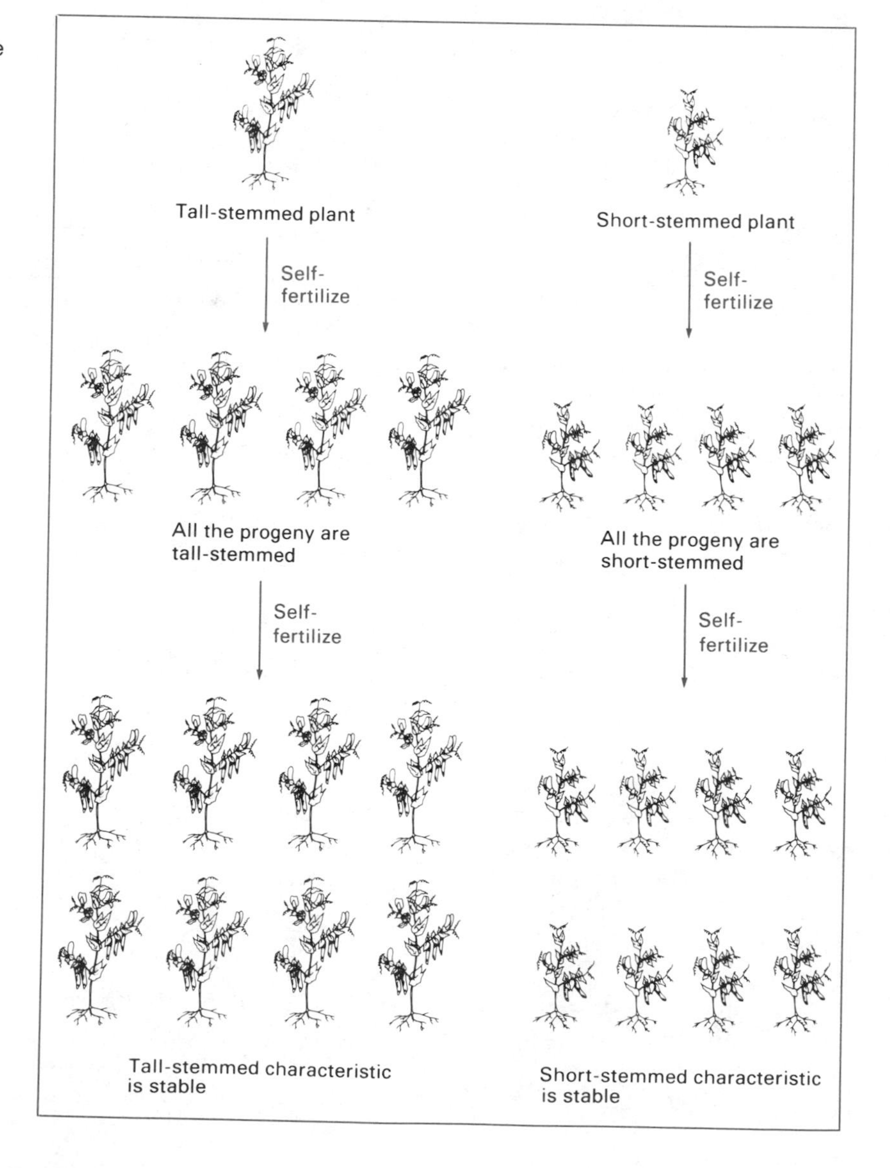

Figure 17.1 The characteristics chosen by Mendel were stable. The progeny plants produced from several rounds of self-fertilization displayed only the parental characteristics.

characteristic; that is to say, the plant could display one form of the characteristic or another, but not both at once. The best-known example of the contrasting pairs of characteristics chosen by Mendel is stem height, which he classified as tall or short.

The second consideration in choosing characteristics to study was a purely experimental one. Mendel anticipated that he would need to determine which characteristic of a pair was displayed by each of several hundred progeny plants in order to be able to distinguish a pattern of inheritance, should one exist. Consequently, his characteristics had to be scored easily. It would be tedious and difficult to score characteristics that were not obvious and needed dissection or chemical analysis before they could be assigned.

In the end Mendel selected seven pairs of characteristics, each of which had two distinct forms (Table 17.1). As well as stem height, he decided to study inheritance of three pairs of characteristics concerned with pod morphology, two pairs involving pea morphology, and a single pair of contrasting flower colours. Each of these characteristics was stable, so if plants were **self-fertilized** then all the progeny would be the same as the parents (Fig. 17.1).

17.1.1 The monohybrid crosses

At first, Mendel studied the inheritance of each pair of characteristics by **cross-fertilizing** plants with alternative forms of the same character (the **monohybrid crosses:** for example, tall-stemmed × short-stemmed). A clear and consistent pattern emerged from these seven monohybrid crosses. In each case the first generation (F_1) of plants arising from the cross displayed only one of the parental characteristics. For example, in the cross between tall-stemmed and short-stemmed plants (Fig. 17.2(a)), all the F_1 plants had tall stems and none had short stems. However, if members of the F_1 generation were self-fertilized, giving rise to a second generation (F_2) of plants, the characteristic which had been absent in the F_1 generation reappeared. In the case of tall-stemmed × short-stemmed the F_2 generation contained 787 plants with tall stems and 277 plants with short stems which, Mendel noted, was a ratio of 2.84:1.

Each of the seven monohybrid crosses produced a similar pattern. Let us take as a second example the cross involving flower colour (violet petals × white petals; Fig. 17.2(b)). In this case all the F_1 plants had violet petals, and in the F_2 generation 705 plants had violet petals and 224 had white petals, a ratio of 3.15:1. The results of each of the monohybrid crosses are summarized in Table 17.2. In each cross only one of the pairs of characteristics is displayed in the F_1 generation, but both appear in the F_2 generation. When all of the monohybrid crosses are added together, the F_2 ratio that is

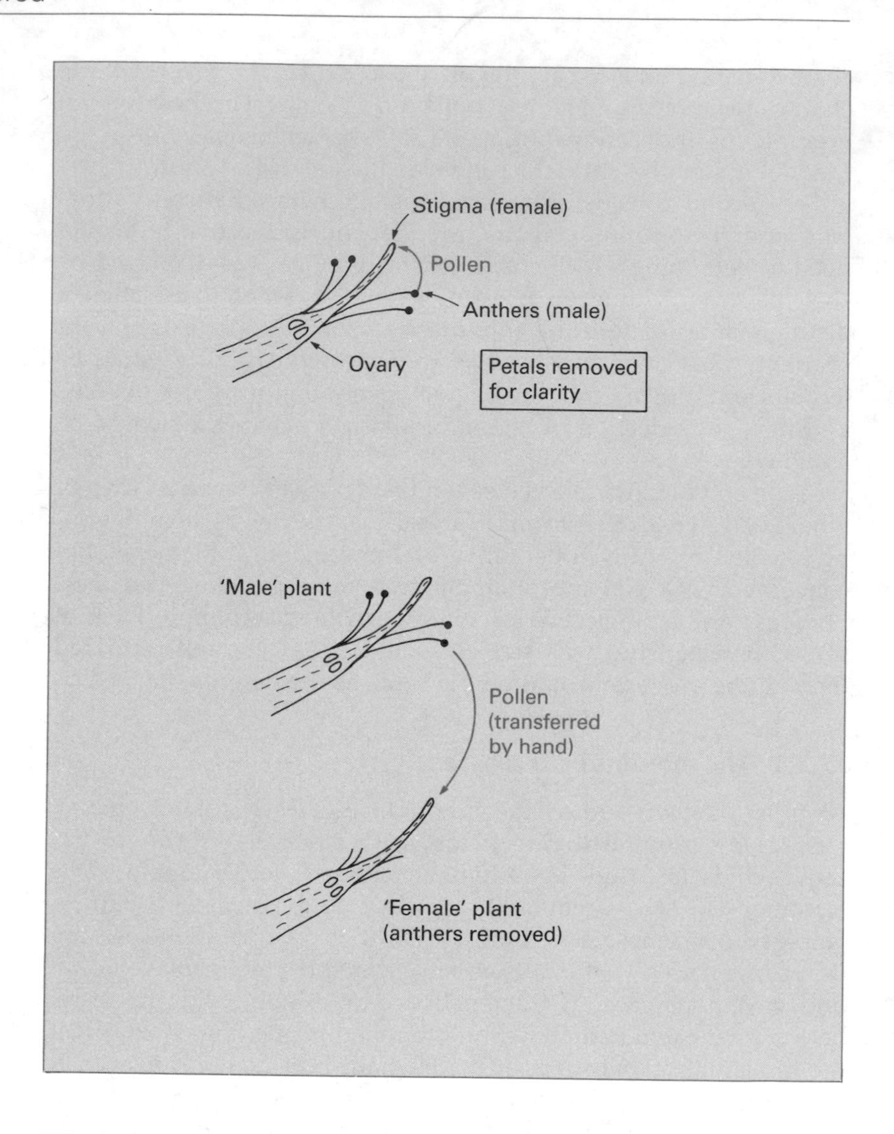

Table 17.2 The results of Mendel's monohybrid crosses

Cross	F_1	F_2	
		Numbers	*Ratio*
Tall stems × short stems	All tall	787 tall : 277 short	2.84 : 1
Axial pods × terminal pods	All axial	651 axial : 207 terminal	3.15 : 1
Full pods × constricted pods	All full	882 full : 299 constricted	2.95 : 1
Green pods × yellow pods	All green	428 green : 152 yellow	2.82 : 1
Round peas × wrinkled peas	All round	5474 round : 1850 wrinkled	2.96 : 1
Yellow peas × green peas	All yellow	6022 yellow : 2001 green	3.01 : 1
Violet flowers × white flowers	All violet	705 violet : 224 white	3.15 : 1
	Total:	14 949 : 5010	2.98 : 1

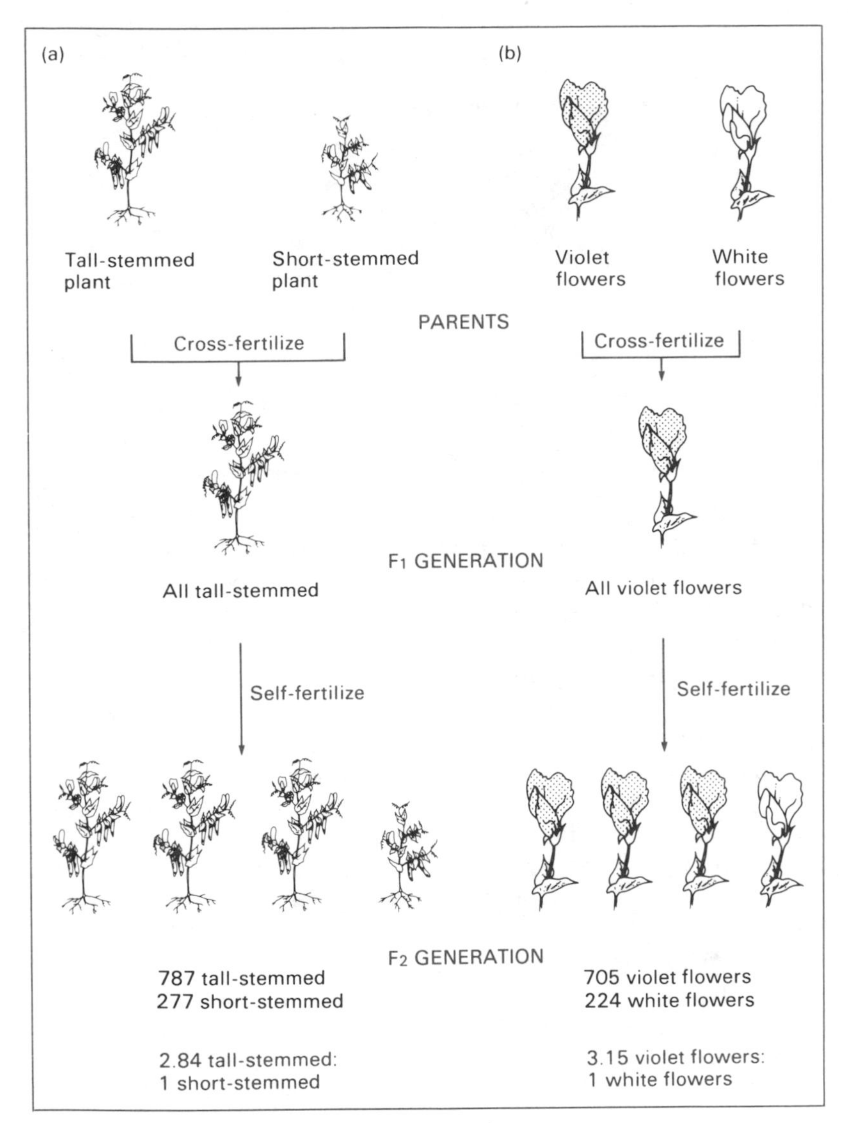

Figure 17.2 Two examples of the monohybrid crosses carried out by Mendel.

obtained works out to 2.98 plants with the F_1 characteristic to every one plant with the second characteristic. Mendel decided that within the limits of experimental error this was a real ratio of 3:1.

(a) Mendel's interpretation of the results of the monohybrid crosses. To explain the consistent pattern of inheritance that he had discovered, Mendel proposed that each trait (for example, plant height) is controlled by a **unit factor**, and that each unit factor can exist in more than one form, with different forms responsible for different characteristics. According to this interpretation, the unit factor for plant height can exist in two alternative forms

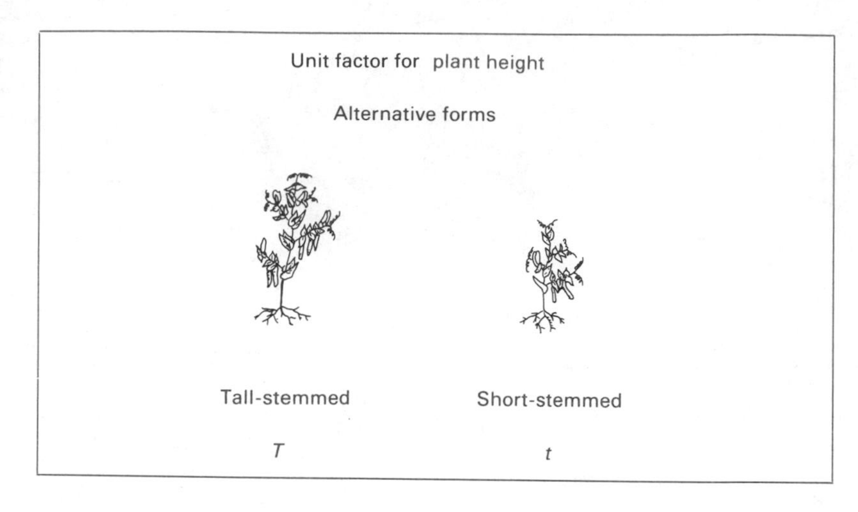

Figure 17.3 Mendel proposed that each trait is controlled by a unit factor that can exist as either of two alternative forms.

(Fig. 17.3), one responsible for tall stems (Mendel designated this version of the unit factor as '*T*') and one responsible for short stems (designated as '*t*'). Mendel imagined the unit factors to be physical structures within the cell. In fact they are exactly the same things that we now call genes: in effect Mendel proposed the existence of genes. As we discovered in Chapter 16, the alternative forms of a gene are now called **alleles**.

Mendel realized that the most important aspect of the monohybrid crosses was that, in each cross, one of the parental characteristics disappeared in the F_1 generation and then reappeared in the F_2 generation. In the crosses between plants with tall stems and plants with short stems, for example, the short-stemmed characteristic was absent in the F_1 generation, but present again in the F_2 plants. This means that the F_1 plants must contain two alleles, one for tall stems and one for short stems; if this were not the case then how could both alleles be transmitted to F_2 plants? This was the first important conclusion that Mendel made:

Each gene exists as a pair of alleles

In fact we would write the genotype (that is, the genetic constitution – Section 12.1) of the F_2 plants as *Tt*. We would refer to these plants as **heterozygotes**, indicating that they have two different alleles for this gene; a **homozygote** has two alleles the same (that is, *TT* or *tt*). Immediately we are able to understand Mendel's second conclusion:

Alleles can be dominant or recessive

Although the genotype of the F_1 plants is *Tt*, the phenotype (the observable characteristic) is tall-stemmed. Therefore the *T* allele must be able to dominate over the *t* allele, so that only the characteristic that it controls – tall stems – appears in the plant. We say that the *T* allele is **dominant** and the *t* allele **recessive**.

(b) The 3:1 ratios. Having established how one of the parental characteristics could disappear in the F_1 generation and reappear in the F_2 plants, Mendel went on to explain how the 3:1 ratios arise in the F_2 generation. As the F_1 plants have the genotype *Tt*, the F_1 cross can be written as:

$$Tt \times Tt$$

This notation is acceptable even though the F_1 cross is a self-fertilization, meaning that in reality it does not involve two individual plants, but instead different parts of the same plant (see p. 312).

We know that the F_1 cross produces some short-stemmed plants, whose genotypes must be *tt*. These F_2 plants must have obtained a *t* allele from each F_1 'parent': there is no other way in which they could arise. This suggests that during sexual reproduction the alleles of each parent separate (we use the term **segregate**), producing intermediate structures that contain just one allele. These intermediate structures are called **gametes** (Fig. 17.4). Two gametes, one from each parent, fuse to bring together the pair of alleles carried by a member of the next generation.

How do the events shown in Fig. 17.4 result in the 3:1 ratio? The useful construction called the **Punnett square** (Fig. 17.5) helps us to see that the 3:1 ratio arises naturally, as long as the F_1 gametes are able to segregate in an entirely random fashion. If segregation is random, then the genotypes displayed by the F_2 plants will be 1*TT*:2*Tt*:1*tt*. As both *TT* and *Tt* plants are tall-stemmed, the phenotypic ratio will be three tall-stemmed to one short-stemmed,

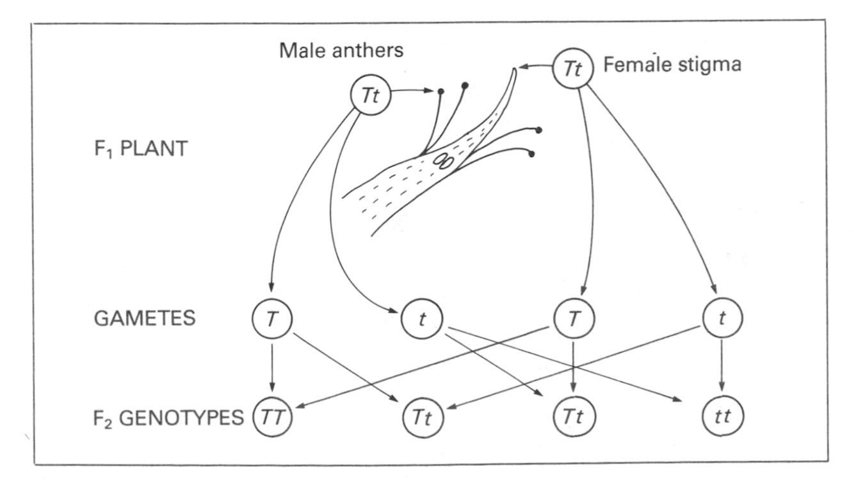

Figure 17.4 The segregation of alleles during the F_1 cross involving tall- and short-stemmed plants. The genotype of the F_1 plant is *Tt*. The male and female reproductive parts produce gametes, each gamete containing just a single allele, *T* or *t* in this case. Two alleles, one from each 'parent', fuse to produce the pair of alleles carried by an F_2 plant.

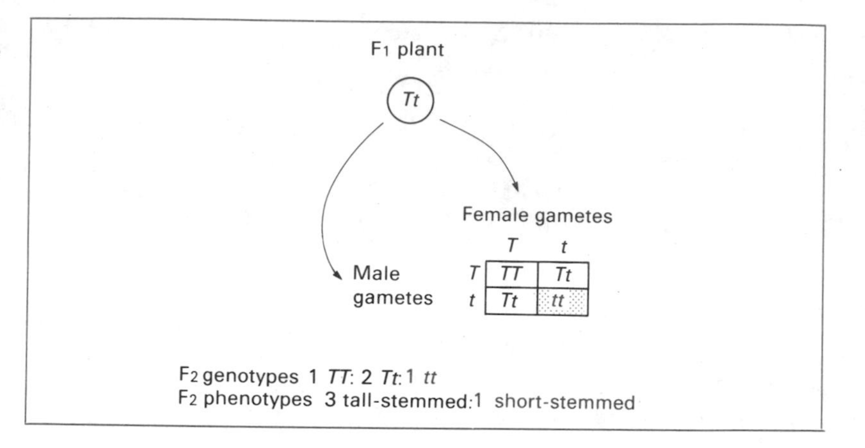

Figure 17.5 The Punnett square treatment of the F_1 cross involving the *T* and *t* alleles. The square enables the possible combinations of alleles to be worked out, and shows the numbers of each genotype that will result. Note that the 3 : 1 ratio will arise only if the gametes of each 'parent' segregate randomly.

exactly as Mendel observed. The 3:1 ratios therefore led Mendel to his First Law:

Alleles segregate randomly

17.1.2 Crosses involving two pairs of characteristics

Next Mendel examined whether different pairs of characteristics are inherited together or independently. He investigated this by carrying out a series of **dihybrid crosses**. As an example we

Lethal genotypes

An important exception to the 3 : 1 ratio occurs if one of the three genotypes turns out to be lethal. This was first shown by Lucien Cuénot in 1904, shortly after Mendel's work was rediscovered. Cuénot was one of the first geneticists to study the inheritance of coat colour with mice, a system that has become very popular as a means of demonstrating the principles of Mendelian genetics. Cuénot obtained the following results when he crossed mice with yellow coats:

Parents:	yellow × yellow
F_1 generation:	2 yellow : 1 agouti

The first thing that these results tell us is that yellow is dominant to agouti. This is clear from the fact that two yellow parents can give rise to an F_1 generation containing agouti mice. Also we can deduce that both yellow parents must be heterozygotes (*Yy*) as otherwise the homozygous recessive genotype (*yy* = agouti) could not occur in the F_1 generation. So the cross can be rewritten as:

Yy × *Yy*

In the F_1 generation we would therefore expect 1 *YY* : 2 *Yy* : 1 *yy*, giving the standard Mendelian of 3 yellow : 1 agouti. The actual results of 2 yellow : 1 agouti are most easily explained if we assume that for some reason the homozygous dominant mice (*YY*) do not survive:

Parents:	*Yy* × *Yy*
F_1 genotypes:	1 *YY* (dies) : 2 *Yy* : 1 *yy*
F_1 phenotypes:	2 surviving yellow : 1 agouti

Yellow coat colour in mice is just one of many characteristics known to be lethal in the homozygous form.

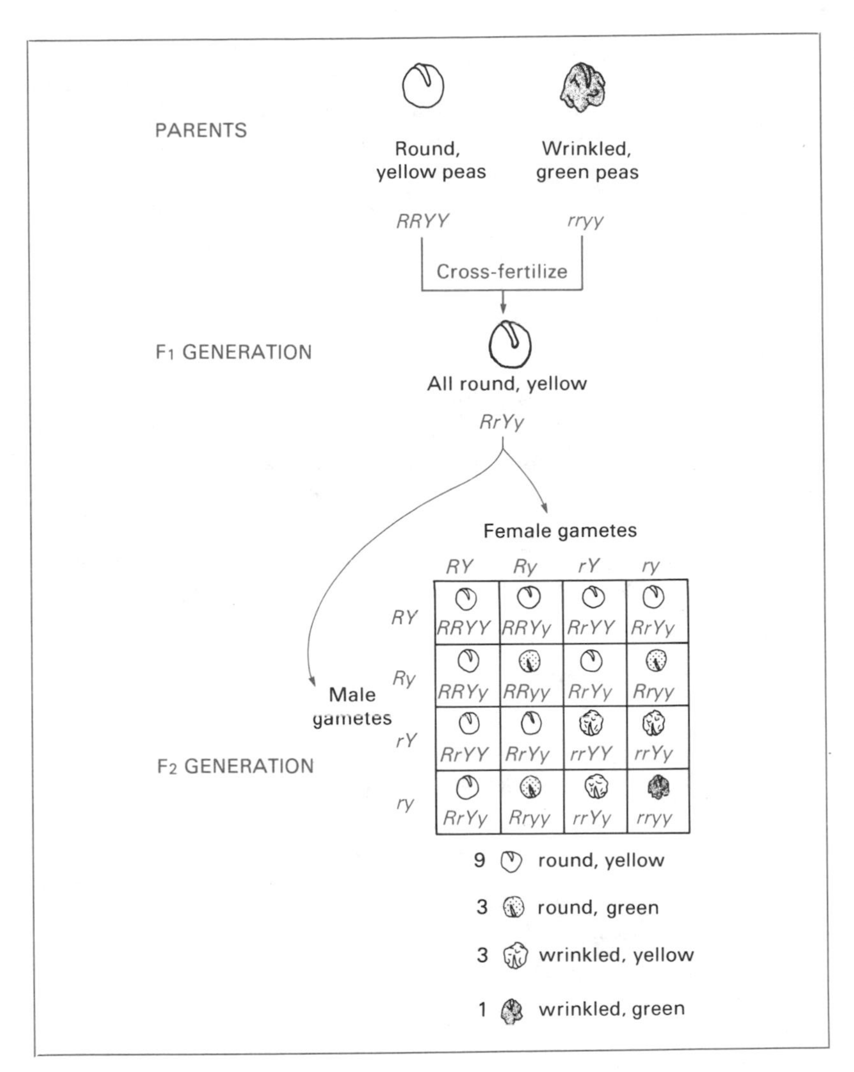

Figure 17.6 The dihybrid cross between a plant with round, yellow peas and one with wrinkled, green peas.

will consider the cross between plants with round, yellow peas (homozygous dominant *RRYY*) and plants with wrinkled, green peas (homozygous recessive *rryy*). This cross and its outcome is illustrated in Fig. 17.6. The F_1 generation all have the same genotype (*RrYy*) so all are double heterozygotes and have round, yellow peas. In the F_2 generation each possible combination of characteristics is represented, although in a regular ratio:

9 plants with round, yellow peas – a parental combination
3 plants with round, green peas } two non-parental
3 plants with wrinkled, yellow peas } combinations
1 plant with wrinkled, green peas – a parental combination

The first observation to make is that the parental characteristics (round, yellow peas and wrinkled, green peas) can be split, and non-parental combinations (round, green peas and wrinkled, yellow peas) can be obtained in the F_2 generation. Different pairs of characteristics are therefore not inherited together. But are they inherited completely independently? The Punnett square treatment of the dihybrid cross tells us that the answer is 'yes': the 9:3:3:1 ratio will arise only if the two pairs of characteristics segregate in an independent fashion. This is Mendel's Second Law:

Different pairs of alleles segregate independently

The dihybrid cross is most easily interpreted if it is looked on simply as two concurrent monohybrid crosses. Taking just the round:wrinkled component of the cross, the expected ratio of 3:1 in the F_2 generation holds true, although it is written above as 12:4. If we now examine just the F_2 plants with round peas, and bring into play Mendel's Second Law, then we would expect three-quarters of the round peas to be yellow and one-quarter to be green, and this is indeed the case – 9/12 yellow, 3/12 green. Now consider the plants with wrinkled peas. Three-quarters of these peas should be yellow and one-quarter green. Again this is the case – 3/4 yellow and 1/4 green.

17.2 How molecular genetics relates to Mendel

It might appear that a series of breeding experiments carried out with pea plants grown in an out-of-the-way part of Central Europe over a hundred years ago has little relevance to the topics now central to molecular genetics. This is a short-sighted view. Mendel's work anticipated several aspects of molecular genetics and provides a convincing demonstration of the potential contribution that genetic analysis can make to the understanding of molecular biological processes. Students of molecular genetics must attempt to develop a true appreciation of the contribution genetic breeding experiments have made and can make in forming hypotheses about gene behaviour at the molecular level. We will therefore examine exactly how Mendel's results fit in with what we now know about the gene.

17.2.1 Each gene exists as a pair of alleles

There is no difficulty in relating Mendel's first important conclusion to what has subsequently been discovered about genes and chromosomes. The chromosome theory proposed by Sutton in 1903 linked Mendel's unit factors to the chromosomes present in the nuclei of cells. The fact that genes occur in pairs is therefore explained by the fact that most cells of higher organisms are

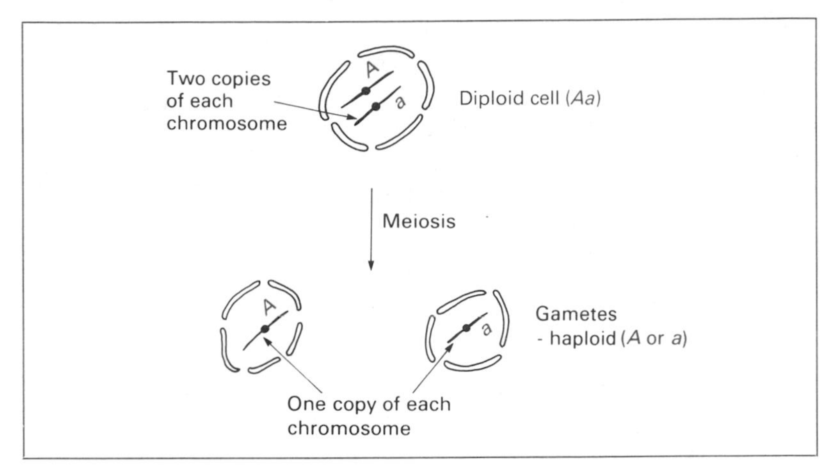

Figure 17.7 Mendel proposed that each cell contains two alleles (*Aa* in this example) but that alleles segregate during sexual reproduction so that each gamete contains just one (*A* or *a*). We now know that this is because in higher organisms such as pea plants, most cells are diploid but the gametes, produced by meiosis, are haploid.

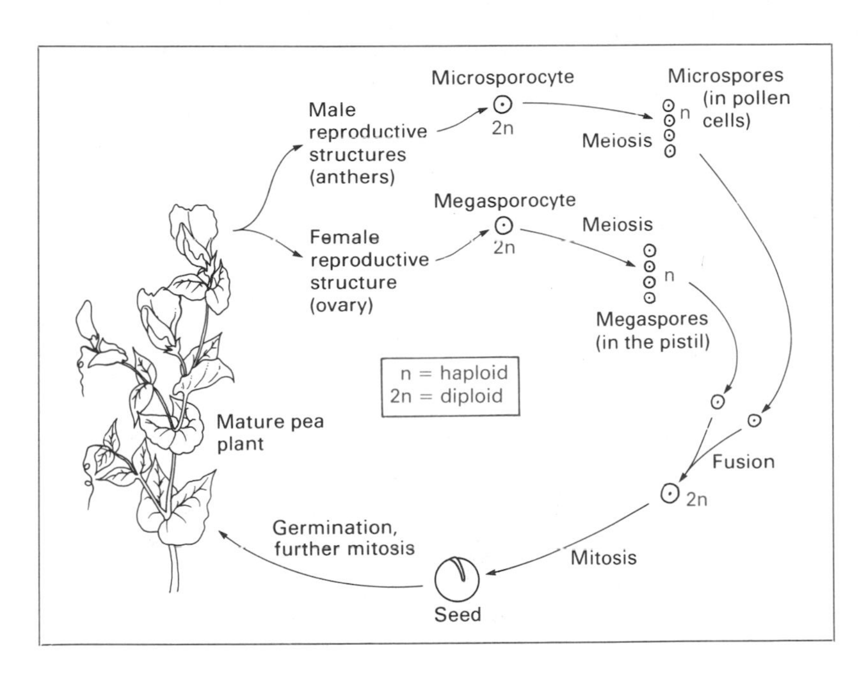

Figure 17.8 The life cycle of a pea plant. As with most higher organisms, in a pea plant the vegetative cells are diploid and the reproductive cells are haploid.

diploid (Section 15.1) meaning that each cell contains two copies of each chromosome, and hence two copies of each gene. During sexual reproduction, meiosis (Fig. 15.17) reduces this double set of chromosomes by half, producing cells (the gametes) that contain just one copy of each chromosome and hence one copy of each gene (Fig. 17.7). This is exactly as Mendel argued: alleles segregate during sexual reproduction.

Mendel's conclusion that alleles occur in pairs holds for the **vegetative** (i.e. non-reproductive) cells of most higher organisms. In pea plants, for example, only the reproductive cells (the male pollen and the female egg cells) are haploid (Fig. 17.8). However, things are different with many lower organisms, including all bacteria as well as some lower eukaryotes, notably certain strains of the yeast *Saccharomyces cerevisiae*. With these organisms, the vegetative cells themselves may be haploid and contain only a single allele for each gene. This fact is of considerable relevance to the methods used for genetic analysis of bacteria and yeast, and will be returned to in Chapter 18.

17.2.2 Dominance and recessiveness

In a heterozygote only one allele, the dominant one, contributes to the phenotype; the recessive allele is, in effect, silent. This is quite easily understood when we remember that genes code for RNA or protein molecules that have particular functions in the cell. Let us envisage a simple system in which a gene can exist in one of two forms (that is, alleles): functional and non-functional (Fig. 17.9(a)). The functional version of the gene we designate '*G*': it is the correct version that can be expressed by transcription and translation to produce a functional gene product. The non-functional form we will designate '*g*': this version contains a mutation that destroys the biological information carried by the gene so it no longer codes for the functional gene product.

Now let us look at the different possible genotypes and deduce what phenotypes will result (Fig. 17.9(b)). First, there is the homozygote, *GG*. An organism with this genotype will carry two functional copies of the gene and so will be able to synthesize the

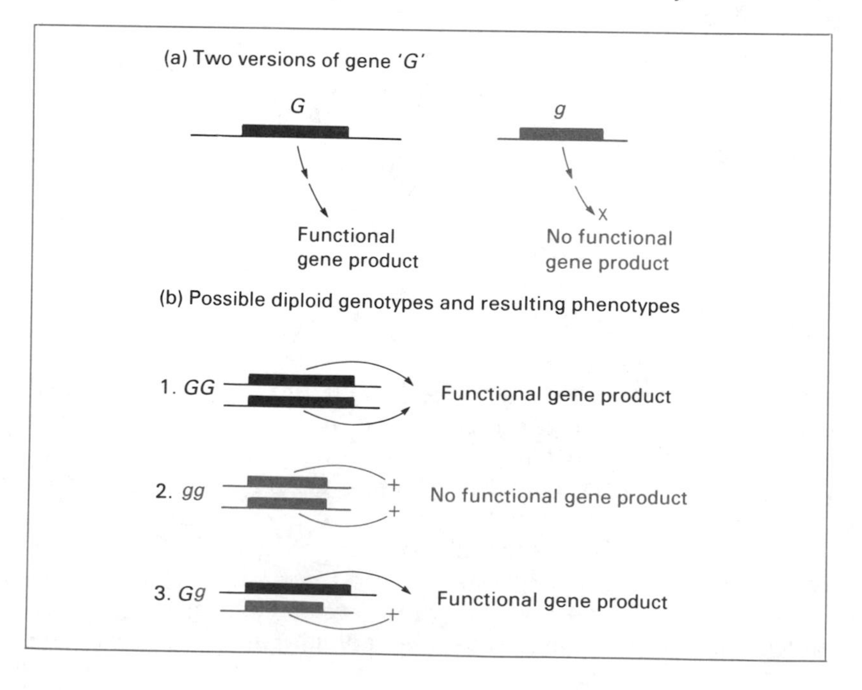

Figure 17.9 A molecular explanation for dominance and recessiveness. In this example *G* is dominant and *g* recessive.

gene product. The characteristic that the gene product is responsible for will therefore be displayed by the organism. Second, there is the other homozygote, *gg*. Now there are two non-functional versions of the gene: the organism cannot synthesize the gene product and the characteristic it is responsible for will be absent. Finally, there is the heterozygote, *Gg*, with one functional and one non-functional copy of the gene. This organism contains a functional gene and so will be able to synthesize the gene product: the characteristic will be displayed. *G* will be dominant and *g* recessive, precisely as Mendel deduced.

(a) Molecular explanations for the traits studied by Mendel. How does our molecular explanation of dominance and recessiveness correlate with the characteristics actually studied by Mendel? For some traits we now understand the allele pair quite well. Round peas and wrinkled peas provide a good example.

A wrinkled pea has a higher sucrose content than a round pea, and as a consequence absorbs more water while it is developing in the pod. When the pea reaches maturity some of this water is lost, causing the pea to collapse and become wrinkled. Round peas, which have less sucrose but more starch, do not take up so much water when they are developing and so become less dehydrated when they mature: as a result they stay round (Fig. 17.10(a)). The key difference between wrinkled and round peas therefore lies with the relative amounts of sucrose and starch that they contain. This in turn depends on the activities of a series of enzymes that in the 'normal' (round) pea convert the bulk of the sucrose into starch (Fig. 17.10(b)). Each of these enzymes is of course coded by a gene. The *r* allele is a mutated version of one of these genes, so a plant with the genotype *rr* does not produce the enzyme coded by the gene. This means that *rr* peas convert less sucrose into starch, so they become wrinkled. In the heterozygous state, *Rr*, the *R* allele directs synthesis of the enzyme in sufficient amounts to reduce the sucrose level. *Rr* plants therefore have round seeds: *R* is dominant to *r*.

One or two of the traits studied by Mendel are more difficult. The reason why tall stems are dominant over short stems is more tricky to explain in terms of single genes. But what is important here is that the demonstration of a single pair of alleles for a trait, and the determination of which allele is dominant and which recessive, provides important information on which to base additional experiments to find the molecular explanation for the phenomenon.

(b) Complications not observed by Mendel. Mendel was fortunate in that each of the seven pairs of contrasting characteristics that he studied display a simple dominant–recessive rela-

Figure 17.10 The molecular basis to the round–wrinkled pea phenotype controlled by the *R* and *r* alleles.

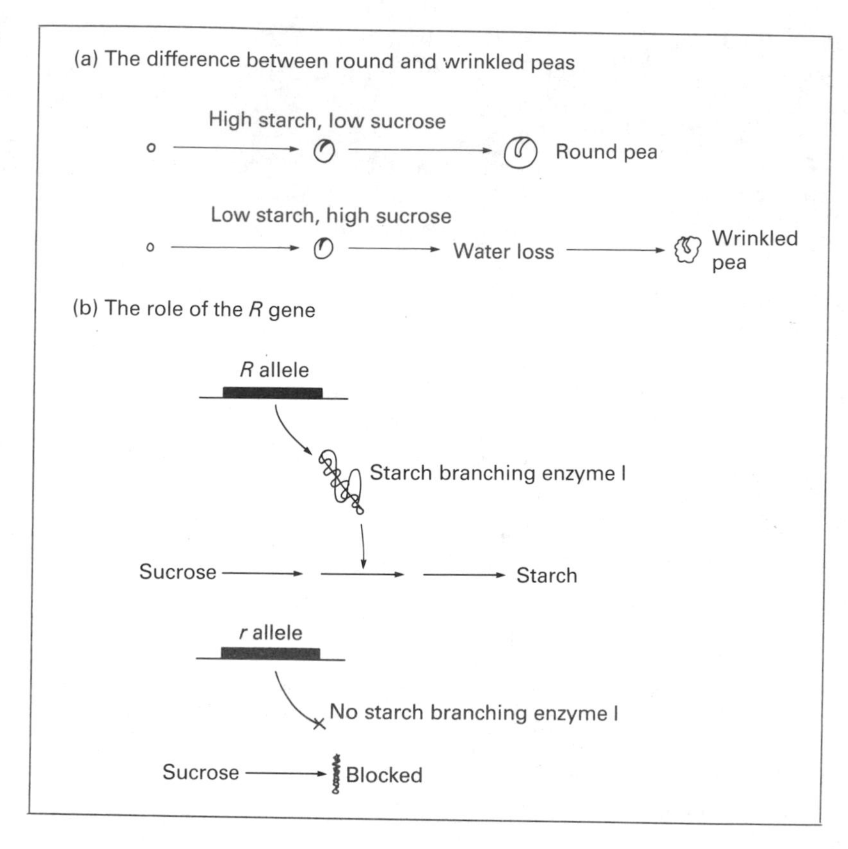

tionship, and can be explained in terms of dominant = active gene and recessive = inactive gene. This is not always the case and variations on this basic theme are known; three examples follow:

1. ***Incomplete dominance***, where the heterozygous form displays a phenotype intermediate between the two homozygous forms. Flower colour in plants such as carnations (but not peas) is an example, with *RR* being red and *rr* white, but with the heterozygous *Rr* pink. An explanation is that the heterozygous form, although possessing a functional gene for pigment synthesis, cannot produce sufficient quantities of the relevant enzyme and so only limited amounts of the pigment are made (Fig. 17.11).
2. ***Codominance***, where both alleles are active in the heterozygous state. Normally, to observe codominance it is necessary to examine the biochemistry of the organism. This is the case with the best understood example, the MN blood group series in man. The MN series centres on glycoprotein molecules in the blood which exist in two forms, M and N. These are coded by a pair of alleles designated L^{M} and L^{N} (Fig. 17.12). Homozygous $L^{M}L^{M}$ produce just M glycoproteins, and the blood group is

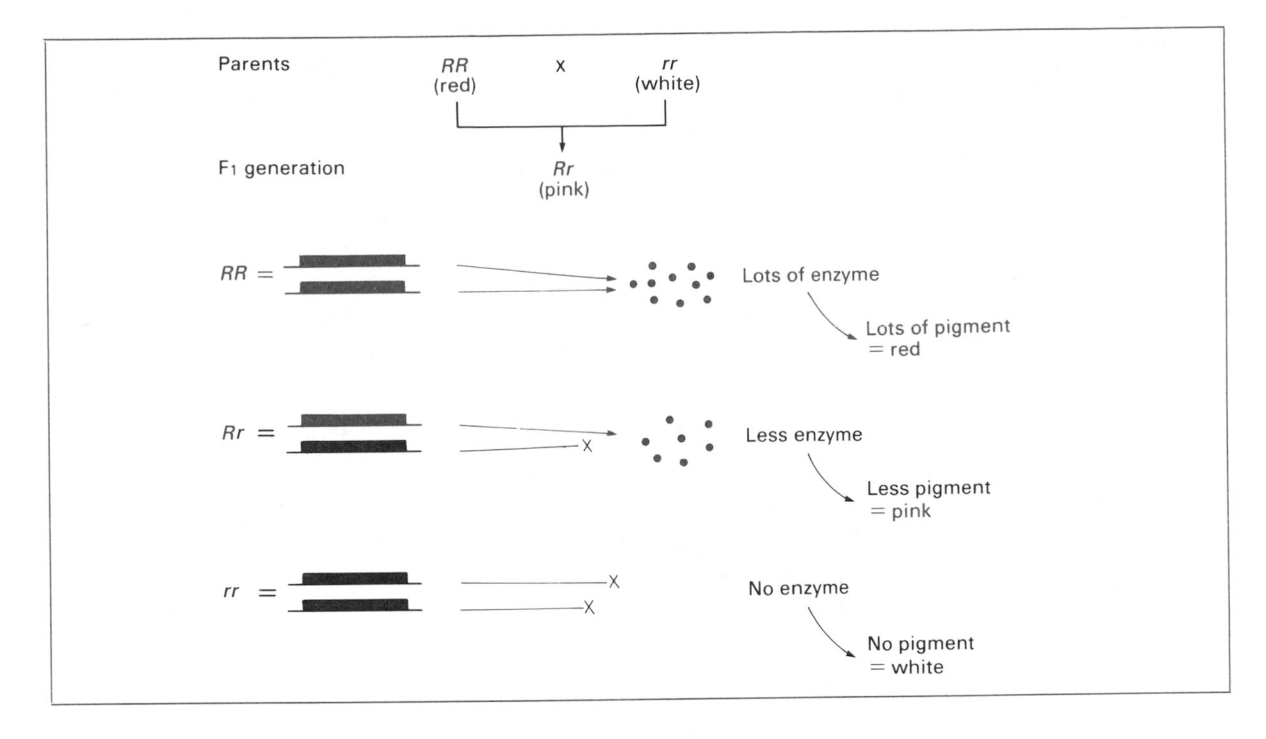

Figure 17.11 Incomplete dominance with alleles for red and white flower colour in plants such as carnations.

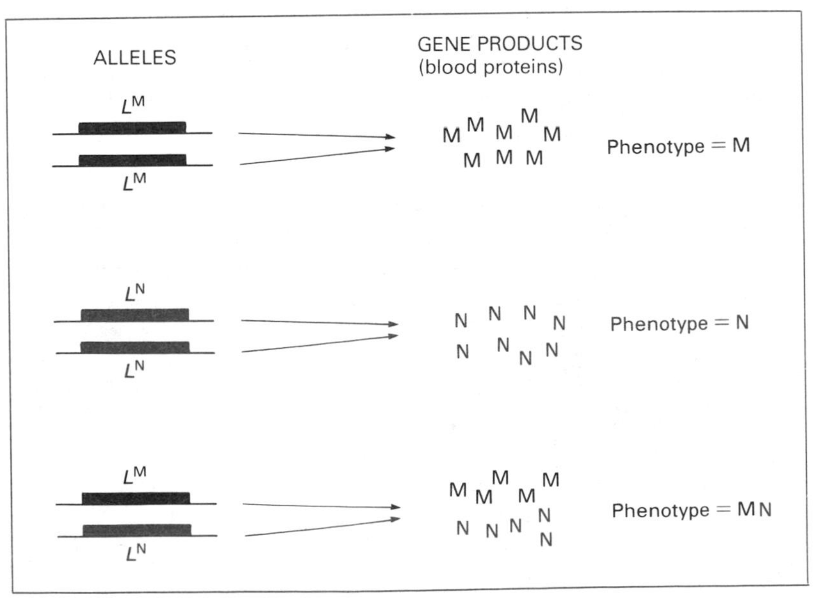

Figure 17.12 Codominance: the MN blood group in man.

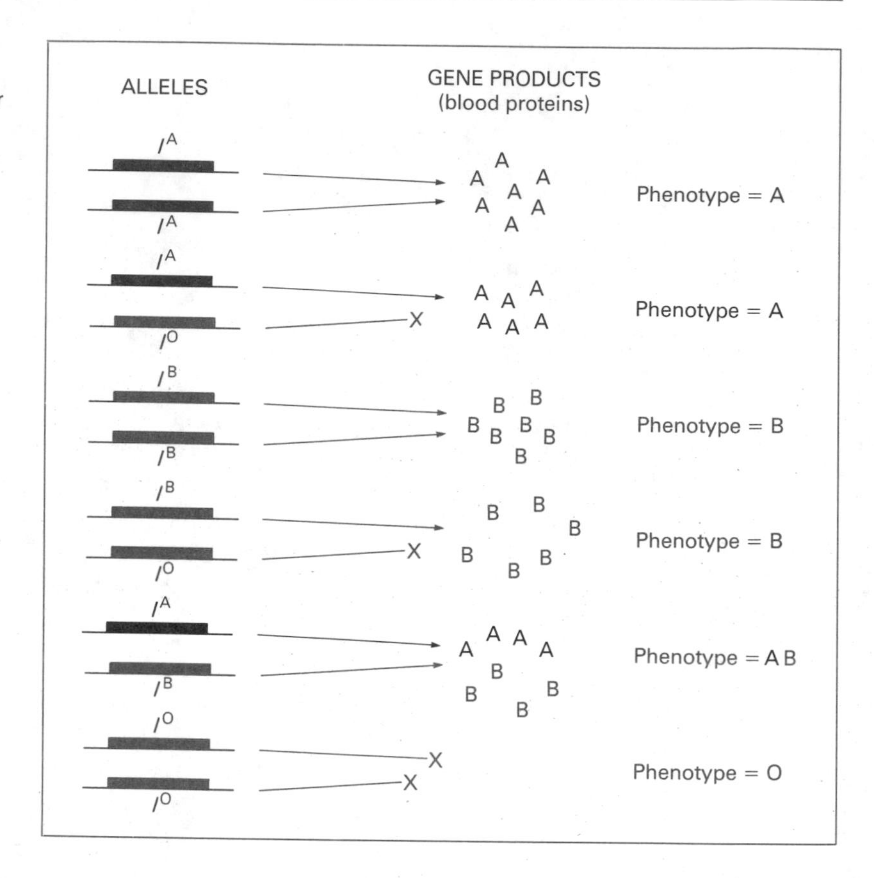

Figure 17.13 Multiple alleles: the ABO blood series in man. I^A is dominant over I^O; I^B is dominant over I^O; I^A and I^B display codominance.

therefore M. Similarly, homozygous L^NL^N produce N glycoproteins and give rise to blood group N. However, in the heterozygote, both alleles are present (L^ML^N) and both still direct synthesis of glycoproteins. The blood type is MN: neither allele is dominant. Several other examples of codominance are known.

3. ***Multiple alleles*** would probably not have complicated Mendel's data but provide still another variation on the simple dominant–recessive pattern. Some genes can exist as more than just two alleles, and instead have multiple allelic forms. A single organism has only two alleles, but different individuals can have quite different combinations of alleles. A second human blood grouping, ABO, is an example. Three different alleles, I^A, I^B and I^O, contribute to the blood type. The relationships between these alleles and the phenotypes that arise from the possible genotypes are shown in Fig. 17.13. Note that this system includes pairs of alleles that display standard dominance–recessiveness and pairs that display codominance.

Blood groups

Blood type is determined by the nature of the different proteins present on the surface of red blood cells. These proteins are genetically specified as described in the text. The first human blood group to be discovered was the ABO system by Landsteiner in 1900. Other important systems are Rhesus, Hh, Lewis and Sese. Although interesting to the geneticist, the MN system is less important in physiological and medical terms. Blood groups have been the subject of research because of the importance of blood transfusion in surgery, and the disastrous consequences of transferring blood of the wrong type.

17.2.3 Mendel's Laws – molecular explanations

We have not had too much difficulty in explaining how Mendel's deductions about allele pairs and dominance–recessiveness relate to the molecular interpretation of the gene. Can we provide similar explanations for Mendel's two Laws?

(a) Mendel's First Law – alleles segregate randomly. The First Law does not cause us any problems. During meiosis (see Fig. 15.17), each pair of chromosomes replicates and the daughters segregate into the gametes. Each gamete then has an equal chance of fusing with a gamete derived from the other parent: there are no stipulations regarding which of the four gametes can give rise to a member of the next generation (see Fig. 17.4). The entire process occurs at random, just as postulated by Mendel.

(b) Mendel's Second Law – different pairs of alleles segregate randomly. Of all of Mendel's deductions, it is the Second Law that has caused geneticists most concern because it is by no means inviolate. In fact it is the exception rather than the rule, especially when several different pairs of alleles are being studied.

To segregate independently, two pairs of alleles must be on different chromosomes: if they are on the same chromosome then they would be expected to segregate together. When the chromosome theory was proposed, it was realized that there must be many more genes than there are chromosomes in a cell, and so some genes must display **linkage**, meaning that they are on the same chromosome and so will not segregate in accordance with Mendel's Second Law. It was assumed that Mendel had been very lucky in that each of the pairs of alleles he studied was on a different chromosome of the pea, and so none displayed linkage. As Mendel studied seven pairs of alleles, and peas have seven pairs of chromosomes, the chances of each pair of alleles being on a different chromosome are quite low. In fact, there are other explanations for why Mendel did not see linkage.

Linkage was, however, observed in some of the crosses carried out by the rediscoverers of Mendel's work. It is to these experiments that we must now turn our attention.

Mendel and linkage

In fact Mendel's seven genes did not all lie on different chromosomes. Blixt (see Reading) has pointed out that two of the genes probably lie on chromosome 1 and three on chromosome 4. However, most of the linked genes are so distantly separated on their chromosomes that they will segregate more or less independently, because of the processes that will be described in Chapter 18. The only dihybrid cross that should have displayed linkage (stem height and full/constricted pods) was apparently not carried out by Mendel.

Reading

Anon (1950) The birth of genetics. *Genetics*, **35** (Suppl.); C. Stern and E. Sherwood (1966) *The Origins of Genetics: A Mendel Source Book*, W. H. Freeman, San Francisco; R. Olby (1985) *Origins of Mendelism*, University of Chicago Press, London – describe the background to Mendel's experiments.

J. A. Peters (1959) *Classic Papers in Genetics*, Prentice Hall, New Jersey, USA – includes Mendel's paper.

J. R. S. Fincham (1990) Mendel – now down to the molecular level. *Nature*, **343**, 208–9 – describes the molecular basis to the round–wrinkled phenotype.

W. M. Watkins (1966) Blood-group substances. *Science*, **152**, 172–81 – provides some information on blood types.

S. Blixt (1975) Why didn't Gregor Mendel find linkage? *Nature*, **256**, 206.

Problems

1. Define the following terms:

monohybrid cross	gamete
self-fertilization	homozygous
cross-fertilization	heterozygous
allele	Punnett square
dominant	dihybrid cross
recessive	vegetative cell
segregation	linkage

2. Explain what is meant by pairs of contrasting characteristics. Why were they important for Mendel's experiments?
3. Describe the outcome of Mendel's cross between tall-stemmed and short-stemmed pea plants.
4. Carefully explain how Mendel was able to deduce that alleles occur in pairs and that they display dominance and recessiveness.
5. Explain how Mendel derived his First and Second Laws.
6. What is the molecular explanation for dominance and recessiveness?
7. Explain what is meant by incomplete dominance and codominance.
8. What are multiple alleles?
9. Why is Mendel's Second Law not universally true?
10. Green pods are dominant to yellow pods in pea plants. What will be the phenotypes and genotypes of the F_1 and F_2 plants from a green pod (*GG*) and yellow pod (*gg*) cross?
11. What phenotypes and ratios will be obtained in the F_2 generation from a dihybrid cross between pea plants with full green pods (*FFGG*) and plants with constricted yellow pods (*ffgg*)?

12. Deduce the F_2 ratios for the trihybrid cross *AABBCC* × *aabbcc*.
13. What would be the outcome of crossing an F_1 plant from Mendel's tall-stemmed and short-stemmed cross with the short-stemmed parent? (This is called a **backcross**.)
14. Fig. 17.11 shows a cross between carnations with red flowers and carnations with white flowers. What would be the result of allowing the F_1 plants to self-fertilize?

18 Using Mendelian genetics to study eukaryotic genes

Linkage – the development of gene mapping • Gene mapping with microbial eukaryotes – gene mapping with *S. cerevisiae* • Beyond gene mapping – understanding complex phenotypes

The rediscovery of Mendel's experiments in 1900 provided the new generation of geneticists with conclusive proof that inheritance could be studied in a scientific manner. This stimulated a burst of activity and the next 10 years witnessed a number of important developments that built on Mendel's work and interpreted it in terms of other new ideas in biology. Gradually the power of genetic crosses became apparent.

The most important development was the proposal by Sutton in 1903 that genes reside on chromosomes. This immediately raised a question about Mendel's Second Law, that of independent segregation, because it was realized that the number of chromosomes possessed by an organism must be considerably less than the total number of inheritable traits displayed by that organism. This being so then clearly a single chromosome must carry a large number of genes, and different genes on the same chromosome would not be inherited independently, but instead would be transmitted together; that is, they would display linkage. This clear appreciation of the concept of linkage led directly to techniques for mapping genes on eukaryotic chromosomes, techniques that today still form one of the cornerstones of molecular genetics.

18.1 Linkage

One of the first experimental demonstrations of linkage was provided in 1905 by W. Bateson, E. Saunders and R. Punnett. They carried out a series of dihybrid crosses along the same lines as Mendel, although with a different plant, the sweet pea. In their experiments plants with purple flowers and long pollen grains were crossed with plants with red flowers and round pollen grains

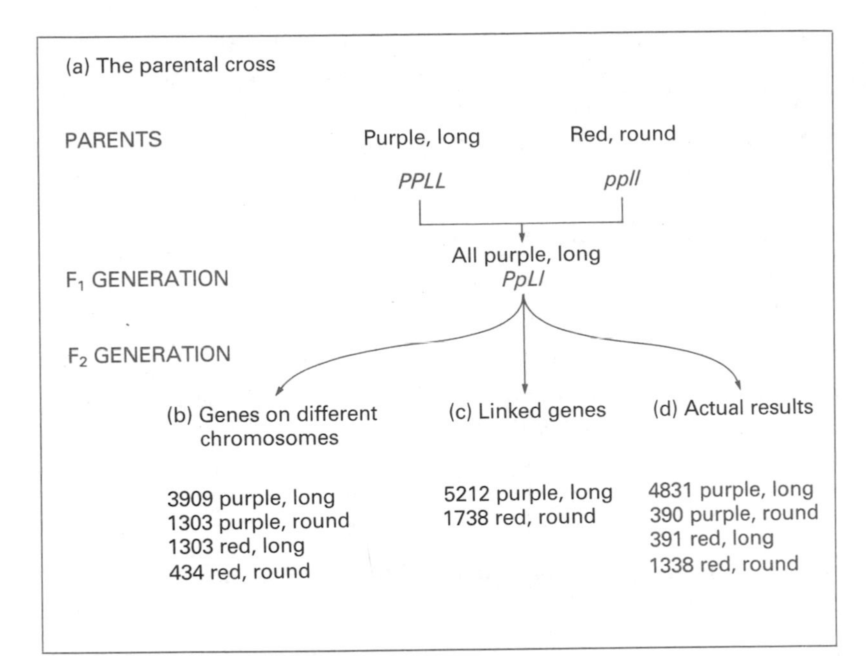

Figure 18.1 The Bateson, Saunders and Punnett cross. The genes are for flower colour (alleles P = purple, p = red) and pollen grain shape (L = long, l = round). The predicted phenotypes of the 6950 F_2 plants are given for: (b) a typical dihybrid cross between unlinked genes, and (c) genes showing complete linkage. The actual results of the cross are given in (d).

(Fig 18.1(a)). In the F_1 generation the plants, as expected, were all of the same phenotype, with purple flowers and long pollen grains. These two traits are therefore dominant. The F_1 plants were then allowed to self-fertilize and 6950 F_2 plants obtained. Now that we understand Mendel's Laws we can attempt to predict the ratios of phenotypes that these 6950 plants should display. First, if the genes for flower colour and pollen grain shape lie on different chromosomes then we would expect the 9:3:3:1 ratio of a typical dihybrid cross (Fig. 18.1(b)). This would give us:

3909 purple, long
1303 purple, round
1303 red, long
434 red, round

On the other hand, if the two genes are linked on the same chromosome then they will segregate together, and in effect we will have carried out a monohybrid cross between the alleles 'purple-flowers-and-long-pollen-grains' and 'red-flowers-and-round-pollen-grains' (Fig. 18.1(c)). We would therefore expect a 3:1 ratio:

5212 purple, long
1738 red, round

The important thing is that we would not expect to see any non-parental combinations (purple, round and red, long) in the F_2

generation, as these would require independent segregation of two genes that are linked together on the same chromosome.

Which of our predictions is correct? Surprisingly, it is neither. The results obtained by Bateson, Saunders and Punnett were (Fig. 18.1(d)):

4831 purple, long
390 purple, round
391 red, long
1338 red, round

This is neither a 9:3:3:1 nor a 3:1 ratio. Although the majority of the F_2 plants have the parental combination of characteristics, there are nonetheless a few with the non-parental features, although not as many as we would expect if the genes were unlinked. It is a half-way house: the genes display only **partial linkage**.

18.1.1 Thomas Hunt Morgan and the development of gene mapping

Partial linkage was puzzling for the early geneticists. Several hypotheses to explain it were put forward (including the correct one, by Hugo De Vries, one of Mendel's rediscoverers) but the breakthroughs in understanding the behaviour of linked genes during crosses were made almost exclusively by Thomas Hunt Morgan (Section 1.3) and his colleagues at Columbia University in New York. Morgan initially set out to study ways of inducing mutations in fruit flies, but the chance discovery of a single male fly with white eyes in a population of wild-type, red-eyed flies prompted him in 1910–11 to carry out his first Mendelian-type crosses. The results were very illuminating.

(a) Crosses involving the white-eye gene. Morgan's first white-eyed fly was male so he crossed it with a wild-type female. At first glance, the results appeared to agree with the predicted outcome of a monohybrid cross (Fig. 18.2(a)). All the F_1 flies had red eyes (red eyes are dominant to white eyes), and in the F_2 generation a 3:1 red eye to white eye ratio was seen. However, Morgan noticed one peculiarity: in the F_2 generation all the white-eyed flies were male. This was not as expected as the 3:1 ratio should hold for both sexes. This result may seem puzzling, but it can be explained if we assume that the white-eye gene lies on the X chromosome. The Punnett square analysis of the cross makes this clear (Fig. 18.2(b)).

To test the hypothesis that the white-eye gene is on the X chromosome, Morgan carried out a series of additional crosses. We will look at just one of these, between a white-eyed male fly and a female from the F_1 generation of the previous cross. The outcome of this new cross was a 1:1:1:1 ratio of phenotypes: equal numbers

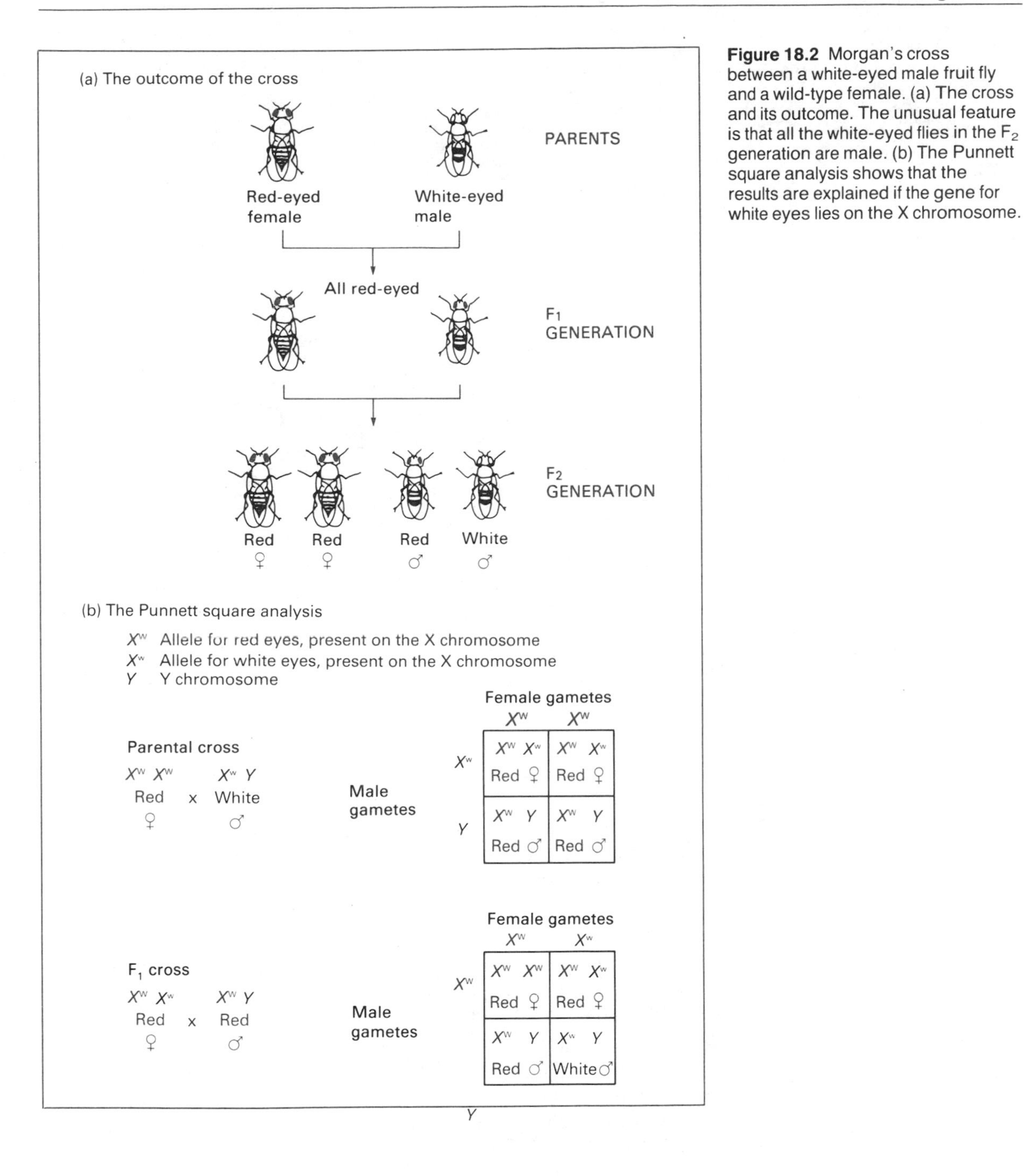

Figure 18.2 Morgan's cross between a white-eyed male fruit fly and a wild-type female. (a) The cross and its outcome. The unusual feature is that all the white-eyed flies in the F_2 generation are male. (b) The Punnett square analysis shows that the results are explained if the gene for white eyes lies on the X chromosome.

of red- and white-eyed flies, half of each type being male and half female (Fig. 18.3(a)). Again the Punnett square is the best way to analyse the results of the cross (Fig. 18.3(b)). The hypothesis is confirmed: the gene for red and white eyes is **sex-linked**.

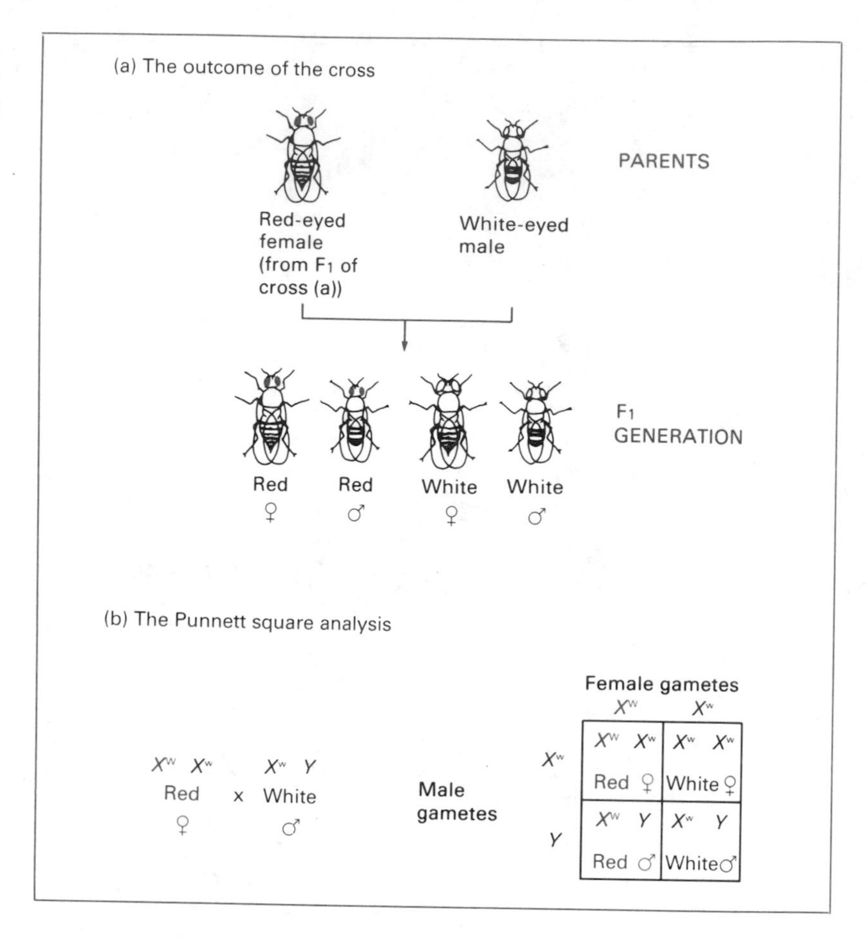

Figure 18.3 Morgan's cross between a red-eyed female fruit fly, from the F_1 generation of the cross shown in Fig. 18.2, and a white-eyed male. In (b) the notation used is the same as in Fig. 18.2.

(b) Crossing-over. Soon after completing the white-eye crosses Morgan discovered another fruit fly variant, this one with miniature wings. A series of crosses established that miniature wings are inherited in the same way as white eyes and that the relevant gene lies on the X chromosome. So what would happen during a cross involving both of these two genes: would the linkage be complete or only partial? To find out, Morgan crossed a female fly with white eyes and miniature wings (genotype X^{wm} X^{wm}) with a wild-type male (genotype X^{WM} Y). The F_1 generation, as expected, consisted of wild-type females (genotypes X^{wm} X^{WM}, i.e. heterozygotes) and males with white eyes and miniature wings (genotypes X^{wm} Y). Check with Fig. 18.4(a) and make sure that you can follow what has happened so far.

Now we come to the important part of the cross. If the two genes display complete linkage then the F_2 generation should be made up of half white-eyed flies with miniature wings (half of these flies will be male and half female) and half wild-type flies (again half

The testcross

The cross between white-eyed male fruit flies and wild-type females is an example of a **testcross**. The standard testcross is between a homozygous recessive organism and one displaying the dominant phenotype. The testcross establishes whether the organism with the dominant phenotype is homozygous or heterozygous for that allele. If homozygous (example 1 below) the progeny will all have the dominant phenotype; if heterozygous (example 2) then half the progeny will have the dominant phenotype and half the recessive phenotype.

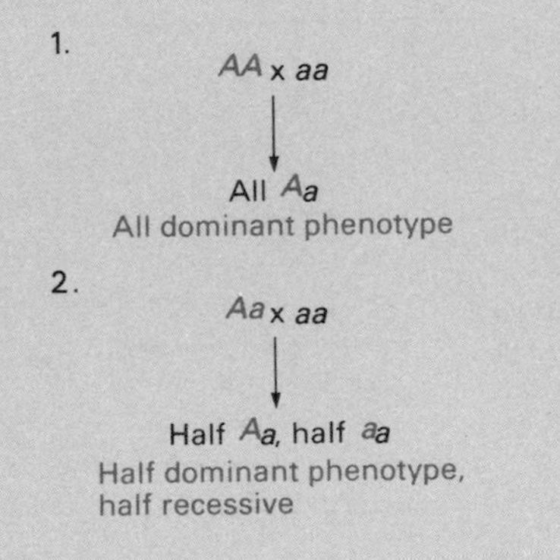

As the white-eyed allele is on the X chromosome, the crosses shown on Figs 18.2 and 18.3 are specialized versions of the testcross, slightly different in concept to the standard version. Male fruit flies are **hemizygous** for alleles on the X chromosome. This is because there is only a single allele present, as the Y chromosome is genetically distinct and does not carry alleles equivalent to those on the X chromosome. An allele on the X chromosome is therefore always expressed as the phenotype of the male, even if recessive. The outcomes of testcrosses involving alleles on the X chromosome are as follows.

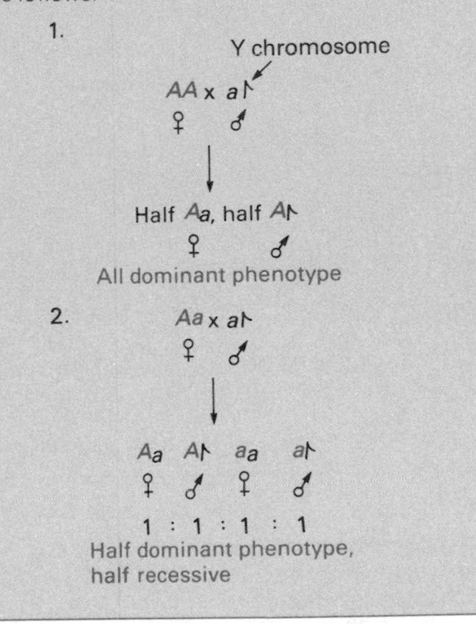

male and half female). The scheme is outlined in Fig. 18.4(b). The important point is that only the parental combinations of characteristics will be seen in the F_2 generation. How does this compare with the actual results obtained by Morgan? Again only partial linkage was observed: of the 2441 flies in the F_2 generation, 900 (that is 36.9%) possessed a non-parental combination of characteristics, being either red-eyed flies with miniature wings, or white-eyed flies with normal wings.

Morgan's results mirrored those from the earlier experiments of Bateson, Saunders and Punnett. However, Morgan was already sure that the two genes lay on the same chromosome and so was forced to provide an explanation for his results. He did so by referring to observations made of meiosis in salamanders by a Belgian cytologist F. A. Janssens in 1909. Janssens had made a very close examination of the way pairs of chromosomes line up during nuclear division and had suggested that the close proximity of the chromatids of the two chromosomes of a pair could lead to breakage and transfer of segments between chromosomes (Fig. 18.5(a)). We now understand that **crossing-over** indeed occurs and is a manifestation of recombination between DNA molecules

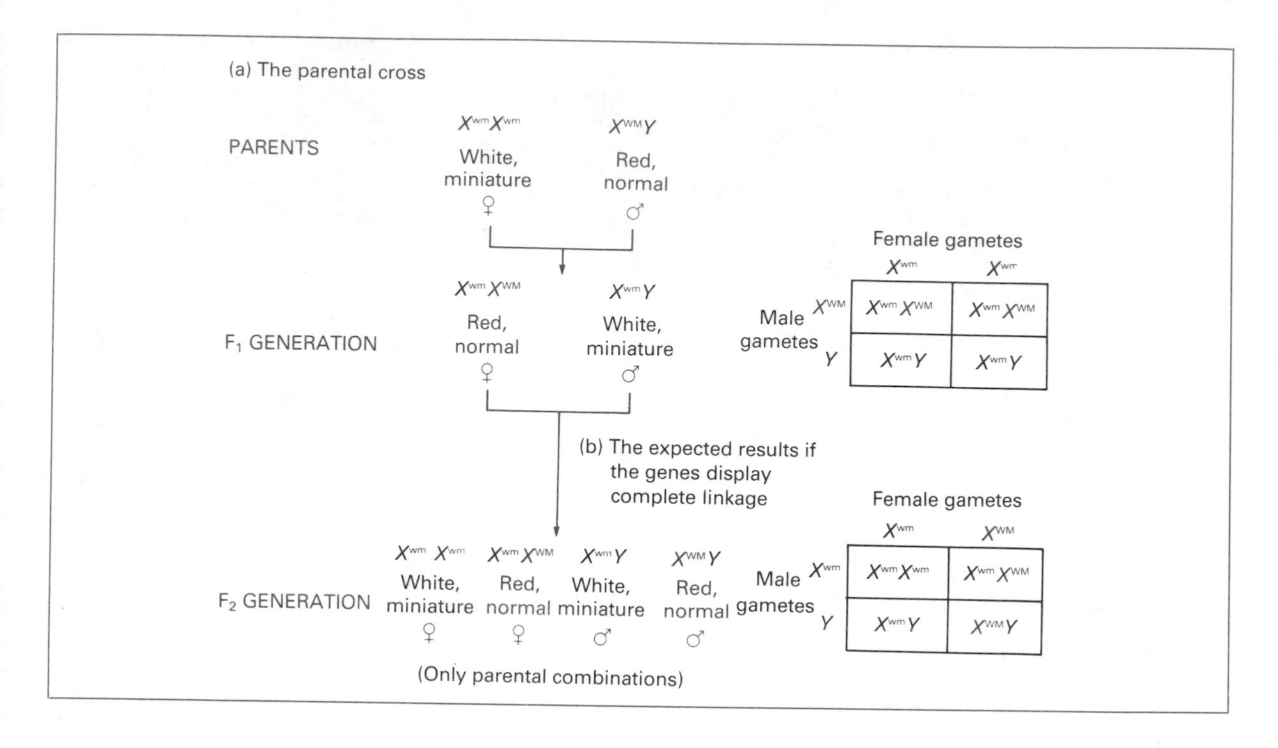

Figure 18.4 The cross involving eye colour and miniature wings in *Drosophila*. In (b) the predicted outcome is shown if the two genes display complete linkage. In fact, of the 2441 flies in the F_2 generation, 900 possessed non-parental combinations of alleles. Key: X^{WM}, alleles for red eyes and normal wings (wild-types) present on the X chromosome; X^{wm}, alleles for white eyes and miniature wings, present on the X chromosome; *Y*, Y chromosome.

(Section 12.3.2). Morgan built on Janssens' hypothesis by proposing that crossing-over could lead to the uncoupling of linked genes (Fig. 18.5(b)), and hence could explain how non-parental forms (more correctly called **recombinants**, as they result from recombination) arise in a dihybrid cross even when the genes involved occupy different positions on the same chromosome.

(c) The mapping technique. Very soon after performing the white eye–miniature wing cross Morgan obtained a third sex-linked gene, this one for yellow body. He carried out a new cross involving yellow body and white eyes. Again he found only partial linkage, although, in this case, the linkage seemed to be tighter as only 1.3% of the F_2 flies were recombinants. Morgan explained this by proposing that the gene for white eyes is closer on the X chromosome to the gene for yellow body than it is to the gene for miniature wings. So, the first two genes (white eyes and yellow body) are less likely to be uncoupled by crossing-over. There would simply be less chance of a crossover occurring in the limited region separating these two genes on the X chromosome (Fig. 18.6).

The next phase of the work was carried out by Arthur Sturtevant, one of Morgan's students. Two more sex-linked genes became available – for vermilion eyes and rudimentary wings – and when

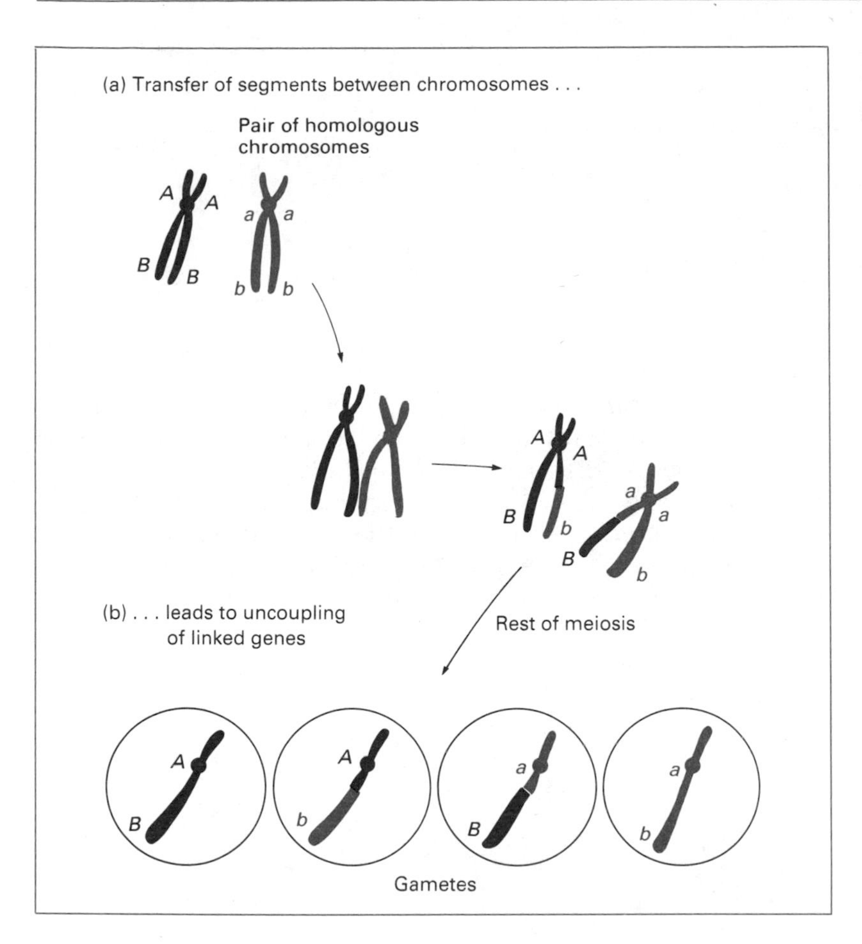

Figure 18.5 Morgan's proposition that crossing-over could lead to linked genes becoming uncoupled.

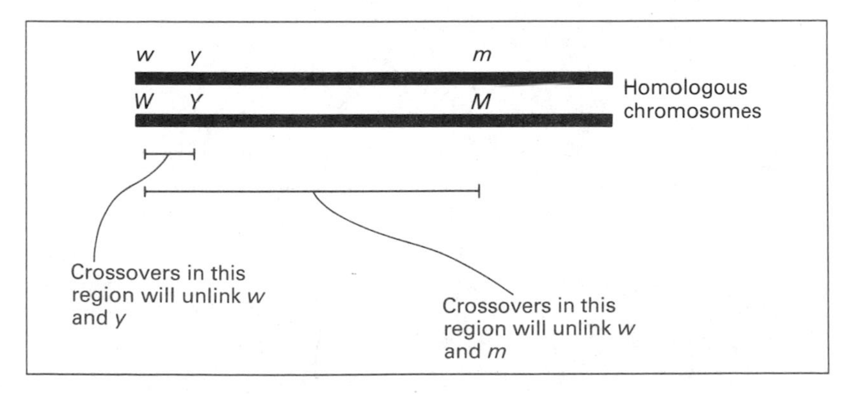

Figure 18.6 Two genes that are relatively close together on a chromosome are less likely to be uncoupled by crossing-over than two that are further apart. White eyes (*w*) and yellow body (*y*) will therefore recombine less frequently than white eyes and miniature wings (*m*).

all five genes were combined in a variety of crosses a consistent set of results was obtained. Not only was the **recombination frequency** different for each cross, but the differences could be related to the relative distances between the genes on the X

Figure 18.7 Sturtevant's five-gene map for the *Drosophila* X chromosome. Abbreviations: *y*, yellow body; *w*, white eyes; *v*, vermilion eyes; *m*, miniature wings; *r*, rudimentary wings.

X chromosome						
Map position	0	1.3	30.7	33.7	57.6	
Gene	*y*	*w*	*v*	*m*	*r*	

chromosome. A map for these five genes could be drawn (Fig. 18.7).

(d) Linkage groups. Once the way to map genes had been discovered Morgan's group set about mapping as many fruit fly genes as possible and by 1915 had assigned locations for 85 of them. These genes fall into four **linkage groups**, the sex-linked group and three others, corresponding to the four pairs of chromosomes seen in the fruit fly nucleus. The distances between genes are expressed in **map units**, with one map unit being the distance between two genes that recombine with a frequency of

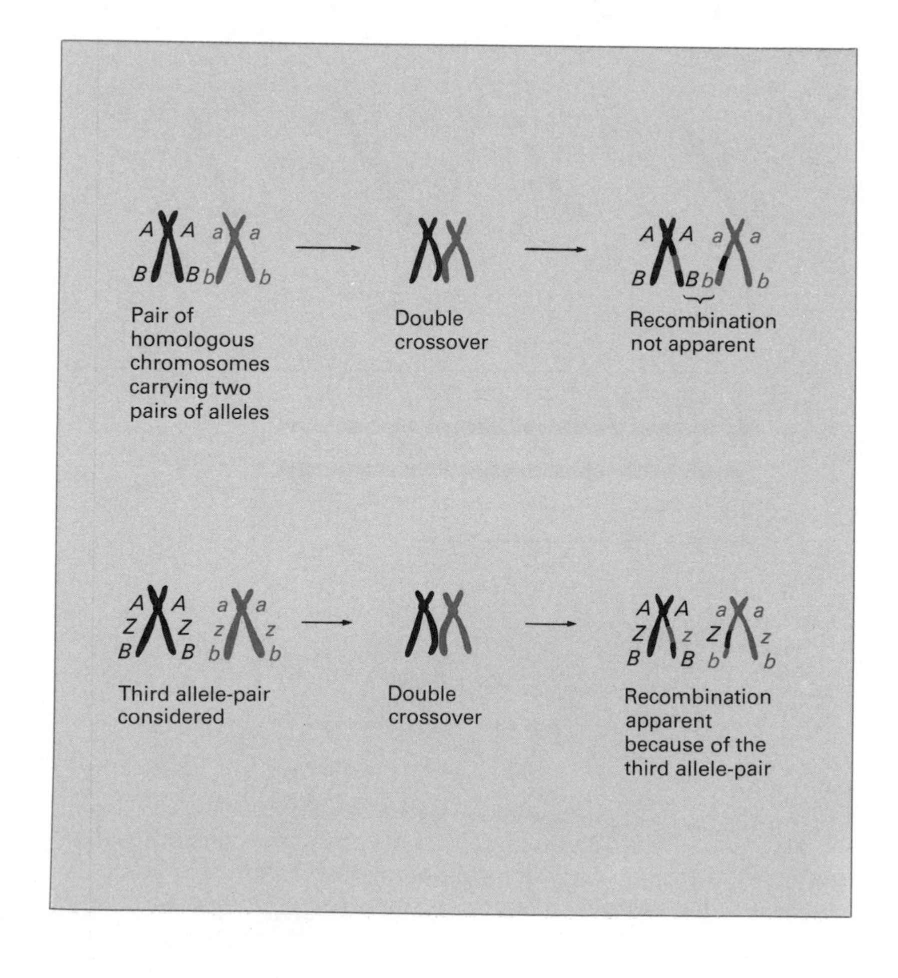

1% (so the distance between the genes for white eyes and yellow body, which recombine with a frequency of 1.3%, is 1.3 map units, see Fig. 18.7). More recently the name **centiMorgan** (cM) has begun to replace the map unit.

18.2 Gene mapping with microbial eukaryotes

Gene mapping techniques are basically the same with most eukaryotes although with many higher vertebrates lengthy generation times or ethical considerations prevent the experiments from being carried out. However, with microbial eukaryotes it is necessary to use a radically different approach. The rationale behind the experiments remains the same – to determine how frequently recombination occurs between two linked genes – but with microbial eukaryotes the crosses are set up and the results analysed in a different way. The microbial eukaryotes include organisms such as the yeast *Saccharomyces cerevisiae*, the fungus *Neurospora crassa* and the photosynthetic alga *Chlamydomonas reinhardtii*. These are the eukaryotes that are easiest to work with in the laboratory (p. 59), and much of the fundamental research on eukaryotic molecular biology is being carried out with them. Clearly we must understand how they are studied.

18.2.1 Gene mapping with *S. cerevisiae*

We will use the yeast *S. cerevisiae* to illustrate the principles of gene mapping with a microbial eukaryote. The first point to appreciate is

Table 18.1 Some typical genetic markers for *S. cerevisiae*. Each of these gives rise to a characteristic that is easily recognized by testing cells on a suitable agar medium. Markers are usually mutant versions of normal genes

Marker	*Phenotype*	*Method by which cells carrying the marker are distinguished*
ade2	Requires adenine	Grows only when adenine is present in the medium
can1	Resistant to canavanine	Grows in the presence of canavanine
cup1	Resistant to copper	Grows in the presence of copper
cyh1	Resistant to cycloheximide	Grows in the presence of cycloheximide
leu2	Requires leucine	Grows only when leucine is present in the medium
suc2	Able to ferment sucrose	Grows when sucrose is the only carbohydrate present
ura3	Requires uracil	Grows only when uracil is present in the medium

that the nature of the inherited characteristics studied with a microorganism will be different from those studied with *Drosophila* and other multicellular organisms. Wing size, for example, is clearly inapplicable for yeast! Rather than examining morphological traits, biochemical characteristics are generally used. A list of typical characteristics, called **genetic markers**, for yeast is given in Table 18.1.

(a) The yeast life cycle. A typical yeast life cycle is shown in Fig. 18.8. Note that in the example shown the vegetative cells are haploid rather than diploid (compare this with the pea life cycle shown in Fig. 17.8). Not all strains of yeast have haploid vegetative cells – many diploid strains exist – but haploids are generally used in gene mapping experiments. This is because genetic analysis with diploids is always complicated by the fact that a cell with the dominant phenotype may be either homozygous or heterozygous.

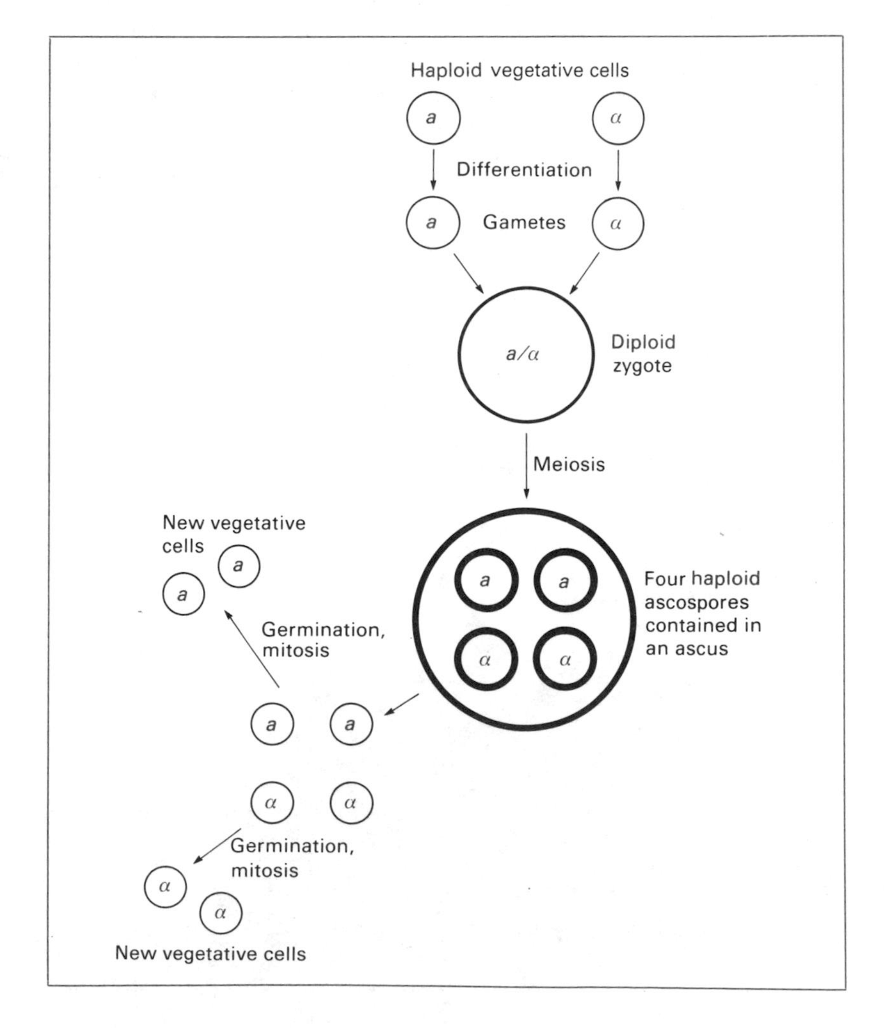

Figure 18.8 Sexual reproduction with a haploid strain of yeast. Vegetative cells of opposite mating type (*a* and *α*) differentiate into gametes, which then fuse to produce a diploid zygote. Meiosis gives four haploid ascospores, contained in a structure called an ascus. The ascus bursts, releasing the ascospores, each of which will germinate and divide by mitosis to produce new vegetative cells.

This problem does not arise with haploid cells (each cell has just one copy of each chromosome, so all cells are 'homozygous' for all genes), making genetic analysis much more straightforward.

Sexual reproduction occurs when two strains of yeast, of opposite mating types, are mixed together (see Fig. 18.8). The haploid vegetative cells undergo a subtle morphological and physiological change, resulting in gametes that fuse to produce a diploid **zygote**. Meiosis occurs within the zygote, giving rise to a **tetrad** of four haploid **ascospores**, contained in a structure called an **ascus**. The ascus bursts open, releasing the ascospores, which then divide by mitosis to produce new haploid vegetative cells.

(b) Carrying out a cross with *S. cerevisiae*. We will carry out a cross between the following two strains of yeast:

mating type α, wild-type
mating type a, $leu2^-$

The $leu2^-$ strain carries a mutation that inactivates the *leu2* gene, which codes for β-isopropylmalate dehydrogenase, one of the enzymes involved in biosynthesis of leucine. The strain is therefore a leucine auxotroph (Section 12.1.1), and can survive only if leucine is supplied in the growth medium. The wild-type strain will of course be a prototroph. We could write the cross as:

$leu2^- \, a \times \alpha \, leu2$

Actually, when setting out the genetic analysis in this way, we assume that the strains are of opposite mating types and are not concerned with which is which. Also, the wild-type organism

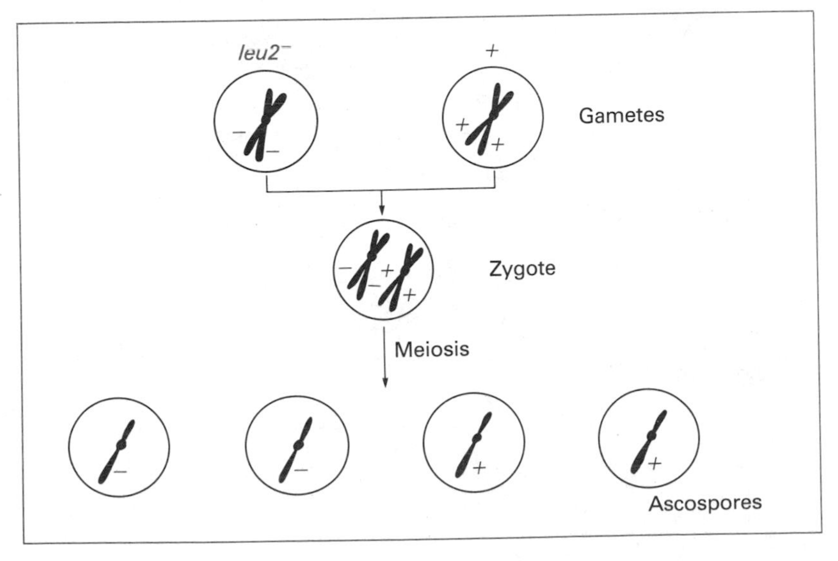

Figure 18.9 A cross between a $leu2^-$ and a wild-type strain of yeast. The gametes are in the G2 phase of the cell cycle (Section 15.4.1). Fusion between two gametes of opposite mating types results in a diploid zygote, which then undergoes meiosis (check with Fig. 15.17) to produce four haploid ascospores.

would be designated as '+'. So we would write the cross simply as:

leu2$^-$ × +

Fusion of the *leu2*$^-$ and wild-type gametes produces diploid zygotes that subsequently undergo meiosis, giving four haploid ascospores per zygote. According to Mendel's Laws the result of the cross should be equal numbers of *leu2*$^-$ and wild-type ascospores (Fig. 18.9). To verify this result we must determine the phenotypes of individual ascospores. Either of two methods could be used:

1. **Random spore analysis** (Fig. 18.10). A sample of ascospores is spread on to an agar medium and the ability of each to survive without leucine is tested. This is done by first spreading on to a complete medium (containing leucine), on which all ascospores,

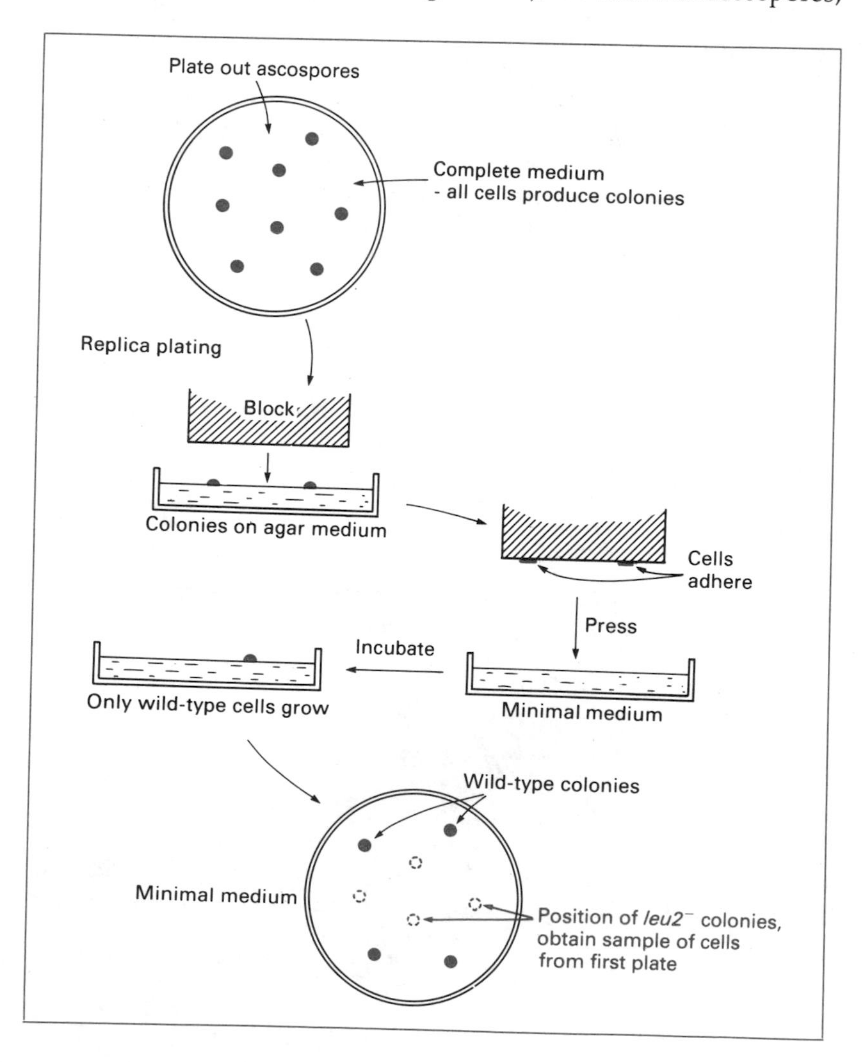

Figure 18.10 Random spore analysis and replica plating.

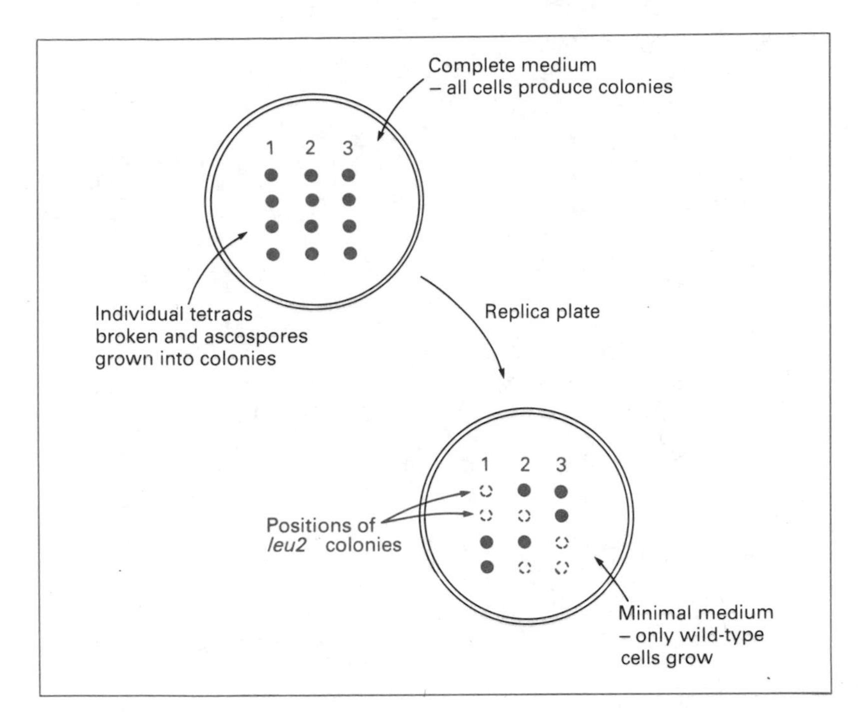

Figure 18.11 Tetrad analysis.

whether *leu2*$^-$ or wild-type, will be able to divide and produce colonies. The number of colonies tells you how many ascospores are in the sample tested. Each colony is then transferred by **replica plating** on to minimal medium (containing no leucine), on which *leu2*$^-$ cells will not be able to grow. The number of leucine auxotrophic ascospores in the original sample can then be worked out.

2. **Tetrad analysis** (Fig. 18.11). This technique makes use of the fact that the four ascospores that result from meiosis of a single zygote are initially retained in the ascus (see Fig. 18.8). Individual asci can be separated from one another, broken open, and the four ascospores grown on agar medium. The number of leucine auxotrophic ascospores that arise from a single gamete fusion event can therefore be determined, allowing the results of the cross to be examined in detail.

(c) Mapping genes. How are these techniques used to map genes in *S. cerevisiae*? To illustrate the strategy we will look at a second cross, carried out in the same way as the one we have just described, but with parent yeast strains that carry markers for two genes rather than one. We will use hypothetical markers, a^- and b^-, so our cross can be written as:

$$a^-b^- \times ++$$

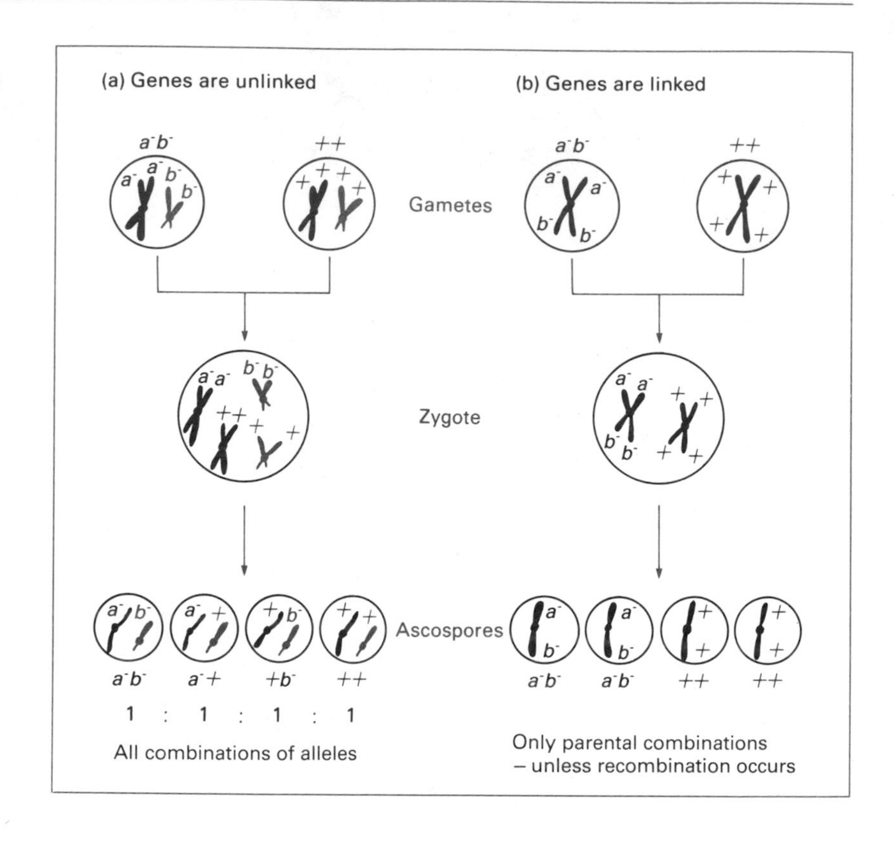

Figure 18.12 Distinguishing between (a) unlinked and (b) linked genes in a yeast cross.

Ascospores arising from the cross could have any of four genotypes:

a^-b^-, $++$ } parental combinations

a^-+, $+b^-$ } non-parental (recombinant) combinations

Are the genes on the same chromosome? If the genes are unlinked then we would expect equal numbers of each ascospore (Fig. 18.12(a)). Random spore analysis will therefore reveal a 1:1:1:1 ratio for the four possible genotypes. If, on the other hand, a^- and b^- are present on the same chromosome, then ascospores with the parental combinations (a^-b^- and $++$) will be more frequent (Fig. 18.12(b)). By now you will appreciate that if the genes are on the same chromosome then recombinant ascospores will arise only if crossing-over occurs between a^- and b^-.

What is the map distance between the genes? An approximate estimation of the map distance between a^- and b^- can be obtained from the number of recombinant ascospores detected by random spore analysis. By analogy with the strategy for gene mapping in fruit flies, the recombination frequency (i.e. the number of recombinant

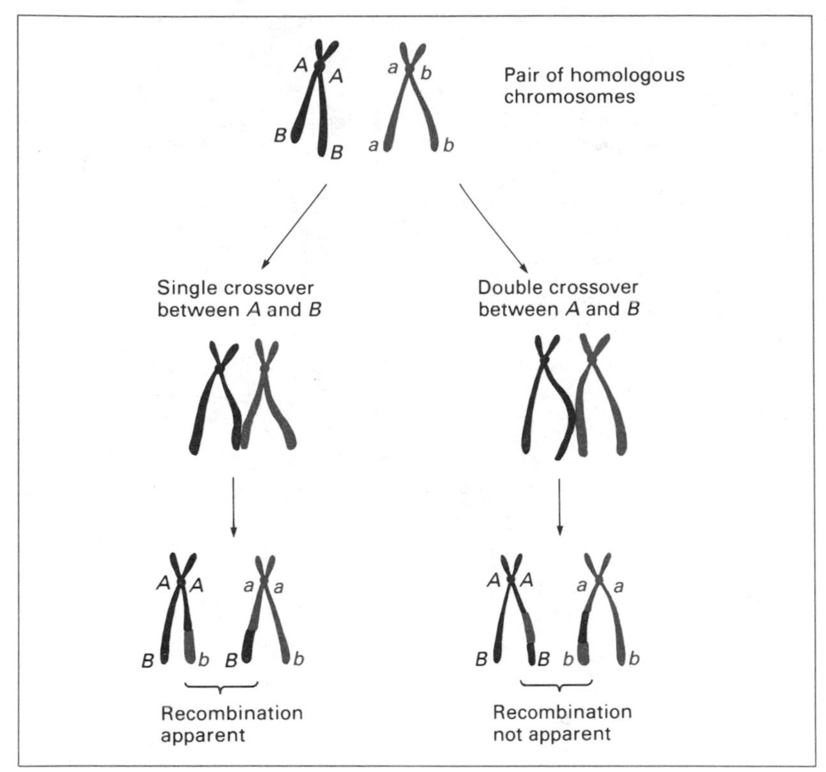

Figure 18.13 Double crossover events are not detected by random spore analysis as the genes under study do not display recombination.

ascospores divided by the total number of ascospores) will correspond to the distance between the genes in centiMorgans.

Unfortunately, measurement of map distances by random spore analysis is only approximate. The method does not take account of double crossover events, as these will not result in a recombinant ascospore (Fig. 18.13). Because these crossovers are missed, the apparent recombination frequency between a^- and b^- will be less than the real value. The map distance calculated by random spore analysis will therefore be shorter than the true distance between the genes.

This is where tetrad analysis comes in. Tetrad analysis is important because it can determine the number of crossover events and so give a more correct map distance. Before explaining how, we must look at the three different ascospore combinations that are possible in a tetrad (Fig. 18.14):

1. A **parental ditype** (PD) is a tetrad that contains two ascospores with the genotype of one parent, and two ascospores with the genotype of the other parent (a^-b^-, a^-b^-, $++$, $++$).
2. A **non-parental ditype** contains two of each of the non-parental combinations (a^-+, a^-+, $+\,b^-$, $+\,b^-$).
3. A **tetratype** contains four different ascospores, one of each possible genotype (a^-b^-, a^-+, $+\,b^-$, $++$).

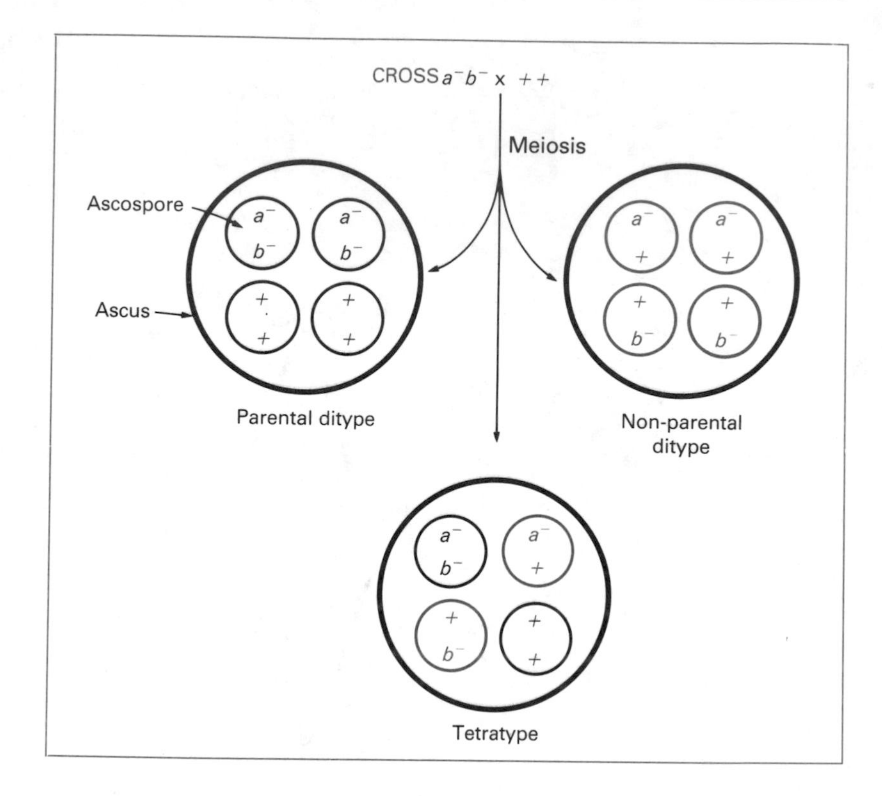

Figure 18.14 The three types of tetrad that can arise from a yeast cross involving two genes.

To understand how a map distance is arrived at, look carefully at each of the following points:

1. If in a particular zygote no crossovers occur between a^- and b^-, then the tetrad that results will be a parental ditype, PD (Fig. 18.15(a)).
2. If a single crossover occurs between a^- and b^-, then a tetratype, T, will result (Fig. 18.15(b)). Note that crossovers can occur between any two chromatids of a pair of homologous chromosomes.
3. If two crossovers occur between a^- and b^-, then a PD, T or NPD tetrad will result (Fig. 18.15(c)):

   If two chromatids are involved in the double crossover then the tetrad will be PD.

   If three chromatids are involved then the tetrad will be T.

   If four chromatids are involved then the tetrad will be NPD.

   As illustrated in Fig. 18.15(c), double crossovers will occur in the ratio 1 two-chromatid event:2 three-chromatid events: 1 four-chromatid event.

Now remember that we are trying to work out the number of crossover events that have occurred between a^- and b^-. The number of single crossover events is determined as follows:

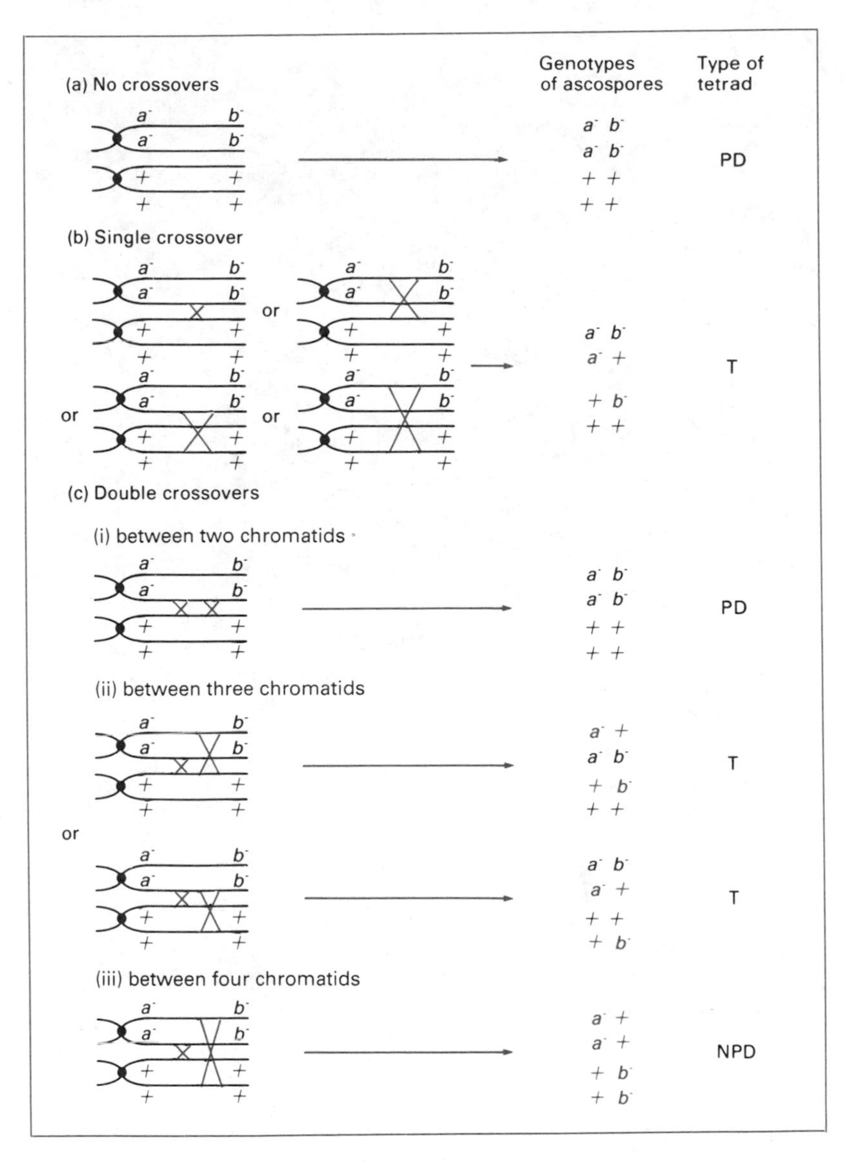

Figure 18.15 The events that give rise to the different types of tetrad during a cross involving two linked genes in yeast.

All single crossovers give a T (see point 2), but some Ts (equal to twice the number of NPDs) result from double crossovers (see point 3).
Therefore, the number of single crossover events = T − 2NPD.

The number of double crossovers can also be calculated:

The number of tetrads representing double crossovers = 4NPD (see point 3).
Therefore, the number of actual crossovers (two per each double crossover) = 8NPD.

Gene mapping in yeast – a worked example

Consider the cross:

$a^-b^- \times ++$

The following results are obtained by tetrad analysis:

PD tetrads	142
NPD tetrads	3
T tetrads	61
Total:	206

The first conclusion is that these genes are linked. This is evident from the very low number of NPD tetrads that are seen. For unlinked genes the PD:NPD ratio should be about one.

The crossover frequency can be calculated from the formula:

$$CF = \frac{T + 6NPD}{2(T + NPD + PD)}$$

$$= \frac{61 + 18}{2 \times 206} = 0.192 = 19.2\%$$

The two genes represented by the markers a^- and b^- are therefore 19.2 centiMorgans apart.

Ordered tetrads

Many microbial eukaryotes retain their meiotic products (e.g. the ascospores of *S. cerevisiae*) within a structure equivalent to the ascus, and so are amenable to tetrad analysis. In a few cases the analysis can be taken further as the ascus is constructed in a such a way that the ascospores derived from individual chromatids of the gametes can be identified.

Neurospora crassa provides an example. With this fungus the ascospores are contained within a linear ascus that develops as follows:

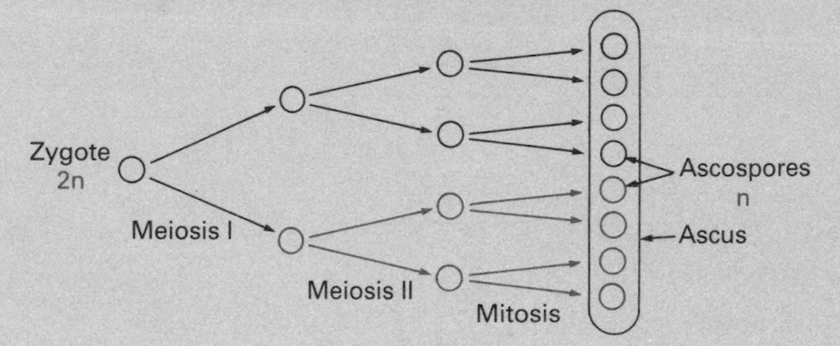

We still call this a tetrad even though the mitotic division means that the ascus contains eight ascospores.

The ordered nature of the *N. crassa* ascus does not change the way in which crossover frequencies are calculated by tetrad analysis. The advantage of the ordered tetrad is that it enables us to examine more directly the recombination events that give rise to the genotypes of the individual ascospores. For example, it is possible to confirm that double crossovers during meiosis I can involve any two pairs of chromatids:

(a) Double crossovers involving two chromatids

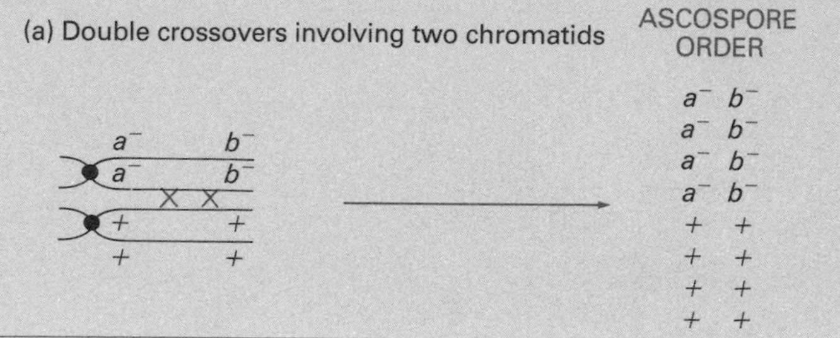

The total number of crossovers occurring in the tetrads analysed is therefore:

$$T - 2NPD + 8NPD = T + 6NPD$$

The crossover frequency is therefore:

$$\frac{T + 6NPD}{2 \times \text{Total number of tetrads}} = \frac{T + 6NPD}{2(T + NPD + PD)}$$

Note that to obtain the crossover frequency we divide the number of crossovers by twice the total number of tetrads. This is because each ascospore contains four chromatids, but only two of these participate in an individual crossover event.

Crossover frequency, as measured by tetrad analysis, provides a direct estimation of map distance and is accurate for genes up to

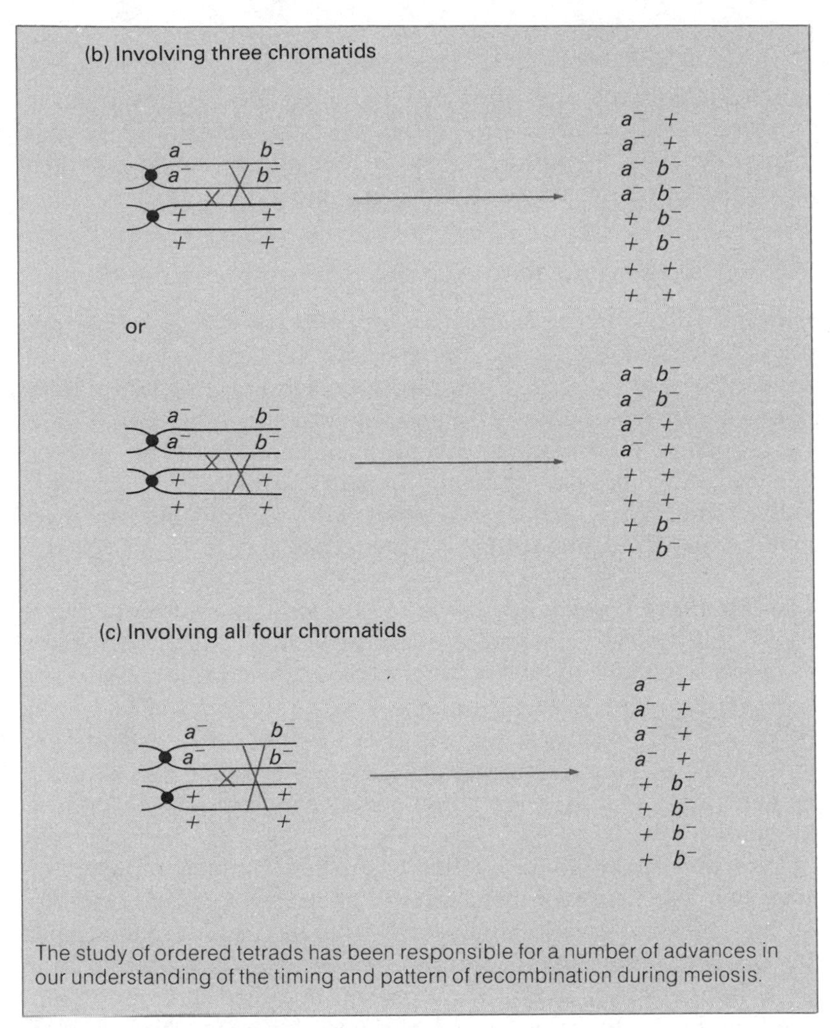

The study of ordered tetrads has been responsible for a number of advances in our understanding of the timing and pattern of recombination during meiosis.

about 40 centiMorgans apart. At greater distances the possibility of triple crossovers provides further complications that lead to underestimations of the true distances between genes.

18.3 Taking Mendelian genetics beyond gene mapping

The Mendelian approach to genetic analysis can do much more than simply tell us the relative positions of genes on chromosomes. By examining the relationships between genotypes and phenotypes we can deduce a great deal about the functions of individual genes and the ways in which different genes interact with one another. This is important as the molecular approach to genetics, which typically involves gene isolation and DNA sequencing as the initial steps, is often not able to identify the function of a gene, or to determine which sets of genes work together in a cell or organism (see Section 21.4). The molecular approach will therefore be successful only if it is backed up by the information provided by Mendelian experiments. In the remainder of this chapter we will therefore look at a few of the ways in which Mendelian genetics can be used to examine gene activities.

18.3.1 A single gene may give rise to a complex phenotype

Students encountering Mendelian genetics for the first time are often misled into believing that the link between genotype and phenotype is always clearcut and unambiguous. In multicellular organisms this is not always the case and a single gene may give rise to a complex or variable phenotype. Analysis of this phenotype can often provide valuable information about the function of the gene. To illustrate this point we will consider the two phenomena called **pleiotropy** and **penetrance**.

(a) Pleiotropic genes give rise to complex phenotypes. Pleiotropy refers to the fact that a single gene may affect more than one trait. Mendel's gene for flower colour (see Table 17.1) is an example as it also influences the colour pattern of the leaves of the plant. This indicates that the trait interpreted by Mendel as 'flower colour' is more correctly 'pigment synthesis'. As a result the gene affects all the pigmented parts of the pea plant, not just the petals.

Pleiotropy is common in higher organisms, as the complexity of eukaryotic biochemistry means that a genetic defect that leads to the loss of an enzyme activity rarely has a clearcut phenotypic effect. Cystic fibrosis, for example, results from a mutation in a single human gene, coding for a protein involved in chloride ion secretion by epithelial cells. This defect leads to the complex phenotype that we call 'cystic fibrosis', which manifests itself as a variety of symptoms including lung disease, a deficiency in pancreatic enzymes, and excessive loss of salt in the sweat (Fig. 18.16). In extreme cases the sheer complexity of a pleiotropic phenotype can obscure the nature of the primary biochemical defect, but once the latter has been identified the pleiotropic effects are a bonus as they enable the function of the gene in different tissues to be assessed.

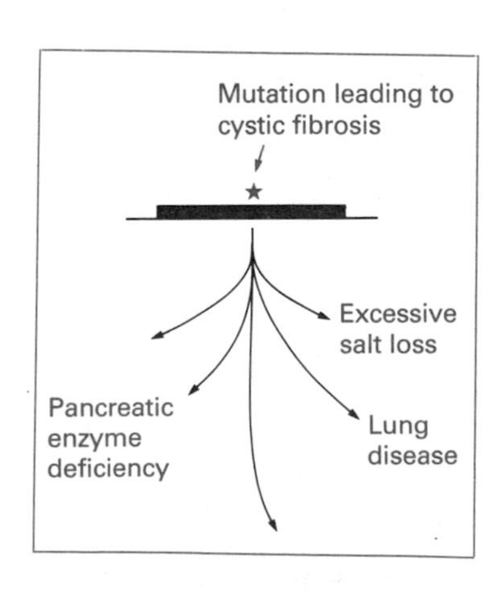

Figure 18.16 A single gene can give rise to a complex, multifactorial phenotype.

(b) Incomplete penetrance leads to a variable phenotype. Genes that are incompletely penetrant are ones that do not always give rise to their associated phenotype. An example is provided by the human disease retinoblastoma, which is a cancer of the eye. Retinoblastoma is inherited, indicating that it is a genetic disease and is controlled by a gene or set of genes. However, the disease is incompletely penetrant and of every ten individuals that, according to their genotype, should have retinoblastoma, on average only nine actually develop the disease (a penetrance of 90%). This observation provides an important clue to the genetic basis of retinoblastoma. It indicates that the relevant gene or genes do not directly specify the disease, but merely provide vulnerability towards it.

We now know that retinoblastoma is associated with a single gene (*Rb*) that regulates division of cells in the retina. Inactivation of the gene, for example by environmental radiation or other mutagenic agents, results in over-proliferation of retinal cells, leading to retinoblastoma (Fig. 18.17). Normal individuals are homozygous for *Rb*, so each retinal cell has two copies of the gene. It is very unlikely that inactivating mutations will occur in both genes of a single cell, so homozygotes virtually never develop the disease. Heterozygotes, on the other hand, are more vulnerable to retinoblastoma. They have just one active copy of *Rb* per cell, and so a single inactivating mutation in any retinal cell could lead to retinoblastoma. The incomplete penetrance of retinoblastoma tells us that there is a 90% chance that such a mutation will occur during the individual's lifetime, which makes those of us who are homozygotes thankful for our good fortune.

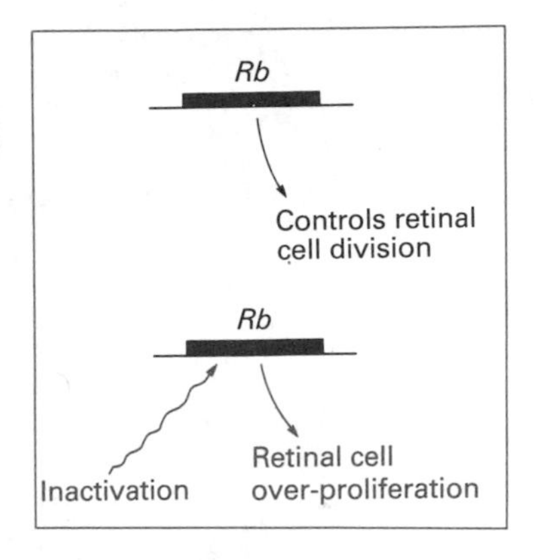

Figure 18.17 Retinoblastoma results from inactivation of the *Rb* gene, which regulates division of cells in the retina.

18.3.2 Complex phenotypes that result from interactions between genes

Although many phenotypes can be linked to the activity of just a single gene, there are numerous others that result from interactions between two or more separate genes. Understanding these interactions is crucial to our awareness of how the genome as a whole functions in living cells.

To a certain extent, the number of genes involved in a phenotype is a function solely of the way in which the phenotype is defined. The *lac* operon provides an example (Section 10.3.1). The phenotype could be defined as 'utilization of lactose', in which case it would be multigenic, as the entire operon is required. On the other hand, the more specific phenotype 'synthesis of β-galactosidase' is monogenic as it depends only on *lacZ*. Many phenotypes can be minimalized to single genes if we break them down far enough. The important point is that by first defining a multigene phenotype, and then breaking it down into a set a monogenic phenotypes, we are taking the first steps towards understanding the relevant gene interactions.

(a) Gene interactions in flower colour. Flower pigmentation illustrates various ways in which different genes can interact within a single phenotype. With some plant species more than one

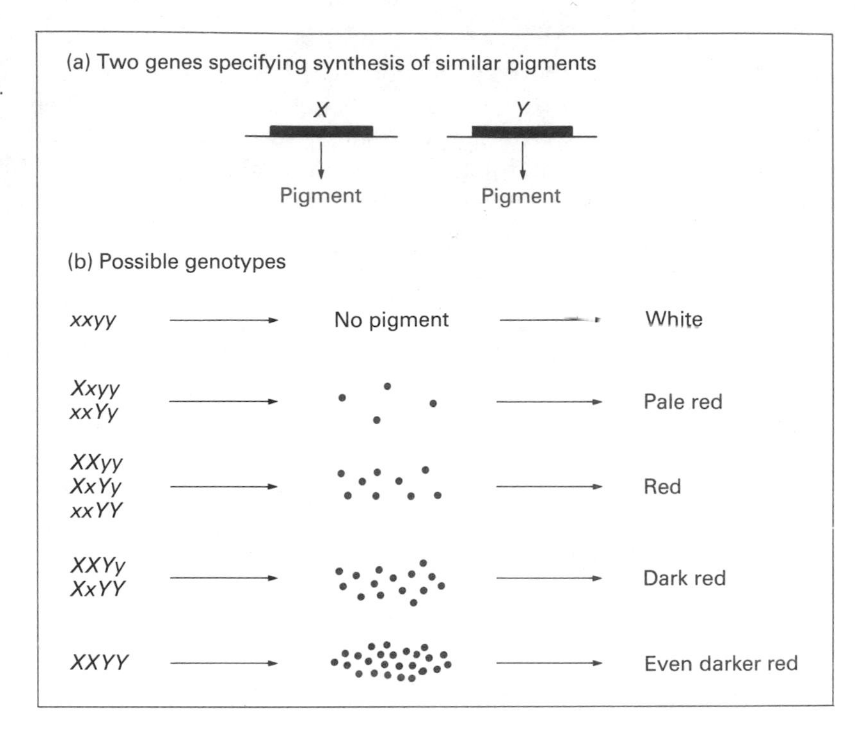

Figure 18.18 The additive effects of two genes *X* and *Y*, whose dominant alleles specify synthesis of a pigment.

gene specifies the synthesis of pigments of the same or similar colours (Fig. 18.18(a)). In our example there are two genes, *X* and *Y*, both responsible for synthesis of the same pigment. The two recessive alleles, *x* and *y*, are inactive and do not contribute to pigment synthesis. The possible genotypes are (Fig. (18.18(b)):

No dominant alleles	*xxyy*
1 dominant allele	*Xxyy* or *xxYy*
2 dominant alleles	*XXyy*, *XxYy* or *xxYY*
3 dominant alleles	*XXYy* or *XxYY*
4 dominant alleles	*XXYY*

X and *Y* are incompletely dominant and so interact in an additive fashion, with each *X* or *Y* allele providing one unit of pigmentation. This results in a complex phenotype in which the flowers can take on any of five different colour shades determined by the number of dominant alleles that are present.

A more puzzling interaction between genes for flower colour was discovered by Bateson and Punnett in 1905. They crossed two white-flowered sweet peas and obtained an F_1 generation comprised entirely of plants with purple flowers. When these plants were allowed to self-fertilize they gave rise to a F_2 ratio of 9 purple : 7 white (Fig. 18.19(a)). At first glance this cross may seem difficult to understand but it is easy enough if we deal with it step

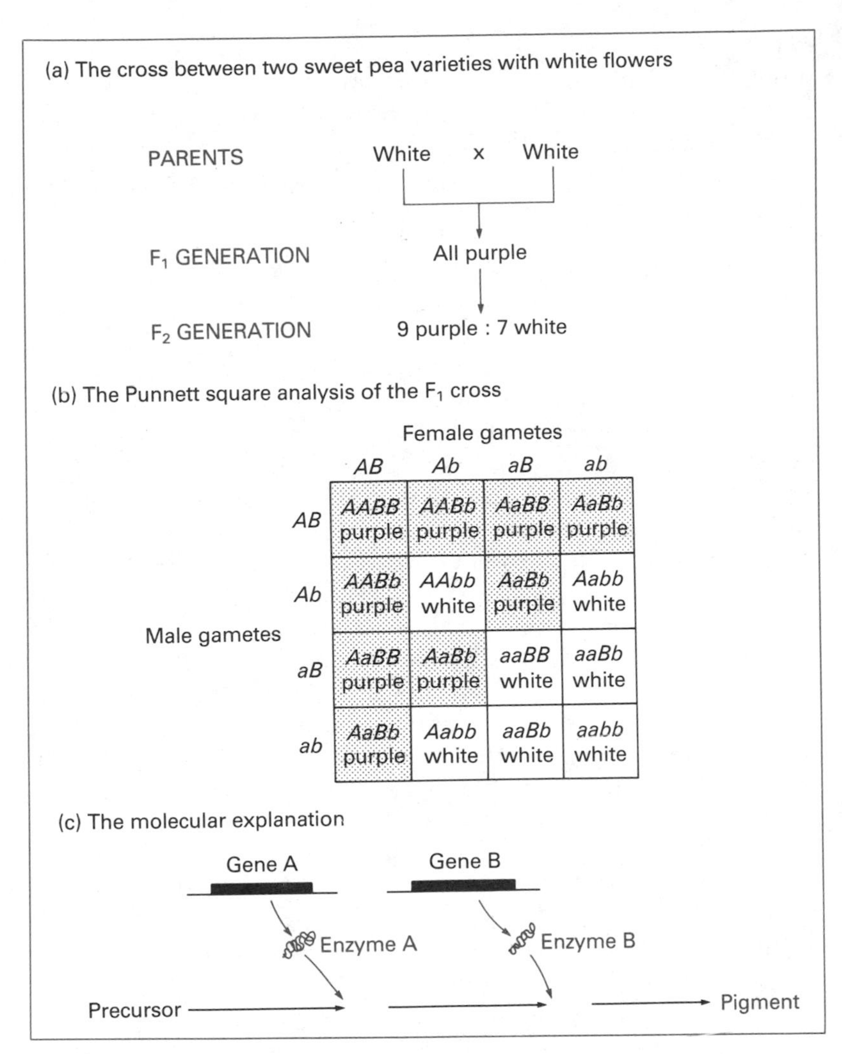

Figure 18.19 Bateson and Punnett's sweet pea cross that demonstrates epistasis between genes for flower colour. (a) The outcome of the cross. (b) The Punnett square explanation of the F_1 cross. The F_1 plants are assumed to be double heterozygotes, *AaBb*. (c) Genes *A* and *B* code for two enzymes that catalyse different steps in the biochemical pathway that results in pigment synthesis.

by step. First, the 9:7 ratio is clearly a modified 9:3:3:1, indicating that we are dealing with two genes. Recall that a standard 9:3:3:1 ratio is made up of the following genotypes (refer back to Fig. 17.6):

9 *A – B –*
3 *A – bb*
3 *aaB –*
1 *aabb*

The 9:7 ratio could be explained if we assume that the phenotype is purple only if both dominant alleles are present. Pigment synthesis does not occur if either pair of alleles is homozygous recessive, so seven of the genotypes result in white flowers (Fig. 18.19(b)). This is an example of **epistasis**, an interaction that

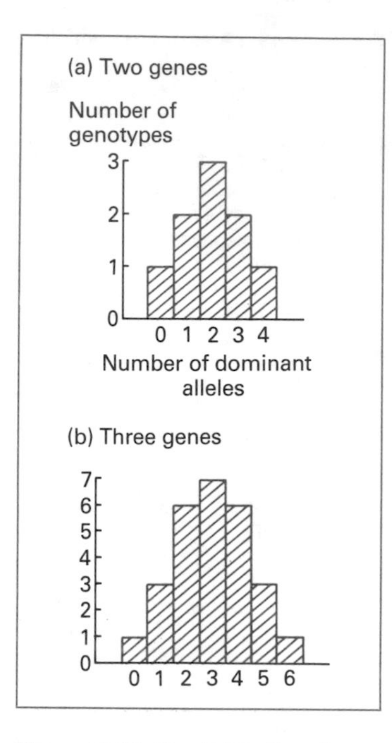

Figure 18.20 The distribution of phenotypes that will occur with (a) two and (b) three genes acting in an additive fashion and displaying incomplete dominance.

requires a particular combination of alleles at two separate genes to produce a phenotype. It is often an indication of two genes involved in a single biochemical pathway, in this case pigment synthesis (Fig. 18.19(c)). If either gene is present only as its recessive alleles then the enzyme it codes for will not be synthesized, and pigment production will be blocked.

(b) Interactions between multiple genes. The examples we have used to illustrate the complexity of flower pigmentation phenotypes have involved just two genes in either case. What if more than two genes contribute to a particular phenotype?

An indication of the effect that multiple genes might have was provided by our example of additive pigmentation genes (see Fig. 18.18). The five different levels of pigmentation that result from interactions between *X* and *Y* will not occur with equal frequencies in a random population of plants. Instead, the intermediate phenotype will predominate, as it can be produced by three separate genotypes, *XXyy*, *XxYy* and *xxYY*. The distribution of phenotypes is most clearly described in histogram form (Fig. 18.20(a)). Note that if the interaction involved three genes rather than two then the overall distribution pattern would be the same, although the curve would be smoother (Fig. 18.20(b)). If enough genes are involved, then the different phenotypes will form a continuous gradation, and intermediates will be impossible to distinguish.

A continuous distribution pattern is precisely what is found if **quantitative phenotypes** such as height or skin colour are examined in the population as a whole. These phenotypes result from interactions between large numbers of genes, and often are complicated further by environmental effects. Height, for example, is affected by diet. Quantitative phenotypes can be analysed in some detail by Mendelian genetics, but it has proved difficult to identify the individual genes involved in many of these traits. As a result we have not made much progress in examining them at the molecular level.

Reading

A. H. Sturtevant (1965) *A History of Genetics*, Harper & Row, New York, USA; J. A. Peters (1959) *Classic Papers in Genetics*, Prentice Hall, New Jersey, USA – good sources for the early developments in genetic analysis.

T. H. Morgan (1910) Sex limited inheritance in *Drosophila*. *Science*, **32**, 120–2; A. H. Sturtevant (1913) The linear arrangement of six sex-linked factors in *Drosophila* as shown by mode of association. *Journal of Experimental Zoology*, **14**, 39–45 – two of the original papers.

H. B. Creighton and B. McClintock (1931) A correlation of cytological and genetical crossing-over in *Zea mays*. *Proceedings of the National Academy of Sciences USA*, **17**, 492–7.

J. R. S. Fincham, P. R. Day and A. Radford (1979) *Fungal Genetics*, 4th edn, Blackwell, Oxford – the classic text on genetic analysis of eukaryotic microorganisms.

Problems

1. Define the following terms:

partial linkage	zygote
autosome	ascospore
sex chromosome	replica plating
crossing-over	ascus
linkage group	tetrad
map unit	pleiotropy
centiMorgan	penetrance
genetic marker	epistasis
mating type	quantitative phenotype

2. Explain why Morgan was able to deduce that the gene for white eyes lies on the X chromosome of *Drosophila*.
3. What is crossing-over and how does it explain recombination between linked genes?
4. How were the first genes mapped on the *Drosophila* X chromosome?
5. Describe what is meant by the term 'genetic marker' and provide some examples of genetic markers for *S. cerevisiae*.
6. Explain how a cross involving *S. cerevisiae* is set up.
7. Distinguish between random spore analysis and tetrad analysis.
8. What are the limitations of random spore analysis when used to map genes?
9. Distinguish between the different types of tetrad that arise from *S. cerevisiae* crosses.
10. Explain how the results of tetrad analysis are used to determine a recombination frequency for two linked genes in *S. cerevisiae*.
11. Provide examples to distinguish between pleiotropy and penetrance.
12. What can flower pigmentation tell us about interactions between genes?

13. What is currently known about the genetic basis to height?

14. *Explain why heterozygosity complicates gene mapping experiments with eukaryotes.

15. Vermilion eyes (*v*) and rudimentary wings (*r*) are two genes on the X chromosome of *Drosophila*. The cross *VVRR* (red eyes and normal wings) × *vvrr* (vermilion eyes and rudimentary wings) produced the following F_2 generation:

359 flies with red eyes and normal wings
381 flies with vermilion eyes and rudimentary wings
131 flies with red eyes and rudimentary wings
139 flies with vermilion eyes and normal wings

What is the map distance between the genes for vermilion eyes and rudimentary wings?

16. Genes, *a*, *b*, *c* and *d* lie on the same chromosome of *Drosophila*. In a series of crosses the following recombination frequencies were observed:

Genes	Recombination frequency (%)
a,c	40
a,d	25
b,d	5
b,c	10

Draw a map of the chromosome showing the positions of these genes.

17. An *S. cerevisiae* cross involving the markers a^- and b^- is set up:

$a^-b^- \times ++$

Random spore analysis reveals the following data:

Genotype	Number of spores
a^-b^-	566
a^-+	548
$+b^-$	533
$++$	578

What do these results tell us?

18. A second *S. cerevisiae* cross is set up with markers x^- and y^-. Tetrad analysis gives the following results:

Tetrad type	Number
PD	101
NPD	1
T	56

What can be deduced from these results?

19 Genetic analysis of bacteria

Basic features • Mapping by conjugation – the interrupted mating experiment • Mapping by transduction • Mapping by transformation • Relative merits of the three methods

It is rather surprising that despite the establishment by Morgan and his colleagues of genetic analysis techniques of general applicability to eukaryotic organisms, very few biologists during the early decades of the century attempted to develop analogous procedures for the study of bacterial genes. Indeed, as late as 1940 a significant number of biologists were not even convinced that bacteria actually possessed genes, but instead retained traces of the nineteenth-century awe for biological processes and considered that certain phenomena simply could not be unravelled by scientific methodology.

The person primarily responsible for placing the genetic analysis of bacteria on the same level as that of eukaryotes was Joshua Lederberg (p. 253). Lederberg's first contribution was the demonstration in 1946 that genes can be passed between bacteria during conjugation. Over the next 20 years he played a part in most of the important advances in bacterial genetics.

Three major techniques for genetic analysis of bacteria became available during the 1950s. These are **conjugation analysis, transduction analysis** and **transformation analysis**. Each can provide valuable information on bacterial genes, although with many species only one or two of the three procedures can be used. Importantly, each technique forms the basis to a method for gene mapping in bacteria. In this chapter we will look at each of the techniques in turn, with our emphasis on how they are used, individually and in combination, to obtain comprehensive gene maps. Before we start you should review Sections 13.1 and 14.3, as bacteriophages and plasmids will be central to our discussion.

19.1 Basic features of gene mapping in bacteria

The gene mapping techniques for eukaryotes, described in the previous chapter, are based on recombination (as a result of

The importance of gene mapping in bacteria

Gene maps are very interesting but how do they aid our understanding of gene activity? With a eukaryote the huge size of the genome means that it can be very difficult to isolate a gene if its map position is not known. Gene mapping is therefore an important prerequisite for detailed analysis of eukaryotic gene structure and activity by recombinant DNA techniques (Chapters 20 and 21).

With bacteria this argument does not apply. Bacterial genomes are much smaller than eukaryotic ones, so isolation of an individual gene is a less awesome task. Bacteria can be manipulated in ways that are not possible with most eukaryotes, again aiding gene isolation. Knowing the map position of a bacterial gene is therefore not a requirement for its successful isolation. However, this does not mean that mapping is 'old-fashioned' or a waste of time.

An example will help us understand the value of gene mapping. Let us say we are interested in streptomycin resistance in *E. coli*. Wild-type *E. coli* is sensitive to streptomycin but resistant mutants can be obtained by treating with a mutagen and plating onto agar containing streptomycin:

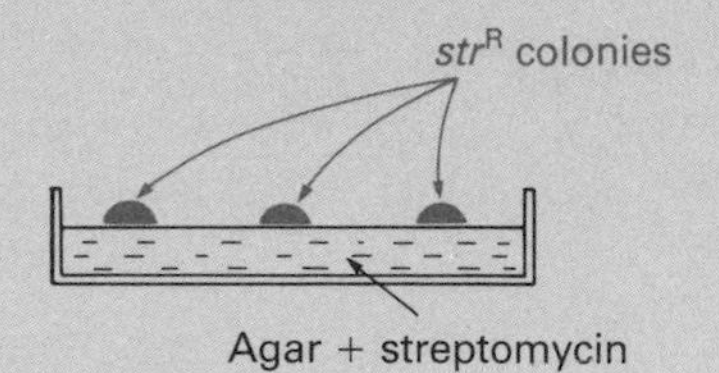

The first thing we would want to know is whether each of these three str^R colonies have mutations in the same gene. We would therefore use cells from each colony in a series of gene mapping experiments, thereby identifying the map position of the gene that is mutated in each one:

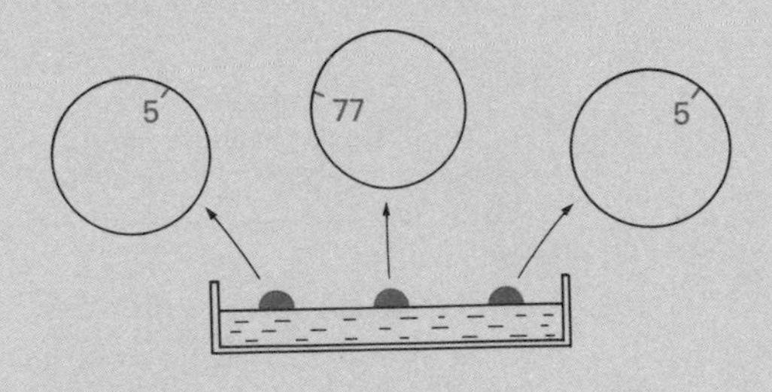

The str^R markers map to two positions on the *E. coli* DNA molecule, showing that mutations in two different genes can confer streptomycin resistance.

Working out the number of genes involved in a phenotype is just one of the values of gene mapping. In addition, gene mapping can:

1. Determine if a group of related genes are clustered into an operon or dispersed around the genome.
2. Identify the insertion sites for plasmids, lysogenic bacteriophage genomes and transposable elements.
3. Reveal DNA rearrangements such as inversions and deletions.

crossing-over) between homologous chromosomes from organisms with slightly different genetic constitutions. The recombination frequency displayed by linked genes provides a measure of their distances apart on the eukaryotic chromosome.

Each of the techniques used to map genes in bacteria is similar in that recombination has to occur between homologous pieces of DNA from two cells of different genetic constitutions. As with eukaryotic microorganisms, the genes are followed by means of genetic markers, which are biochemical characteristics generally obtained by mutagenesis, and easily distinguished by studying growth on agar media. The basic procedure, shared by each of the three different mapping techniques, is as follows (Fig. 19.1):

1. Two strains of the bacterium are required, one to act as a **donor** and one as a **recipient**. The recipient bacterium must possess mutated versions of the genes under study; the donor bacterium must carry unmutated wild-type copies of these genes.

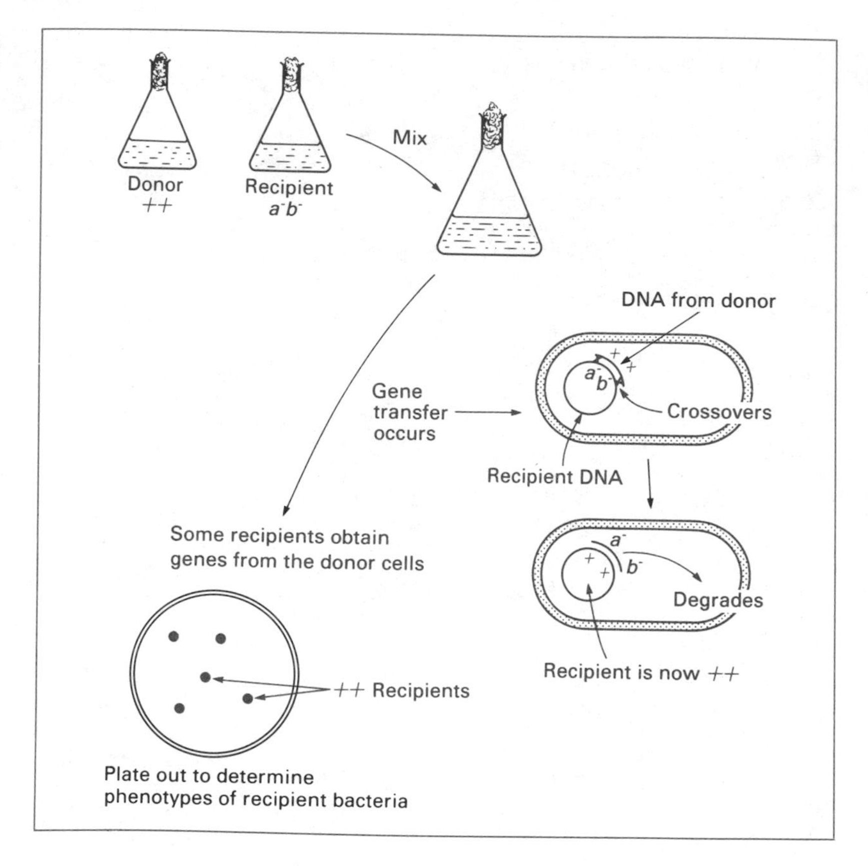

Figure 19.1 The basic features of a gene mapping experiment with a bacterium.

2. **Gene transfer** is effected, and the DNA molecule (or more likely a small part of it) of the donor bacterium is transferred into the recipient bacterium. Exactly how this is achieved is the key distinction between the three mapping techniques.
3. Recombination occurs between the DNA from the donor bacterium and the intact, functioning DNA molecule of the recipient cell. Usually, to be of value for gene mapping, two individual recombination events must occur in such a way as to insert a piece of the donor cell DNA, containing one or more of the genes under study, into the recipient cell's DNA molecule. This **double recombination** event will result in the original copies of the relevant genes present in the recipient cell being replaced by the versions obtained from the donor cell.
4. The resulting change in genotype of the recipient bacterium is assessed by spreading samples on to a **selective medium**, on which only recipient cells containing the unmutated wild-type version(s) of the gene(s) obtained from the donor parent are able to survive.

Although the three gene mapping techniques share the common features outlined above, each method is different in detail. The main distinction between them lies with the way in which transfer between donor and recipient cells is achieved. The nature of this gene transfer is the main limitation on the amount of information that can be obtained from each technique, and also dictates how the results of the experiment are interpreted. We will look at each technique in turn, and then briefly examine their strengths and weaknesses.

19.2 Mapping by conjugation

The discovery of bacterial conjugation by Lederberg and Tatum in 1946 (see Chapter 14) was the key breakthrough that enabled the first gene mapping experiments to be carried out with *E. coli*. The important observation was that transfer of some bacterial genes could occur, along with the F plasmid, during certain types of mating (see Fig. 14.10). In particular, a mating involving an Hfr strain, as studied by Hayes, Cavalli-Sforza and others, almost always results in transfer of bacterial genes from the Hfr cell to the F^- recipient (see Fig. 14.10(b)). The transferred genes may then integrate into the recipient cell's DNA to produce recombinants. Hfr in fact stands for high frequency of recombination. Can recombination brought about in this way be used to map genes? The answer is yes, as was demonstrated first by Elie Wollman and François Jacob at the Pasteur Institute early in 1955, using the **interrupted mating** technique that is still the basis for conjugation mapping.

19.2.1 The interrupted mating experiment

Wollman and Jacob's aim was to chart the transfer of bacterial genes from an Hfr cell into an F^- cell. To do this they needed an F^- strain that carried several mutated genes and an Hfr strain that possessed correct copies of these genes. Transfer of each wild-type gene from the Hfr cell to the F^- cell would be signalled by a change in genotype of the recipient cell.

The series of markers studied by Wollman and Jacob are listed in Table 19.1. The cross could be described as:

$$\text{Hfr } str^S \times F^- \ thr^- \ leu^- \ azi^R \ ton^A \ lac^- \ gal^- \ str^R$$

The experiment was carried out by mixing the Hfr and F^- cells together and then removing samples from the culture at various time intervals (Fig. 19.2). Each sample was immediately agitated in a Waring blender to snap the conjugation tubes linking Hfr and F^- cells and so interrupt the mating. The cells were spread on to an agar medium containing streptomycin, on which only the recipient cells would be able to grow (note the differences between the

Table 19.1 The series of *E. coli* genetic markers used by Wollman and Jacob in the first interrupted mating experiment

Marker	*Phenotype*
thr$^-$	Requires threonine
leu$^-$	Requires leucine
azi^R	Resistant to azide
ton^A	Resistant to colicin and T1 phage
lac$^-$	Unable to utilize lactose
gal$^-$	Unable to utilize galactose

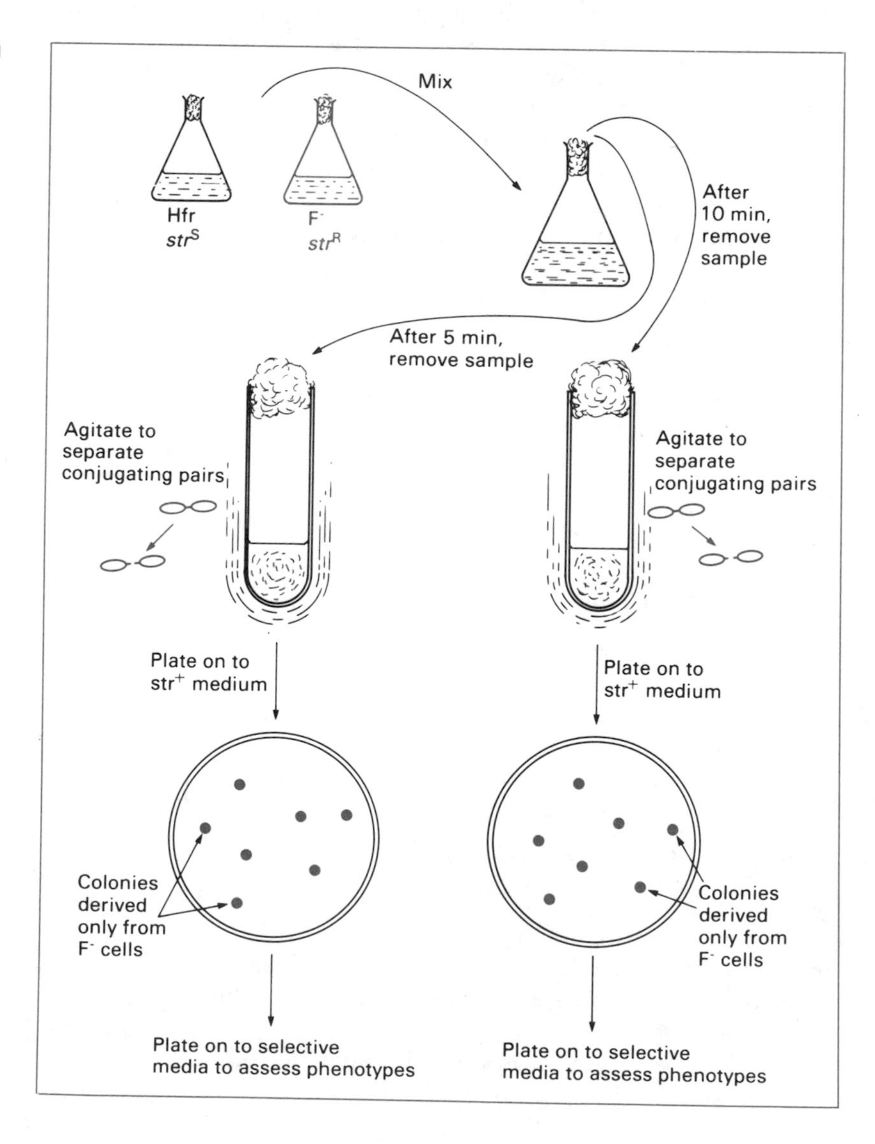

Figure 19.2 The interrupted mating experiment. See the text for details.

genotypes of the Hfr and F^- cells with respect to streptomycin resistance). The genotypes of the resulting colonies, which are derived only from recipient cells, were then determined by replica plating (Section 18.2.1) on to suitable media.

(a) Gene transfer occurs sequentially. Wollman and Jacob were not sure exactly what to expect from the interrupted mating experiment because, at that time, the precise details of Hfr conjugation had not been worked out. Although they had their own ideas about what might happen, the actual results came as a complete surprise. Rather than all the genes being transferred at once, the genes were passed from the Hfr to F^- cells at different times. First,

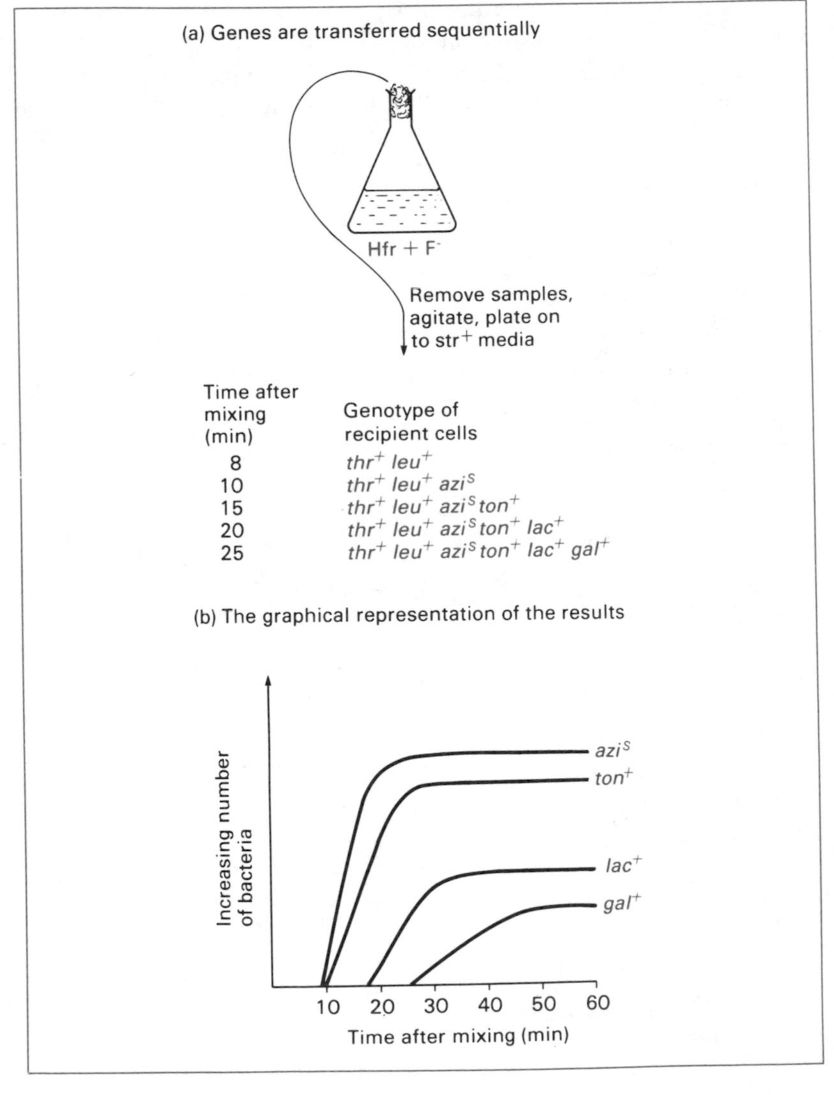

Figure 19.3 The results of the interrupted mating experiment carried out by Wollman and Jacob.

The kinetics of gene transfer during conjugation

The kinetics of gene transfer during conjugation are illustrated by the graph shown in Fig. 19.3(b). Two additional points about this graph should be noted.

The first point is that there is a slope to each line on the graph. This indicates that conjugation is not absolutely synchronous in the population of cells, and that a particular gene is transferred in some matings before it is in others. For this reason, conjugation mapping is not completely accurate and is unable to distinguish the relative positions of two genes less than about 2 minutes apart.

The second point to note is that each gene has a plateau value which indicates the maximum proportion of the population of F^- cells that can obtain that particular gene. This plateau value is always less than 100% and becomes gradually smaller for genes transferred later in the mating. The plateau indicates that not all F^- bacteria become involved in conjugation and that many matings terminate naturally during the course of the experiment.

after about 8 minutes of mating, recipient cells that had lost their auxotrophic requirements for threonine and leucine began to appear (Fig. 19.3(a)). As the mating continued, the proportion of cells displaying leucine and threonine prototrophy increased until, after a few more minutes, almost all the recipient cells were thr^+ leu^+. Next, after about 10 minutes, azide sensitivity began to appear in the recipient cells, gradually increasing in frequency until about 80% of the population were azi^S. Eventually, each of the genes was transferred to the recipient cells, Wollman and Jacob portrayed the results in the form of a graph illustrating the kinetics of transfer of each gene (Fig. 19.3(b)).

(b) Interpretation of the interrupted mating experiment. The Wollman–Jacob experiment is easily explained once we remember that Hfr cells transfer a portion, or possibly all, of the host DNA as a linear molecule into F^- cells. During this transfer the F region is the first to enter the recipient cell, followed by the rest of the *E. coli*

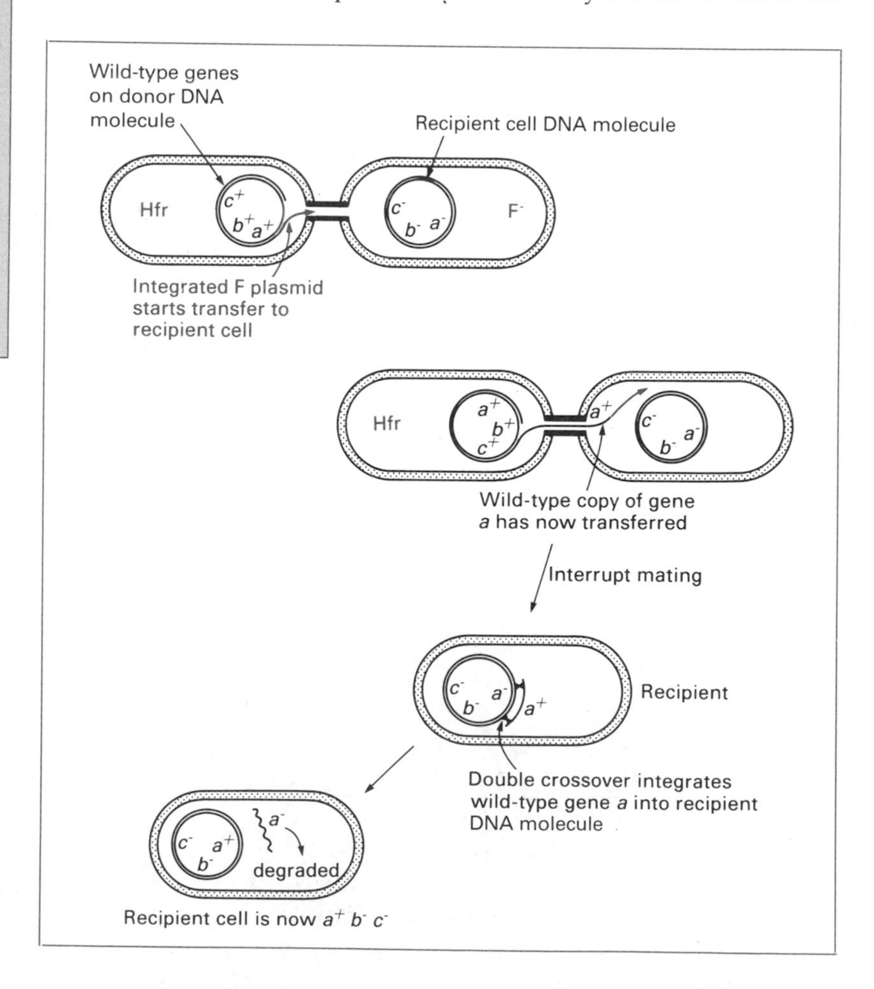

Figure 19.4 The events that occur during an interrupted mating between an Hfr bacterium and an F^- recipient cell.

DNA molecule. If mating is interrupted, only those genes already transferred will be able to recombine with the recipient cell's DNA and give rise to recombinants (Fig. 19.4). By measuring the times at which different genes appeared in the recipient cells, Wollman and Jacob were in effect mapping the relative positions of the genes on the *E. coli* DNA molecule. Indeed, map distances on *E. coli* DNA are now expressed in minutes, with 1 minute representing the length of DNA that takes 1 minute to transfer. There are 100 minutes on the map because transfer of the entire *E. coli* DNA molecule takes about 100 minutes.

Transfer of the entire DNA molecule is not very likely, however, as many bacteria abandon mating naturally well before 100 minutes have passed. To map the entire *E. coli* DNA it has been necessary to carry out a series of interrupted mating experiments, with a variety of Hfr strains, each with the F plasmid inserted at a different position. The regions studied by the individual experiments overlap, enabling the complete map to be built up (Fig. 19.5). It soon became apparent that although genes are transferred in a linear fashion, the *E. coli* genetic map is actually circular.

Conjugation mapping is a powerful technique for determining the relative positions of genes on the *E. coli* DNA molecule. However, it is not tremendously accurate, and genes that are less than 2 minutes apart cannot be separated by conjugation mapping. We know that 2 minutes corresponds to about 80 kb of DNA, enough space for quite a few genes. Furthermore, conjugation mapping is only applicable to those species of bacteria that have F plasmids or their equivalent: many do not. We must therefore look at other methods for gene mapping in bacteria.

(a) A series of linear gene maps from experiments with different Hfr strains

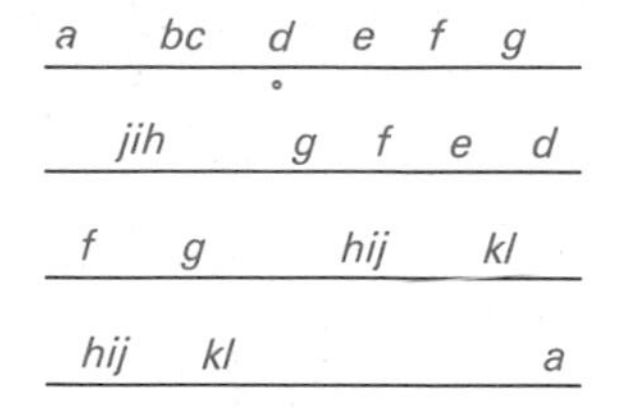

(b) Combining the linear maps gives a circular molecule

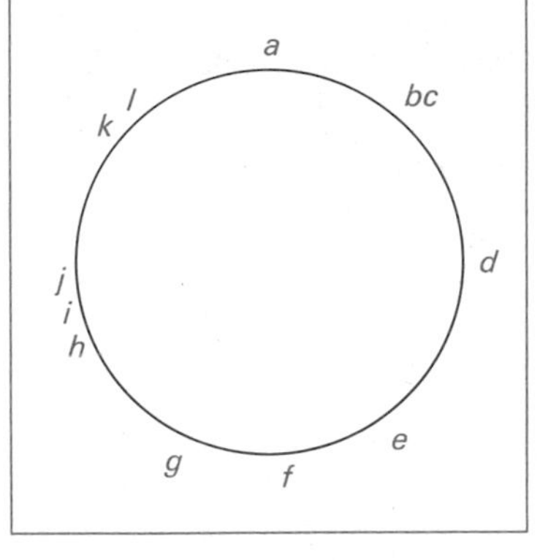

Figure 19.5 Interrupted mating experiments with a variety of Hfr strains provided a series of linear maps (a) which, when joined together, proved that the *E. coli* DNA molecule is circular (b).

19.3 Mapping by transduction

Conjugation and transduction (as well as transformation) are based on totally different mechanisms for gene transfer from donor bacterium to recipient bacterium. Instead of physical contact between cells, with transduction gene transfer is effected by a bacteriophage.

19.3.1 The discovery of transduction

Transduction was discovered by Joshua Lederberg and Norton Zinder in 1952, not with *E. coli* but with the related enterobacterium *Salmonella typhimurium*. They mixed together two auxotrophic strains of this organism, one called LA22 (with the genotype phe^- trp^- met^+ his^+) and one called LA2 (phe^+ trp^+ met^- his^-), and obtained a small number of prototrophs (phe^+ trp^+ met^+ his^+), about one in every 100 000 cells. By now, the reader should appreciate that prototrophs can arise only by way of gene transfer between bacteria, albeit in this case with a very low frequency (Fig. 19.6(a)).

At first it was thought that these results must be due to a conjugative process similar to that characterized in *E. coli*, but further examination suggested that this was not the case. In particular a modification of the basic experiment proved that cell-to-cell contact is not needed for gene transfer between LA2 and LA22: prototrophs were still obtained when individual cultures were placed in opposite arms of a U-tube, separated from each other by a sintered glass filter (Fig. 19.6(b)). Media and small particles can pass through this filter, but not cells. Now it became apparent that gene transfer is unidirectional and that only strain

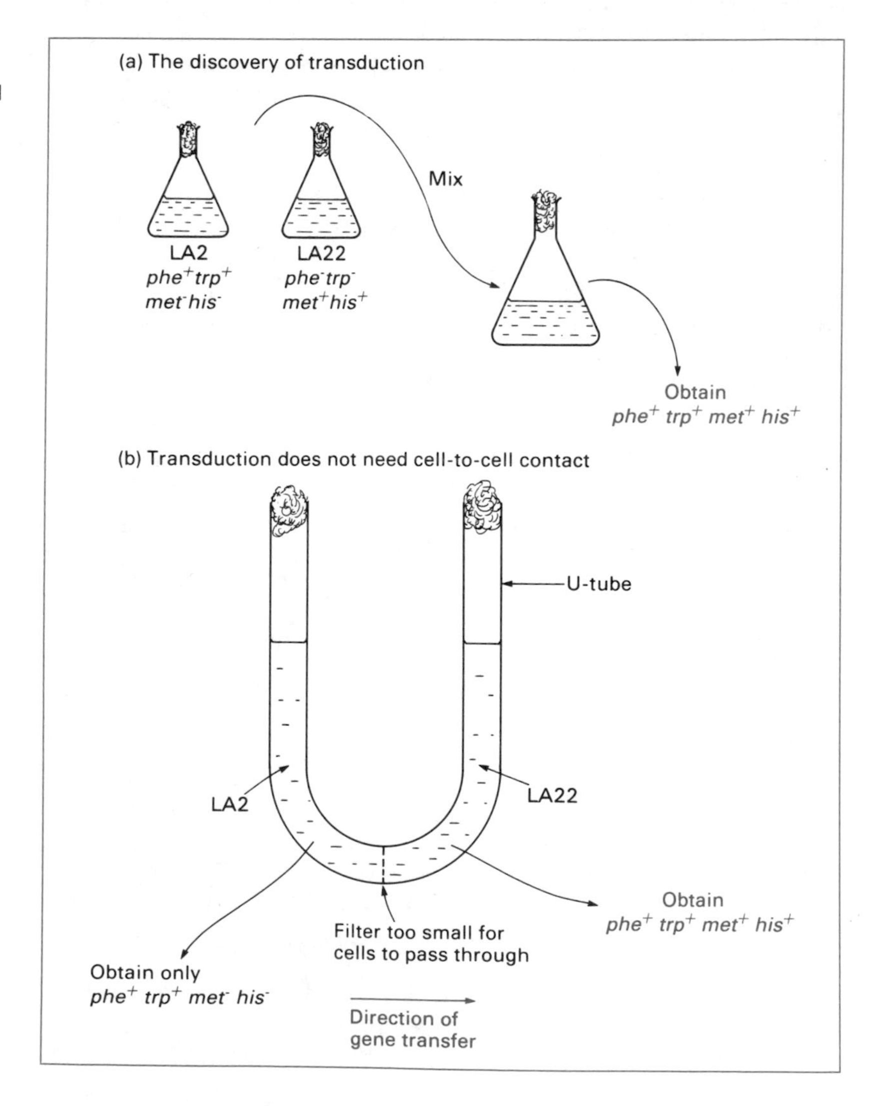

Figure 19.6 Transduction. (a) The discovery, and (b) the first indication that cell-to-cell contact is not needed in order for gene transfer to occur.

LA22 was becoming prototrophic. These results suggested that this type of genetic exchange is mediated by a filterable agent that is produced by LA2 cells and transfers to LA22 cells.

(a) Transduction is mediated by a bacteriophage. The explanation of transduction turned out to be based on a familiar system, bacteriophage infection, but in an unexpected way. It was discovered that strain LA2 contains the prophage version of a phage called P22. Every now and then a P22 prophage is induced and enters into the lytic stage of its infection cycle. During the lead up to lysis the bacterial DNA molecule is broken into fragments, some of which by chance happen to be about the same length as the P22 phage DNA molecule. By mistake a portion of bacterial DNA can be packaged into a P22 phage coat, although this happens only very infrequently (Fig. 19.7). The resulting phage particle is still infective, as infection is a function purely of the protein components of the phage coat, but the infection is **abortive** because of course the bacterial DNA fragment injected into a new cell cannot itself direct synthesis of new phages.

The bacterial DNA fragment transferred into a new cell can,

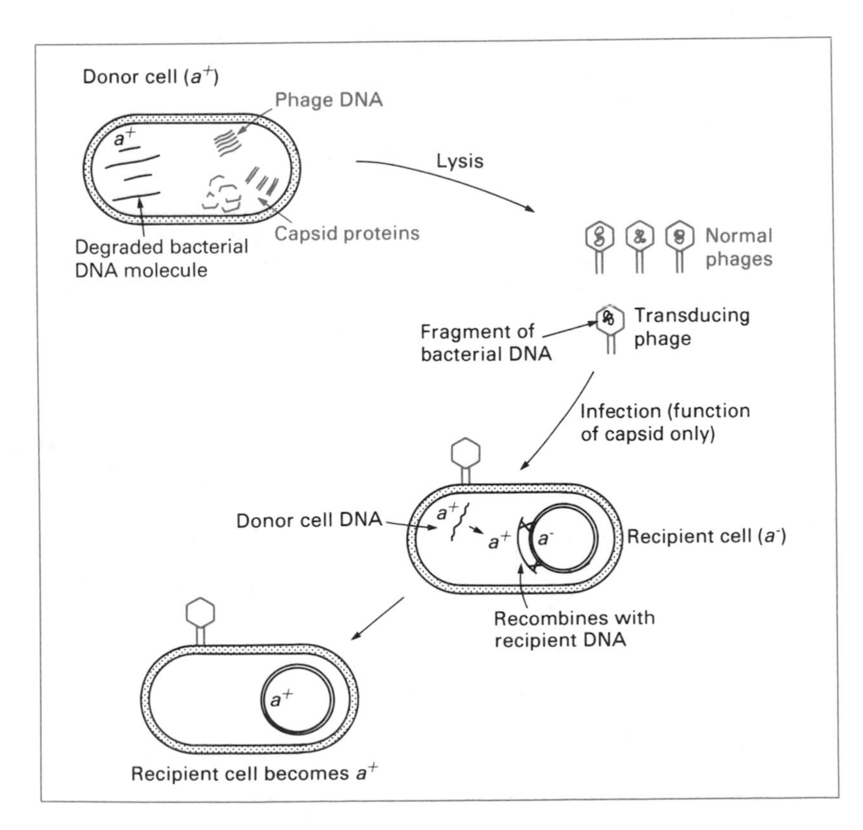

Figure 19.7 The way in which gene transfer occurs during transduction.

however, recombine with the new cell's DNA. This means that gene transfer can occur: the donor cell is the one that was lysed by the phage, and the recipient cell is the one that is infected by the transducing phage particle.

19.3.2 Mapping genes by transduction

The amount of DNA carried by a transducing P22 phage particle is about 40 kb, or about 1% of the total *S. typhimurium* genome. That is sufficient to carry a dozen or more genes. **Cotransduction**, the transfer of two or more genes from donor to recipient, will occur if the genes are relatively close together on the *S. typhimurium* DNA molecule and hence can be contained in a single 40 kb fragment (Fig. 19.8). A transduction mapping experiment, therefore, uses a wild-type donor and double mutant recipient and tests for cotransduction of the two wild-type genes into the recipient cell. This type of experiment is very useful for establishing a relatively close proximity between two genes on a bacterial DNA molecule.

Transduction mapping can also be taken a stage further as the

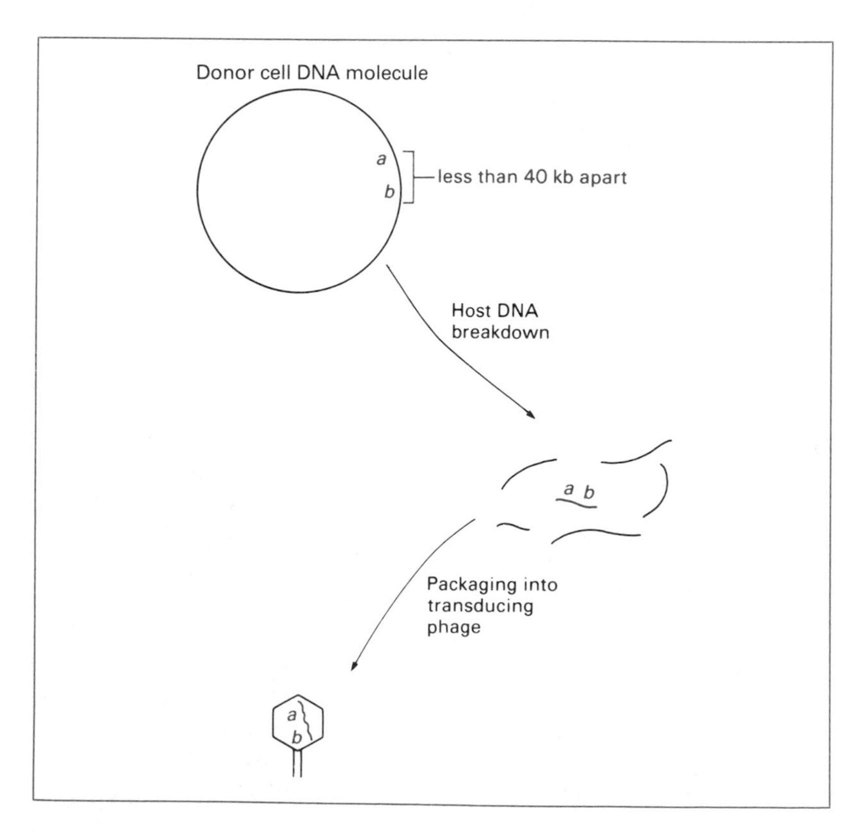

Figure 19.8 Cotransduction of two genes may occur if they are less than 40 kb apart on the donor cell DNA molecule.

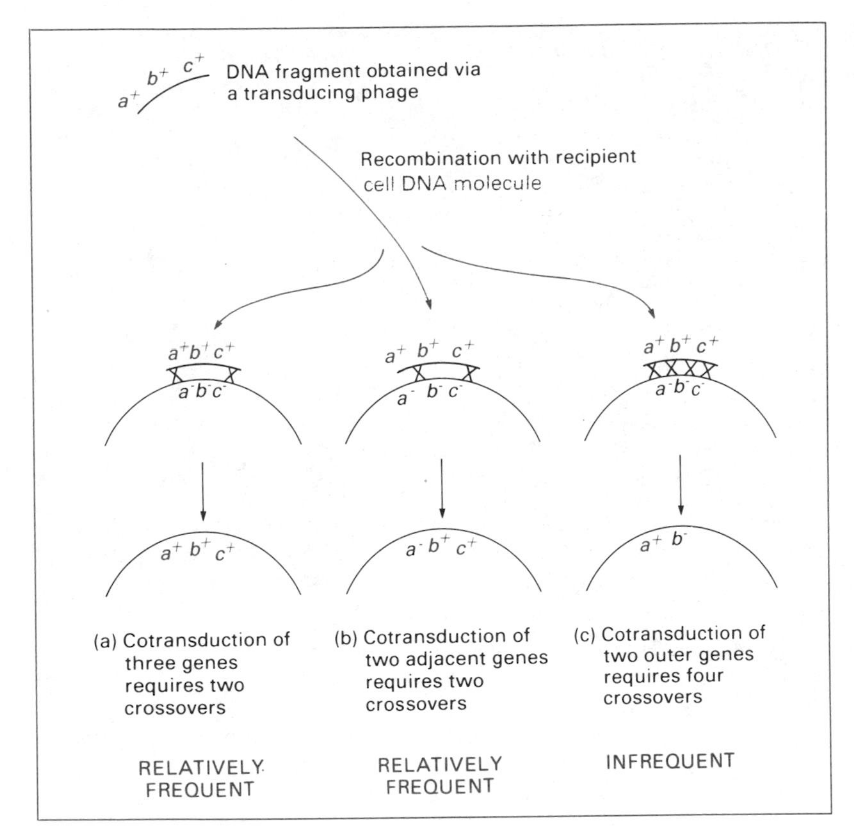

Figure 19.9 Cotransduction can be used to work out the order of genes on a DNA fragment as adjacent genes will cotransduce at a higher frequency than non-adjacent genes.

order of genes can be determined if three or more markers are used. Say, for instance, genes *a*, *b* and *c* lie in that order in the same 40 kb region of the DNA molecule. Cotransduction of the genes would be studied with a donor bacterium carrying wild-type copies ($a^+b^+c^+$) and a triple mutant recipient ($a^-b^-c^-$). Cotransduction of all three genes would be signalled by the recipient becoming $a^+b^+c^+$. For this to occur, the two crossovers that result in recombination must occur outside the three genes (Fig. 19.9(a)). However, crossing-over could also occur between the genes, so the genes may be cotransduced as pairs, or even transduced individually (Fig. 19.9(b)). The crucial detail is that only adjacent genes (*ab* and *bc* in the example shown) will be cotransduced as a pair with any great frequency. This is because cotransduction of the adjacent genes requires just two crossover events, whereas cotransduction of the outer genes alone (*ac* in this example) requires four crossovers (Fig. 19.9(c)). The cotransduction frequencies for the different pairs of genes therefore allow the gene order to be determined.

Specialized transduction

The events described in the text are more properly called generalized transduction to distinguish the phenomenon from **specialized transduction**, a feature of several lysogenic phages including λ. Specialized transduction arises when excision of an integrated prophage occurs slightly inaccurately, so that a short piece of bacterial DNA remains attached to the viral genome. This piece is then transferred via the phage particle to the next infected cell.

The critical difference between generalized and specialized transduction is that the former involves transfer of random pieces of bacterial DNA, whereas the latter results only in transfer of genes immediately adjacent to the prophage integration site.

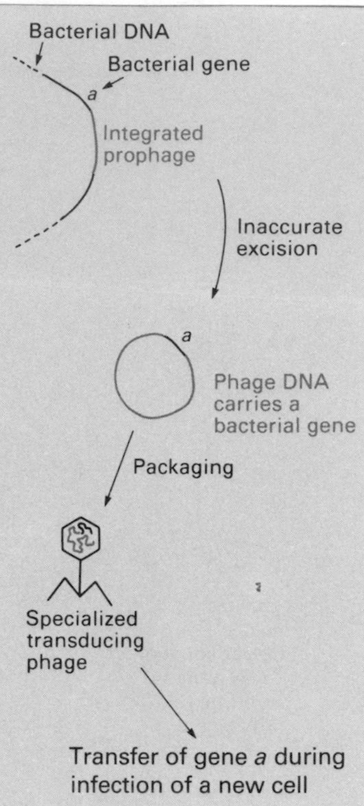

Bacterial DNA
Bacterial gene
a
Integrated prophage
Inaccurate excision
a
Phage DNA carries a bacterial gene
Packaging
Specialized transducing phage
Transfer of gene *a* during infection of a new cell

19.4 Mapping by transformation

Transformation is similar to transduction in the way it is used in gene mapping, but the basis of the genetic exchange is different. With transformation, 'naked' DNA molecules are taken up by the recipient cell – recall the transforming principle used to show that genes are made of DNA (see Section 2.2.1).

Exactly how naked DNA enters a bacterial cell is not known. It is clear, however, that some bacteria, such as *Bacillus* and *Haemophilus* species, have efficient mechanisms for DNA uptake whereas others, including *E. coli*, are transformed only with difficulty, although certain experimental treatments can improve the transformation efficiency by rendering the cells more **competent** (Fig. 19.10).

To carry out a transformation mapping experiment a sample of DNA from the donor strain, wild-type for the genes under study, must be prepared (Fig. 19.11). The DNA is mixed with competent recipient cells and uptake takes place followed by recombination. If two genes are close enough together then **cotransformation** may

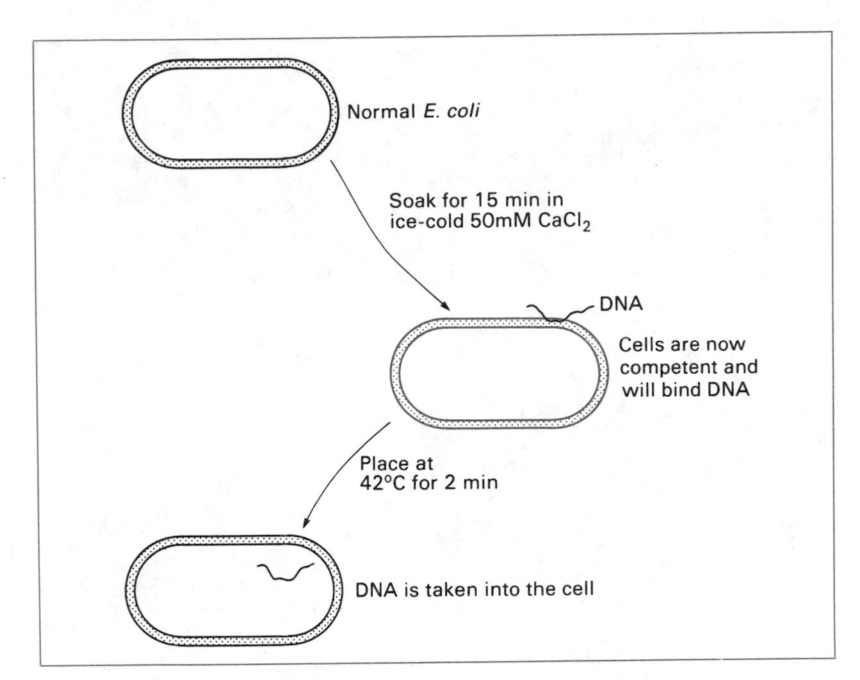

Figure 19.10 Preparation of *E. coli* cells that are competent for DNA uptake.

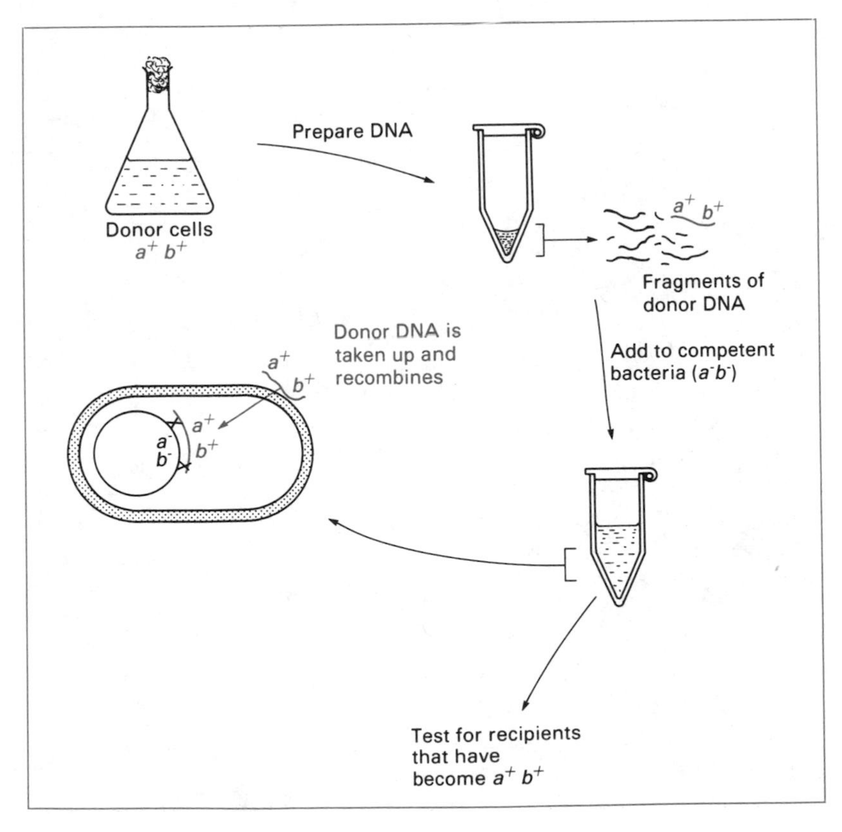

Figure 19.11 Cotransformation of wild-type genes a^+ and b^+ into an a b recipient bacterium.

F-duction

An important additional means of gene mapping is provided by **F-duction**. This depends on the fact that F′ plasmids carry a small portion of the *E. coli* DNA molecule (Section 14.3.2). If an F′ plasmid is introduced into an F⁻ cell, then the cell will become diploid for any genes that the F′ carries:

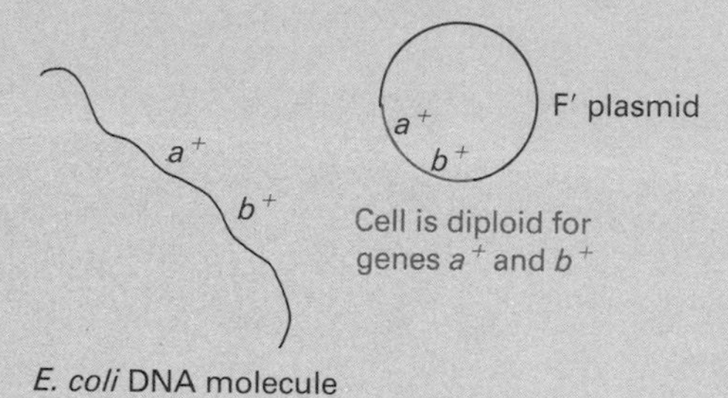

This provides a novel means of gene transfer and allows genes to be mapped in a manner analogous to transduction and transformation mapping. Note though that as recombination does not occur the gene order cannot be determined. Despite this limitation F-duction has a number of plus points. First, transfer of the F′ plasmid from an F′ to F⁻ cell is very efficient, so a high proportion of partial diploids (**merodiploids**) can be obtained. Second, the gene transfer is not random as different F′ plasmids carrying pieces of *E. coli* DNA of varying sizes from all regions of the genome are available:

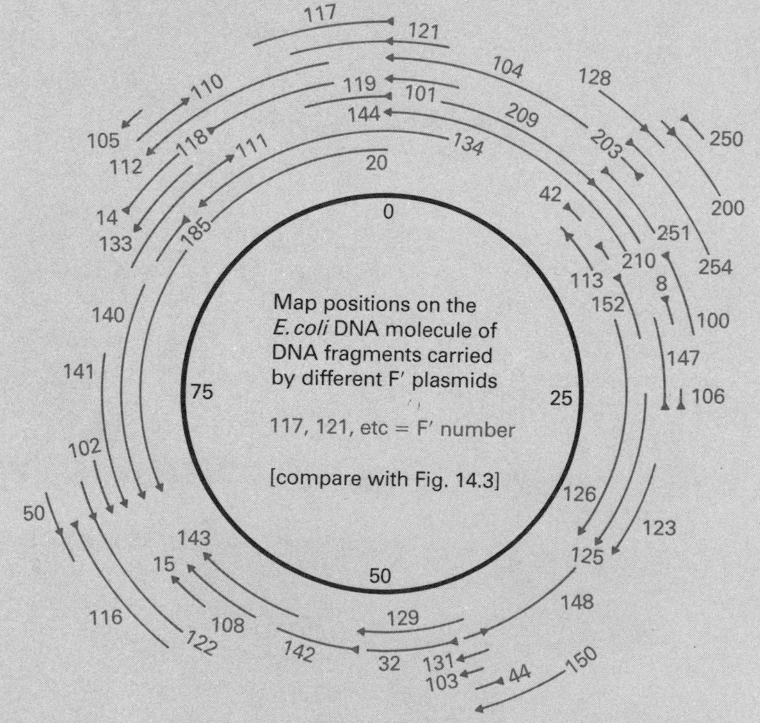

The mapping experiment can therefore be targeted directly at the region of the genome in which the gene is thought to lie.

The *cis-trans* complementation text

An additional advantage of F-duction is that the merodiploids are stable. This expands the repertoire of genetic analyses that can be carried out with bacteria. Among other things, it enables us to distinguish individual genes in an operon, something that can be difficult to achieve by standard gene mapping. Consider the *trp* operon for instance:

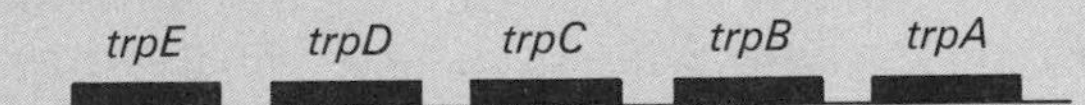

Here we have five genes, very close together, each involved in the same phenotype. If we use a series of tryptophan auxotrophs as recipients in gene mapping experiments then all the *trp*⁻ markers will map to the same position on the *E. coli* DNA molecule. It might prove impossible to distinguish the individual genes even by transduction mapping. We could, however, use the ***cis–trans* complementation test** to determine if two tryptophan auxotrophs carry mutations in the same or different genes. A merodiploid is set up with a host bacterium carrying one marker (*trp1*) and an F′ plasmid carrying a second marker (*trp2*):

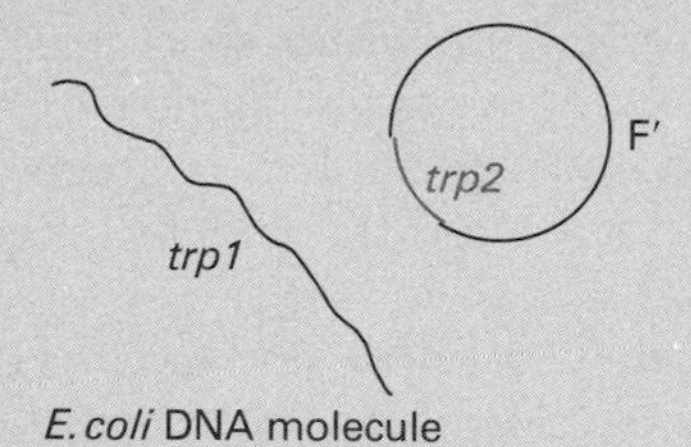

If *trp1* and *trp2* represent the same gene then the merodiploid will still be auxotrophic for tryptophan:

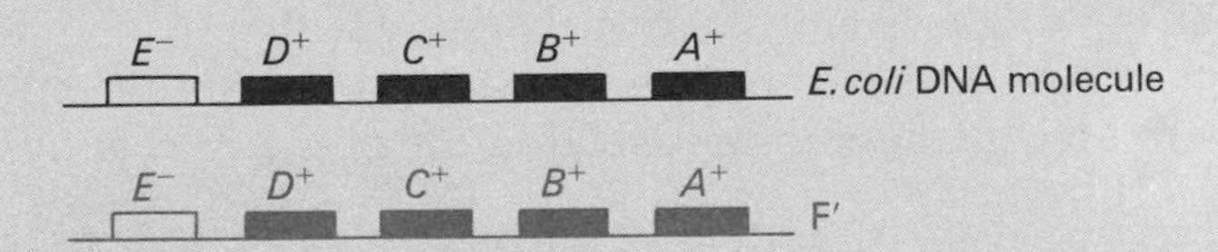

If, on the other hand, *trp1* and *trp2* represent different genes, then the two mutations will complement one another, and the merodiploid will be a prototroph:

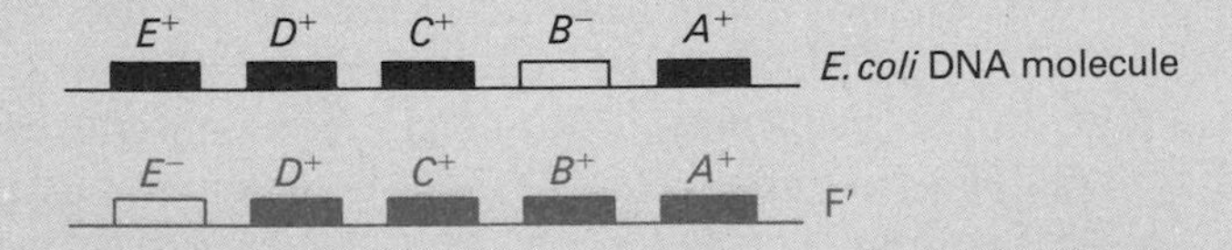

The *cis-trans* test can therefore tell us if a locus identified by gene mapping represents a single gene or a cluster of related genes.

occur. Gene orders can be worked out by cotransformation, just as they can by cotransduction.

Maps of bacterial genomes

Species	Approximate number of mapped genes
Escherichia coli	1000
Salmonella typhimurium	850
Bacillus subtilis	700
Pseudomonas aeruginosa	250
Caulobacter crescentus	115
Pseudomonas putida	85
Neisseria gonorrhoeae	60
Staphylococcus aureus	55
Proteus mirabilis	22

Data taken from *Genetic Maps*, 5th edn (ed. S.J. O'Brien), Cold Spring Harbor Laboratory Publications, New York, 1990.

19.5 Relative merits of the three methods

The existence of three different mapping techniques has brought about great advances in understanding gene organization in bacteria. To map a newly discovered gene in *E. coli* a good approach would be first to perform an interrupted mating experiment, which will position the gene with an accuracy of about 2 minutes, equivalent to about 80 kb. The exact position of the gene can then be established with greater accuracy by testing for cotransformation or cotransduction with genes known to lie in the same region of the genome. Finally, the gene order can be worked out by cotransduction and/or cotransformation.

This situation does not hold for all bacteria. Conjugation mapping is clearly limited to those species that possess F plasmids or equivalents. This means that it is not generally applicable to the Gram-positive group of bacteria, which includes several important genera (notably *Streptomyces*, from which 80% of all antibiotics are obtained). Transduction mapping is also limited in scope because transducing phages are known for only a few species. For many bacteria, transformation mapping is the only technique that can be used satisfactorily.

These limitations are not so important today because recombinant DNA techniques have provided new gene mapping procedures that do not depend on gene transfer in the conventional sense, and so are applicable to virtually all species, including higher organisms. Recombinant DNA technology also enables the researcher to go several stages beyond gene mapping and obtain detailed information on gene structure and expression. These powerful techniques are described in the next two chapters.

Reading

D. Freifelder (1987) *Microbial Genetics*, Jones and Bartlett, Boston, USA – Chapters 13, 14 and 15 cover gene mapping.

E. L. Wollman and F. Jacob (1955) Sur la mecanisme du transfert de materiel genetique au cours de la recombinaison chez *Escherichia coli* K12. *Comptes Rendus de l'Academie des Sciences*, **240**, 2449–51; F. Jacob and E. L. Wollman (1955) Etapes de la recombinaison genetique chez *Escherichia coli* K12. *Comptes Rendus de l'Academie des Sciences*, **240**, 2566–8 – the first interrupted mating experiment.

N. Willetts and R. Skurray (1980) The conjugation system of F-like plasmids. *Annual Review of Genetics*, **14**, 47–76 – an advanced review.

N. D. Zinder and J. Lederberg (1952) Genetic exchange in *Salmonella. Journal of Bacteriology*, **64**, 679–99; N. D. Zinder (1958) 'Transduction' in bacteria. *Scientific American*, **199**(5), 38–43 – the first report and a review.

R. D. Hotchkiss and J. Marmur (1954) Double marker transformations as evidence of linked factors in desoxyribonucleate transforming agents. *Proceedings of the National Academy of Sciences USA*, **40**, 55–60 – the first attempt at transformation mapping.

B. W. Holloway (1979) Plasmids that mobilize bacterial chromosome. *Plasmid*, **2**, 1–19 – includes a description of F plasmids from bacteria other than *E. coli*.

S. H. Goodgal (1961) Studies on transformation of *Hemophilus influenzae. Journal of General Physiology*, **45**, 205–28; D. Dubnau, C. Goldthwaite, I. Smith and J. Marmur (1967) Genetic mapping in *Bacillus subtilis. Journal of Molecular Biology*, **27**, 163–85 – mapping with bacteria other than *E. coli*.

Problems

1. Define the following terms:

donor bacterium	interrupted mating
recipient bacterium	abortive infection
gene transfer	cotransduction
double recombination	competence
selective medium	cotransformation

2. Outline the basic features of an experiment to map genes in a bacterium.

3. Distinguish between the ways by which gene transfer is effected in conjugation mapping, transduction mapping and transformation mapping.

4. Describe how Wollman and Jacob carried out the first interrupted mating experiment.

5. Explain how interrupted mating experiments proved that the *E. coli* DNA molecule is circular.

6. Describe how transduction was discovered.

7. Explain how transduction mapping can be used to determine gene order.

8. Explain what transformation is and how it can be used in gene mapping experiments.

9. Compare the relative merits and limitations of the three gene mapping techniques for bacteria.

10. Deduce the methods used by Jacob and Wollman to test for transfer of each of the markers they studied in the first interrupted mating experiment. (The markers are listed in Table 19.1.)

11. A series of interrupted mating experiments with different Hfr strains produces the following linear gene maps:

1. *A J F C*
2. *D H E B*
3. *I B E H*
4. *J F C I*
5. *A G D H*

Draw a circular map of the bacterial DNA molecule showing the positions of these genes.

12. In a series of transduction experiments, involving genes *A*, *B*, *C* and *D*, the following combinations were found to cotransduce relatively frequently:

A and *B*
C and *D*
A and *D*

Deduce the gene order.

Cloning genes

20

What is gene cloning? • Constructing recombinant DNA molecules – restriction endonucleases – ligases • Cloning vectors and the way they work – cloning with pBR322 – other types of cloning vector for *E. coli* • Cloning vectors for eukaryotes – *Saccharomyces cerevisiae* – plants and animals

Many spectacular advances were made in genetics and molecular biology during the 75 years that followed the rediscovery of Mendel's work in 1900. These advances were in large part due to the gene analysis techniques described in the previous two chapters but were of course also supplemented by additional experiments that made use of a wide range of biochemical, biophysical and crystallographic procedures. The reader needs only to recall some of the momentous experiments described in this book – the Avery, Hershey–Chase and Meselson–Stahl experiments for instance – to appreciate the diversity of the methodology that was brought to bear on the gene and its mode of action.

Towards the end of the 1960s limitations in the available technology started to become apparent. By this time the genetic code had been elucidated and the basic features of transcription and translation worked out. Molecular geneticists wished to examine the gene in increasing detail, but were frustrated by their inability to design ways of carrying out the critical experiments. The situation was eventually retrieved by the development in the early 1970s of a whole new methodology, referred to as **recombinant DNA technology** or **genetic engineering**, and having at its core the technique called **gene cloning**. In this chapter we will look at gene cloning and what it involves and then, in the final chapter, examine the information that can be obtained by studying cloned genes.

20.1 What is gene cloning?

The basic events in a gene cloning experiment are as follows (Fig. 20.1):

1. A fragment of DNA containing the gene to be cloned is inserted into a second (usually circular) DNA molecule, called a **cloning vector**, to produce a **recombinant DNA molecule**.

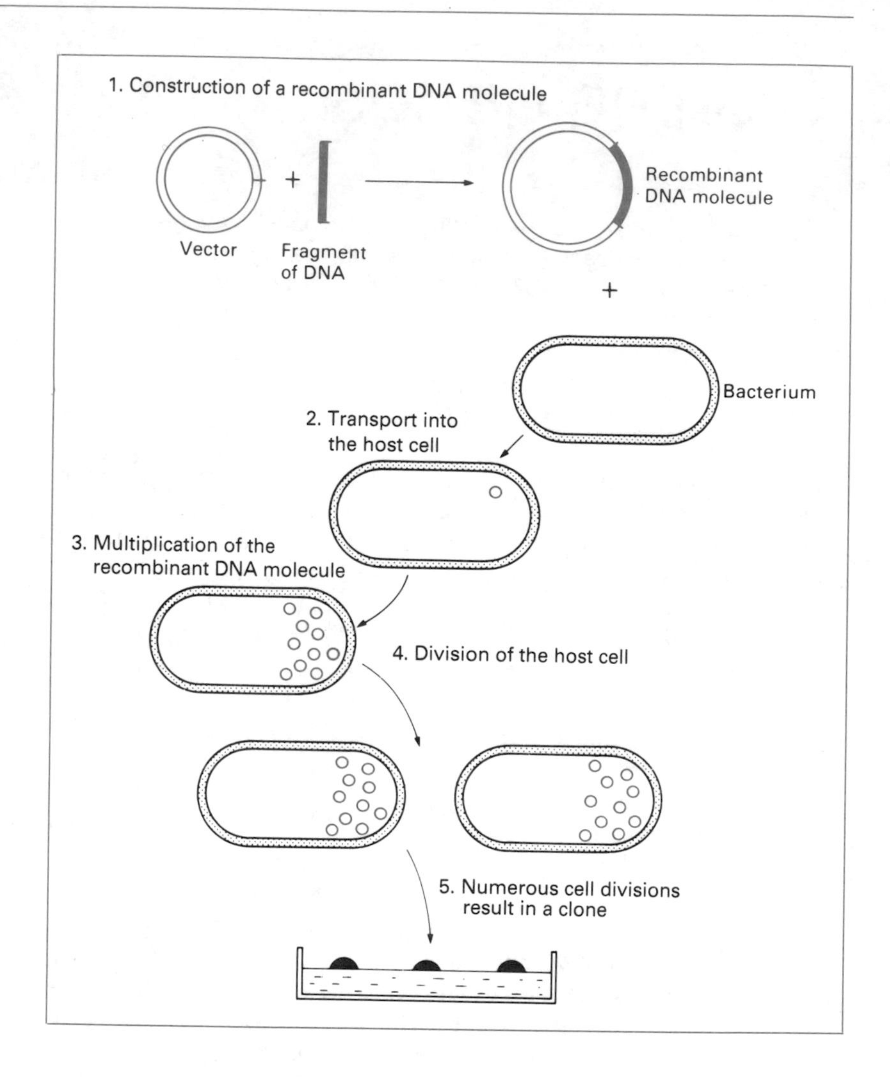

Figure 20.1 The basic events in a gene cloning experiment.

2. The recombinant DNA molecule is introduced into a host cell (often but not always *E. coli*) by transformation or an equivalent procedure.
3. Within the host cell the vector directs multiplication of the recombinant DNA molecule, producing a number of identical copies.
4. When the host cell divides, copies of the recombinant DNA molecule are passed to the progeny and further vector replication takes place.
5. A large number of cell divisions give rise to a **clone**, a colony of cells each containing multiple copies of the recombinant DNA molecule.

Gene cloning is therefore a relatively straightforward procedure; why then has it assumed such importance in biology? The answer

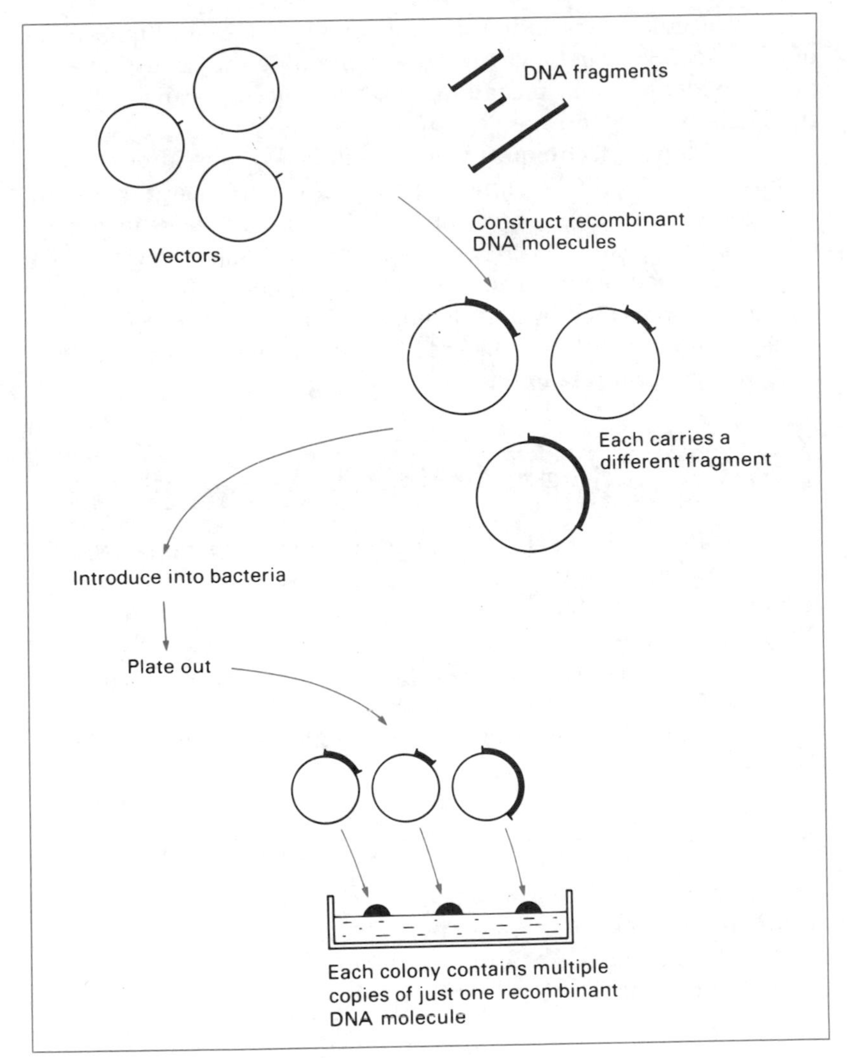

Figure 20.2 Cloning allows individual fragments of DNA to be purified.

is largely because cloning can provide a pure sample of an individual gene, separated from all the other genes with which it normally shares the cell. To understand exactly how this arises, let us take a second look at a gene cloning experiment, drawn in a slightly different way (Fig. 20.2). The DNA fragment to be cloned can be part of a mixture of many different fragments, each carrying different genes or parts of genes. This mixture could indeed be the entire genetic complement of a higher organism: man for example. All these fragments will be inserted into different vector molecules to produce a family of recombinant DNA molecules, one of which carries the gene of interest. During transformation each host cell takes up just a single recombinant DNA molecule, so that although the final set of clones may contain many different recombinant

DNA molecules, each individual clone will contain multiple copies of just one molecule. Once the clone containing the gene of interest has been identified, its recombinant DNA molecules can be purified from cell extracts and the gene recovered.

Before cloning techniques were developed it was impossible to obtain purified samples of individual genes. This meant that work on gene structure and expression had to deal with genes in general terms, rather than concentrating on the particular features of a single specified gene. There was no way of building on the information provided by Mendelian analysis and progress in many areas of genetics became blocked. Cloning provided the spur that enabled genetical research to get back in gear.

20.2 Constructing recombinant DNA molecules

The first step in a gene cloning experiment, construction of recombinant DNA molecules, involves cutting DNA molecules at specific points and joining them together again in a controlled manner. DNA manipulative techniques of this type make use of purified enzymes that, in the cell, participate in processes such as DNA replication. The two main types of DNA manipulative enzyme used in gene cloning are **restriction endonucleases** and **DNA ligases**.

20.2.1 Enzymes for cutting DNA: restriction endonucleases

During construction of a recombinant DNA molecule the circular vector must be cut at a single point into which the fragment of DNA to be cloned can be inserted. Not only must each vector be cut just once, but all the vector molecules must be cut at precisely

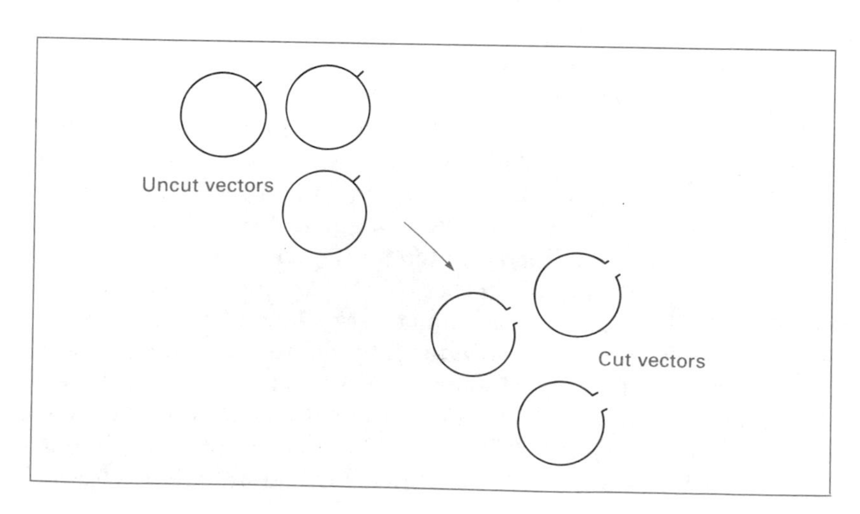

Figure 20.3 All vector molecules must be cut at exactly the same position.

Table 20.1 The recognition sequences of some commonly used restriction endonucleases

Enzyme	*Organism*	*Recognition sequence**	*Type of end*
*Pvu*I	*Proteus vulgaris*	CGATCG	Sticky
*Pvu*II	*Proteus vulgaris*	CAGCTG	Blunt
*Eco*RI	*Escherichia coli*	GAATTC	Sticky
*Bam*HI	*Bacillus amyloliquefaciens*	GGATCC	Sticky
*Hin*fI	*Haemophilus influenzae*	GANTC	Sticky
*Sau*3A	*Staphylococcus aureus*	GATC	Sticky
*Alu*I	*Arthrobacter luteus*	AGCT	Blunt
*Hae*III	*Haemophilus aegyptius*	GGCC	Blunt
*Not*I	*Nocardia otitidis-caviarum*	GCGGCCGC	Sticky

* The sequence shown is that of one strand, given in the 5′ → 3′ direction. Note that almost all recognition sequences are palindromes: when both strands are considered they read the same in each direction, e.g.

```
         5′-GAATTC-3′
EcoRI:      ||||||
         3′-CTTAAG-5′
```

the same position (Fig. 20.3). Clearly, a very special type of nuclease is needed.

The relevant enzymes are called Type II restriction endonucleases. Each recognizes a specific nucleotide sequence and cuts a DNA molecule at this sequence and nowhere else. For example, the restriction endonuclease called *Pvu*I (isolated from *Proteus vulgaris*) cuts DNA only at the hexanucleotide CGATCG. In contrast, a second enzyme from the same bacterium, called *Pvu*II, cuts at a different hexanucleotide, in this case CAGCTG. Many restriction endonucleases recognize hexanucleotide target sites, but others cut at four-, five- or even eight-nucleotide sequences (Table 20.1). There are also examples of restriction endonucleases with degenerate recognition sequences, meaning that they cut DNA at any of a family of related sites. *Hin*fI, for example, recognizes GANTC, so cuts at GAATC, GATTC, GAGTC and GACTC.

(a) Blunt ends and sticky ends. The exact nature of the cut produced by a restriction endonuclease is very important in the design of a gene cloning experiment. Many restriction endonucleases make a simple double-stranded cut in the middle of the recognition sequence (Fig. 20.4(a)), resulting in a **blunt** or flush end. *Pvu*II and *Alu*I are examples of blunt-end cutters.

Other restriction endonucleases cut DNA in a slightly different way. With these enzymes the two DNA strands are not cut at exactly the same position. Instead the cleavage is staggered,

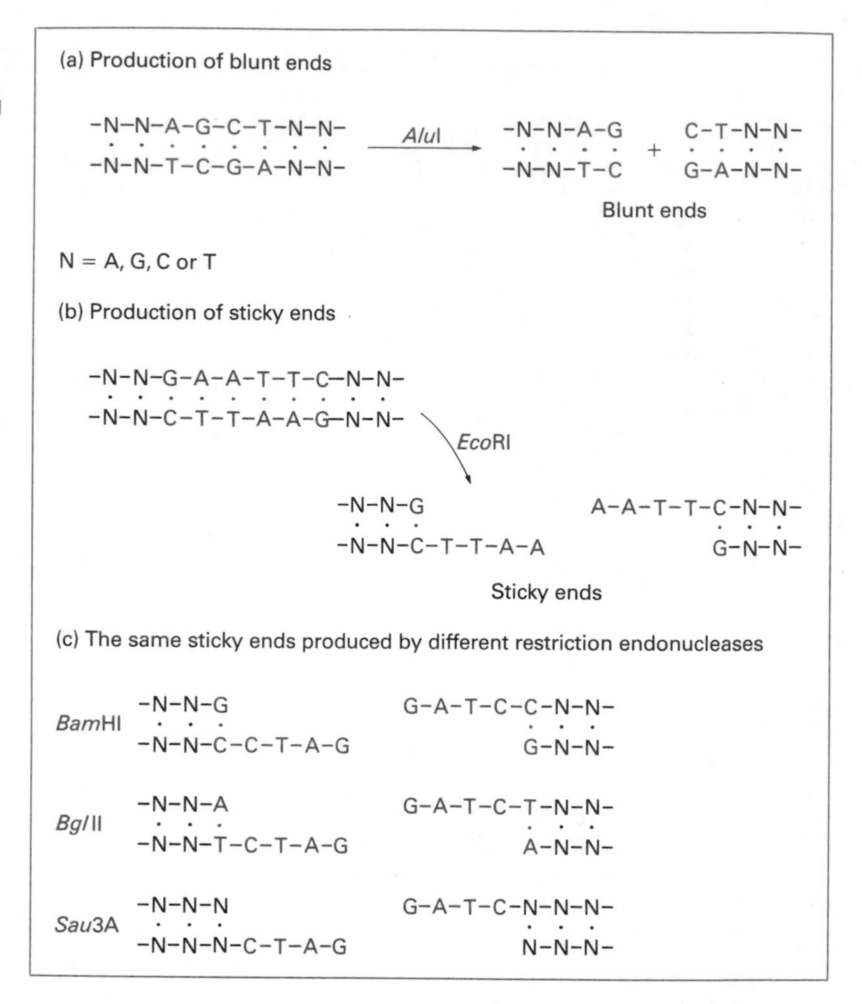

Figure 20.4 The ends produced by cleavage of DNA with different restriction endonucleases. (a) A blunt end produced by *Alu*I. (b) A sticky end produced by *Eco*RI. (c) The same sticky ends produced by *Bam*HI, *Bgl*II and *Sau*3A.

usually by two or four nucleotides, so that the resulting DNA fragments have short single-stranded overhangs at each end (Fig. 20.4(b)). These are called **sticky** or **cohesive** ends, as base pairing between them can stick the DNA molecule back together again. One important feature of sticky-end enzymes is that restriction endonucleases with different recognition sequences may produce the same sticky ends. *Bam*HI (recognition sequence GGATCC) and *Bgl*II (AGATCT) are examples: both produce GATC sticky ends (Fig. 20.4(c)). The same sticky end is also produced by *Sau*3A, which recognizes just the tetranucleotide GATC. Fragments of DNA produced by cleavage with either of these enzymes can be joined to each other, as each fragment will have a complementary sticky end.

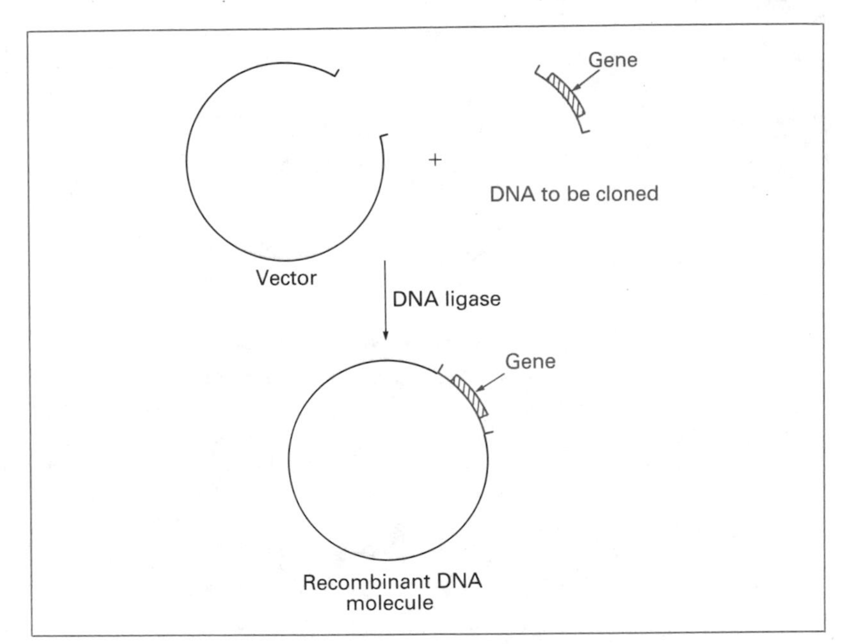

Figure 20.5 Ligation: the final step in construction of a recombinant DNA molecule.

20.2.2 Ligases join DNA molecules together

The final step in construction of a recombinant DNA molecule is the joining together of the vector molecule and the DNA to be cloned (Fig. 20.5). This process is referred to as ligation, and the enzyme that catalyses the reaction is called DNA ligase.

All living cells produce DNA ligases but the enzyme most frequently used in gene cloning is one involved in replication of T4 phage, and is purified from infected *E. coli* bacteria. Within the cell the enzyme synthesizes phosphodiester bonds between adjacent nucleotides during replication of the phage DNA molecules (see Section 11.2.2 for the role of ligase in DNA replication). In the test-tube, purified DNA ligases carry out exactly the same reaction and can join together restricted DNA fragments in either of two ways:

1. By repairing the discontinuities in the base-paired structure formed between two sticky ends (Fig. 20.6(a)). This is relatively efficient because the ends of the molecule are held in place, at least transiently, by sticky-end base pairing.
2. By joining together two blunt-ended molecules (Fig. 20.6(b)). This is a less efficient process because DNA ligase cannot 'catch hold' of the molecules to be ligated and has to wait for chance associations to bring the ends together.

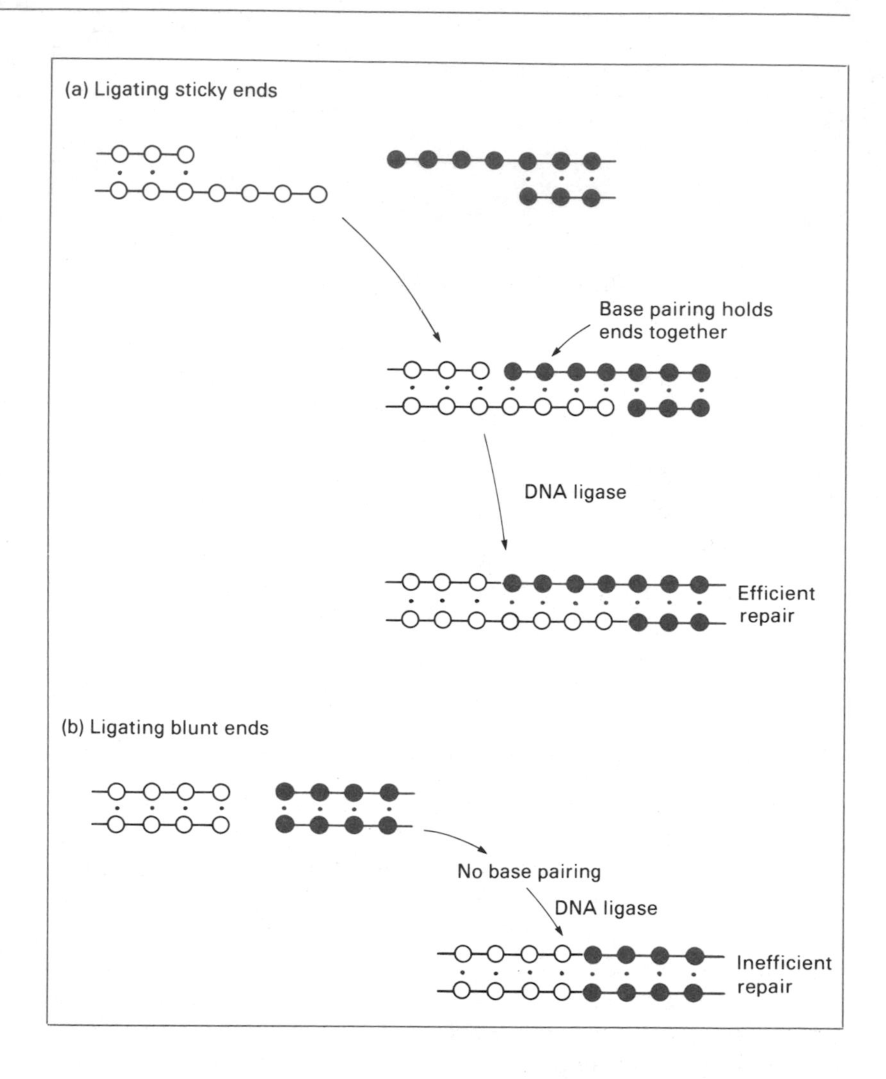

Figure 20.6 Ligation of (a) sticky ends, and (b) blunt ends.

20.3 Cloning vectors and the way they work

The vector is the central component of a gene cloning experiment. It provides the replicative function that enables the recombinant DNA molecule to multiply within the host cell. We have already encountered two types of naturally occurring DNA molecule that can replicate inside cells – plasmids and bacteriophage genomes – and both are used extensively as cloning vectors.

Literally hundreds of different cloning vectors are now available for use with different types of host cell. The largest number exists for *E. coli*, and the best known of these is the plasmid vector **pBR322**. We will use pBR322 as a typical example of a vector and see how it works in a cloning experiment.

20.3.1 Cloning with pBR322

The genetic and physical map of pBR322 (Fig. 20.7) gives an indication of why it has become such a popular cloning vector. Its first useful feature is its size, which, at 4363 bp, means that it and recombinant DNA molecules derived from it can be purified from *E. coli* cells with ease. Larger molecules (especially those of 50 kb and above) are much more difficult to purify intact.

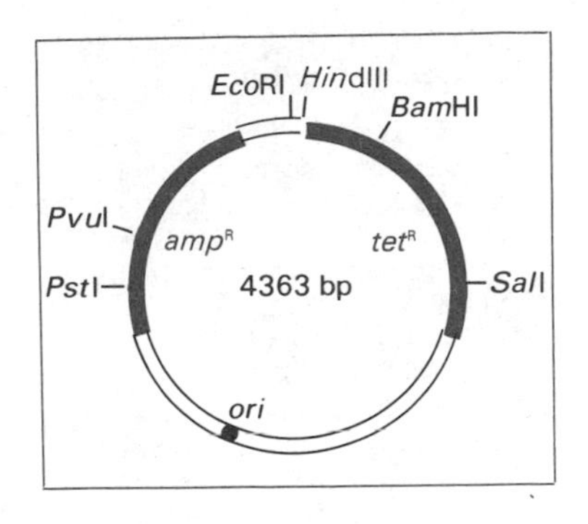

Figure 20.7 The cloning vector pBR322, showing the positions of the genes it carries and a few of the restriction sites that can be used for construction of recombinant molecules. Abbreviations: amp^R, ampicillin resistance; tet^R, tetracycline resistance.

The second feature of pBR322 is that it carries two antibiotic resistance genes, one coding for a β-lactamase (which modifies ampicillin into a form that is non-toxic to *E. coli*), and the second (actually a set of genes) coding for enzymes that detoxify tetracycline. The presence of these genes means that a bacterium transformed with pBR322 will be able to grow on a medium containing ampicillin and/or tetracycline, whereas cells lacking the plasmid will not. Transformed bacteria can therefore be selected by checking for growth on the appropriate medium (Fig. 20.8).

A third advantage of pBR322 is that it is a relaxed plasmid (Section 14.3.3) and has a reasonably high copy number. Generally there about 15 molecules present in a transformed *E. coli* cell, but this number can be increased, up to 1000–3000, by **plasmid amplification** in the presence of a protein synthesis inhibitor such as chloramphenicol (Fig. 20.9). Incubation with chloramphenicol prevents replication of the bacterial DNA molecule but not replication of pBR322. Cell division therefore ceases but the plasmid continues to multiply until large numbers are present in each cell.

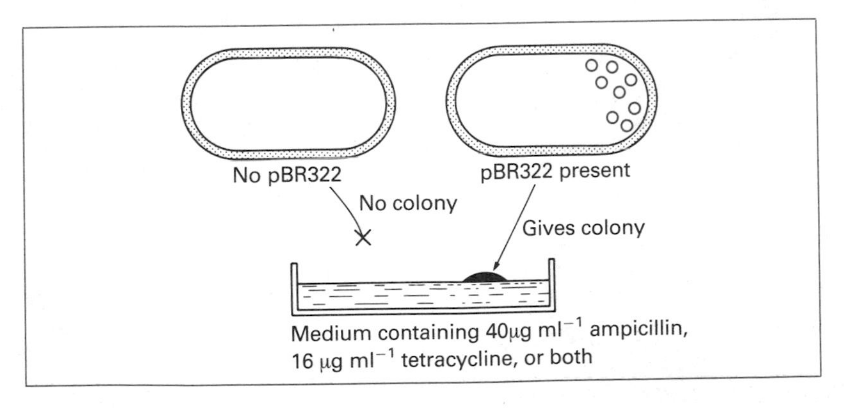

Figure 20.8 Cells containing pBR322 can be selected by plating on an agar medium that includes normally toxic amounts of ampicillin and/or tetracycline.

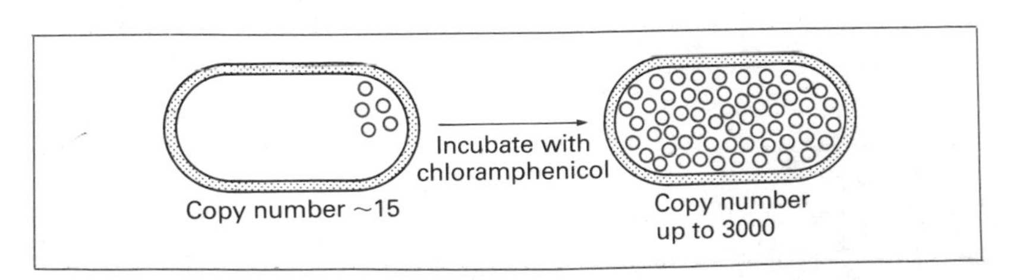

Figure 20.9 Plasmid amplification.

The pedigree of pBR322

The remarkable convenience of pBR322 as a cloning vector did not arise by chance. In fact the plasmid is not 'natural' at all: it was constructed in the laboratory by ligating together segments obtained from three other plasmids. The *amp*^{R} gene originally resided on the plasmid R1, a typical antibiotic-resistance plasmid that occurs in natural populations of *E. coli* (Section 14.3.1). The *tet*^{R} gene is derived from pSC101, and the replication origin from pMB1, which is related to the colicin-producing plasmid ColE1.

An amplified *E. coli* culture will therefore provide a good yield of recombinant pBR322 molecules.

(a) The cloning experiment. A typical cloning experiment with pBR322 is illustrated in Fig. 20.10. First, the vector molecule must be restricted to open the circle; pBR322 has several unique recognition sites for different restriction endonucleases (see Fig. 20.7), and any one of these could be used for this purpose. In our example we have used *Bam*HI, which will produce sticky ends at its cut site within the *tet*^{R} gene cluster.

The DNA to be cloned must also be cut with a restriction endonuclease, either the same one or another which will produce the same sticky ends. The DNA molecules are then joined together by ligation and mixed with competent *E. coli* cells (Section 19.4) so that transformation will take place. The bacteria are then spread on to agar medium and colonies grown. The composition of the medium is designed with the following points in mind:

1. Cells that have not been transformed and so contain no plasmid molecules are *amp*^{S}*tet*^{S}.
2. Cells that have been transformed with a pBR322 molecule that

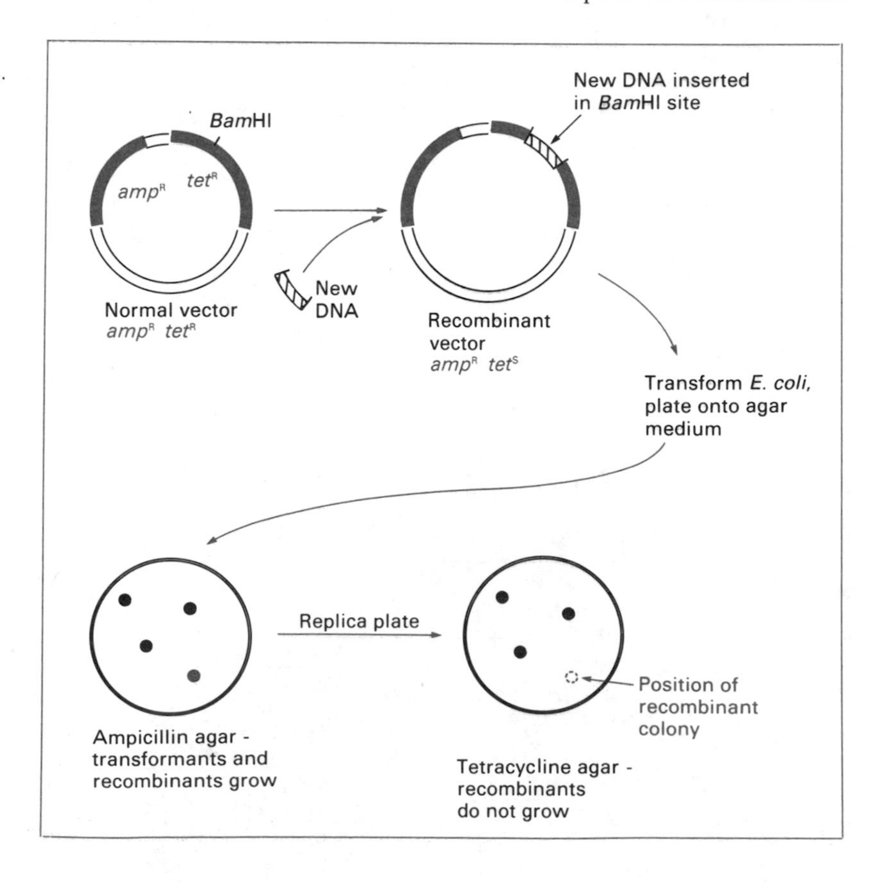

Figure 20.10 A cloning experiment with pBR322. See the text for details.

has recircularized without insertion of a DNA fragment (we call these cells **transformants**) will be *amp*^{R}*tet*^{R}.
3. Cells that contain a recombinant DNA molecule (we call these **recombinants**) will have lost tetracycline resistance because of the presence of the inserted DNA fragment in the middle of the tetracycline resistance gene cluster. Recombinants are therefore *amp*^{R}*tet*^{S}.

By checking for growth on ampicillin agar we can exclude untransformed cells, but both transformants and recombinants will produce colonies. However, the recombinants can be identified by replica plating them on agar that contains both ampicillin and tetracycline. Colonies that do not grow are recombinants, and cells for DNA purification can be recovered from the ampicillin plate.

20.3.2 Other types of cloning vector for *E. coli*

There are now many different types of *E. coli* cloning vector. Some have properties similar to pBR322, but others make use of alternative methods for identifying recombinant clones, and there are also several that are based on bacteriophages rather than plasmids. We must therefore examine a few additional vectors in order to get a complete picture of how genes are cloned in *E. coli*.

(a) Plasmids that carry the *lacZ'* gene. The main drawback with pBR322 is that the transformed colonies must be replica plated before recombinants can be identified. This takes quite a bit of time and vectors that enable recombinants to be identified at the first plating-out stage offer a distinct advantage; pUC8 (Fig. 20.11(a)) is an example of such a vector.

pUC8 carries two genes: the *amp*^{R} gene that we met with pBR322, and a new gene called *lacZ'*. The latter does not code for resistance to an antibiotic but instead specifies synthesis of part of the β-galactosidase enzyme. You will recall that this is the enzyme that splits lactose into glucose and galactose (Section 10.3.1). A recombinant pUC8 molecule is constructed by inserting new DNA into one of the restriction sites that are clustered near the start of *lacZ'* (Fig. 20.11(b)). This means that:

1. Cells harbouring normal pUC8 plasmids are ampicillin resistant and able to synthesize β-galactosidase.
2. Cells harbouring recombinant pUC8 plasmids are ampicillin resistant but unable to synthesize β-galactosidase.

The two types of cell are distinguished by plating on to agar that contains ampicillin plus two additional compounds. The first of these is called IPTG (isopropyl-thiogalactoside), which is an inducer of the *lac* genes and makes sure that *lacZ'* expression is not

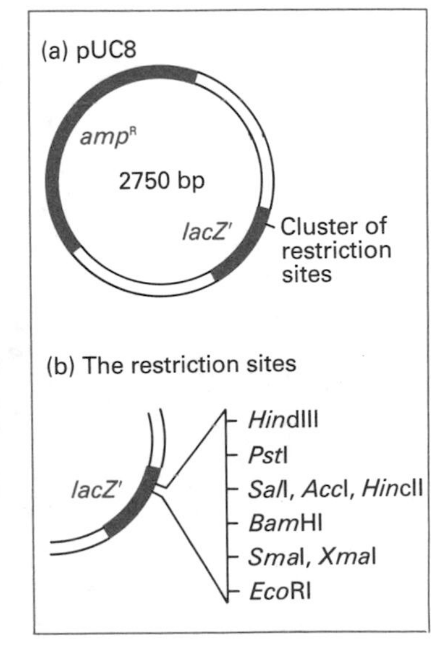

Figure 20.11 pUC8.

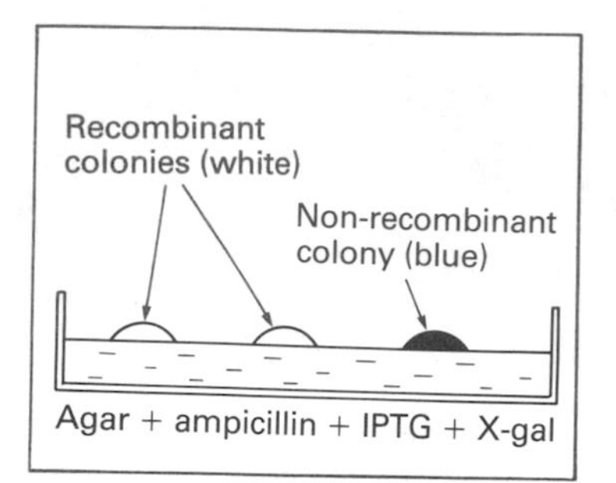

Figure 20.12 Recombinant selection with pUC8.

repressed. The second compound is X-gal (5-bromo-4-chloro-3-indolyl-β-D-galactopyranoside). X-gal is a substrate for β-galactosidase and is broken down by the enzyme into a product that is coloured deep blue. This means that colonies comprising non-recombinant cells (*amp*^{R} *lacZ*$'^+$) are blue, whereas recombinant colonies (*amp*^{R} *lacZ*$'^-$) are white (Fig. 20.12). Recombinants can therefore be selected without the need for replica plating.

(b) Cloning vectors based on bacteriophages. Bacteriophage genomes are the second type of independently replicating DNA molecule that can be used as the basis for a cloning vector. Often a bacteriophage vector is introduced into the host *E. coli* cells by **transfection**, which is exactly the same as transformation but involves phage rather than plasmid DNA. The transfectants are mixed with normal bacteria and poured on to an agar plate, giving

***In vitro* packaging**

With λ vectors, transfection is not a very efficient process when compared with the infection of an *E. coli* culture with phage particles. It would therefore be useful if recombinant λ molecules could be packaged into the λ head-and-tail structures in the test-tube. These particles could then be added directly to a culture of bacteria, resulting in a much higher number of recombinants.

Constructing phage particles in the test-tube may sound difficult but in fact is a relatively easy thing to do. The proteins required for packaging are purified from infected cells and added to vectors that have been linearized and joined to each other through their terminal *cos* sites:

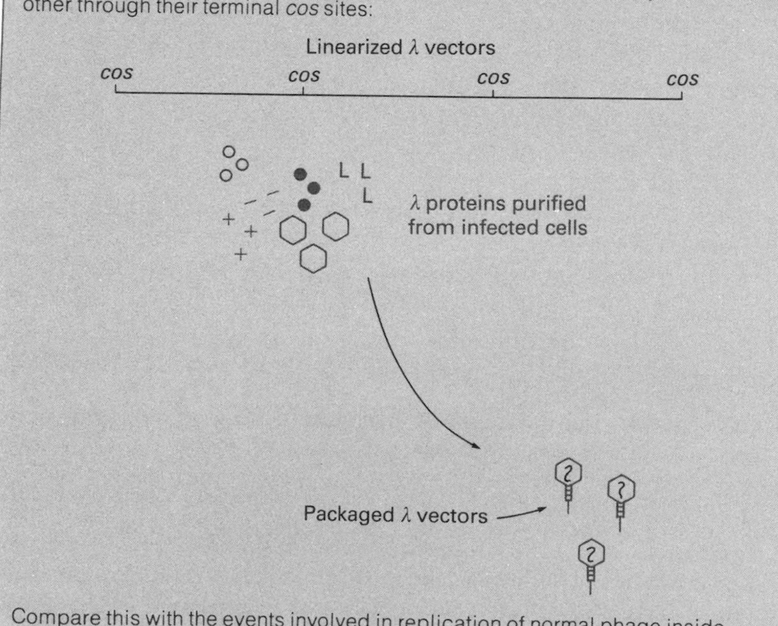

Compare this with the events involved in replication of normal phage inside bacterial cells (see p. 226).

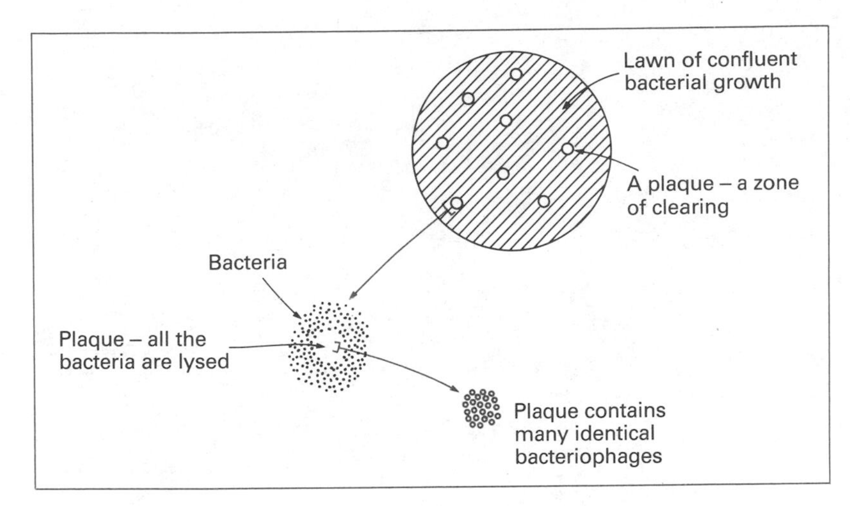

Figure 20.13 Transfection of *E. coli* with a bacteriophage vector leads to plaques on a lawn of bacteria. Each plaque is a clone made up of many identical bacteriophages.

rise to a set of plaques on a lawn of bacteria (p. 222), each plaque being a clone consisting of the bacteriophages produced by a single transfected cell (Fig. 20.13).

Virtually all bacteriophage cloning vectors are based on either λ or M13 phages. Their important features are as follows:

1. ***λ vectors***. With most λ vectors the DNA fragments to be cloned are ligated into the non-essential region of the phage genome (see Fig. 13.6(b)), as this region can be disrupted without harming the genes for the important replicative and structural proteins of the phage. Usually the phage genome has been modified, by deletions and mutations, so that it is unable to insert into the *E. coli* DNA, forcing the bacteriophage to follow the lytic cycle, and ensuring that transfected cells immediately give rise to plaques. A number of strategies are used for identifying recombinant plaques. Some vectors carry the *lacZ'* gene and so are plated onto X-gal agar; others rely on subtle differences in plaque morphology (Fig. 20.14).

   The main advantage of a λ-based vector is that the size of the DNA fragment that can be cloned is much larger than with plasmids such as pBR322 and pUC8. The largest piece of DNA that can be cloned with a plasmid vector is about 8 kb, whereas some λ vectors can handle fragments up to 25 kb. The ability to clone such large fragments is important as it allows construction of a **genomic library**, a set of recombinant clones that contain all of the DNA present in an individual organism. A human genomic library, for example, contains all the human genes cloned in *E. coli*, so any desired gene can be withdrawn from the library and studied. The problem is that a large number of clones may be needed if the library is to be rep-

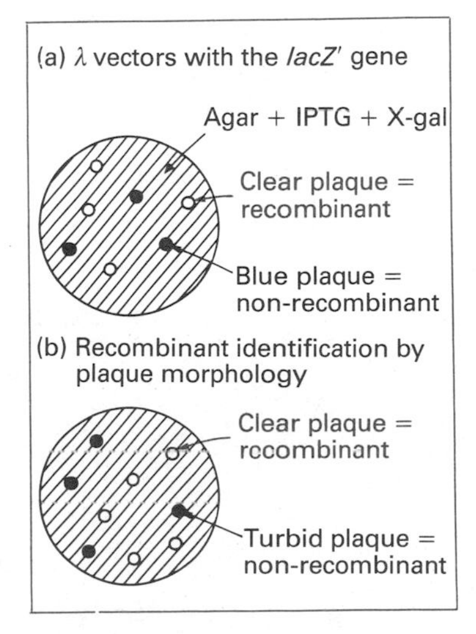

Figure 20.14 Two alternative strategies for recombinant selection with a λ vector.

Packaging constraints

The limitation on the size of DNA fragment that can be cloned with a λ vector results from the need to package the recombinant vector into the λ phage head. The phage head can accommodate about 52 kb of DNA. If the recombinant molecule is longer than 52 kb then it will not be packaged into infective particles. The normal λ genome is 49.5 kb, but up to 22 kb of this can be deleted without affecting the ability of the phage to infect via the lytic cycle. This means that a λ vector can be used to clone DNA fragments up to about 25 kb in length.

Table 20.2 Numbers of clones needed for genomic libraries

Species	*Genome size (kb)*	*Number of clones**	
		17 kb fragments	*35 kb fragments*
Escherichia coli	4 000	700	340
Saccharomyces cerevisiae	20 000	3 500	1 700
Fruit fly	165 000	29 000	14 500
Man	3 000 000	535 000	258 250
Maize	15 000 000	2 700 000	1 350 000
Salamander	90 000 000	16 000 000	8 000 000

* Calculated from the formula:

$$N = \frac{\ln(1 - P)}{\ln(1 - a/b)}$$

where N = number of clones required; P = probability that any given gene is present, set at 0.95 (i.e. 95%) in these calculations; a = average size of the DNA fragments in the library; b = total size of the genome.

resentative of the genome as a whole (Table 20.2). The number required depends on the length of DNA that each clone contains, so vectors able to handle long pieces of DNA are used to keep the size of the library as small as possible. We will return to gene libraries in the next chapter.

2. ***M13 vectors***. M13 vectors are important as they provide a means of obtaining a single-stranded version of a cloned gene. This is because the M13 phage particles that are continuously released by an infected bacterium (see Fig. 13.4) contain the single-stranded form of the M13 genome. A new gene that has been ligated into an M13 vector can therefore be obtained as single-stranded DNA from the phage particles released from recombinant cells (Fig. 20.15). Single-stranded genes are needed for several techniques, notably DNA sequencing (Section 21.2) and are difficult to obtain in any other way.

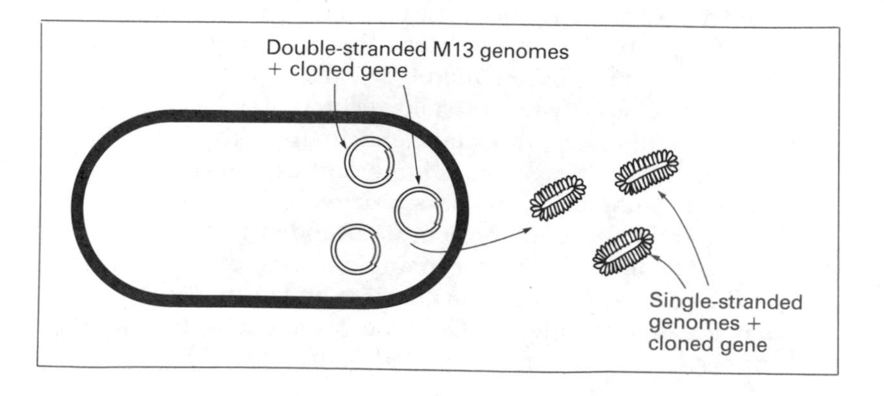

Figure 20.15 M13 vectors enable single-stranded versions of cloned genes to be obtained.

Cosmids

Cosmids combine features of both plasmid and λ vectors. A cosmid has all the normal features of a plasmid vector, including one or more antibiotic-resistance genes, but also carries a λ *cos* site:

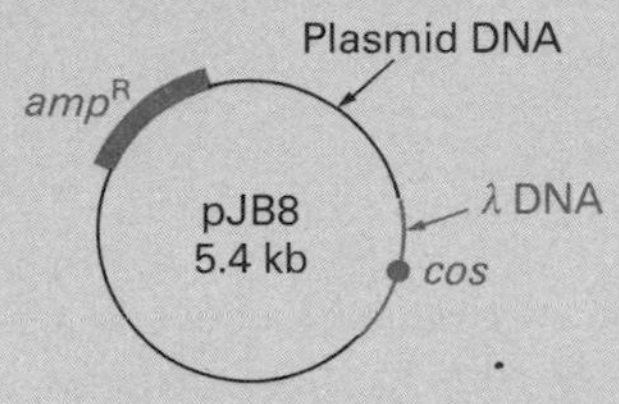

A cloning experiment with a cosmid is carried out as follows. The cosmid is opened at its unique restriction site and a new DNA fragment inserted. The ligation step is carried out in such a way that individual cosmid molecules link together to produce linear chains that are suitable for packaging into λ phage heads. A λ *in vitro* packaging mix is then added to the ligated cosmids, which are cleaved at the *cos* sites and inserted into phage particles. It does not matter that the cosmids do not contain any λ genes as the packaging reaction is dependent solely on the presence of *cos* sites:

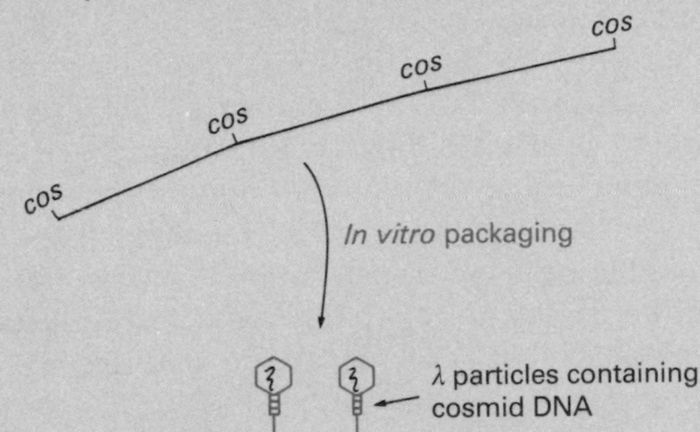

The λ phages that have been constructed are then used to infect an *E. coli* culture, although of course plaques are not formed. Instead, infected cells are plated on to antibiotic agar and resistant colonies grown.

The cosmid itself can be very small, just 8 kb or less, so up to 44 kb of new DNA can be added before the packaging limitation of the λ phage head is reached.

Paul Berg b. 30 June 1926, New York City

Paul Berg was responsible for several of the discoveries that led to the development of recombinant DNA techniques in the 1970s. During these years he also acted as the spokesman and motivator for those biologists who were concerned about the potential risks of gene cloning experiments, in particular the harmful effects that an engineered bacterium might have if it 'escaped' from the laboratory into the environment. The concerns were expressed in the 'Berg letter', published in *Science* on 24 July 1976. Berg organized a conference in February 1975 at Asilomar, California, at which the scientists who were carrying out gene cloning research discussed the safeguards that they should employ. At the heart of these safeguards was the notion of **biological containment**, which means that the host bacterium for a gene cloning experiment carries mutations that will prevent it from surviving outside of the laboratory. The Asilomar recommendations still form the basis to regulations for gene cloning experiments. Berg graduated from Penn State University in 1948 and obtained his doctorate from Western Reserve University 4 years later. He worked on protein synthesis and tRNAs at Washington University, St. Louis, before moving to the Stanford Medical Center in 1959. He was awarded the 1980 Nobel Prize for Chemistry with W. Gilbert and F. Sanger.

20.4 Cloning vectors for eukaryotes

The ease with which *E. coli* can be grown in culture (p. 59) makes this bacterium a popular host for gene cloning and it is generally used if the aim of the experiment is simply to isolate a particular gene. However, under some circumstances it may be desirable to use a eukaryote as a host. This is especially true in **biotechology** (Section 21.5), where the aim may not be to study a gene, but to use cloning to control or improve synthesis of an important metabolic product (e.g. a hormone such as insulin), or to change the properties of the host organism (e.g. to introduce insect resistance into a crop plant). We must therefore think about cloning vectors for eukaryotes.

20.4.1 Cloning vectors for *Saccharomyces cerevisiae*

Yeast can be grown in culture in just the same way as *E. coli* and not surprisingly is the eukaryote that is most frequently used in gene cloning experiments. There is a wide range of yeast cloning vectors, but three examples are sufficient to illustrate the main points.

(a) Cloning vectors based on the 2 μm plasmid. Yeast is unusual for a eukaryote in that it possesses a plasmid. The 2 μm plasmid, as it is called, is 6 kb in size and has a copy number of between 70 and 200. Vectors based on it are called **yeast episomal plasmids** or **YEps**.

YEps are very similar to *E. coli* plasmid vectors, the main difference being the way in which transformed cells are selected. A few YEps carry genes conferring resistance to inhibitors such as copper or methotrexate, but most make use of a radically different selection system. A normal yeast gene is used, generally one that codes for an enzyme involved in amino acid biosynthesis. An example is *leu2*, which we met in Section 18.2.1. To use *leu2* as a selectable marker we need a host yeast strain that has a non-functional *leu2* gene and so will survive only if this amino acid is supplied as a nutrient in the growth medium (Fig. 20.16(a)). Selection is possible because transformants contain the plasmid-borne copy of the *leu2* gene, enabling them to produce colonies on a minimal medium that lacks the amino acid (Fig. 20.16(b)).

(b) Integrative yeast vectors. One disadvantage with a YEp vector is that the 2 μm plasmid is notoriously unstable, and recombinant cells often lose their cloned genes if they are grown in culture for a long time. If the cloned gene is directing synthesis of an important product then the aim will be to grow the recombinants for as long as possible, making a YEp a poor choice. This leads us to the second group of yeast cloning vectors, the **yeast integrative plasmids** or YIps, which are designed to produce recombinants that are highly stable as a result of the vector having integrated into the host chromosomal DNA.

Integration is due to homologous recombination (Section 12.3) between the yeast gene carried by the YIp as the selectable marker and the inactive version of the gene present in one of the yeast chromosomes. A YIp will not replicate as a plasmid as it does not contain any parts of the 2 μm plasmid, and so must integrate via its yeast gene in order to survive. Transformation with a YIp is therefore a very inefficient process, but is worth the effort as the resulting recombinants will be very stable.

(c) Artificial chromosomes. The final type of yeast vector that we will consider is the **YAC** or **yeast artificial chromosome**. This is

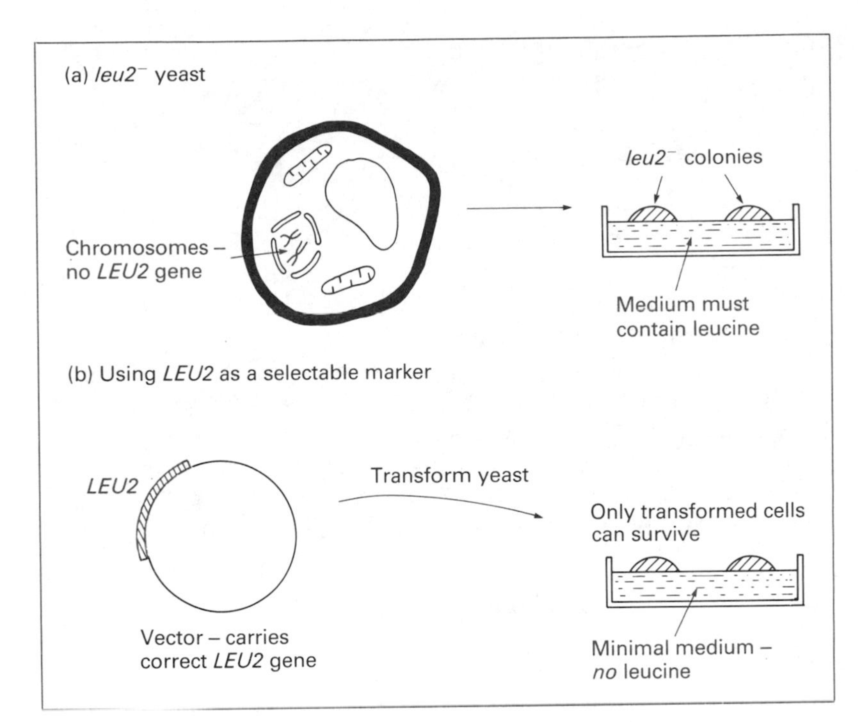

Figure 20.16 Using the *leu2* gene as a selectable marker in a yeast cloning experiment.

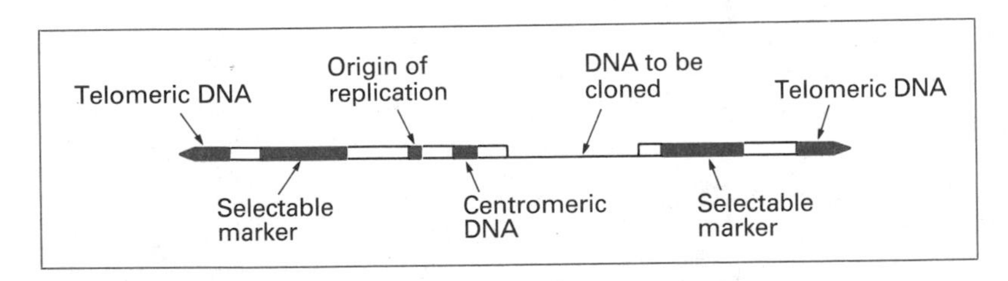

Figure 20.17 A yeast artificial chromosome (YAC).

a totally new approach to gene cloning that has come along as a spin-off from research into chromosome structure. A typical YAC (Fig. 20.17) consists of a centromere, two telomeres, an origin of replication, and one or more genes that act as selectable markers. In essence the YAC is a mini-chromosome.

The initial stimulus in designing artificial chromosomes came from yeast geneticists who wanted to use them to study various aspects of chromosome structure and behaviour, for example to examine chromosome segregation during meiosis. The applications of YACs in gene cloning experiments have only recently been explored. An important use of these vectors will be in obtaining intact clones of very long pieces of DNA, as many important animal genes are much too long to be cloned with a conventional vector. YAC vectors are also being used to reduce the number of clones needed to make a genomic library. A complete human genomic library in a λ vector requires half a million clones (see Table 20.2), but with a YAC vector this number can be reduced

Transforming organisms other than bacteria

E. coli cells are made competent for DNA uptake by soaking in calcium chloride (Section 19.4). In general terms, this treatment is not effective with organisms other than bacteria, although soaking in lithium chloride or lithium acetate does enhance uptake by yeast.

With most organisms the main barrier to DNA uptake is the cell wall. Cultured animal cells, which usually lack cell walls, are easily transformed, especially if the DNA is precipitated on to the cell surface. For other types of cell the answer is often to remove the cell wall. Enzymes that degrade yeast, fungal and plant cell walls are available, and under the right conditions intact **protoplasts** can be obtained. Protoplasts generally take up DNA quite readily: alternatively transformation can be stimulated by subjecting the protoplasts to a short electrical pulse.

There are also physical methods for introducing DNA into cells. Microinjection makes use of a very fine pipette to inject DNA molecules directly into the nucleus of a cell. This technique was initially applied to animal cells but has also been successful with plants. It is also possible to bombard cells with high velocity microprojectiles, usually particles of gold or tungsten that have been coated with DNA. The microprojectiles penetrate the cell wall, taking the DNA molecules with them.

immensely, to just 60000 clones if 150 kb fragments are used, making it much easier to work with the library.

20.4.2 Vectors for plants and animals

There are important potential benefits of gene cloning using plants and animals as the host organisms. For example, there is real optimism that new varieties of crops, with improved nutritional qualities and able to grow under adverse conditions, will be developed as a result of gene cloning.

Most cloning vectors for plants are based on the **Ti plasmid**, which is not a natural plant plasmid but instead belongs to a soil bacterium called *Agrobacterium tumefaciens*. This bacterium invades plant tissues causing a cancerous growth called a crown gall. During infection a part of the Ti plasmid, called the T-DNA, is integrated into the plant chromosomal DNA (Fig. 20.18(a)). Clon-

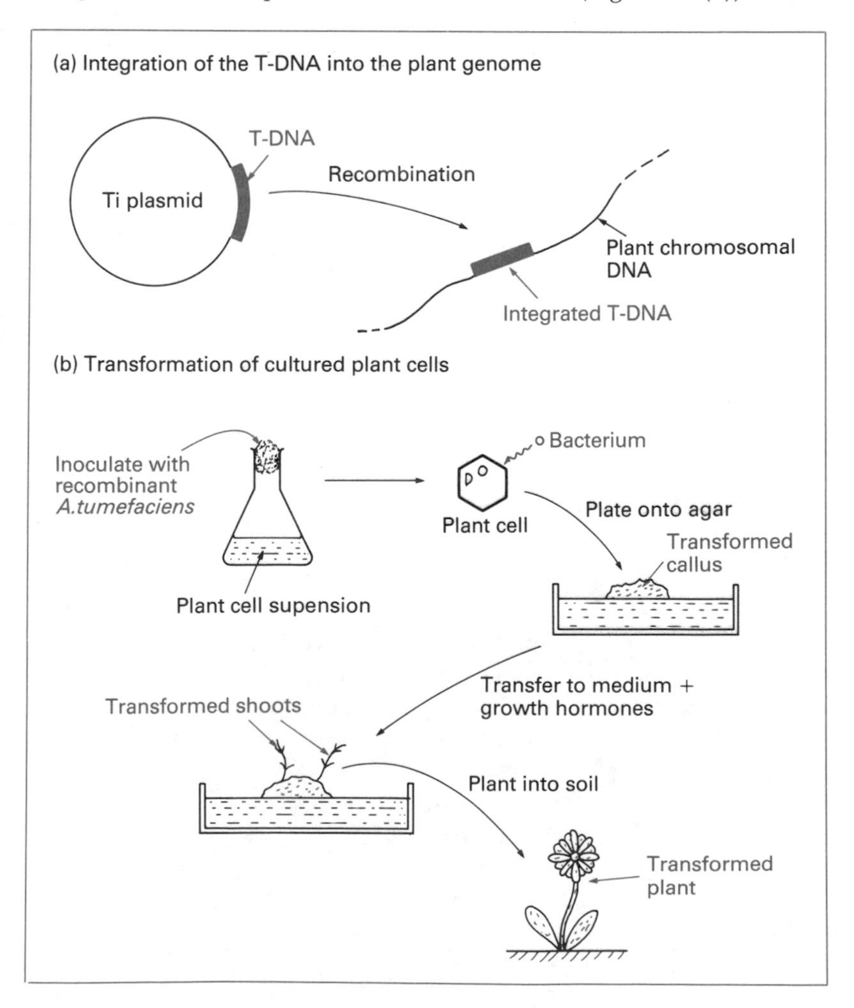

Figure 20.18 Gene cloning with plants. (a) The T-DNA integrates into plant chromosomal DNA. (b) Transformation of plant cells with a Ti plasmid vector.

ing vectors based on the Ti plasmid are therefore designed so that new genes can be inserted into the T-DNA region, which is then transferred into the plant genome by way of the normal *Agrobacterium* infection process. In practice a single plant cell or protoplast is usually transformed, and then grown via a callus into an intact plant (Fig. 20.18(b)).

The simplest mammalian cloning vectors are based on bacterial plasmids such as pBR322, which can integrate into chromosomal DNA by non-homologous recombination (Section 12.3). There are also a number of vectors derived from mammalian viruses. Simian virus 40 (SV40) has received a lot of attention, but suffers from an inability to handle more than a few hundred nucleotides of inserted DNA. Anything more and the vector DNA will not be packaged into the viral capsid. Also an SV40 vector usually kills the host cell shortly after infection, so stable transformants cannot be obtained. A second type of virus, the bovine papillomaviruses, are more suitable in this respect as in some hosts the viral genome takes the form of a multicopy plasmid. Infected cells containing these plasmids are unharmed and continue to divide, with viral genomes being passed on to the daughter cells.

Reading

T. A. Brown (1990) *Gene Cloning: An Introduction*, 2nd edn, Chapman and Hall, London – covers these topics in more detail.

H. O. Smith and K. W. Wilcox (1970) A restriction enzyme from *Haemophilus influenzae*. *Journal of Molecular Biology*, **51**, 379–91.

T. A. Brown (1991) *Molecular Biology Labfax*, BIOS, Oxford – a comprehensive databook providing details of restriction enzymes and other tools used in gene cloning.

P. Lobban and A. D. Kaiser (1973) Enzymatic end-to-end joining of DNA molecules. *Journal of Molecular Biology*, **79**, 453–71 – ligation.

F. Bolivar, R. L. Rodriguez, P. J. Greene *et al.* (1977) Construction and characterisation of new cloning vectors. II. A multi-purpose cloning system. *Gene*, **2**, 95–113 – pBR322.

A. M. Frischauf, H. Lehrach, A. Poutska and N. Murray (1983) Lambda replacement vectors carrying polylinker sequences. *Journal of Molecular Biology*, **179**, 827–42 – λ vectors.

D. T. Burke *et al.* (1987) Cloning of large segments of exogenous DNA into yeast by means of artificial chromosome vectors. *Science*, **236**, 806–12 – a description of YACs and the way they are used in gene cloning.

M. D. Chilton (1983) A vector for introducing new genes into plants. *Scientific American*, **248**(6), 50–9.

J. W. Davies and J. Stanley (1989) Geminiviruses, genes and vectors. *Trends in Genetics*, **5**, 77–81 – a different type of cloning vector for plants.

Problems

1. Define the following terms:

recombinant DNA technology	sticky end
genetic engineering	plasmid amplification
gene cloning	transfection
cloning vector	cosmid
recombinant DNA molecule	genomic library
clone	biotechnology
restriction endonuclease	yeast episomal plasmid
DNA ligase	yeast integrative plasmid
blunt end	yeast artificial chromosome
	Ti plasmid

2. Outline the basic steps in a gene cloning experiment.
3. Explain why gene cloning is important.
4. Outline the roles of restriction endonucleases and DNA ligases in gene cloning.
5. What is the difference between a blunt-ended DNA molecule and a sticky-ended molecule?
6. Outline the useful features of pBR322.
7. Describe how a gene cloning experiment with pBR322 is carried out.
8. How is the *lacZ'* gene used in gene cloning experiments?
9. Explain what a genomic library is and outline the relative merits of λ vectors and yeast artificial chromosomes in genomic library construction.
10. What is the main advantage of an M13 cloning vector?
11. Outline the reasons for carrying out a gene cloning experiment with a eukaryote as the host.
12. Distinguish between a YEp and YIp.
13. What are the key features of a YAC vector?
14. Outline the types of vector that are available for cloning genes in plants and animals.

21 Studying cloned genes

Identifying a gene in a genomic library – hybridization probing – chromosome walking • DNA sequencing • Studying gene expression – transcript analysis – studying gene regulation • Determining the function of a protein coded by a cloned gene • Cloned genes in biotechnology • The future for molecular genetics

Once a gene has been cloned there is almost no limit to the information that can be obtained about its structure and expression. The development of gene cloning techniques in the 1970s stimulated the parallel development of analytical methods for studying cloned genes, and new methods and strategies have been introduced at regular intervals. In this chapter we will look at a selection of the most important of these methods, and will also survey the ways in which cloned genes are being used in the biotechnology industry. First though we must tackle what is often the first problem to arise in a molecular biology project: how to isolate from a genomic library a clone that carries the gene in which you are interested.

21.1 Identifying a gene in a genomic library

Once a genomic library has been prepared (Section 20.3.2), a number of procedures can be employed to attempt identification of a clone carrying the desired gene. Most of these are based on the important technique called **hybridization probing**.

21.1.1 Hybridization probing

Any two single-stranded nucleic acids have the potential to form base pairs with one another. With most pairs of molecules the resulting hybrid structures will be unstable, as only a small number of individual interstrand bonds will be formed (Fig. 21.1(a)). However, if the polynucleotides are complementary then extensive base

Enzymes used in recombinant DNA experiments

Techniques for studying cloned genes make use of purified enzymes which enable DNA and RNA molecules to be manipulated in precisely defined ways. We have already seen how restriction endonucleases and DNA ligases are used to cut and join DNA molecules during gene cloning experiments. Other important enzymes include:

1. **DNA polymerases**, which are enzymes that synthesize a new strand of DNA complementary to an existing DNA or RNA template. Purified DNA polymerase I (Section 11.2.1) is used to synthesize labelled polynucleotides, and a modified version is used in DNA sequencing procedures (Section 21.2.1). Reverse transcriptase (Section 13.2.2) is used to make DNA copies of RNA molecules (see p. 402).
2. **RNA polymerases** are used to make RNA transcripts from cloned genes, for example to study intron splicing.
3. **Nucleases** degrade DNA and/or RNA molecules by breaking the phosphodiester bonds that link one nucleotide to the next. **Exonucleases** remove nucleotides one at a time from the ends of DNA and RNA molecules, whereas **endonucleases** break internal phosphodiester bonds. Restriction endonucleases are the most important type of nuclease in recombinant DNA technology, but there are also many others. Bal31, for instance, is able to progressively shorten double-stranded DNA molecules and so can remove unwanted regions from the ends of fragments to be cloned. S1 nuclease degrades just the RNA component of a double-stranded DNA–RNA hybrid (Section 21.3.1).
4. **Modifying enzymes** add or remove specific chemical groups. Alkaline phosphatase removes the phosphate groups from the 5′ termini of a DNA molecule, preventing it from re-ligating; polynucleotide kinase adds the phosphates back on again. Terminal deoxynucleotidyl transferase adds one or more deoxynucleotides to the 3′-ends of a DNA molecule, producing 'tails' that can be base-paired to tails of the complementary nucleotide when recombinant molecules are being constructed.

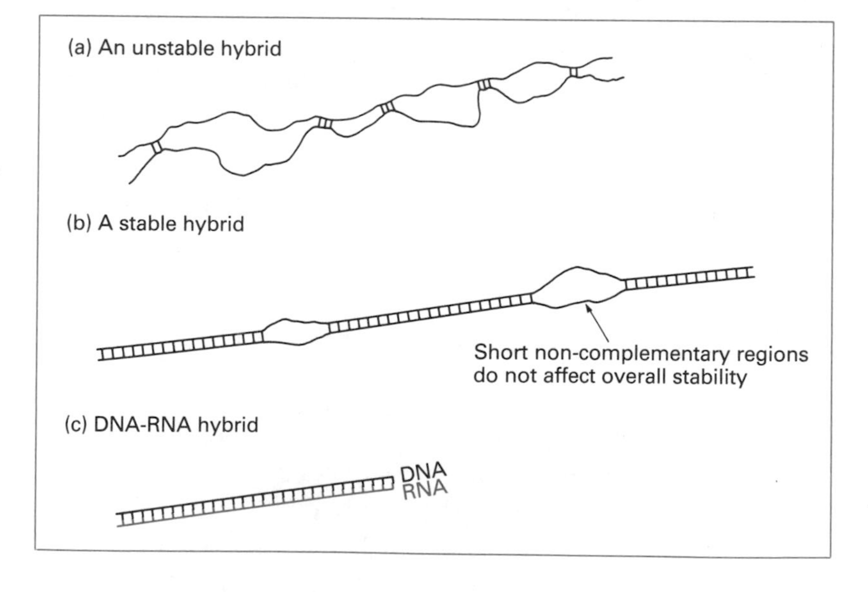

Figure 21.1 Nucleic acid hybridization. (a) An unstable hybrid molecule formed between two non-homologous DNA polynucleotides. (b) A stable hybrid formed between two complementary polynucleotides. (c) A DNA–RNA hybrid, such as may be formed between a gene and its transcript.

pairing will occur to form a stable double-stranded molecule (Fig. 21.1(b)). Not only can this occur between single-stranded DNA molecules to form the DNA double helix, but also between single-stranded RNA molecules and between combinations of one DNA and one RNA strand (Fig. 21.1(c)).

Nucleic acid hybridization can be used to identify a particular recombinant clone if a DNA or RNA probe, complementary to the desired gene, is available. We will think about the probe itself in a few minutes; first we must look at how a hybridization probing experiment is carried out.

(a) Colony and plaque hybridization probing. Hybridization probing can be used to identify recombinant DNA molecules contained in either bacterial colonies or bacteriophage plaques. First the colonies or plaques are transferred to a nitrocellulose or nylon membrane (Fig. 21.2(a)), and then treated to remove all con-

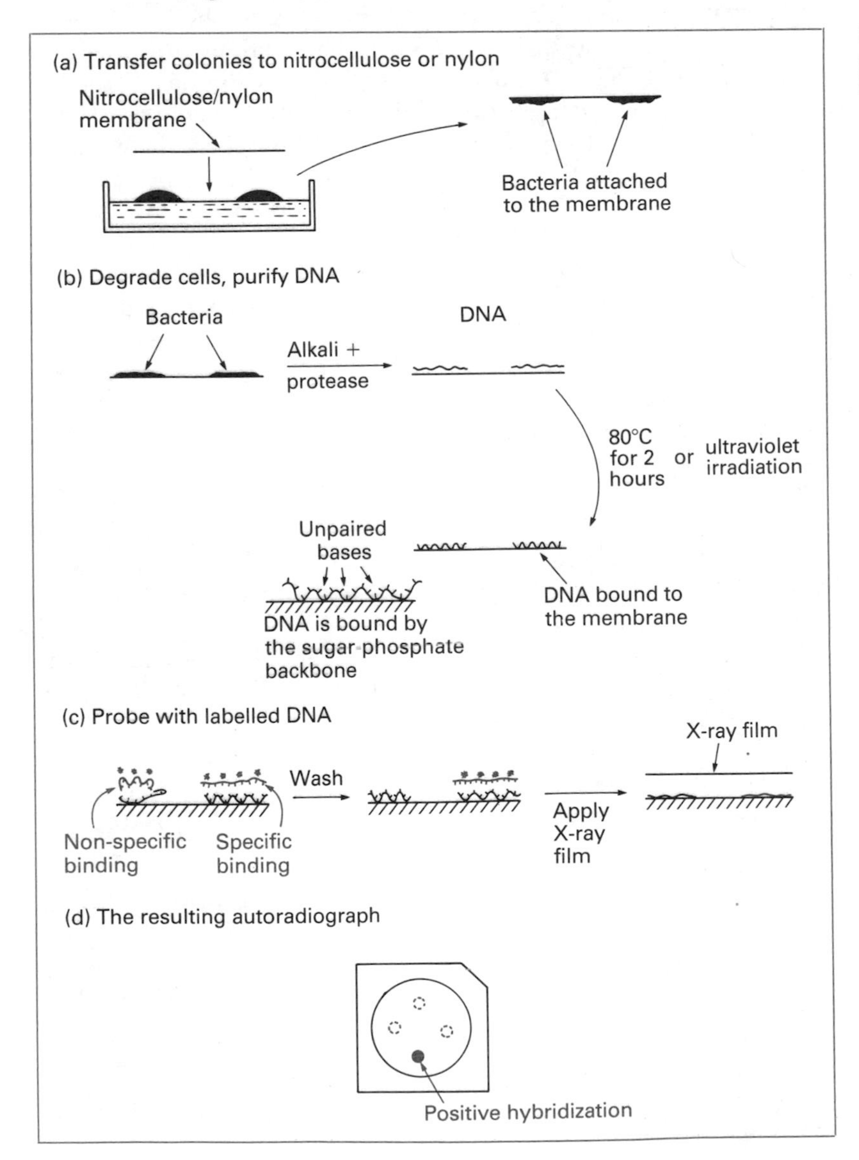

Figure 21.2 Colony hybridization probing with a radioactively labelled probe.

taminating material, leaving just DNA (Fig. 21.2(b)). Usually this treatment also results in denaturation of the DNA molecules, so that the hydrogen bonds between individual strands in the double helix are broken. These single-stranded molecules can then be bound tightly to the membrane by a short heat treatment or by ultraviolet irradiation. The molecules will be attached to the membrane through their sugar-phosphate backbones, so the bases will be free to pair with complementary nucleic acid molecules.

The probe must now be labelled, denatured by heating, and applied to the membrane in a solution of chemicals that promote nucleic acid hybridization (Fig. 21.2(c)). After a period to allow hybridization to take place, the membrane is washed to remove unbound probe molecules, dried, and the positions of the bound probe detected (Fig. 21.2(d)). Traditionally the probe is labelled with radioactive phosphorus, ^{32}P, and the positions of the hybridization signals visualized by **autoradiography**, in which a sheet of X-ray-sensitive film is placed over the membrane, as shown in Fig. 21.2(d). The problem with radioactive labelling is that it presents a health hazard to the person doing the experiment, so a number of methods for labelling DNA with non-radioactive markers have now been developed. In one method the probe DNA is complexed with the enzyme horseradish peroxidase, which can be detected through its ability to degrade a special substrate (called luminol) with the emission of chemiluminescence. The signal can be recorded on normal photographic film in a manner similar to autoradiography.

(b) Choosing a hybridization probe. Clearly the success of colony or plaque hybridization as a means of identifying a particular recombinant clone depends on the availability of a DNA molecule that can be used as a probe. This probe must share at least a part of the sequence of the cloned gene. If the gene itself is not available (which presumably will be the case if the aim of the experiment is to provide a clone of it) then what can be used as the probe?

In practice the nature of the probe is determined by the information available about the desired gene. We will consider two possibilities:

1. ***Oligonucleotide probes for genes whose translation products have been characterized.*** Often the gene that is being sought will be one that codes for a protein that has already been studied in some detail. In particular the amino acid sequence of the protein may have been determined. If this is the case, then it will be possible to use the genetic code to predict the nucleotide sequence of the relevant gene. This prediction will always be an approximation, as only tryptophan and methionine can be

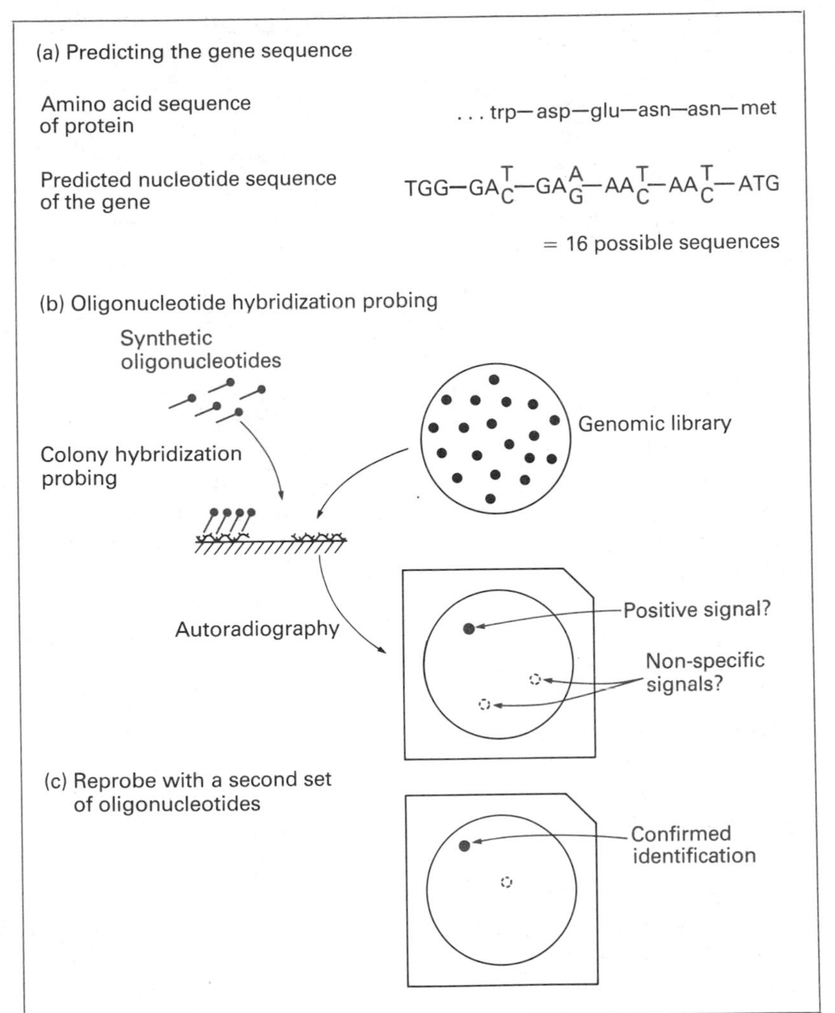

Figure 21.3 The use of two pools of synthetic oligonucleotides to identify a clone from a genomic library.

assigned unambiguously to triplet codons: all other amino acids are coded by at least two codons each (Section 8.2.3). However, these codons are generally related (for example, alanine is coded by GCA, GCT, GCG and GCC), so two of the three nucleotides in a triplet can still be assigned with certainty (Fig. 21.3(a)).

Once the sequence of the gene has been predicted, a mixture of oligonucleotides that will hybridize to the gene can be synthesized. Usually a region of five or six codons where the prediction is fairly unambiguous is chosen, and a set of synthetic oligonucleotides including all the possible sequences is used as a mixed probe (Fig. 21.3(b)). One of these oligonucleotides will have the correct sequence and so will hybridize to the gene being sought. There is also the possibility that one or more of

Figure 21.4 The rabbit and human β-globin genes are sufficiently similar to form a stable base-paired hybrid.

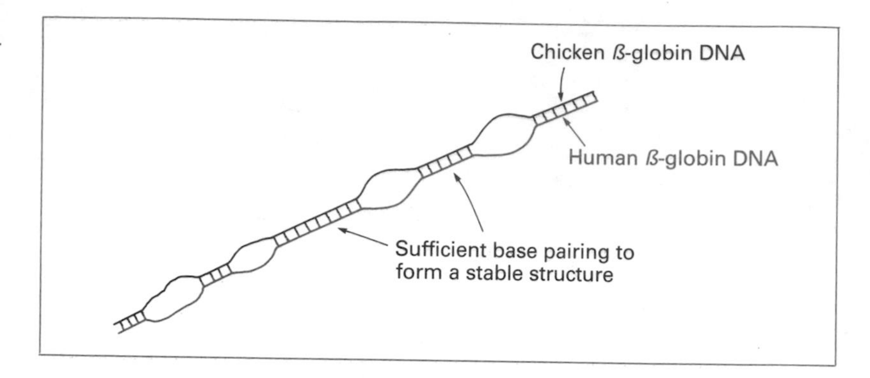

the other oligonucleotides in the mixture will, by chance, cross-hybridize with other genes, resulting in a few unwanted clones being selected as well, but things can usually be sorted out by checking the initial identification with a second set of oligonucleotides, synthesized in accordance with a different region of the predicted gene sequence (Fig. 21.3(c)).

2. ***Using a related gene as a probe.*** Often a substantial amount of nucleotide sequence similarity is seen when two genes for the same protein, but from different organisms, are compared. For instance, the nucleotide sequences of the chicken and human β-globin genes are 75% similar (on average seven of every ten nucleotides are the same). These similarities mean that a single-stranded probe prepared from the gene of one organism will form a hybrid with the gene from the second organism: although the two molecules will not be entirely complementary, enough base pairs will be formed to produce a stable structure (Fig. 21.4). The chicken β-globin gene could therefore be used as a hybridization probe for the equivalent human gene.

   This strategy is called **heterologous probing** and it has proved very useful in screening genomic libraries for genes that have already been identified in other organisms. One important result has been the discovery of mammalian homeobox genes (Section 10.5.2), which were identified after mouse and human genomic libraries were probed with a *Drosophila* gene containing the homeobox sequence.

21.1.2 Chromosome walking

It is not always possible to obtain a hybridization probe for the gene being sought. In fact the most interesting genes are often those about which we know very little, ones whose translation products have not been identified for example. We need to use a more subtle approach in order to identify a clone of a gene of this type.

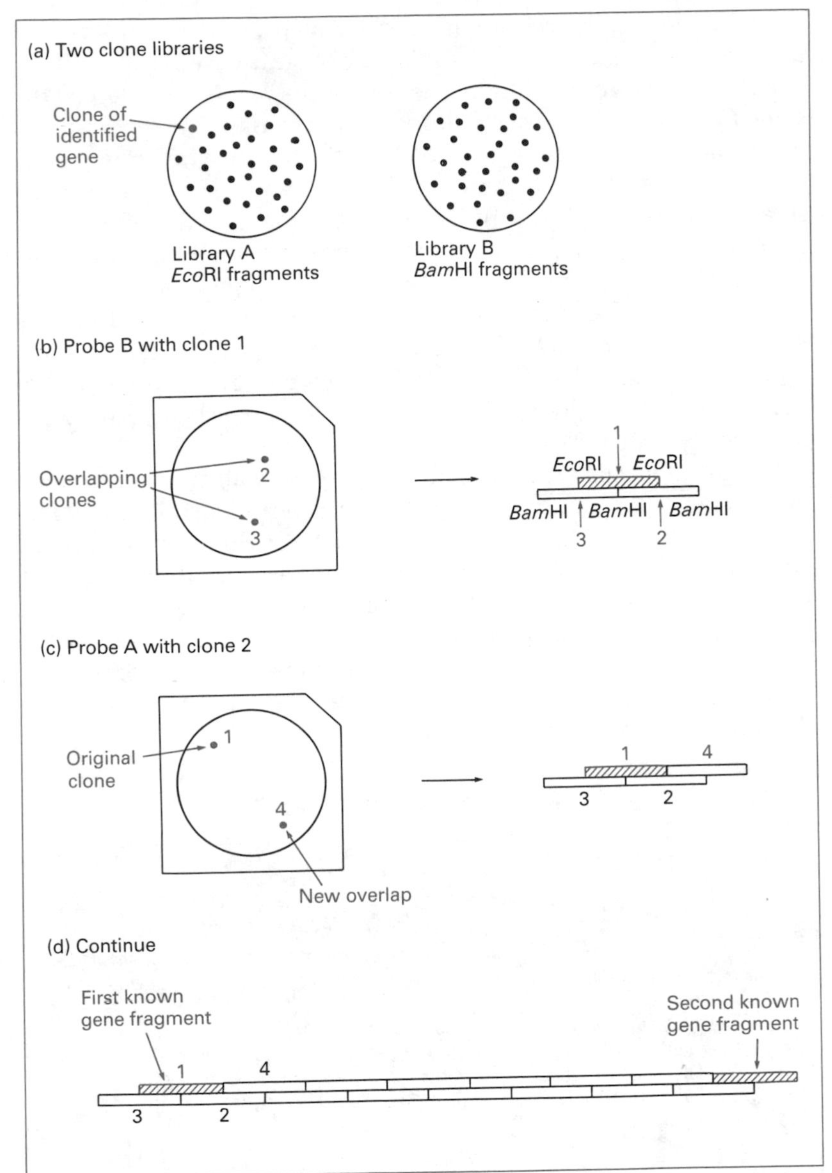

Figure 21.5 Chromosome walking.

Frequently the first information to be obtained about a gene is its map position in the genome. This will have been determined by one or more of the techniques described in Chapters 18 and 19. The mapping experiments may have indicated that the gene we are interested in lies very near to another gene that has already been obtained as an identified clone. This would enable us to use **chromosome walking** to progress along the chromosome from one gene to the other.

Two genomic libraries are needed for a chromosome walk, each prepared with a different restriction endonuclease, say *Eco*RI for one and *Bam*HI for the other. The fragments of DNA carried by the clones in the two libraries will therefore overlap (Fig. 21.5(a)). To begin the chromosome walk the clone containing the known gene is taken from the first library (library A) and used to probe the second library (Fig. 21.5(b)). One or more clones from library B will give positive hybridization signals, showing that the fragments carried by these clones overlap with the fragment carried by the probe. One of these clones from library B is now used to probe library A (Fig. 21.5(c)). The original clone, and possibly some others, will hybridize. The cycle is repeated several times, gradually progressing along the DNA until the gene being sought is reached (Fig. 21.5(d)).

This is a tedious business and the longest walks that have ever been achieved are only 200 to 250 kb in length, meaning that the two genes have to be very close together for this to be a realistic approach. There is also the problem of knowing when you have reached the gene you are looking for: if you know nothing about the translation product then you may not recognize the gene when you see it. Despite these drawbacks, chromosome walking has had some notable successes, its greatest triumph being the isolation of the gene responsible for cystic fibrosis.

cDNA libraries

For many plants and animals a complete genomic library contains such as vast number of clones (see Table 20.2) that identification of the desired one is a mammoth task. With these organisms a **cDNA (complementary DNA) library**, constructed from mRNA, may be more useful. mRNA cannot itself be ligated into a cloning vector, but it can be converted into DNA by cDNA synthesis. The key to this method is the enzyme reverse transcriptase (Section 13.2.2), which will synthesize a DNA polynucleotide complementary to an existing RNA molecule. Once the cDNA strand has been synthesized the RNA component of the hybrid is replaced with DNA, producing a double-stranded DNA molecule that can be ligated into the vector:

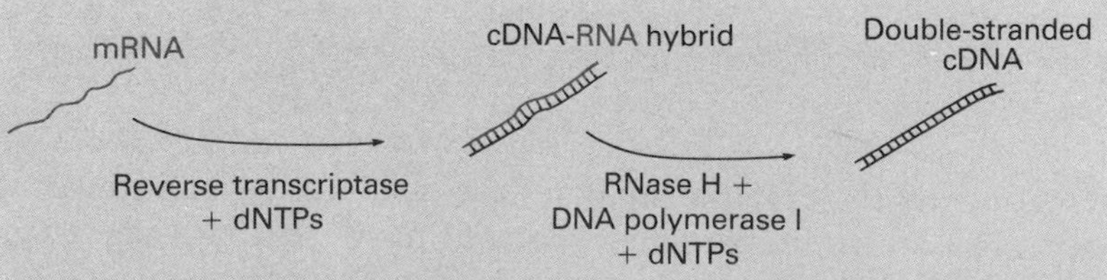

cDNA libraries are used when it is known that the gene of interest is expressed in a particular tissue or cell type. Not all genes are expressed at the same time, so relatively few clones will be needed to make a cDNA library that contains copies of all of the genes that are active in the tissue or cells from which the mRNA is obtained. It will then be a relatively easy task to screen the cDNA library for a clone carrying a particular expressed gene.

21.2 DNA sequencing: working out the structure of a gene

Probably the most important technique available to the molecular biologist is DNA sequencing, by which the precise order of nucleotides in a piece of DNA can be determined. DNA sequencing methods have been around for several years, but only since the late 1970s has rapid and efficient sequencing been possible. Two different techniques were developed almost simultaneously: the chain termination method by Frederick Sanger and Andrew Coulson at Cambridge, and the chemical degradation method by Walter Gilbert and his associates at Harvard. We will look at the Sanger–Coulson method, which has now emerged as the most popular of the two techniques.

21.2.1 The Sanger–Coulson method: chain-terminating nucleotides

The chain termination method requires single-stranded DNA and so the molecule to be sequenced is usually cloned into an M13 vector (Section 20.3.2). This is because chain termination sequencing involves the enzymatic synthesis of a second strand of DNA, complementary to an existing template.

(a) Carrying out a sequencing experiment. The first step in chain termination sequencing is to anneal a short oligonucleotide to the recombinant M13 molecule (Fig. 21.6(a)). This oligonucleotide acts as the primer for the complementary strand synthesis reaction, which is carried out by a modified form of DNA polymerase I. Remember that this enzyme needs a double-stranded region from which to begin strand synthesis (Section 11.2.2).

The strand synthesis reaction is started by adding the enzyme plus each of the four deoxynucleotides (dATP, dTTP, dGTP and dCTP). In addition, a single modified nucleotide is also included in the reaction mixture. This is a **dideoxynucleotide** (e.g. dideoxyATP) which can be incorporated into the growing polynucleotide just as efficiently as the normal nucleotide, but which blocks further strand synthesis. This is because the dideoxynucleotide lacks the hydroxyl group at the 3′ position of the sugar component (Fig. 21.6(b)). This group is needed for the next nucleotide to be attached; chain termination therefore occurs whenever a dideoxynucleotide is incorporated by the enzyme.

If dideoxyATP is added to the reaction mix, then termination will occur at positions opposite thymidines in the template (Fig. 21.6(c)). But termination will not always occur at the first T as normal dATP is also present and may be incorporated instead of

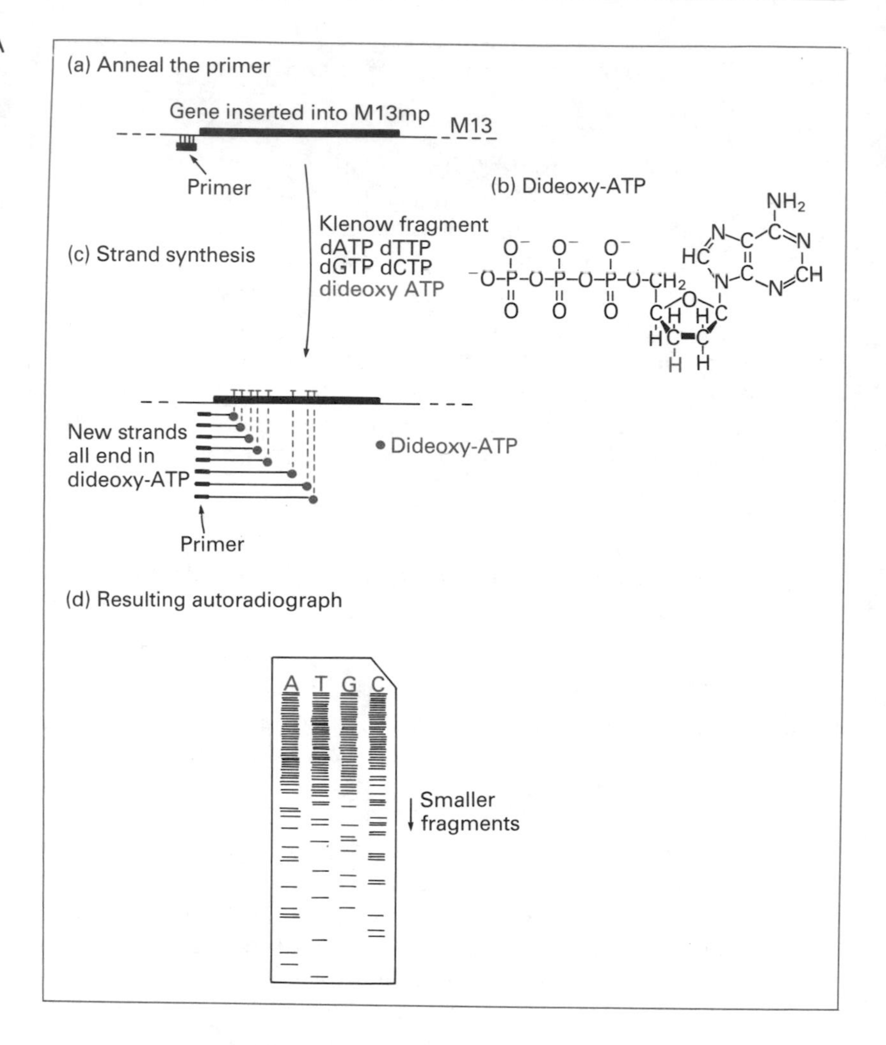

Figure 21.6 Chain termination DNA sequencing.

the dideoxynucleotide. The ratio of dATP to dideoxyATP is such that an individual strand may be polymerized for a considerable distance before a dideoxyATP molecule is added. The result is that a family of new strands is obtained, all of different lengths, but each ending in dideoxyATP. In fact four strand synthesis reactions are carried out in parallel, one with each dideoxynucleotide (dideoxyATP, dideoxyTTP, dideoxyGTP and dideoxyCTP). The result will be four distinct families of newly synthesized polynucleotides, one family containing strands all ending in dideoxyATP, one of strands ending in dideoxyTTP, etc.

The next step is to separate the components of each family so the lengths of each strand can be determined. This can be achieved by polyacrylamide gel electrophoresis (p. 19), which results in a series of bands, each representing a chain-terminated DNA molecule of a different length. Usually the DNA molecules are

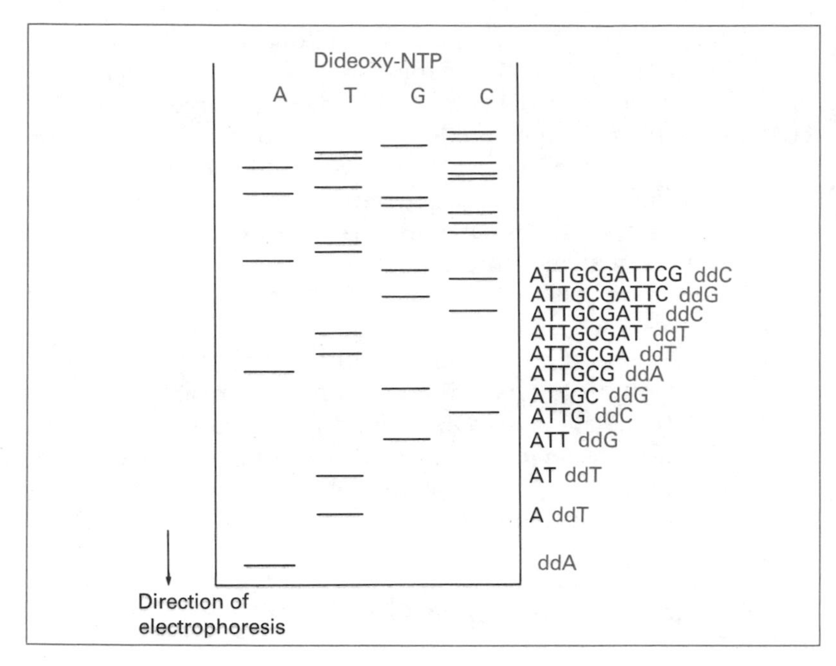

Figure 21.7 Interpreting the autoradiograph produced by a chain termination sequencing experiment. Each track contains the fragments produced by strand synthesis in the presence of one of the four dideoxyNTPs. The sequence is read by identifying the track in which each fragment lies, starting with the one that has moved the furthest, and gradually progressing up through the autoradiograph.

radioactively labelled, so autoradiography is used to visualize the results (Fig. 21.6(d)).

(b) Reading the DNA sequence from the autoradiograph. Reading the sequence is very easy (Fig. 21.7). First the band that has moved the furthest is located. This will represent the smallest piece of DNA, the strand terminated by incorporation of the dideoxynucleotide at the first position in the template. The track in which this band occurs is noted. Let us say it is track A; the first nucleotide in the sequence is therefore A.

The next most mobile band will correspond to a DNA molecule one nucleotide longer than the first. The track is again noted, T in the example shown in Fig. 21.7; the second nucleotide is therefore T and the sequence so far is AT.

The process is continued along the autoradiograph until the individual bands become so bunched up that they cannot be separated from one another. Generally it is possible to read a sequence of about 400 nucleotides from one autoradiograph. This sequence can then be combined with those of overlapping DNA fragments to build up a much longer master sequence.

21.2.2 The power of DNA sequencing

The first DNA molecule to be completely sequenced was the 5386 nucleotide genome of the bacteriophage ϕX174, which was completed in 1975. This was quickly followed by sequences for SV40 virus (5243 bp) in 1977 and the cloning vector pBR322 (4363 bp) in 1978. Gradually sequencing was applied to larger molecules. Frederick Sanger's group published the sequence of the human mitochondrial genome (16.6 kb) in 1981 and of bacteriophage λ (49.5 kb) in 1982. Nowadays sequences of 5–10 kb are routine and most research laboratories have the necessary expertise to generate this amount of sequence information.

The pioneering projects today are the massive genome sequencing initiatives, each aimed at obtaining the nucleotide sequence of the entire genome of a particular organism. The Human Genome Project (Section 16.1.3) is underway and a similar project has begun for the plant *Arabidopsis thaliana*. However, the leader at the moment, as in many areas of molecular biology, is the yeast *Saccharomyces cerevisiae*, with the entire sequence of one chromosome already completed.

21.3 Studying gene expression

Once the DNA sequence of a gene has been obtained the next question is likely to concern the RNA molecules transcribed from the gene. Two questions will probably be asked: which parts of the sequence are transcribed, and how is expression regulated? The first of these two questions can be answered by **transcript analysis**.

21.3.1 Transcript analysis

Most methods of transcript analysis are based on hybridization between the RNA transcript and a fragment of DNA containing the relevant gene. The resulting DNA–RNA hybrid can be analysed by electron microscopy or by treatment with nuclease enzymes.

(a) Electron microscopy of nucleic acid molecules. Electron microscopy can be used to examine nucleic acid molecules, so long as the polynucleotides are first treated with chemicals that will increase their apparent diameter. Untreated molecules are simply too thin to be seen. Usually the DNA molecules are mixed with a protein that coats the polynucleotides in a thick shell, and then stained with an electron-dense compound such as uranyl acetate (Fig. 21.8(a)).

Electron microscopy is particularly useful for determining if a gene contains introns. Consider the appearance of a hybrid between a DNA polynucleotide, containing a gene, and its RNA transcript. If the gene contains introns then these regions of the DNA strand will have no homology with the RNA transcript and so will not base pair. Instead they will loop out, giving a charac-

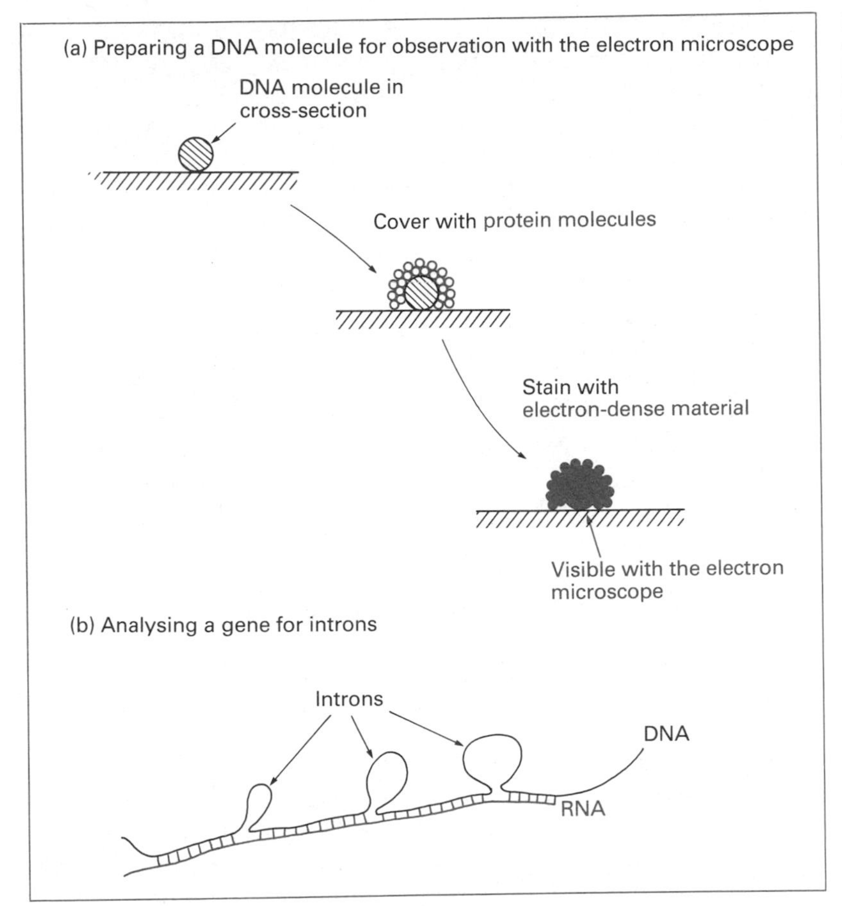

Figure 21.8 Transcript analysis by electron microscopy. (a) Preparing a DNA molecule for observation with the electron microscope. (b) The appearance of a DNA–RNA hybrid formed between a gene containing three introns and its spliced mRNA.

teristic appearance when examined with the electron microscope (Fig. 21.8(b)). The number and positions of these loops will correspond directly to the number and positions of the introns in the gene.

(b) Analysis of DNA–RNA hybrids by nuclease treatment. The second method for studying a DNA–RNA hybrid involves the use of S1 nuclease. This enzyme degrades single-stranded DNA or RNA polynucleotides, including single-stranded regions in predominantly double-stranded molecules, but has no effect on double-stranded DNA or on DNA–RNA hybrids. If a DNA molecule containing a gene is hybridized to its RNA transcript, and then treated with S1 nuclease, the non-hybridized single-stranded DNA regions at each end of the transcription unit will be digested, along with any looped-out introns (Fig. 21.9). The result will be a completely double-stranded hybrid. The single-stranded DNA fragments protected from S1 nuclease digestion can then be recovered if the DNA strand is degraded by treatment with alkali.

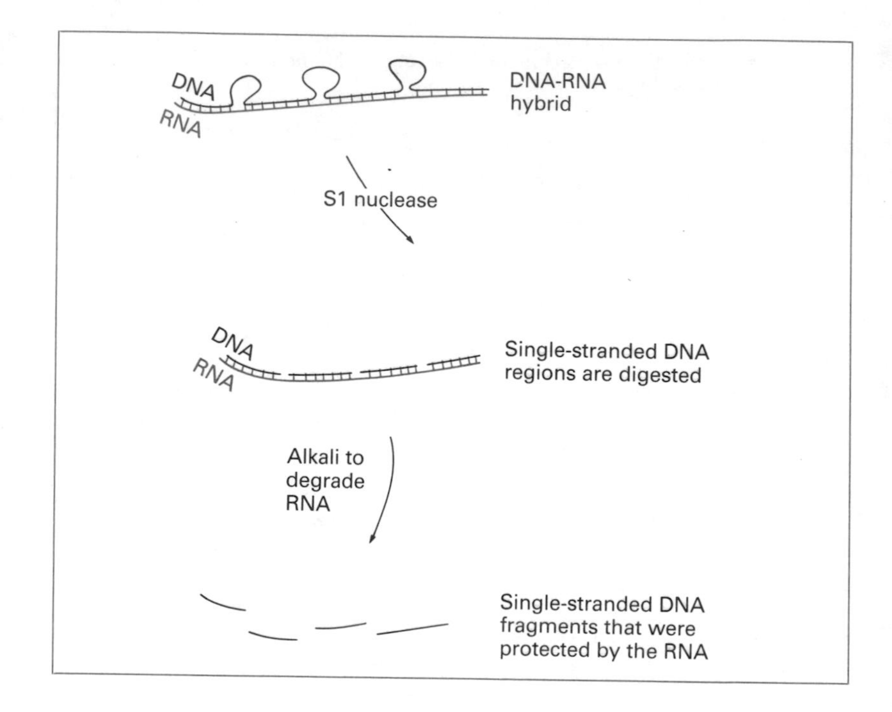

Figure 21.9 The effect of S1 nuclease on a DNA–RNA hybrid.

The manipulations shown in Fig. 21.9 have to be carried out in a special way if they are to be informative. The trick is to use a restriction fragment that spans the start or end of the transcription unit or one of the exon–intron boundaries (Fig. 21.10). This fragment is inserted into an M13 vector and a single-stranded DNA version obtained and hybridized with the transcript. Digestion with S1 nuclease will now produce a protected fragment, one end of which is marked by a restriction site, the other by the transcription unit terminus or the exon–intron boundary. We can now determine the size of this molecule by gel electrophoresis, and then use the restriction site as a marker to position the transcription unit on the DNA sequence.

21.3.2 Studying gene regulation

Once we have mapped the transcription unit on to the DNA sequence we will start to ask questions about how the gene is expressed. You will remember from Chapter 10 that regulatory sequences upstream of genes are usually binding sites for proteins that modulate gene expression. It should therefore be possible to identify a regulatory sequence by searching for a protein-binding site. There are several ways of doing this, the most popular method being **gel retardation analysis**.

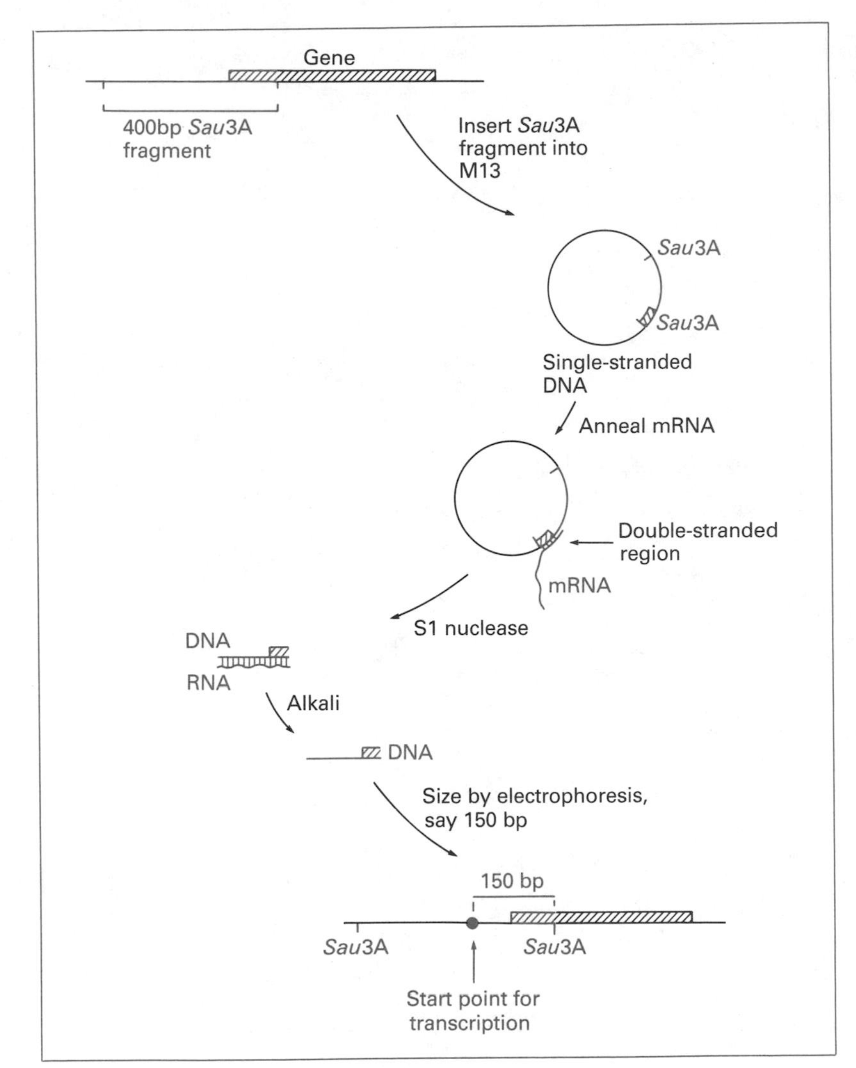

Figure 21.10 Locating a transcription start point by S1 nuclease mapping.

(a) Gel retardation of DNA–protein complexes. Proteins are quite substantial structures and a protein attached to a DNA molecule will result in a large increase in overall molecular mass. If this increase can be detected then a DNA fragment containing a protein binding site will have been identified. In practice a DNA fragment carrying a bound protein is identified by gel electrophoresis, as it will have a lower mobility than the uncomplexed DNA molecule (Fig. 21.11(a)). This is called gel retardation.

To carry out a gel retardation experiment the region containing the regulatory sequence must be digested with a restriction endonuclease and then mixed with DNA-binding protein (Fig. 21.11(b)). The location of the regulatory sequence is then deter-

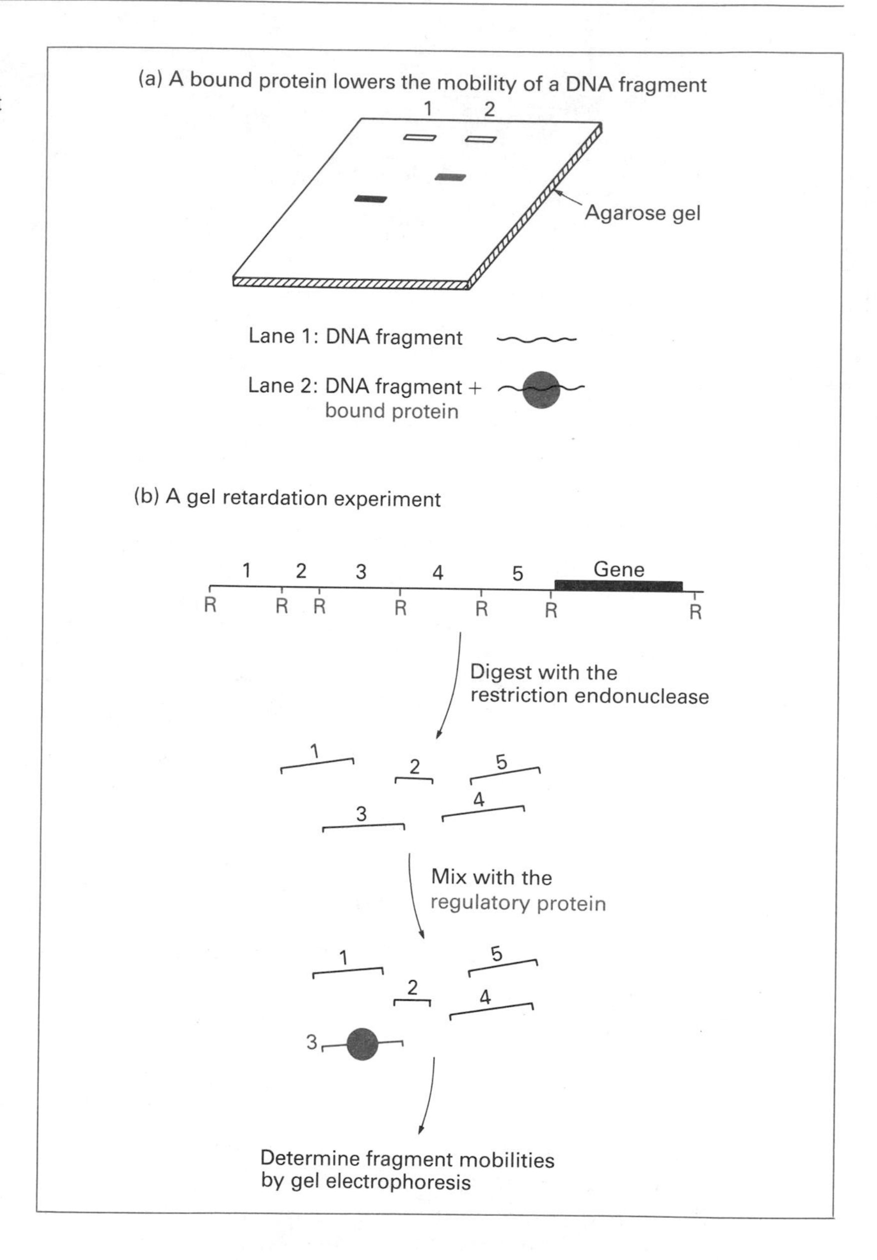

Figure 21.11 Gel retardation analysis. (a) A bound protein will lower the mobility of a DNA fragment during gel electrophoresis.
(b) Carrying out a gel retardation experiment. Each 'R' indicates a restriction site.

mined by identifying which restriction fragment binds to the protein. Often it is not possible to obtain the regulatory protein in pure form, so a crude extract of nuclear protein is used instead. This complicates things as the extract may contain proteins that bind to DNA in a non-specific fashion, resulting in spurious retardations. However, a series of control experiments will distinguish the real regulatory protein from this background 'noise'.

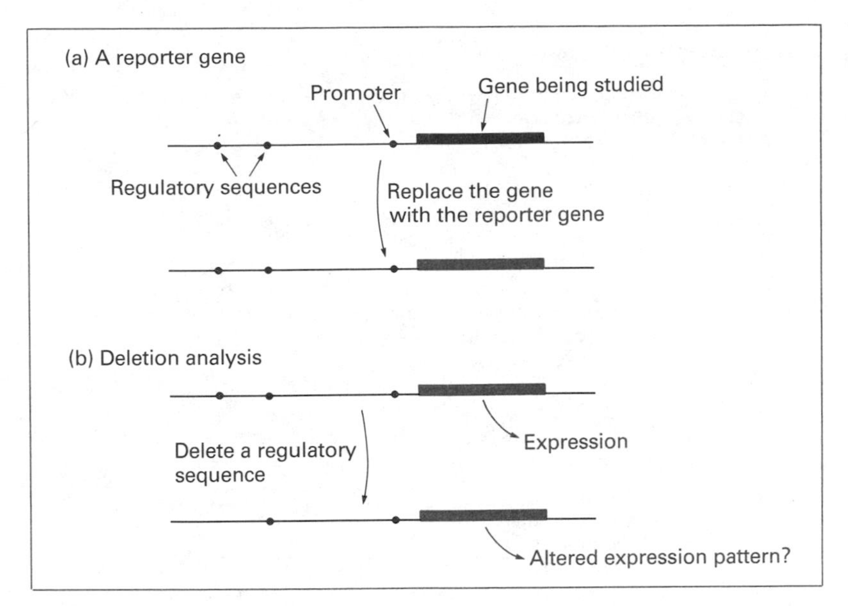

Figure 21.12 A reporter gene and its use in deletion analysis.

(b) Determining the function of a regulatory sequence. A gel retardation experiment will locate the position of a protein-binding site upstream of a cloned gene but will provide no information on the function of the regulatory sequence. A second type of analysis is needed to determine if the sequence is a repressor, enhancer or response element of some description.

Deletion analysis depends on the assumption that removal of the regulatory sequence will result in a change in the way in which expression of the cloned gene is regulated. For example, deletion of a repressor should result in the gene being expressed at a higher level, and removal of a response element might lead to the target gene failing to respond to the regulatory molecule. A deletion experiment therefore requires that expression of the cloned gene be monitored in some way. This may present a problem as it is often difficult to measure the extent to which a gene is expressed, simply because there is no convenient assay for its translation product. The answer is to use a **reporter gene**, a test gene whose expression results in a quantifiable phenotype (Table 21.1). The reporter gene is fused to the upstream region of the gene under study (Fig. 21.12(a)), so that when it is cloned in the host organism its expression pattern exactly mimics that of the original gene. Deletions can then be made in the upstream region of the construct, and the effects assessed by looking for changes in the pattern and extent of expression of the reporter gene (Fig. 21.12(b)).

The results of deletion analyses have to be interpreted with care as complications can arise if a single deletion removes more than one regulatory sequence, or if two different sequences cooperate to

The polymerase chain reaction

The **polymerase chain reaction (PCR)** is a new technique that has many applications in recombinant DNA research. PCR results in the selective amplification of a chosen region of a DNA molecule. The only requirement is that the sequences at the borders of the selected region must be known, so that two short oligonucleotides that will anneal to the target DNA molecule can be synthesized. These oligonucleotides delimit the region that will be amplified.

Amplification is carried out by the DNA polymerase I enzyme from *Thermus aquaticus*, a bacterium that lives in hot springs. This enzyme is thermostable, meaning that it is resistant to denaturation by heat treatment. To begin a PCR amplification, the enzyme and oligonucleotides are added to the target DNA. The oligonucleotides anneal to the template DNA and prime the synthesis of new complementary polynucleotides:

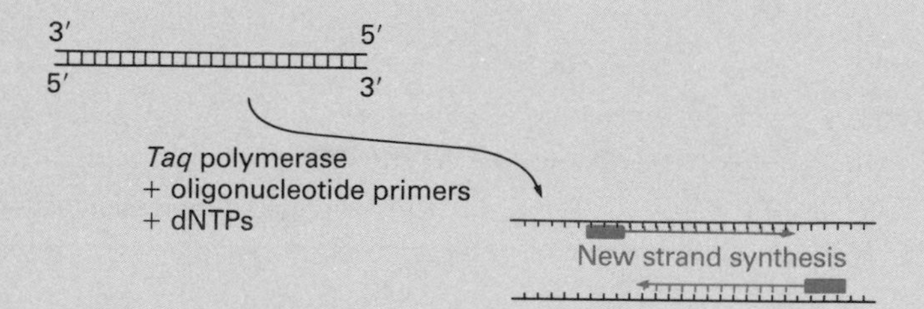

With most types of DNA polymerase this single reaction step would be all that could be carried out, but the thermostability of the *Taq* polymerase means that the reaction mixture can now be heated, to 80° C or more, so that the newly synthesized strands detach from the templates. On cooling, more primers anneal at their respective positions (including positions on the newly synthesized strands), and the *Taq* polymerase, unaffected by the heat treatment, carries out a second round of DNA synthesis:

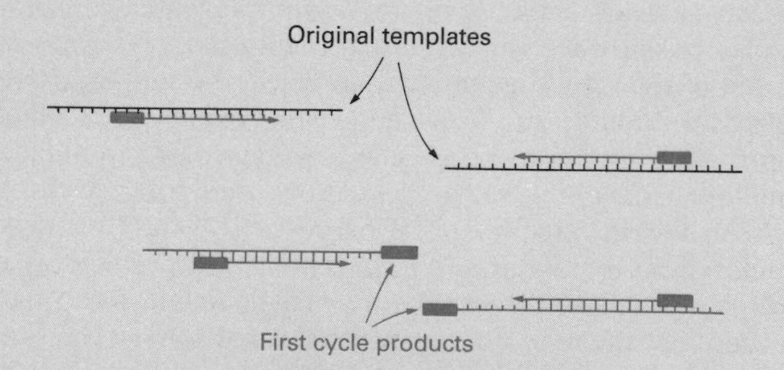

The reaction can be continued through 30–40 cycles, with the DNA amplification proceeding in an exponential fashion.

PCR is proving to be the most important technical breakthrough since the development of gene cloning itself. Like gene cloning it can produce large amounts of identical DNA molecules from a minute amount of starting material. It has the advantages that it is rapid and is successful even if there is just a single DNA molecule with which to begin the amplification. Several kb of DNA can be amplified in one reaction, and the amplified DNA can be studied by any of the methods described in this chapter. The only drawback is the need for oligonucleotides to prime the reaction, which means that the sequence of at least the boundary regions must already be known, thus limiting PCR to the study of genes that have already been characterized in part by cloning methods.

Table 21.1 A few examples of reporter genes used in studies of gene regulation in higher organisms

Gene	*Gene product*	*Assay*
lacZ	β-galactosidase	Histochemical test
uidA	β-glucuronidase	Histochemical test
neo	Neomycin phosphotransferase	Kanamycin resistance
cat	Chloramphenicol acetyltransferase	Chloramphenicol resistance
dhfr	Dihydrofolate reductase	Methotrexate resistance
aphIV	Hygromycin phosphotransferase	Hygromycin resistance
lux	Luciferase	Bioluminescence

produce a single response. Despite these potential difficulties the approach, in combination with studies of protein-binding sites, has provided important information about the regulation of individual genes.

21.4 Determining the function of a protein coded by a cloned gene

Practising molecular biologists quickly come to realize that there are many more genes than there are characterized proteins, and often a cloning experiment will produce a gene whose translation product is unknown. The DNA sequence of the gene will allow the amino acid sequence to be deduced, and rules can be followed to predict the secondary structure that might be taken up by the polypeptide. This is only of limited value though, as neither the amino acid sequence nor the secondary structure provides interpretable information on the function of the protein.

The first option in these circumstances is to enter the amino acid sequence into a computer and search the databases containing all the protein sequences that have ever been published. The computer will not only look for a perfect match but will also seek proteins that show a limited degree of homology. This may allow a hypothesis to be proposed, and subsequently tested by further experimentation, about the function of the unknown protein.

The computer analysis can be supplemented by two techniques that identify the gene product in extracts of cells from which the gene is derived. These are called **hybrid-arrest translation** (HART) and **hybrid-release translation** (HRT). However, at best these methods provide only a sample of the relevant protein; functional characterization may still be difficult. These problems underline one of the key facts of genetics: classical and recombinant experiments are not mutually exclusive as neither can stand on their own. Mendelian genetics normally starts with a phenotype and works backwards towards the gene or genes that are responsible for it. On its own this approach can become stalled as it may not be possible to gain detailed information on the structure and expression of these genes. Gene cloning can provide details on individual genes, but on its own may not be able to put those details in context within the organism as a whole. The great beauty of modern genetics is the way in which the two branches of genetical investigation complement one another.

In vitro mutagenesis

Many of the questions asked by molecular biologists centre on the relationship between the structure of a protein and its mode of activity. The best way of tackling this problem is by introducing a mutation into the gene coding for the protein and then determining what effect the change in amino acid sequence has on the properties of the translation product. Under normal circumstances mutations occur randomly and a large number may have to be screened before one that provides useful information is found. *In vitro* mutagenesis is a more powerful technique as it allows specific alterations to be made at predetermined positions in a cloned gene.

An almost unlimited variety of DNA manipulations can be used to introduce mutations into cloned genes. The simplest are as follows:

1. A restriction fragment can be deleted.
2. The gene can be opened up at a restriction site, a few nucleotides removed to make a short deletion, and the gene re-ligated.
3. A short oligonucleotide, coding for a few additional amino acids, can be inserted at a restriction site.

As well as these approaches, there are a series of techniques under the heading **oligonucleotide-directed mutagenesis**, which allow a point mutation to be introduced at any position in a cloned gene. These procedures enable very specific questions to be asked about the way protein structure and function are interlinked.

21.5 Cloned genes in biotechnology

Biotechnology – the use of living organisms in industrial processes – is not a new subject, although it has received far more attention during the past few years than ever before. According to archaeologists, the British biotechnology industry dates back 4000 years, when fermentation processes that make use of living yeast cells to produce ale and mead were first introduced into this country. Certainly brewing was well established in Britain by the time of the Roman invasion.

During the twentieth century, biotechnology diversified with the development of a variety of industrial uses for microorganisms. In the 1940s fungi and bacteria were first used for the production of antibiotics, enzymes and other organic chemicals, the ease with which microorganisms can be grown in culture being exploited to obtain large amounts of these products. More recently, gene cloning has expanded the range of compounds that can be obtained in this way. Rather than being limited to natural microbial products, we can now transfer a gene from any organism into a bacterium, yeast or fungus and obtain large amounts of the gene product. At least that is the theory: is it so easy in practice?

21.5.1 Bacteria are not ideal hosts for recombinant protein

The first attempts at synthesis of **recombinant protein** employed bacteria such as *E. coli* as the host organism. The first human protein to be synthesized in this way was the peptide hormone somatostatin (anti-growth hormone), which is important in the treatment of a variety of human growth disorders. Being a very short protein, only 14 amino acids in length, it was relatively easy to synthesize an artificial gene coding for the peptide, rather than attempting to locate the real gene in a human genomic library. The artificial gene was then inserted into an *E. coli* cloning vector, downstream of the *lac* promotor (Fig. 21.13), and introduced into host bacteria which subsequently synthesized the protein.

Similar projects were carried out with insulin, interferon and a

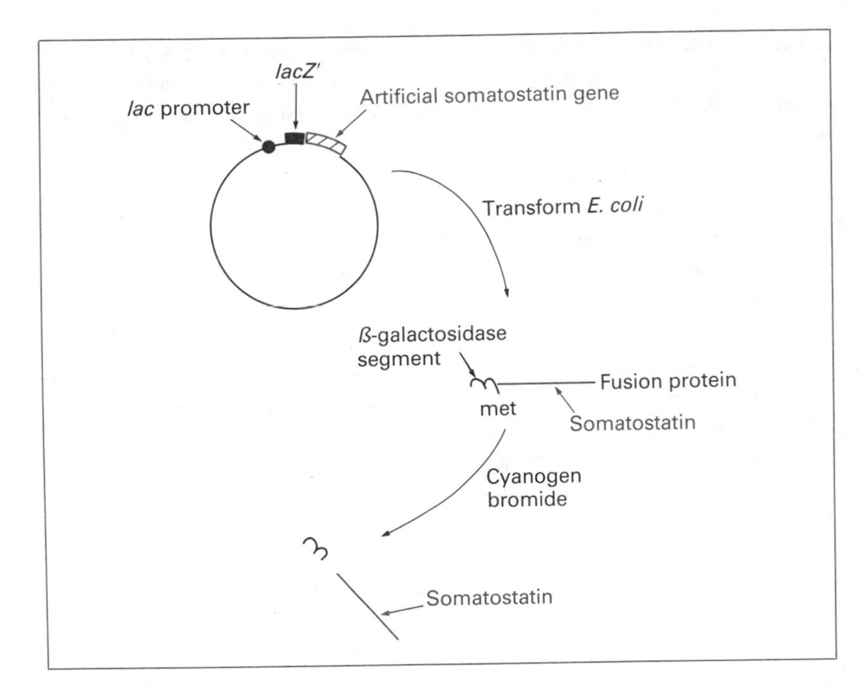

Figure 21.13 Production of somatostatin in recombinant *E. coli*. The synthetic gene coding for somatostatin was fused with the start of the *lacZ'* gene contained within an *E. coli* cloning vector. After transformation, the *E. coli* RNA polymerase transcribed the fused gene, recognizing the *lac* promoter as its binding site. The mRNA was then translated by ribosomes that recognized the *lac* ribosome binding sequence. The fused protein that was synthesized was cleaved with cyanogen bromide, which cuts polypeptides specifically at methionine amino acids. As a result the somatostatin peptide was obtained in pure form.

variety of other human proteins, but in each case the experiment was complicated by the need to modify the gene to get it to function in *E. coli*. A gene from a higher organism will normally be inactive in *E. coli* as the bacterium will not recognize the eukaryotic promoter, and so will not transcribe the cloned gene. Human genes introduced into *E. coli* must therefore be ligated downstream of an authentic *E. coli* promoter before they will be expressed. This is not too much of a problem, but is not the end of the story. A ribosome binding site must also be put in place, and it is no good if the gene contains introns because *E. coli* does not have the U-RNAs and proteins needed for their removal. In fact the list of problems is quite extensive and before long biotechnologists started looking for eukaryotes that might be suitable as alternative hosts for recombinant protein synthesis.

21.5.2 Synthesis of recombinant proteins in eukaryotes

The argument is that a microbial eukaryote, such as yeast or a fungus, is more closely related to an animal, and so should be able to deal with recombinant protein synthesis more efficiently than *E. coli*. Yeast and fungi can be grown just as easily as bacteria in large-scale culture vessels, and ought to express a cloned animal gene in a manner more akin to that occurring in the animal itself. *Saccharomyces cerevisiae*, for example, should be able to deal with introns in animal genes.

Human proteins synthesized in recombinant eukaryotes

Some of the human proteins that have been synthesized from genes cloned in cultured eukaryotic cells

Protein	*Host*
Insulin	Mouse cells
Somatostatin	Monkey cells
Growth hormone	Mouse cells
Parathyroid hormone	Rat cells
Chorionic gonadotropin	Monkey cells
Interferons	Yeast *Aspergillus nidulans* Mouse cells
Interleukins	*Aspergillus nidulans* Insect cells
Tissue plasminogen activator	*Aspergillus nidulans*
Factor VIII blood protein	Hamster cells

To a certain extent these hopes have been realized and microbial eukaryotes are now used for the routine production of some animal proteins. Unfortunately there are still problems, especially with yeast, which tends not to secrete proteins into the growth medium, preferring to package them within the cell vacuole. This makes the purification process more difficult and, importantly, more expensive. The continuing problems are forcing biotechnologists to look more carefully at higher eukaryotes as hosts for recombinant protein. The rapid progress made in the development of cloning vectors for plants and animals (Section 20.4.2) has been allied with similar advances in cell culture systems for higher organisms. These cultures have relatively long generation times, limiting the yield of recombinant protein obtained from them, but this is more than offset by the fact that the recombinant protein is almost always synthesized correctly.

The final possibility is to use an intact organism rather than a cell culture. The seeds of higher plants are very efficient at protein synthesis, as they accumulate large quantities of storage proteins and other compounds that the young seedling uses as a nutrient supply during the early stages of germination. Many crop plants have been bred specifically for the protein content of their seeds and the genes involved in seed development are quite well understood. What if a gene for a natural seed protein were to be replaced by a gene coding for some useful animal protein? The answer is that under some circumstances the animal protein will accumulate in the seeds, as has been demonstrated by the synthesis of pharmaceutical compounds called enkephalins in the seeds of engineered oilseed rape plants. Enkephalins are small proteins, only a few amino acids in length, and it still has to be established that larger animal proteins can also be synthesized in the special environment found within the developing seed. Further success would open up an exciting new area of biotechnology.

21.6 The future for molecular genetics

Genetics is one of the most living of the life sciences. New discoveries follow one after another and the pace of research is, if anything, getting more rapid every year. Since the First Edition of this book was published there have been major advances in our understanding of RNA editing (Section 7.5) and the non-universality of the genetic code (Section 8.2.3), topics that are fundamental to our appreciation of how the biological information contained within a gene is released and made use of by the cell. There have been equally exciting breakthroughs in research concerning the role of genes in development (Section 10.5), and the viral basis to cancer (Section 13.2.3). Molecular genetics is moving into new fields with the establishment of molecular anthropology (Section 16.2) and promises to yield new approaches to disciplines as disparate as archaeology and palaeontology. On top of all this is the phenomenal amount of information that will come out of the Human Genome Project before the turn of the century.

These are indeed exciting days to be a molecular geneticist. Understanding DNA is looked on as the key to combating the problems presented by AIDS, cancer, and genetic diseases such as cystic fibrosis and muscular dystrophy. The genetic basis to mental illnesses such as schizophrenia are becoming known, raising hopes that our powerful investigative tools may soon be ranged against these devastating diseases. Gregor Mendel would be surprised at how far genes have come since he discovered them in his vegetable garden in Brno; one hopes that he would be pleased with what he started.

Reading

T. A. Brown (1990) *Gene Cloning: An Introduction*, 2nd edn, Chapman and Hall, London – a more detailed description of recombinant DNA techniques.

M. Grunstein and B. J. Hoffman (1975) Colony hybridization: a method for the isolation of cloned cDNAs that contain a specific gene. *Proceedings of the National Academy of Sciences USA*, **72**, 3961–5 – the first description of this important method.

W. Bender *et al.* (1983) Chromosome walking and jumping to isolate DNA from the *Ace* and *rosy* loci and bithorax complex in *Drosophila melanogaster*. *Journal of Molecular Biology*, **168**, 17–33 – one of the first examples of chromosome walking.

F. Sanger, S. Nicklen and A. R. Coulson (1977) DNA sequencing with chain terminating inhibitors. *Proceedings of the National Academy of Sciences USA*, **74**, 5463–7; A. Maxam and W. Gilbert (1977) A new method of sequencing DNA. *Proceedings of the National Academy of Sciences USA*, **74**, 560–4 – the two original descriptions of modern DNA sequencing techniques.

T. J. White *et al.* (1989) The polymerase chain reaction. *Trends in Genetics*, **5**, 185–9 – a useful review of PCR.

M. M. Garnier and A. Revzin (1981) A gel electrophoretic method for quantifying the binding of proteins to specific DNA regions. *Nucleic Acids Research*, **9**, 3047–60 – gel retardation analysis.

K. Itakura *et al.* (1977) Expression in *Escherichia coli* of a chemically synthesized gene for the hormone somatostatin. *Science*, **198**, 1056–63 – one of the first successes with recombinant protein in *E. coli*.

G. Saunders *et al.* (1989) Heterologous gene expression in filamentous fungi. *Trends in Biotechnology*, **7**, 283–7 – fungi as hosts for recombinant protein synthesis.

J. Vandekerckhove *et al.* (1989) Enkephalins produced in transgenic plants using modified 2S storage proteins. *Biotechnology*, **7**, 929–32 – recombinant protein synthesis in seeds.

Problems

1. Define the following terms:

   hybridization probing
   nucleic acid hybridization
   autoradiography
   heterologous probing
   chromosome walking
   dideoxynucleotide
   transcript analysis
   gel retardation analysis
   deletion analysis
   reporter gene
   hybrid-arrest translation
   hybrid-release translation
   recombinant protein

2. Describe how a colony hybridization experiment is carried out.
3. Explain how the amino acid sequence of a protein can be used to design a hybridization probe for its gene.
4. Give an example of the use of heterologous probing.
5. Outline the way in which a chromosome walk is conducted.
6. Describe how a chain termination DNA sequencing experiment is carried out.
7. How is a DNA sequence read from a sequencing autoradiograph?
8. Describe two ways of mapping the positions of introns within cloned genes.
9. What is the basic assumption behind deletion analysis?
10. Explain how a reporter gene is used in a deletion analysis experiment.
11. Outline the problems associated with determining the function of a cloned gene.
12. To what extent is *E. coli* a suitable host for the synthesis of recombinant protein?
13. What alternatives are there to bacteria as the hosts for recombinant protein synthesis?

Answers to selected problems

Chapter 3

15. 5′-TTGCATTGCTAT-3′

16. (a) 30%
(b) 40%
(c) 20%
(d) 20%

17. 54%

Chapter 4

12. 4^{95}

13. 29 out of 36 = 81%

14. 5′-AUGGGUACUAAAGGGCGCUAUAUAUACGGACGUUAG-3′
5′-AUGGGAACCAAAGCGCGCUACAUAUAUGGUCCUUAG-3′

Chapter 5

12. The sequences can be summarized as follows:

1st nucleotide	6 As	1 T	
2nd nucleotide	6 Ts	1 G	
3rd nucleotide	6 As	1 T	
4th nucleotide	5 Gs	1 C	1 A
5th nucleotide	6 As	1 C	
6th nucleotide	5 Cs	1 G	1 T
7th nucleotide	6 As	1 T	
8th nucleotide	6 Ts	1 A	
9th nucleotide	4 As	3 Ts	

So a consensus sequence would be 5′-ATAGACAT$^{A}_{T}$-3′. Only the last nucleotide is doubtful.

13. Each of them.

```
    5′
     AC          C
       GUUUGG     A
       CAAACC     G
      U       U
    3′

     5′
      AGCUAGCU ACAGCU
                     A
                      G
      UCGAUCGA      C
  GUGU        GUGUUU
3′

     5′
      GGCUAGGUCG A
                   A
      CCGAUCCAGC  A
     3′
```

Chapter 6

10.

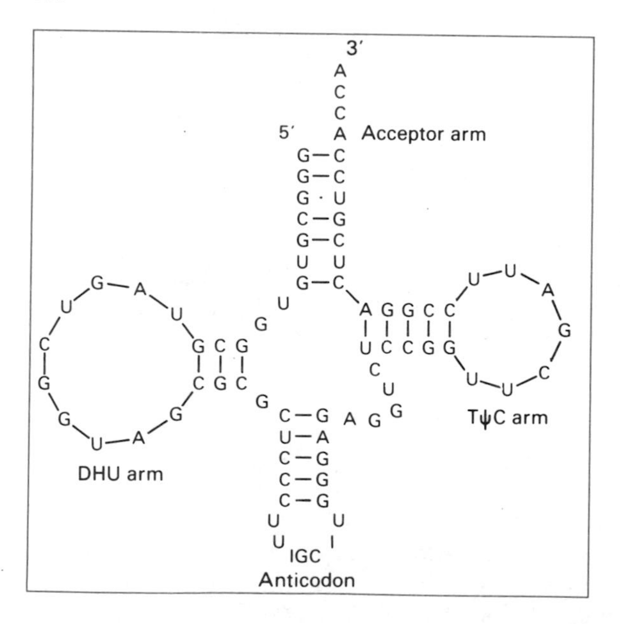

11.

```
                  5′
                   AUU        G
                      AGCAU    U
   CAACGGUCUA UCGUA        A
3′                        A
```

12.

```
5'          AA        UUC         AUC     CC
  CUCCUC      AUGA       AGUGC       UAA    C
  ::::::      ::::       :::::       :::
  GAGGAG      UACU       UCACG       AUU    U
3'        AG        CGA         CAA       UA
```

Chapter 8

15. $4^4 = 256$

16. Codons = AAA, AAG, AGA, GAA, AGG, GGA, GAG, GGG.
Amino acids = lys, arg, glu, gly.

17. met-leu-ala-asp-pro-glu-met-met-leu-tyr-ile-ile-tyr-ala-gln.

Chapter 10

16. (a) Absence of galactose, repressor protein binds to the operator, there is no transcription.
(b) Presence of galactose, galactose binds to the repressor protein, the repressor–inducer complex cannot bind to the operator, transcription occurs.

17. (a) If the repressor cannot bind to the operator, then the operon will be expressed all the time, even in the absence of lactose.
(b) If there is no repressor protein, then the operon will be expressed all the time, even in the absence of lactose.
(c) If RNA polymerase cannot attach to the promoter then no transcription will occur so the operon will not be expressed at all.
(d) If the repressor is not able to bind lactose, then the effect of the inducer will be lost. The operon will not be expressed at all, as the repressor will always be bound to the operator.
(e) *lacZ* will still be transcribed correctly, but only an incomplete transcript of *lacY* will be made. *lacA* will not be transcribed at all. The result will be that only β-galactosidase will be synthesized, although its synthesis will be regulated correctly.
(f) If the *lac* mRNA is not degraded, then any mRNA that is synthesized in response to the presence of lactose will be retained in the cell, even after all the lactose is used up. The result is that the enzymes will be present all the time, and the ability to regulate the operon will be lost.

Chapter 11

10.

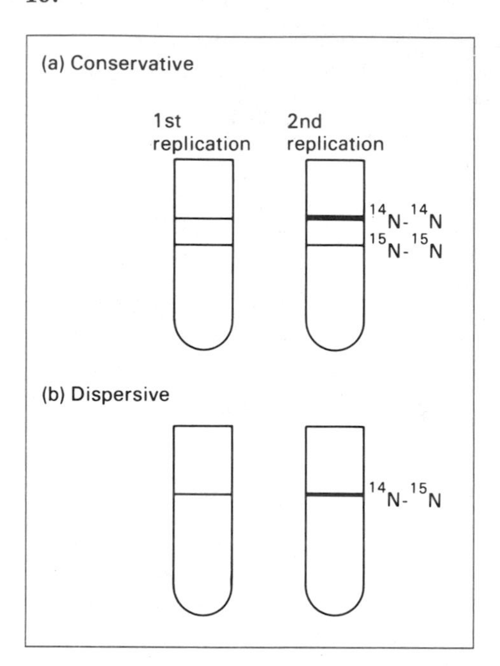

11.

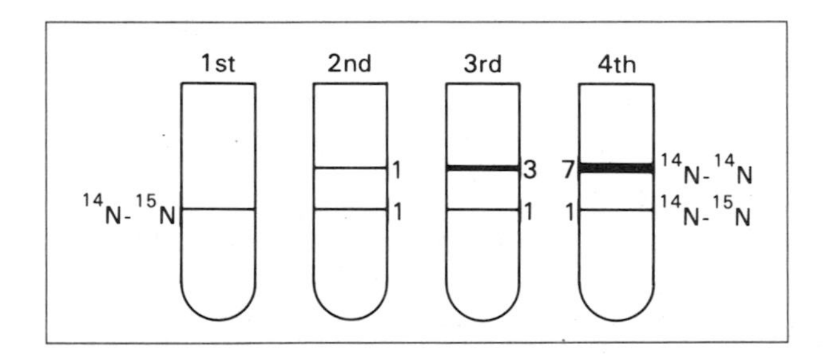

Chapter 12

17. Amino acids = met-gly-arg-thr-thr-gly-ser-tyr-trp-phe-ser.
For (a) (b) and (d) there are several alternatives. For (c) the only possibilities are:-
nucleotide 24 C to A (TAC → TAA) or G (TAC → TAG)
nucleotide 26 G to A (TGG → TAG)
nucleotide 27 G to A (TGG → TGA)

18. (a) GGA, GGU, GGG or GGC to GCA, GCU, GCG or GCC
(b) UGG to UGA or UAG
(c) UGU or UGC to CGU or CGC
(d) AGU or AGC to AUU or AUC

(e) AGU or AGC to ACU or ACC
(f) ACU or ACC to UCU or UCC

19. See p. 357.

Chapter 14

15. You will require an F^- strain that is streptomycin sensitive and also carries a genetic marker that distinguishes it from the F^+ strain. A strain $F^- \ str^S \ amp^R$ (amp^R = ampicillin-resistant) would be suitable. The F^+ and F^- cells would then be mixed together so conjugation can take place. The cells are then spread on an agar medium containing both streptomycin and ampicillin. Only if R100 is mobilizable will there be any $str^R \ amp^R$ cells, able to grow on the selective medium.

Chapter 17

10. F_1: All *Gg*, all green pods.
F_2: 1 *GG* : 2 *Gg* : 1 *gg*; 3 green pods : 1 yellow pod.

11. 9 full green : 3 full yellow : 3 constricted green : 1 constricted yellow.

12. Phenotypes will be 27 *ABC*:9 *ABc* : 9 *AbC* : 9 *aBC* : 3 *Abc* : 3 *aBc*: 3 *abC* : 1 *abc*.

13. Equal numbers of heterozygous tall-stemmed plants and homozygous short-stemmed plants.

14. F_2 genotypes will be 1 *RR* : 2 *Rr* : 1 *rr*.
F_2 phenotypes will be 1 red : 2 pink : 1 white.

Chapter 18

15. Number of recombinant progeny = 131 + 139 = 270

Total number of progeny = 1010
Recombination frequency = 26.7%
Map distance = 26.7 cM

16.

a	*d*	*b*	*c*
0	25	30	40 cM

17. That a^- and b^- are on different chromosomes.

18. (i) As the number of PD tetrads greatly exceeds the number of NPD tetrads, we know the markers are linked.

(ii) The crossover frequency can be calculated:

$$CF = \frac{T + 6NPD}{2 \times (T + NPD + PD)}$$

$$= \frac{56 + 6}{2 \times (56 + 1 + 101)}$$

$$= \frac{62}{2 \times 158} = \frac{62}{316}$$

$$= 19.6\%$$

Chapter 19

10. *thr*$^-$ Test for growth on minimal medium.
leu$^-$ Test for growth on minimal medium.
azi^R Test for growth on a medium containing azide.
ton^A Mix with a sample of T1 phage, spread onto agar, check for plaques.
lac$^-$ Test for growth on a medium with lactose as the only sugar.
gal$^-$ Test for growth on a medium with galactose as the only sugar.

11.

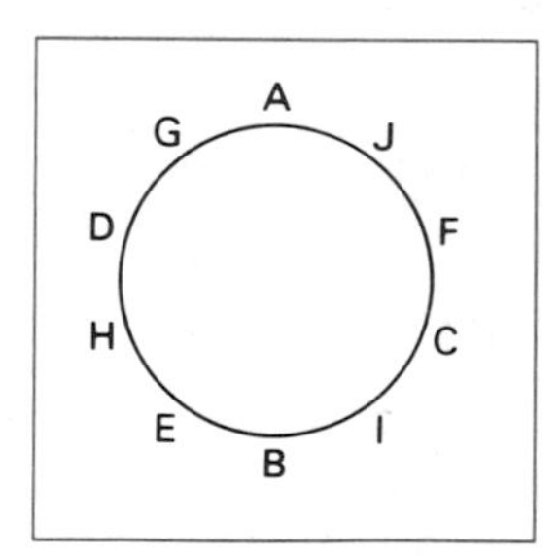

12. C–D–A–B

Glossary

Abortive infection Infection of a bacterium with a transducing bacteriophage that carries a segment of bacterial DNA, derived from the previous host, instead of the phage genome. The infection cycle is said to 'abortive' as no new phage particles are produced.

Acridine dye A member of a family of chemical compounds that cause frameshift mutations by intercalating between adjacent base pairs of the double helix.

Acute transforming retrovirus A retrovirus that has captured an oncogene and which therefore is a very efficient inducer of cell transformation.

Adenine A purine, one of the nitrogenous bases found in DNA and RNA.

Adenylate cyclase The enzyme that converts ATP to cyclic AMP and which is inhibited by glucose during catabolite repression.

A form One of the structural configurations of the double helix, although not the one thought to be the commonest form of DNA in the cell.

Allele One of two or more alternative forms of a gene.

Alpha DNA The clustered repeat sequences of DNA located in the centromeric regions of human chromosomes.

α-helix A helical configuration taken up by some segments of a polypeptide; one of the commonest types of protein secondary structure.

***Alu* family** A type of repetitive DNA found in the genomes of several species of mammal.

Amino acid One of the monomeric units that polymerize to give a protein molecule.

Aminoacylation The attachment of an amino acid to the acceptor arm of a tRNA molecule; one of the prerequisite steps for protein synthesis.

Aminoacyl- or A-site The site on the ribosome to which the aminoacyl-tRNA attaches during translation.

Aminoacyl-tRNA synthetase One of the diverse family of enzymes responsible for aminoacylation.

Ancient DNA DNA that is present in preserved biological material.

Anticodon The triplet of nucleotides in a tRNA molecule that is complementary to and base pairs with a codon in an mRNA molecule.

Archaea A group of organisms once thought to be related to bacteria but now looked on as a distinct taxon.

Ascospore One of the haploid products of meiosis in an ascomycete such as the yeast *Saccharomyces cerevisiae*.

Ascus The structure which contains the four ascospores produced by a single meiotic cell division in the yeast *Saccharomyces cerevisiae*.

Attenuation A mechanism by which expression of an amino acid biosynthetic operon (e.g. the *trp* operon) is regulated by the levels of the amino acid in the cell.

Autoradiography A method of detecting radioactively labelled molecules through exposure of an X-ray-sensitive photographic film.

Autosome Any chromosome that is not a sex chromosome.

Auxotroph A mutant microorganism that will grow only if supplied with a nutrient not required by the wild-type.

Back mutation A mutation that reverses the effect of a previous mutation by restoring the original nucleotide sequence.

Bacterial transformation The transformation of a bacterium from one form to another, or the acquisition of a new characteristic by a bacterium, by the uptake of genetic material.

Bacteriophage or phage A virus whose host is a bacterium.

Bacterium In general terms, any prokaryote.

Base analogue A chemical compound that is structurally similar to one of the bases in DNA and which may act as a mutagen.

Base pair The hydrogen-bonded structure formed between two complementary nucleotides.

Beads-on-a-string The substructure of chromatin in which individual nucleosomes can be visualized by electron microscopy as beads on a 'string' of DNA.

β-sheet A hydrogen-bonded sheet configuration taken up by some segments of a polypeptide; one of the commonest types of protein secondary structure.

B form The standard configuration of the double helix, believed to be the predominant form in the cell, the structure of which was deduced by Watson and Crick in 1953.

Biological containment One of the precautionary measures taken to prevent replication of recombinant DNA molecules in micro-organisms in the natural environment. Biological containment involves the use of vectors and host organisms that have been modified so that they will not survive outside of the laboratory.

Biological information The information that is carried by the genes of an organism and which directs the development of a living example of that organism.

Biotechnology The use of living organisms, often but not always microbes, in industrial processes.

Blunt end An end of a DNA molecule, at which both strands terminate at the same nucleotide position with no single-stranded extension.

Buoyant density The density possessed by a molecule or particle when suspended in an aqueous salt or sugar solution.

CAAT box A component of the nucleotide sequence that makes up the eukaryotic promoter.

Capsid The protein coat that encloses the DNA or RNA molecule of a virus.

CAP site A nucleotide sequence upstream of some bacterial genes and operons; the attachment point for the catabolite activator protein.

Cap structure The modified 5′-end of a eukaryotic mRNA molecule.

Capsule A covering, often of polysaccharide, that surrounds certain types of bacterial cell.

Cascade system A system in which the completion of one event triggers the initiation of a second event. In genetics, usually a system for controlling the order in which genes are expressed.

Catabolite activator protein The regulatory protein that mediates catabolite repression of gene activity by responding to glucose levels in the cell.

Catabolite repression The process by which glucose is able to repress activity of genes and operons involved in utilization of alternative sugars as energy sources.

cDNA (complementary DNA) library A library of clones that have been prepared from mRNA after conversion into double-stranded DNA.

Cell cycle The period between one cell division and the next.

Cell-free protein synthesizing system A cell extract containing all the components required for protein synthesis (i.e. ribosomal subunits, tRNAs, amino acids, enzymes and cofactors) and able to translate added mRNA molecules.

Cell transformation The alteration in morphological and biochemical properties that occurs when an animal cell is infected by an oncogenic virus.

CentiMorgan The unit used to describe the distance between two genes on a chromosome. 1 cM is the distance that corresponds to a 1% probability of recombination in a single meiotic event.

Central Dogma The key hypothesis in molecular genetics, proposed by Crick in 1958, which states that DNA makes RNA makes protein.

Centromere The structural feature of a chromosome that is the point at which the pair of chromatids of the metaphase chromosome are held together.

Chi form An intermediate structure in recombination between DNA molecules.

Chromatid A chromosome arm.

Chromatin Originally, the deeply staining material present in the nuclei of cells and corresponding to the chromosomes. Now used more specifically to refer to the structural association between DNA and protein in chromosomes.

Chromosomes The structures, comprising DNA and protein, that are found in eukaryotic nuclei and which carry the bulk of the cell's genes.

Chromosome theory The theory, stated in its most convincing form by Sutton in 1903, that genes reside on chromosomes.

Chromosome walking A recombinant DNA technique that allows a series of overlapping restriction fragments to be identified, often used to determine the relative positions of genes on large DNA molecules.

***Cis–trans* complementation test** A genetic analysis that tests whether two mutations lie in the same or different genes, and which can also provide information on dominance and recessiveness. The test involves introducing the two mutated genes into a single cell, for example by introducing an F′ plasmid carrying one mutated gene into a recipient bacterium with a chromosomal copy of the second mutated gene.

Clone A population of identical cells, generally those containing identical recombinant DNA molecules.

Clone library A collection of clones that contain a number of different genes. See **Genomic library**.

Cloning vector A DNA molecule, capable of replication in a host organism, into which a gene is inserted to construct a recombinant DNA molecule.

Closed promoter complex The initial complex formed between RNA polymerase and a promoter, in which the double helix is still completely base paired.

Cloverleaf A convenient two-dimensional representation of the structure of a tRNA molecule.

Codominance The situation whereby both members of a pair of alleles contribute to the phenotype.

Codon A triplet of nucleotides that code for a single amino acid.

Cohesive end An end of a double-stranded DNA molecule where there is a single-stranded extension.

Colinearity Refers to the fact that a gene and the polypeptide for which it codes are related in a direct linear fashion, with the 3′-end of the template strand of the gene corresponding to the amino-terminus of the polypeptide.

Competent Refers to a culture of bacteria that have been treated to enhance their ability to take up DNA molecules.

Complementary Refers to nucleotides or nucleotide sequences that are able to base pair, e.g. A and T are complementary; the sequence 5′-ATGC-3′ is complementary to 5′-GCAT-3′.

C-*onc* The cellular version of a gene carried by a transforming retrovirus.

Concatamer A DNA molecule that comprises a series of smaller DNA molecules linked head-to-tail. Concatamers are formed during the replication of some viral and phage genomes.

Conditional-lethal mutation A mutation that results in a cell or organism able to survive only under certain conditions.

cI repressor One of the regulatory proteins responsible for controlling expression of λ genes.

Conjugation Physical contact between two bacteria, usually associated with transfer of DNA from one cell to the other.

Conjugation mapping The technique that allows the relative positions of bacterial genes to be mapped by determining the time it takes for each gene to be transferred during conjugation.

Consensus sequence A nucleotide sequence used to describe a large number of related although non-identical sequences. Each position of the consensus sequence represents the nucleotide most often found at that position in the real sequences.

Constitutive mutant A mutant in which a gene or set of genes normally subject to regulation is expressed all the time.

Conventional pseudogene A DNA sequence that resembles a gene but which has become inactive owing to the accumulation of mutations.

Copy number Usually, the number of molecules of a plasmid contained in a single cell. Also, the number of copies of a gene, transposon or repetitive element in a genome.

Core enzyme The version of *E. coli* RNA polymerase that has the subunit composition $\alpha_2\beta\beta'$ and is able to transcribe DNA although is unable to locate promoters efficiently.

Core octamer The structure comprising two subunits each of histones H2A, H2B, H3 and H4, which forms the central component around which DNA is wound to form a nucleosome.

Corepressor A small molecule that must bind to a repressor protein before the latter is able to attach to its operator site.

Cosmid A cloning vector consisting of the λ *cos* site inserted into a plasmid, used to clone DNA fragments up to 40 kb in size.

***cos* site** The sequence within the λ DNA molecule that forms single-stranded overhangs at the ends of the linear version of the genome.

Cotransduction The transfer of two or more genes from one bacterium to another via a transducing phage.

Cotransformation The uptake of two or more genes on a single DNA molecule during transformation of a bacterium.

Cro repressor One of the regulatory proteins responsible for controlling expression of λ genes.

Cross-fertilization Fertilization of a female gamete with a male gamete derived from a different individual.

Crossing-over The exchange of polynucleotides between chromosomes during meiosis.

Cruciform structure The cross-shaped structure that can arise by intramolecular pairing within a double-stranded DNA molecule that contains an inverted repeat.

Cyclic AMP A modified version of AMP in which an intramolecular phosphodiester bond links the 5′ and 3′ carbons.

Cyclobutyl dimer A dimeric structure formed between two

adjacent pyrimidine bases in a polynucleotide, resulting from ultraviolet irradiation.

Cytochemistry The analysis, often by determination of staining properties, of the chemical composition of cell components.

Cytoplasm A general name for the complex mixture of molecules that comprises the ground substance of the extranuclear regions of a living cell.

Cytosine A pyrimidine, one of the bases found in DNA and RNA.

Defective virus A viral genome that has lost essential nucleotide sequences and so cannot replicate on its own. A defective virus may give rise to progeny by making use of proteins synthesized from a non-defective virus genome present in the same cell.

Degeneracy Refers to the genetic code and the fact that most amino acids are coded for by more than one triplet codon.

Deletion analysis The identification of regulatory sequences for a gene by determining the effects on gene expression of specific deletions in the upstream region.

Deletion mutation A mutation that arises through deletion of a nucleotide from a DNA sequence.

Denaturation Breakdown by chemical or physical means of the non-covalent interactions that are responsible for the secondary and higher levels of structure of proteins and nucleic acids.

Density gradient centrifugation Separation of molecules and particles on the basis of buoyant density, usually by centrifugation in a concentrated sucrose or caesium chloride solution.

Deoxyribonuclease An enzyme that breaks a DNA polynucleotide by cleaving phosphodiester bonds.

Dideoxynucleotide A modified nucleotide that lacks the 3′ hydroxyl group and so prevents further chain elongation when incorporated into a growing polynucleotide.

Dihybrid cross A sexual cross in which the inheritance of two pairs of alleles is followed.

Diploid Having two copies of each chromosome.

Direct repeats Two or more identical nucleotide sequences present in a single polynucleotide.

Discontinuous gene A gene in which the biological information is divided between two or more exons, separated by introns.

Disulphydryl bridge A particular type of covalent bond, incorporating two sulphur atoms, that can form between two cysteines in a single or on different polypeptides.

DNA Deoxyribonucleic acid, the genetic material.

DNA bending A topological state in which a DNA molecule is bent so that two distant sites on the double helix are brought close together.

DNA-binding protein Any protein that attaches to DNA as a part of its normal function, e.g. histones, RNA polymerase, *lac* repressor.

DNA gyrase A Type II topoisomerase found in *E. coli*.

DNA ligase An enzyme that repairs single-stranded discontinuities in double-stranded DNA molecules. In the cell, ligases are involved in DNA replication. Purified ligases are used in construction of recombinant DNA molecules.

DNA polymerase I The *E. coli* enzyme that completes synthesis of individual Okazaki fragments during DNA replication.

DNA polymerase III The main DNA replicating enzyme of *E. coli*.

DNA repair The correction of nucleotide errors introduced during DNA replication or resulting from the action of mutagenic agents.

DNA sequencing Determination of the order of nucleotides in a DNA molecule.

DNA topoisomerase An enzyme that introduces or removes turns from the double helix by transient breakage of one or both polynucleotides.

DNA tumour virus A virus containing a DNA genome and able to cause cancer after infection of an animal cell, e.g. SV40, adenovirus.

Dominant Refers to the allele that is expressed in a heterozygote.

Donor cell The cell that donates genes during a genetic cross between two bacteria.

Double helix The base-paired structure comprising two polynucleotides that is the natural form of DNA in the cell.

Double recombination Two recombination events between the same DNA molecules, which may result in insertion of a short linear molecule into a longer linear or circular molecule.

Downstream Towards the 3′-end of a polynucleotide.

Early genes Bacteriophage genes that are expressed during the early stages of infection of a bacterial cell; usually the products of early genes are involved in replication of the phage genome.

Electrophoresis Separation of molecules on the basis of their net electrical charge.

Elongation factor One of the proteins that play an ancillary role in the elongation step of translation.

Endonuclease An enzyme that breaks phosphodiester bonds within a nucleic acid molecule.

Enhancer A special type of eukaryotic regulatory sequence that can increase the rate of transcription of a gene located some distance away in either direction.

Episome A plasmid capable of integration into the host cell's chromosome.

Epistasis The ability of one gene to mask the phenotype derived from a second gene.

Euchromatin The regions of a eukaryotic chromosome that appear less condensed and stain less deeply with DNA-specific dyes.

Eukaryote An organism whose cells are characterized by the presence of membrane-bound nuclei.

Excision repair A repair system that is able to correct various types of DNA damage by excising a region of polynucleotide and resynthesizing the correct nucleotide sequence.

Exon One of the coding regions of a discontinuous gene.

Exon shuffling A theory that supposes exons to contain discrete subunits of genetic information that may be 'shuffled' to produce new functional genes.

Exonuclease An enzyme that sequentially removes nucleotides from the ends of a nucleic acid molecule.

Extrachromosomal gene Any gene that is not carried by the cell's chromosome(s). For example, genes present on mitochondrial or chloroplast genomes, genes carried by plasmids.

Extragenic DNA Any part of the genome that is not a gene, a decayed gene (e.g. pseudogenes) or a gene-related sequence (e.g. promoters, introns).

F^+ cell A bacterium that carries an F plasmid.

F^- cell A bacterium that does not carry an F plasmid.

F' cell A bacterium that carries a modified F plasmid, one which carries a small piece of DNA derived from the host bacterial

DNA molecule.

F-duction The transfer of bacterial genes from donor to recipient bacterium via an F′ plasmid, used as the basis to a gene mapping technique.

5′-P terminus The end of a polynucleotide that terminates with a mono-, di- or triphosphate attached to the 5′-carbon of the sugar.

fMet N-formylmethionine, the modified amino acid carried by the tRNA that initiates translation in bacteria.

F plasmid A fertility plasmid carrying genes that direct conjugal transfer of DNA between bacteria.

Frameshift mutation A mutation that results from insertion or deletion of a group of nucleotides that is not a multiple of three, and which therefore changes the frame in which the altered gene is translated.

G1 phase The first gap period of the cell cycle.

G2 phase The second gap period of the cell cycle.

Gamete A reproductive cell, usually carrying the haploid chromosome complement, that can fuse with a second gamete to produce a new cell during sexual reproduction.

Gap period One of two intermediate periods within the cell cycle. Gap 1 occurs between mitosis and the DNA synthesis phase, gap 2 between DNA synthesis and the next mitosis.

GC box A component of the nucleotide sequence that makes up the eukaryotic promoter.

Gel electrophoresis Electrophoresis performed in a gel matrix so that molecules of similar electrical charge can be separated on the basis of size.

Gel retardation analysis A technique that identifies a DNA fragment that has a bound protein by virtue of its decreased mobility during gel electrophoresis.

Gene A segment of DNA that contains biological information and codes for an RNA and/or polypeptide molecule.

Gene amplification The production of multiple copies of a DNA segment in order to increase the rate of expression of a gene carried by the segment.

Gene cloning Insertion of a fragment of DNA, containing a gene, into a cloning vector, and subsequent propagation of the recombinant DNA molecule in a host organism.

Gene expression The process by which the biological information carried by a gene is released and made available to the cell, through transcription possibly followed by translation.

Generalized recombination Recombination between two double-stranded DNA molecules which share extensive nucleotide sequence similarity.

Genetic code The rules that determine which triplet of nucleotides codes for which amino acid during translation.

Genetic engineering The use of experimental techniques to produce DNA molecules containing new genes or new combinations of genes.

Genetic marker An allele whose phenotype is easily recognizable and which can therefore be used to follow the inheritance of its gene during a genetic cross.

Genetic material The chemical material of which genes are made, now known to be DNA in most organisms, RNA in a few.

Genetics The branch of biology devoted to the study of genes.

Gene transfer The passage of a gene or group of genes from a donor to a recipient organism.

Genome The entire genetic complement of a cell.

Genomic library A collection of clones sufficeint in number to include all the genes of a particular organism.

Genotype A description of the genetic composition of an organism.

GT–AG rule Refers to the fact that with introns present in nuclear protein-coding genes, the first two nucleotides of the intron are GT and the last two AG.

Guanine A purine, one of the nucleotides found in DNA and RNA.

Half-life The time needed for half the atoms in a radioactive sample to decay.

Haploid Refers to a cell that contains a single copy of each chromosome.

Helicase The enzyme responsible, during DNA replication, for breaking the hydrogen bonds that hold the double helix together.

Helix-turn-helix A structural motif found in several DNA-binding proteins.

Helper virus A viral genome that enables a defective or modified virus to complete its infective cycle, by providing gene products for which the defective genome does not code.

Hemizygous Refers to a gene that is present in only one copy in a diploid cell, such as those genes on the Y chromosome.

Heredity The passage of characteristics from parents to offspring.

Heterochromatin The regions of a chromosome that appear relatively condensed and stain deeply with DNA-specific stains.

Heteroduplex A base-paired structure formed between two polynucleotides that are not entirely complementary.

Heterogeneous nuclear RNA (hnRNA) The nuclear RNA fraction that comprises the unprocessed transcripts synthesized by RNA polymerase II.

Heterologous probing The use of a labelled nucleic acid molecule to identify related molecules by hybridization probing.

Heteropolymer An artificial nucleic acid molecule made up of a mixture of different nucleotides.

Heterozygous Refers to a diploid cell or organism that contains two different alleles for a particular gene.

Hfr cell A bacterium whose DNA molecule contains an integrated copy of the F plasmid.

Histone One of the basic proteins that make up nucleosomes and have a fundamental role in chromosome structure.

Hogness box A nucleotide sequence that makes up part of the eukaryotic promoter.

Holliday structure An intermediate structure believed to be formed during recombination between two DNA molecules.

Holoenzyme The version of the *E. coli* RNA polymerase that has the subunit composition $\alpha_2\beta\beta'\sigma$ and is involved in efficient recognition of promoter sequences.

Homeobox A conserved sequence element found in a number of genes believed to be involved in development of eukaryotic organisms.

Homeotic gene A gene, mutations in which result in the transformation of one body part into another.

Homologous chromosomes Two or more identical chromosomes.

Homologous genes Two genes, from different organisms and therefore of different sequence, that code for the same gene product.

Homologous recombination Recombination between two double-stranded DNA molecules which share extensive nucleotide sequence similarity.

Homopolymer An artificial nucleic acid molecule comprising just one nucleotide.

Homozygous Refers to a diploid cell or organism that contains two identical alleles for a particular gene.

Housekeeping genes The genes whose products are required by the cell at all times.

Hybrid-arrest translation (HART) A method used to identify the polypeptide coded by a cloned gene.

Hybridization probing A method that uses a labelled nucleic acid molecule to identify complementary or homologous molecules through the formation of stable base-paired hybrids.

Hybrid-release translation (HRT) A method used to identify the polypeptide coded by a cloned gene.

Hydrogen bond A type of chemical bond that is relatively weak but is important in stabilizing the higher orders of structure of many biomolecules.

Illegitimate recombination Recombination between two double-stranded DNA molecules which have little or no nucleotide sequence similarity.

Immunoscreening The use of an antibody to detect a polypeptide synthesized by a cloned gene.

Incompatibility The inability of certain types of plasmid to coexist in the same cell.

Incompatibility group Comprises a group of different plasmids that are unable to coexist in the same cell.

Incomplete dominance The situation where neither of a pair of alleles displays dominance and the phenotype of the heterozygote is intermediate between the phenotypes of the two alternative homozygotes.

Inducer A molecule that induces expression of a gene or operon by binding to a repressor protein and thereby preventing the repressor from attaching to its operator site.

Induction A chemical or physical treatment which results in excision from the host genome of the integrated form of a lysogenic phage, followed by the switch to the lytic mode of infection.

Initiation codon The codon, usually but not exclusively 5′-AUG-3′, which indicates the point at which translation of an mRNA should begin.

Initiation complex The complex which comprises mRNA, a small ribosomal subunit, aminoacylated initiator-tRNA and initiation factors and which forms during the initiation stage of translation.

Initiation factor One of the protein molecules that play an ancillary role in the initiation stage of translation.

Insertion mutation Alteration of the sequence of a DNA molecule by insertion of one or more nucleotides.

Insertion sequence One of a group of relatively small transposable elements found in bacteria.

Intercalating agent A chemical compound which is able to invade the space between adjacent base pairs of a double-stranded DNA molecule, often causing mutations.

Interphase The period between cell divisions.

Interrupted mating The artificial cessation of conjugation between bacteria, often brought about by vigorous shaking, used as part of the procedure for conjugation mapping of bacterial genes.

Intron In a discontinuous gene, one of the segments that does not contain biological information.

Inversion mutation Alteration of the sequence of a DNA molecule by removal of a segment followed by its reinsertion in the opposite orientation.

Inverted repeat Two identical nucleotide sequences repeated in opposite orientations in a DNA molecule, possibly adjacent to one another or possibly some distance apart.

***In vitro* mutagenesis** Any one of several techniques used to produce a specified mutation at a predetermined position in a DNA molecule.

***In vitro* packaging** Synthesis of infective λ particles from a preparation of λ capsid proteins and a concatamer of DNA molecules separated by *cos* sites.

Isoacceptors Two or more tRNA molecules that are specific for the same amino acid.

Isotope One of two or more atoms that have the same atomic number (and are therefore the same element) but have different atomic weights.

Karyotype The entire chromosome complement of a cell, with each chromosome described in terms of its appearance at metaphase.

Kinetochore The part of the centromere to which microtubules of the spindle apparatus attach.

***Lac* operon** The cluster of three structural genes that code for enzymes involved in utilization of lactose by *E. coli*.

Lactose repressor The regulatory protein that controls transcription of the *lac* operon in response to the levels of lactose in the environment.

Lagging strand The strand of the double helix which, during DNA replication, is copied in a discontinuous fashion.

Late genes Bacteriophage genes that are expressed during the later stages of the infection cycle. Late genes usually code for proteins needed for synthesis of new phage particles.

Latent period The period between injection of a phage genome into a bacterial cell and the time when cell lysis occurs.

Leader segment The untranslated segment that lies upstream of the initiation codon on an mRNA molecule.

Leading strand The strand of the double helix which, during DNA replication, is copied in a continuous fashion.

Leaky mutation A mutation that results in only partial loss of a characteristic.

Lethal mutation A mutation that results in a cell or organism that is unable to survive.

Leucine zipper A structural motif found in several DNA-binding proteins.

LINE (long interspersed nuclear element) A type of dispersed repetitive DNA, often with transposable activity.

LINE-1 One of the best characterized human LINEs.

Linkage The physical association between two genes that results from them being present on the same chromosome.

Linkage group A group of genes that display linkage. With eukaryotes a single linkage group usually corresponds to a single chromosome.

Linker DNA The DNA that links nucleosomes together and which makes up the 'string' in the beads-on-a-string model for chromatin structure.

Lysogenic The pattern of bacteriophage infection that involves integration of the phage genome into the host DNA molecule.

Lysozyme An enzyme that weakens the cell walls of certain types of bacteria, and is synthesized by several types of phage in order to burst the host cell at the end of the infection cycle.

Lytic The pattern of infection displayed by a bacteriophage that lyses the host cell immediately after the initial infection. Integration of the phage DNA molecule into the host genome does not occur.

Macromolecule In general terms, any large molecule. In biology, more specifically polymeric compounds such as nucleic acids, proteins and polysaccharides.

Major groove The larger of the two grooves that spiral around the surface of the double helix model for DNA.

Major histocompatibility complex A mammalian multigene family that codes for a series of cell surface proteins. Several members of the gene family are highly polymorphic and have been used extensively in population studies.

Map unit A unit used to describe the distance between two genes on a chromosome, now superseded by centiMorgan.

Marker *See* **Genetic marker, Radioactive marker**.

Mating type The equivalent of gender for a eukaryotic micro-organism. Gametes of opposite mating type fuse to initiate sexual reproduction.

Maturation promoting factor A protein in *Saccharomyces cerevisiae* that promotes mitosis and is required for passage of the cell through the start point of the cell cycle.

Meiosis The series of events, involving two cell divisions, by which diploid cells are converted to haploid cells.

Melting Denaturation of a double-stranded DNA molecule into non-base-paired polynucleotides, possibly although not exclusively by heating.

Messenger RNA (mRNA) A transcript of a protein-coding gene.

Metaphase chromosome A chromosome at the metaphase stage of cell division, when the structure is at its most organized and features such as the banding pattern can be visualized.

Minimal medium A medium that provides only the minimum number of different nutrients needed for growth of a particular bacterium.

Minor groove The smaller of the two grooves that spiral around the surface of the double helix model for DNA.

−25 box A component of the nucleotide sequence that makes up the prokaryotic promoter.

Mismatch DNA repair A DNA repair system that is able to recognize mismatched nucleotide pairs and replace the incorrect nucleotide on the daughter polynucleotide.

Missense mutation An alteration in a nucleotide sequence that converts a codon specifying one amino acid into a codon for a second amino acid.

Mitosis The series of events that result in division of a single cell into two daughter cells.

Model building An experimental approach in which possible structures of biological molecules are assessed by building scale models of them.

Molecular biology An ill-defined term now used to describe the branch of biology devoted to the study of the molecular nature of the gene and its biochemical reactions, such as transcription and translation. According to Chargaff, anything published in the *Journal of Molecular Biology*.

Molecular genetics A synonym for 'Genetics', used to emphasize that genetics involves the study not only of heredity but also of the molecular nature of genes and gene expression.

Monohybrid cross A sexual cross in which the inheritance of just a single pair of alleles is followed.

Monomer One of the structural units that are joined together to form a polymer.

Mosaic gene A discontinuous gene.

M phase The period of the cell cycle when mitosis or meiosis occurs.

Multigene family A group of genes, possibly although not always clustered, that are related either in nucleotide sequence or in terms of function.

Multiple alleles The different alternative states of a gene that has more than two alleles.

Mutagen A chemical or physical agent able to cause a mutation in a DNA molecule.

Mutagenesis Experimental treatment of a group of cells or organisms with a mutagen in order to induce mutations.

Mutant A cell or organism with an abnormal genetic constitution.

Mutation An alteration in the nucleotide sequence of a DNA molecule.

Nitrogenous base One of the purine or pyrimidine compounds that form part of the molecular structure of a nucleotide.

Non-homologous recombination Recombination between two double-stranded DNA molecules which have little or no nucleotide sequence similarity.

Non-permissive host An animal species whose cells may give rise to a tumour when infected with a particular virus.

Non-polar A hydrophobic (water-hating) chemical group.

Nonsense mutation An alteration in a nucleotide sequence that converts a triplet coding for an amino acid into a termination codon.

Non-viral retroelement A mobile DNA element that transposes via an RNA intermediate that is copied into DNA by reverse transcription. Some non-viral retroelements are similar to retroviruses but lack genes for capsid proteins.

Nuclease An enzyme that degrades a nucleic acid molecule.

Nucleic acid Originally, the acidic chemical compound isolated from the nuclei of eukaryotic cells. Now, the polymeric molecules comprising nucleotide monomers: DNA and RNA.

Nucleic acid hybridization Formation of a double-stranded molecule by base pairing between complementary polynucleotides.

Nucleoid The DNA-containing region within a prokaryotic cell.

Nucleolus The region of the nucleus in which rRNA transcription occurs.

Nucleoplasm A general name for the complex mixture of molecules that comprises the ground substance of the nucleus of a living cell.

Nucleoside A chemical compound comprising a purine or pyrimidine base attached to a five-carbon sugar.

Nucleosome The structure comprising histone proteins and DNA that is the basic organizational unit in chromatin structure.

Nucleotide A chemical compound comprising a purine or pyrimidine base attached to a five-carbon sugar, to which a mono-, di-, or triphosphate is also attached. The monomeric unit of DNA and RNA.

Nucleus The membrane-bound structure of a eukaryotic cell within which the chromosomes are contained.

Okazaki fragment One of the short segments of RNA-primed DNA that are synthesized during replication of the lagging strand of the double helix.

Oligonucleotide-directed mutagenesis An *in vitro* mutagenesis technique that involves the use of a synthetic oligonucleotide to introduce a predetermined nucleotide alteration into the gene to be mutated.

Oncogene A gene which when carried by a retrovirus is able to cause cell transformation.

Open promoter complex The complex formed between *E. coli* RNA polymerase and a promoter, in which the double helix is partially unwound in readiness for the start of RNA synthesis.

Open reading frame A series of codons with an initiation codon at the 5′-end. Often considered synonymous with 'gene' but more properly used to describe a DNA sequence which looks like a gene but to which no function has been assigned.

Operator A nucleotide sequence element to which a repressor protein attaches in order to prevent transcription of a gene or operon.

Origin of replication A site on a DNA molecule where unwinding begins in order for replication to occur.

Overlapping genes Two genes whose coding regions overlap, either completely or partially.

Pararetrovirus A viral retroelement whose encapsidated genome is made of DNA.

Partial linkage The property usually displayed by two genes on the same chromosome in a eukaryote, which are not always inherited together because of the possibility of recombination between them.

pBR322 A particular type of artificial plasmid, frequently used as a vector for cloning genes in *E. coli*.

Penetrance The extent to which a phenotype is expressed, measured as the proportion of individuals with a particular genotype that actually display the associated phenotype.

Peptide bond The chemical bond that links adjacent amino acids in a polypeptide.

Peptidyl- or P-site The site on the ribosome to which the tRNA attached to the growing polypeptide is bound during translation.

Peptidyl transferase The enzyme activity responsible for peptide bond synthesis during translation.

Phenotype The observable characteristics displayed by a cell or organism.

Phosphodiester bond The chemical bond that links adjacent nucleotides in a polynucleotide.

Photoreactivation A DNA repair system in which thymine dimers are corrected by a light-activated enzyme.

Plaque A zone of clearing on a lawn of bacteria caused by lysis of the cells by infecting phage particles.

Plasmid A usually circular piece of DNA, primarily independent of the host chromosome, often found in bacterial and some other types of cell.

Plasmid amplification A method involving incubation with an inhibitor of protein synthesis aimed at increasing the copy number of certain types of plasmid in a bacterial culture.

Pleiotropy The ability of a single gene to produce a complex phenotype that consists of two or more distinct characteristics.

Point mutation A mutation that results from a single nucleotide alteration in a DNA molecule.

Polar A hydrophilic (water-loving) chemical group.

Polar mutation A mutation in one gene that affects the expression of other, downstream genes.

Polyadenylation The post-transcriptional addition of a series of A residues to the 3′-end of a eukaryotic mRNA molecule.

Poly(A) polymerase The enzyme responsible for polyadenylation of a eukaryotic mRNA molecule.

Polymer A chemical compound constructed as a long chain of identical or similar units.

Polymerase chain reaction (PCR) A technique that enables multiple copies of a DNA molecule to be generated by enzymatic amplification of a target DNA sequence.

Polymorphism A variable DNA sequence, one that can exist in a number of different, although related forms.

Polynucleotide A polymer consisting of nucleotide units.

Polynucleotide phosphorylase An enzyme able to synthesize polynucleotides from nucleotide subunits without the need for a template. Used for the synthesis of artificial messengers during elucidation of the genetic code.

Polyprotein A translation product that consists of a series of proteins joined into a single polypeptide, and which is processed by proteolytic cleavage to produce the mature proteins.

Polysome An mRNA molecule in the process of being translated by several ribosomes at once.

Pre-mRNA The primary, unprocessed transcript of a protein-coding gene.

Prepriming complex The structure, consisting of a series of proteins attached to the origin of replication, which initiates DNA replication in *E. coli*.

Pre-rRNA The primary, unprocessed transcript of a gene or group of genes specifying rRNA molecules.

Pre-tRNA The primary, unprocessed transcript of a gene or group of genes specifying tRNA molecules.

Pribnow box A component of the nucleotide sequence that makes up the prokaryotic promoter.

Primary transcript The immediate product of transcription of a gene or group of genes, which will subsequently be processed to give the mature transcript(s).

Primase The RNA polymerase enzyme that synthesizes the primer needed to initiate replication of a DNA polynucleotide.

Primer A short oligonucleotide that is attached to a single-stranded DNA molecule in order to provide a site at which DNA replication can begin.

Primosome The structure, comprising several different proteins, that is responsible for initiation of DNA replication, including initiation of synthesis of Okazaki fragments.

Prion An unusual infectious agent that appears to consist purely of protein with no nucleic acid.

Processed pseudogene A pseudogene whose sequence resembles the mRNA copy of a parent gene, and which probably arose by integration into the genome of a reverse transcribed version of the mRNA.

Proflavin A chemical mutagen, one of the acridine dyes, frequently used to induce mutations for experimental purposes.

Prokaryote An organism whose cells are characterized by the absence of a distinct nucleus and a general lack of membranous architecture.

Promiscuous DNA A DNA sequence derived from the chloroplast genome but present in the mitochondrial DNA of a photosynthetic organism.

Promoter The nucleotide sequence, upstream of a gene, to which RNA polymerase binds in order to initiate transcription.

Proofreading The ability of a DNA polymerase to correct misincorporated nucleotides as a result of its 3′ to 5′ exonuclease activity.

Prophage The integrated form of the DNA molecule of a lysogenic phage.

Protease An enzyme that degrades protein.

Protein The polymeric compounds that are made up from amino acid monomers.

Protomer One of the individual polypeptide subunits that combine to make the protein coat of a virus.

Protoplast A cell from which the cell wall has been completely removed.

Prototroph An organism that has no nutritional requirements beyond those of the wild-type and is therefore able to grow on minimal medium.

Provirus The DNA copy of a retroviral genome that is integrated into host chromosomal DNA.

Pseudogene A nucleotide sequence that has similarity to a functional gene but within which the biological information has become scrambled so that the pseudogene is not itself functional.

Punctuation codon A codon that designates either the start or the end of a gene.

Punnett square A tabular treatment used to predict the genotypes of the progeny resulting from a genetic cross in which a number of alleles are followed.

Purine One of the two types of nitrogenous base compounds that are components of nucleotides.

Pyrimidine One of the two types of nitrogenous base compounds that are components of nucleotides.

Quantitative phenotype A phenotype, such as height, which has a continuous distribution pattern within a population, and which is typically determined by the combined effects of a number of genes.

Radioactive marker A radioactive atom which is incorporated into a chemical compound and whose radioactive emissions can be used to detect and follow the compound during a biochemical reaction.

Radiolabelling The technique by which a radioactive atom is attached to a chemical molecule or macromolecule, e.g. incorporation of ^{32}P-dATP into DNA.

Reading frame One of the three overlapping sequences of triplet codons that are contained in any DNA sequence.

Recessive Refers to the allele whose phenotype is not expressed in a heterozygote.

Recipient cell The cell that receives DNA during gene transfer between bacteria.

Recombinant A cell, derived from a genetic cross, that displays neither of the parental combinations for the alleles under study.

Recombinant DNA molecule A DNA molecule created in the test-tube by ligating together pieces of DNA that are not normally contiguous.

Recombinant DNA technology All the techniques involved in the construction, study and use of recombinant DNA molecules.

Recombinant protein A polypeptide that is synthesized in a recombinant cell as the result of expression of a cloned gene.

Recombination A physical process that can lead to exchange of segments of polynucleotides between two DNA molecules and which can result in the progeny of a genetic cross possessing combinations of alleles not displayed by either parent.

Recombination frequency The proportion of recombinant progeny in the total progeny from a genetic cross.

Regulatory gene A gene that codes for a protein, such as a repressor, involved in regulation of expression of other genes.

Relaxed plasmid A plasmid whose replication is not linked to replication of the host genome, and which therefore can exist as multiple copies within the cell.

Release factor One of the protein molecules that play ancillary roles in termination of translation.

Renaturation The return of a denatured molecule back to its natural state.

Repetitive DNA A DNA sequence that is repeated a number of times in a DNA molecule or in a genome.

Replica-plating A technique whereby the colonies on an agar plate are transferred *en masse* to a new plate, on which colonies will grow in the same relative positions as before.

Replication fork The region of a double-stranded DNA molecule that is being unwound to enable DNA replication to occur.

Replication origin A site on a DNA molecule where unwinding begins in order for replication to occur.

Reporter gene A gene whose phenotype can be assayed in a transformed organism, and which can be used in, for example, deletion analysis of regulatory regions.

Restriction endonuclease An enzyme that cuts DNA molecules only at a limited number of specific nucleotide sequences.

Retrovirus A viral retroelement whose encapsidated genome is made of RNA.

Reverse transcriptase An enzyme that synthesizes a DNA copy of an RNA template.

Rho A protein that is required for termination of some prokaryotic transcripts.

Rho-dependent terminator A nucleotide sequence that acts in conjunction with rho to cause termination of transcription in prokaryotes.

Rho-independent terminator A nucleotide sequence that acts independently of rho to cause termination of transcription in prokaryotes.

Ribonuclease An enzyme that degrades RNA.

Ribosomal RNA (rRNA) The RNA molecules that act as structural components of ribosomes.

Ribosome One of the protein–RNA structures on which translation occurs.

Ribosome binding site The nucleotide sequence that acts as the site for attachment of a ribosome to an mRNA molecule.

Ribozyme An RNA molecule that possesses catalytic activity.

RNA Ribonucleic acid, one of the two forms of nucleic acid in living cells.

RNA editing A process by which nucleotides not coded by the gene are introduced at specific positions in an mRNA molecule after transcription.

RNA polymerase An enzyme capable of synthesizing an RNA copy of a DNA template.

RNase D The enzyme responsible for processing pre-tRNA by cleaving at the 3′-termini of the mature tRNA sequences.

RNase P The enzyme responsible for processing pre-tRNA by cleaving at the 5′-termini of the mature tRNA sequences.

RNA transcript An RNA copy of a gene.

Rolling-circle replication A strategy that enables a large number of copies of a circular DNA molecule to be synthesized in a short time.

Satellite DNA DNA comprising clustered repetitive sequences, so-called because it forms a satellite band in a density gradient.

Satellite virus A small viral genome that moves from cell to cell within the capsid of a second virus.

Scanning A proposed system for the initiation of eukaryotic translation, in which the ribosome attaches to the 5′-terminal cap structure of the mRNA and then scans along the molecule until it reaches an initiation codon.

Second site reversion A second mutation that reverses the effect of a previous mutation in the same gene although without restoring the original nucleotide sequence.

Sedimentation coefficient A value used to express the velocity with which a molecule or structure sediments when centrifuged in a dense solution.

Segregation The separation of homologous chromosomes, or members of allele pairs, into different gametes during meiosis.

Selective medium A medium that supports the growth of only cells that carry a certain genetic marker.

Self-fertilization Fertilization of a female gamete with a male gamete derived from the same individual.

Selfish DNA DNA that appears to have no function and apparently contributes nothing to the cell in which it is found.

Semi-conservative replication The mode of DNA replication in which each daughter double helix comprises one polynucleotide from the parent and one newly synthesized polynucleotide.

Serotype A strain of bacterium that is distinguished from other serotypes of the same species by virtue of its different immunological characteristics.

Sex cell A cell that divides by meiosis.

Sex chromosome A chromosome which is involved in sex determination.

Sex-linked Refers to a gene that is located on a sex chromosome, or a phenotype resulting from such a gene.

Sex pilus A tube structure present on the exterior of F^+ bacteria, thought to be involved in transfer of DNA during conjugation.

Shine–Dalgarno sequence The *E. coli* ribosome binding site.

Silent mutation An alteration in a DNA sequence which does not affect the expression or functioning of any gene or gene product.

SINE (short interspersed nuclear element) A type of dispersed repetitive DNA, typified by the *Alu* sequences found in the human genome.

Single-strand binding protein One of the proteins that attaches to single-stranded DNA in the region of the replication fork, preventing reannealing of unreplicated DNA.

Site-specific recombination Recombination between two double-stranded DNA molecules which have only short regions of nucleotide sequence similarity.

Small nuclear ribonucleoprotein (snRNP) The nuclear particles that comprise one or two snRNA molecules complexed with proteins.

Small nuclear RNA (snRNA) The RNA component of the nucleus that comprises relatively small molecules thought to be involved in splicing and other transcript processing events.

Somatic cell A cell that is not a sex cell, i.e. one that divides by mitosis.

S1 nuclease An enzyme that degrades specifically single-stranded molecules or single-stranded regions in predominantly double-stranded nucleic acid molecules.

SOS response The response of *E. coli* to DNA damage induced by ultraviolet irradiation and other agents, involving synthesis of repair systems and other enzymes.

Specialized transduction The result of incorrect excision of a prophage from the host genome, so that the new phage particles contain phage DNA plus some bacterial genes, which will subsequently be transferred to a new bacterial cell during the next round of infection.

S phase The period of the cell cycle when DNA synthesis occurs.

Spliceosome The protein–RNA structure believed to be responsible for splicing.

Splicing The removal of introns from the primary transcript of a discontinuous gene.

Split gene A discontinuous gene.

Spontaneous mutation rate Refers to mutations that occur naturally and in the absence of an artificial mutagenic treatment.

Stable RNA RNA molecules, such as rRNA and tRNA but not mRNA, which are not subject to rapid turnover in the cell.

Stem-loop structure A structure comprising a base-paired stem and non-base-paired loop, which can form in a single-stranded polynucleotide that contains an inverted repeat.

Sticky end An end of a double-stranded DNA molecule where there is a single-stranded extension.

Stringent plasmid A plasmid present at a low copy number in the cell.

Structural gene A gene that codes for an RNA molecule or protein other than a regulatory protein.

Supercoiling A conformational state in which a double helix is overwound or underwound so that superhelical coiling is introduced.

Suppression A mutation in a gene that reverses the effect of a previous mutation in a different gene.

S value The unit of measurement for a sedimentation coefficient.

Tandem repeat Direct repeats that are adjacent to each other with no intervening DNA.

TATA box A component of the nucleotide sequence that makes up the prokaryotic promoter.

Tautomeric shift The spontaneous change of a molecule from one structural isomer to another.

Telomerase The enzyme that maintains the ends of chromosomes by synthesizing telomeric repeat sequences.

Telomere The end of a chromosome.

Temperate phage A bacteriophage that is able to follow a lysogenic mode of infection.

Temperature-sensitive mutation A type of conditional-lethal mutation, one which is expressed only above a certain threshold temperature.

Template strand The polynucleotide of a gene that acts as the template for RNA synthesis during transcription.

Termination codon One of the three codons (5′-UAA-3′, 5′-UAG-3′ and 5′-UGA-3′ in the standard genetic code) that mark the position where translation of an mRNA should stop.

Testcross A genetic cross between a heterozygote and the recessive homozygote.

Tetrad The four haploid cells that result from a single meiotic event in a eukaryotic microorganism.

θ-form An intermediate in certain modes of DNA replication.

30 nm chromatin fibre The substructure of chromatin that consists of a possibly helical array of nucleosomes in a fibre approximately 30 nm in diameter.

3′-OH terminus The end of a polynucleotide which terminates with a hydroxyl group attached to the 3′-carbon of the sugar.

Thymine A pyrimidine, one of the nitrogenous bases found in DNA.

Ti plasmid The large plasmid found in *Agrobacterium tumefaciens* cells and used as the basis for a series of cloning vectors for higher plants.

***Tra* gene** One of a group of genes carried by an F plasmid and coding for proteins involved in DNA transfer during conjugation.

Trailer segment The untranslated region that lies downstream from the termination codon of an mRNA molecule.

Transcript An RNA copy of a gene.

Transcript analysis Experiments designed to determine which regions of a DNA molecule are transcribed into RNA.

Transcription The synthesis of an RNA copy of a gene.

Transcription factor A protein that plays an ancillary role in initiation of a eukaryotic transcript.

Transduction The transfer of bacterial genes from one cell to another by accidental packaging into a phage particle during infection.

Transduction mapping The use of transduction to determine the relative positions of two or more genes on a bacterial DNA molecule.

Transfection The introduction of purified bacteriophage DNA molecules into a bacterial cell.

Transfer RNA (tRNA) One of the small RNA molecules that act as adaptors during translation and are responsible for decoding the genetic code.

Transformant A cell that has become transformed by the uptake of naked DNA.

Transformation The acquisition by a cell of new genes by the uptake of naked DNA.

Transformation mapping The use of transformation to determine the relative positions of two or more genes on a bacterial DNA molecule.

Transforming principle The chemical substance responsible for transforming *Streptococcus* bacteria from one serotype to another, shown by Avery and colleagues to be DNA.

Transition A point mutation that results in a purine being replaced by another purine or a pyrimidine by a pyrimidine.

Translation The synthesis of a polypeptide, the amino acid sequence of which is determined by the nucleotide sequence of an mRNA in accordance with the rules of the genetic code.

Translocation The movement of a ribosome from one codon to the next on an mRNA molecule during translation.

Transposable element A genetic element that is able to move from one site to another in a DNA molecule.

Transposase The enzyme that catalyses transposition of a transposable genetic element.

Transposition The movement of a genetic element from one site to another in a DNA molecule.

Transversion A point mutation that results in a purine being replaced by a pyrimidine, or vice versa.

Triplet binding assay An experimental technique that enables the coding specificity of a triplet of nucleotides to be determined.

tRNA deacylase The enzyme responsible for cleaving the bond between a tRNA molecule and the growing polypeptide during translation.

tRNA nucleotidyl transferase The enzyme responsible for the post-transcriptional attachment of the nucleotide sequence 5′-CCA-3′ to the 3′-end of a tRNA molecule.

Turnover rate The rate at which a population of molecules is degraded in the cell.

Ultracentrifuge A machine used to subject samples to a high gravitational force by spinning them at a very high speed.

Ultraviolet spectroscopy The determination of the amount of ultraviolet radiation absorbed by a sample, used to identify chemical compounds and to measure the concentrations of nucleic acid solutions.

Unit factor The term used by Mendel for the gene.

Upstream Towards the 5′-end of a polynucleotide.

Uracil A pyrimidine, one of the nitrogenous bases found in RNA.

U-RNA A nuclear RNA molecule involved in splicing and other transcript processing events.

U-RNP A nuclear particle, consisting of one or two U-RNAs and several proteins, involved in splicing and other transcript processing events.

Vegetative cell A cell that is not a sex cell, i.e. one that divides by mitosis.

Velocity sedimentation analysis A method that measures the mass of a molecule or particle by determining the rate at which it sediments through a dense solution when subjected to a high centrifugal force.

Viral retroelement A virus whose genome replication process involves reverse transcription.

Virion The genome of a virus.

V-*onc* The version of an oncogene carried by a transforming retrovirus.

Virulent Refers to a bacteriophage that follows the lytic mode of infection.

Virus An infective particle, composed of protein and nucleic acid, that must parasitize a host cell in order to replicate.

Wild-type Refers to a gene, cell or organism that displays the typical phenotype and/or genotype for the species and is therefore adopted as a standard.

Wobble hypothesis The hypothesis proposed by Crick in 1965 to explain how a single tRNA carrying just one anticodon can decode more than one codon.

X-ray diffraction analysis A method for determining the structure of a molecule by analysing the angles at which X-rays are deflected by a crystal of the compound.

X-ray diffraction pattern The pattern of spots on an X-ray sensitive film that is produced by passing X-rays through a crystal.

Yeast artificial chromosome (YAC) A cloning vector comprising the structural components of a yeast chromosome and able to clone very large pieces of DNA.

Yeast episomal plasmid (YEp) A yeast vector carrying the $2\,\mu m$ plasmid origin of replication.

Yeast integrative plasmid (YIp) A yeast vector that relies on integration into a host chromosome for replication.

Z-DNA A structural form of DNA in which the two polynucleotides are wound into a left-handed helix.

Zinc finger A structural motif found in several DNA-binding proteins.

Zygote The cell that results from fusion of gametes during meiosis.

Index

ABBREVIATIONS: 'B' after an entry indicates a Box, 'F' a Figure, 'G' a definition in the Glossary, and 'T' a Table.

Abortive infection 365, 425G
Acceptor site 142, 142F
Acridine dye 121, 425G
Acrocentric 273B
Activation site 166, 425G
Acute transforming virus 240, 240F
Adenine 27F, 29, 425G
Adenylate cyclase 161, 162F, 425G
A-form 38, 38B, 425G
Agarose gel electrophoresis 19B
Agrobacterium tumefaciens 392–3
AIDS virus 212, 238–40
Alkaline phosphatase 396B
Allele 299F, 300, 314, 425G
Alpha DNA 274, 274F 295, 425G
α-helix 116–17, 117F, 425G
Alu family 295, 425G
Alu RNA 77B
Amino acid 13, 114–15, 114F, 114T, 115F, 118B, 425G
Amino acid sequence analysis 118B
Aminoacylation 133–5, 134F, 135B, 135F, 425G
Aminoacyl-site 142, 142F, 425G
Aminoacyl-tRNA synthetase 133–5, 135B, 135F, 426G
Ancient DNA 304–5, 305F, 426G
Ångström 36B
antennapaedia 169F, 170
Antibiotic resistant mutant 196F, 197
Antibody
  in serotype testing 15B
Anticodon 87F, 88, 88F, 426G
Apolipoprotein-B 110
Archaea (archaebacteria)
  distinctive features 245
  genes 250–1
  introns 51
  types 250B
Artificial chromosome 390–2
Ascospore 338F, 339, 426G
Ascus 338F, 339, 426G
A-site 142, 142F, 425G
Astrachan, L. 95
ATP 32
Attenuation 164B, 426G
Autoradiography 398, 399F, 405, 426G
Auxotrophic mutant 196F, 197, 426G
Avery, O. 17–18, 17B, 19B, 41B

Bacillus megaterium 250
Bacillus subtilis 249F, 250, 372B
Back mutation 199, 426G
Bacterial conjugation
  discovery 254
  gene transfer 255–7, 255F, 256F
  kinetics of transfer 362B
  mapping 359–63
  plasmids 252, 254–5
  species other than *E. coli* 257
Bacterial sex 253–7
Bacterial transformation 14–16, 426G
Bacteriophage
  cloning vectors 386–8
  DNA replication 188
  gene expression 230–3
  gene organization 227–9
  genes are DNA 20–3, 41B
  genomes 221–2, 221T
  life cycles 222–7
  sizes 219F
  types 220–1, 220F
  use by Delbrück 9, 9B, 10B
Bacteriophage D108 262
Bacteriophage λ
  cloning vectors 386F, 387–8, 387F, 388B
  DNA sequence 406
  gene organization 227, 228F
  gene regulation 231–3
  genes 221T, 229, 229F
  integration into bacterial DNA 224B
  lysogenic infection cycle 222–6, 225F
  replication of genome 226B
  structure 221, 221T
Bacteriophage MS2 220, 221T
Bacteriophage M13
  cloning vectors 388, 388F
  infection cycle 226–7, 227F
  structure 221, 221T
Bacteriophage Mu 261F, 262
Bacteriophage ϕX174 228, 228B, 228F, 405–6
Bacteriophage PM2 220, 221T
Bacteriophage SPO1 221, 221T
Bacteriophage T2 20–3
Bacteriophage T4
  gene regulation 230–1
  genes 221T, 229, 229F
  lytic infection cycle 222, 223F
  structure 221, 221T
Bal31 396B
Baltimore, D. 236, 237B
Base analogue 201F, 203, 426G
Base pair
  unit of length 38B
  Watson–Crick structures 39F

Base pairing
G–U base pair 136F, 137
importance in genetics 39–40
in double helix 39–40, 39F
inosine base pairs 137F, 138
Base ratio 33–4, 33T, 41B
Bateson, W. 10B, 328, 350
Beadle, G.W. 10B
Beads-on-a-string structure 269–70, 270F, 426G
Berg, P. 389B
β-lactamase 383
β-sheet 117, 117F, 426G
B-form 38, 38B, 427G
bicoid 170F
Biological containment 389B, 427G
Biotechnology 389, 414–16, 427G
Blood groups 322–4, 323F, 324B, 324F
Blunt end 379, 380F, 427G
Bohr, N. 8, 8B, 10B
Bovine papillomavirus 393
Bovine spongiform encephalopathy 241B
Breener, S. 95, 95B, 121, 122–3
Bridges, C. 7
5-bromouracil 201F, 203
Buoyant density 19B, 427G
Burst size 224B

CAAT box 166, 427G
Caesium chloride 19B
Cancer 240–2
CAP 161, 162F
Capping 98–100, 99F, 427G
CAP site 161, 161F
Cap struture 99, 99F, 427G
Capsules 14, 14F, 15B, 427G
Cascade system 230, 427G
Catabolite activator protein 161, 162F, 427G
Catabolite repression 161, 162F, 427G
cDNA library 402B, 427G
Cech, T. 107, 107B
Cell culture 59B
Cell cycle
control 281
definition 277–8
events 278–81
Cell-free protein synthesis 123, 124F, 428G
Cell structure 57–9, 58F
Cell transformation 235–6, 235F, 428G
CENP-B 274, 274B, 274F
CentiMorgan 337, 428G
Centromere 271, 273–4, 273B, 273F, 274B, 274F, 428G
C-form 38, 38B
Chain termination DNA sequencing 403–5
Chargaff, E. 33–4, 33B, 41B
Charging 133–5, 134F, 135B, 135F
Chi form 208F, 209, 428G
Chromatography 34
Chase, M. 22–3, 41B
Chloramphenicol 144B, 383
Chloroplast DNA 265F, 282, 283T, 286–7, 286F
Chromatid 271, 272F, 428G
Chromatin 268
Chromosome
bacterial 248B
carry genes 6–7, 10B
DNA packaging 268–71
during cell cycle 277–81
human 272F, 292F, 291T
location 265F
made of DNA and protein 12–13
morphology 271–7
number of chromosomes 265–8, 266T
number of genes 46B
Chromosome walking 400–2, 401F, 428G
Cis-trans complementation test 371B, 428G
Classical satellite DNA 295–6
Clone 376, 376F, 428G
Clone library 49B, 429G
Cloning vector
animals 393
artificial chromosome 390–2
bacteriophage 386–8
constructing recombinant DNA molecules 378–82
cosmid 389B
eukaryotic 389–92
lacZ' vectors 385–6
outline 49B, 375, 376F
pBR322 383–5
plants 392–3
pUC8 385–6, 385F, 386F
S. cerevisiae 390–2
Closed promoter complex 66, 66F, 429G
Cloverleaf structure 87–8, 87F, 429G
Clustered repetitive DNA 295–6
Codominance 322–3, 323F, 429G
Codon
definition 121, 429G
non-standard codons 127–9
table 127T
Codon recognition 135–8, 137F
Cohesive end 380, 380F, 429G
Colinearity 121, 121F, 429G
Colony hybridization probing 397–8, 397F
Col plasmid 252, 252T
Competent cell 368, 369F, 429G
Complementation text 371B
Complex phenotype 348–52
Composite transposon 261, 261F, 261T
Conditional-lethal mutant 197, 429G
cI repressor 231–2, 231F, 233F, 429G
Conjugation, *see bacterial conjugation*
Conjugation mapping 359–63, 429G
Consensus sequence 64, 429G
Constitutive expression 159B
Constitutive mutant 197–8, 430G
Contrasting characteristics 309–11, 310T
Conventional pseudogene 293, 294F, 430G
Copy number 257–8, 430G
Core octamer 270, 270F, 430G
Corepressor 163B, 430G
Correns, C. 6, 10B
Cosmid 389B, 430G
Cotransduction 366–7, 366F, 367F, 430G
Cotransformation 368–71, 369F, 430G

Coulson, A. 403
Crick, F.
  biography 38B
  Central Dogma 52
  discovery of double helix 32–3, 35–6, 41B
  discovery of triplet code 122–3
  ideas on genetic code 120–1
  postulates existence of mRNA 95
  views on DNA replication 174–5
  wobble hypothesis 136
Cro repressor protein 119, 120F, 232, 232F, 430G
Cross-fertilization 311, 312B, 430G
Crossing-over 332–4, 335F, 430G
Cruciform structure 69, 69F, 430G
CTP 32
Cyclic AMP 161, 162F, 430G
Cyclobutyl dimer 205, 205F, 430G
Cytochemistry 12, 431G
Cytoplasm 246, 246F, 431G
Cytosine 27F, 29, 431G

dATP 28F, 29
dCTP 28F, 29
Defective virus 240, 241F, 431G
deformed 170
Degeneracy 126, 431G
Degradative plasmid 252, 252T
Delbrück, M. 9, 9B, 10B
Deletion analysis 167B, 411–13, 411F, 431G
Deletion mutation 193, 194F, 431G
Denaturation 119, 119F, 431G
Density gradient centrifugation 19B, 177–8, 178F, 297B, 431G
Deoxyribonuclease
  in testing transforming principle 17
  protection experiments 64B
2′-deoxyribose 26–9, 27F
Development 154, 168–71
De Vries, H. 6, 10B
dGTP 28F, 29
Dideoxynucleotide 403, 404B, 431G
Differentiation 153–4
Dihybrid cross 316–18, 431G
Dihydrouridine 87, 91F, 92
Diploid 266, 431G
Direct repair 206
Discontinuous gene 50–2, 431G
Dispersed repetitive DNA 295
Disulphydryl bridge 117, 118F, 432G
DNA
  component of chromosomes 12
  density gradient centrifugation 19B
  electrophoresis 19B
  experimental proof as genetic material
    Hershey–Chase experiment 20–3
    need for experimental proof 13–14
    transforming principle 14–19
    universal acceptance 23
  GC content 34, 35B
  labelling with ^{32}P 21–2, 21B, 21F
  structure
    double helix 32–40
    nucleotides 26–9
    polynucleotides 29–31
DNA-binding protein
  activity in gene regulation 168
  identification of binding sites 409–10, 410F
  interaction with DNA 119, 120F
  types 167–8, 168F
DNA gyrase 186, 432G
DNA ligase 184, 184F, 381, 381F, 432G
DNA packaging
  chromatin 268
  histones 269–70, 269T
  nucleosomes 269–71, 270F, 272B
  30 nm chromatin fibre 271, 271F
DNA polymerase
  DNA polymerase I 179–81, 181B, 184B
  DNA polymerase II 181B
  DNA polymerase III 179–81, 181B, 184B
  in recombinant DNA studies 396B, 402B, 412B
  mammalian 186B
DNA repair 205–7, 432G
DNA replication
  in *E. coli* 179–86
  in eukaryotes 186
  of molecules 187–8
  problems at ends of molecules 275–7, 275F, 277F
DNase protection experiment 64B
DNA sequencing
  chain termination method 403–5
  Human Genome Project 295–9
  outline 49B
  power 405–6
DNA synthesis, in replication 179, 180F
DNA topoisomerase 185–6, 185F, 246B, 432G
Dominance 314–15, 321–2, 432G
Donor cell 357, 432G
Double helix
  different forms 40, 40B
  dimensions 38
  experimental basis 33–5
  importance of base pairing in genetics 39–40
  Watson–Crick structure 35–9, 37F
Double Helix, The 25, 36B, 38B
Double recombination 358, 432G
Drosophila
  developmental mutants 169–71
  5S rRNA genes 48F
  gene mapping 330–7
  genetic map 7F, 10B
  inherited characteristics 7B, 8
  use in genetics 6–8, 6B, 10B
dTTP 28F, 29
Dulbecco, R. 235
Dystrophin gene 290–1

Early gene 230, 433G
Electron microscopy
  detecting introns 102B
  studying ribosome structure 82–3

Electron microscopy (*cont.*)
transcript analysis 406–7, 407F
Electrophoresis 18, 19B, 433G
Elongation factor
E. coli 142, 144B, 146B
eukaryotes 147B
Endonuclease 396B, 433G
Enhancer 165F, 166, 166F, 433G
Epistasis 351–2, 433G
Escherichia coli
cell size 219F
cloning vectors 386–8
gene map 372, 372B
gene nomenclature 158B
gene organization 248–9, 248F
genetic markers 360T
genome size 266T, 267
genomic library 388T
histone-like proteins 247, 247T
infection with bacteriophage T2 20–3, 20F
in molecular genetics 58, 59B
non-standard genetic code 128T
number of genes 46B, 85B
recombinant protein 414–15
ribosome binding sites 139B
RNA polymerase 62, 62F
rRNA genes 83–4, 84F
studying translation in 144B
Ethidium bromide 203, 204F
Eubacteria
definition 245
genes 248–50
Eukaryotes, cell structure 57–9, 58F
Evolution
of genes 52B
of man 302–5
Excision repair 206–7, 206F, 433G
Exon, definition 50, 433G
Exon shuffling 51–2, 51F, 433G
Exonuclease 396B, 433G
Extrachromosomal gene
genetic systems 282–3
genomes 283–7
location 265
origins 287B

F^- cell 254, 255F, 256F, 433G
F′ cell 256, 256F, 370B, 433G
F^+ cell 254, 255F, 256F, 433G
F-duction 370B, 434G
Fertility plasmid 252, 252T
Filamentous bacteriophage 220–1, 220F
5′-terminus 30, 30F
Flower colour 322, 323F, 328–30, 329F, 349–52, 350F, 351F
fmet 141, 141F, 434G
Formate dehydrogenase 128T
N-formylmethionine 141, 141F
F plasmid 252, 252F, 255F, 254–5
Franklin, R.
biography 36B
ideas on DNA structure 39B
X-ray diffraction work 33, 35, 41B
Fox. G. 245
Frameshift mutation 196, 196F, 434G
Fruit fly, *see Drosophila*

G1 phase 278F, 279, 434G
G2 phase 278, 278F, 434G
Gamete 315, 434G
G-banding 272, 272F
GC box 165F, 166, 434G
GC content 34, 35B, 297B
Gel electrophoresis 19B, 49B, 434G
Gel retardation analysis 408–10, 410F, 434G
Gene
archaeal 250–1
eubacterial 248–50
first use of the term 6, 10B
lengths 44
nomenclature 158B
numbers in chromosomes 46B
numbers in organisms 46B
organization on DNA molecules 46–52
segments of DNA 44–6
transcription unit 290B
Gene amplification 85–6, 85F, 434G
Gene cloning
cloning vectors 382–93
constructing recombinant DNA molecules 378–82
description 375–8, 376F
importance 376–8
Gene expression
outline 45, 52–4, 53F
studying 406–13
Gene mapping
accurate gene mapping 336B
bacteria 356–72
Drosophila 330–7
early progress 8
microbial eukaryotes 337–47
S. cerevisiae 337–47
Generalized recombination
definition 207
model 208–9, 208F, 210F
role in genetics 210
Gene regulation
development 168–71
DNA-binding proteins 167–8
E. coli in outline 151–2
eukaryotes in outline 151–2
lactose utilization 157–62
nucleosomes and 272B
possible strategies 154–5
reasons for 151
studying 408–13
upstream binding sites in eukaryotes 165–7
Gene relic 293
Genetic code
colinearity 121
definition 53
early deductions 120–3
elucidation 123–6
features 126–9
function 113
non-standard codes 127–9
punctuation codons 127
table of codons 127T
triplet nature 121–3
Genetic engineering 375, 435G
Genetic marker 337T, 338, 359, 360T, 435G
Genetic material
experimental proof for DNA 13–23
requirements 13
Genetics
definition 3
Gene transfer 358, 358F, 435G

Genome
  definition 217, 435G
  sizes 266T
Genomic library
  identifying genes in 395–402
  sizes 387–8, 388T
Genotype 193, 193B, 435G
Gilbert, W. 51, 51B
Globin genes 47–50, 48F
Globin mRNA 96
Glucocorticoid hormone 166F, 167, 167B
Glucose, effects on *lac* operon 160–2, 162F
Glutathione peroxidase 128T, 129
Griffith, F. 14–16
Group I introns 106F, 107–8
Group II introns 108
GT–AG introns
  consensus sequences 101–3
  splicing 103–5
GT–AG rule 102, 102F, 435G
GTP 32
Guanine 27F, 29, 435G

Haemophilus influenzae 250
Half-life of mRNA 95–6, 96F
Hall, B.D. 95
Halophile 250B
Haploid 266, 435G
Head-and-tail bacteriophage 220F, 221
Heat, mutagenic effect 203, 204F
Helical bacteriophage 220–1, 220F
Helicase 182, 182F, 435G
Helix-turn-helix structure 119, 120F, 167, 168F, 435G
Hemizygous 333B, 436B
Hershey, A.
  biography 22
  Hershey–Chase experiment 20F, 22–3, 41B
Hershey–Chase experiment
  basis to 20–1
  details 20F, 22–3
  use of radiolabels 21–22, 21B
Heteroduplex 208F, 209, 436B
Heterogeneous nuclear RNA 98, 436B
Heterologous probing 400, 436B
Heteropolymer 124–5, 125B, 436B
Heterozygote 314, 436B
Hfr cells 256, 256F, 359–63, 436B
Histone
  genes 290
  histone-like proteins, *E. coli* 247, 247T
  mRNA 104B
  proteins 269–70, 269T
HIV-1, HIV-2, *see AIDS virus*
hnRNA 98
Holley, R. 86, 86B
Holliday, R. 209
Holliday structure 208F, 209, 436B
Homeobox 170–1, 170F, 436B
Homo erectus 303
Homologous chromosomes 280, 436B
Homologous genes 47, 436B
Homologous recombination, *see generalized recombination*
Homopolymer 123, 124F, 437G
Homo sapiens
  cell size 219F
  centromeres 273–4, 274F
  chromosomal DNA sizes 291T
  chromosome banding patterns 272F, 292F
  chromosome number 266T, 267
  coding DNA amount 267B
  evolution studies 302–5
  5S rRNA genes 47, 85B
  gene locations 291–3, 292F
  gene relics 293
  genes and gene families 290–3
  gene sizes 290–1
  genome organization 292F
  genome sequencing project 296–9, 406
  genome size 266T, 267
  genomic library 388T
  globin genes 47–50, 48F
  introns 50T, 290
  metallothionein gene, regulation of 165–7, 165F, 166F
  mitochondrial genome 282T, 283–5, 284F
  multigene families 291–3
  number of genes 46B, 85B
  polymorphic genes 299–300
  population studies 300–2
  repetitive DNA 268, 295–6, 297B
  sex determination 298B
Homozygote 314, 437G
Hormones, on gene regulation 153, 153F, 166F, 167, 167B
Housekeeping genes 151
Human Genome Project 295–9, 406
hunchback 170, 170F
Hybrid-arrest translation 413–14, 437G
Hybridization
  definition 102B, 395–7, 396F
  probing 395–400, 437G
Hybrid-release translation 413–14, 437G
Hydrogen bonds 37–8, 37B, 437G

Icosahedral bacteriophage 220, 220F
Illegitimate recombination 207, 437G
Incompatibility 257–8, 437G
Incomplete dominance 322, 323F, 437G
Incomplete penetrance 348–9
Inducible operon 163B, 165
Induction, of bacteriophage λ 225, 225F, 437G
Inhibitors, studying translation 144B
Initiation codon 127, 127T, 437G
Initiation complex 140–1, 140F, 437G
Initiation factor
  E. coli 141–2, 146B
  eukaryotes 147B
Inosine 91F, 92, 137F, 138
Insertion mutation 193, 194F, 438G
Insertion sequence 259–60, 260F, 438G
Intercalating agent 203, 204F, 438G
Intergenic DNA 44

Interrupted mating experiment 359–63, 438G
Intron
- archaeal 250
- definition 51, 438G
- examining a gene for 102B
- Group I 107–8, 286
- Group II 108, 286
- GT–AG introns 101–5
- in human genes 290
- numbers in human genes 50T
- organellar 286
- *Tetrahymena* self-splicing intron 107–8

Inversion mutation 193, 194F, 438G
Inverted repeats 260, 260B, 438G
In vitro mutagenesis 414B, 438G
In vitro packaging 386B, 438G
Iodothyronine 5′-deiodinase 128T, 129
Isopropyl thiogalactoside 386, 386F, 387F
Isotope 21–2, 176F

Jacob, F.
- biography 156B
- interrupted mating experiment 359
- *lac* operon 155
- mRNA discovery 95

Janssens, F.A. 333
Johannsen, W. 6, 10B

Karyotype 271
Kilobase 38B
Kinetochore 273–4, 273F, 274B, 274F
Kirromycin 144B
Khorana, H.G. 125, 126B
knirps 170F
Koch, R. 14
Kornberg, A. 179, 181B
Krüppel 170F

Labelling 21–2, 21F, 176–7, 176F, 398
lac operon 46–7, 157–9, 248F, 349
lac repressor 159–60, 159F, 160F, 161F
Lactose 157F, 160B
Lactose utilization in *E. coli*
- enzymes 156–7
- glucose effects 160–2, 162F
- *lac* operon 157–9, 158F
- *lac* repressor 159–60, 159F, 160F, 161F

lacZ′ gene 385–6, 385F, 387
Lagging strand 182, 183F, 439G
Late gene 230, 439G
Latent period 222, 223F, 224B, 439G
Leader segment 68, 68F
Leading strand 182, 183F, 439G
Leaky mutation 197B, 439G
Leder, P. 125
Lederberg, J. 253B, 254, 363
Lethal genotype 316B
Lethal mutation 196, 439G
Leucine zipper 168, 168F, 439G
Life cycle
- pea 318–19, 319F
- *S. cerevisiae* 338–9, 338F

Light and Life 8, 8B, 9B, 10B
LINE 295, 295B, 439G
Linkage 325, 325B, 328–37, 439G
Linkage group 336, 439G
Lysogenic infection cycle 222–6, 225F
Lysozyme 230, 439G
Lytic infection cycle 222, 223F

Macleod, C. 17–18, 19B, 41B
Macromolecule 13, 440G
Major groove 37B, 38, 440G
Major histocompatibility complex 300, 300F, 440G
Man, *see Homo sapiens*
Maniatis, T. 48
Map unit 336–7, 440G
Mating type 338F, 339, 440G
Matthaei, H. 123, 124
Maturation promoting factor 281, 440G
McCarty, M. 17–18, 19B, 41B
Meiosis 279–81, 280F, 440G
Mendel, G.
- biography 4B
- characteristics studied 309–11
- dihybrid crosses 316–18
- early experiments in genetics 4–5, 10B, 309–18
- importance 309
- Laws 316, 318, 325
- linkage and Mendel 325
- monohybrid crosses 311–16
- rediscovery of his work 5–6, 10B
- relation to molecular genetics 318–25

Merodiploid 370B
Meselson, M.
- biography 177B
- DNA replication 176
- mRNA discovery 95
- recombination model 209, 210F

Meselson–Radding model 209, 210F
Meselson–Stahl experiment 176–8, 178F
Messenger RNA, *see mRNA*
Metacentric 273B
Metallothionein gene, regulation of 165–7, 165F, 166F
Metaphase chromosome 271–7, 272F, 440G
Methanogens 250B
7-methylguanosine 90, 91F, 98, 99F
Microinjection 392B
Micrometre 36B
Micron 36B
Microprojectile 392B
Microsatellite DNA 296, 299F
Minisatellite DNA 295
Minor groove 37B, 38, 440G
Mismatch 191
Mismatch repair 207, 207F, 440G
Missense mutation 195, 195F, 441G
Mitochondrial DNA
- gene products 286B
- genes 183–6
- genetic code 128, 128T
- human population studies 302–3
- location 265F
- sizes 282, 282T

Mitochondrial Eve hypothesis 303, 304F
Mitosis 279–80, 280F, 441G
Model building 33, 441G

Molecular biology
  advent of 8–9
  definition 3, 441G
Molecular clock 303B
Molecular genetics 3, 441G
Monod, J. 155, 156B
Monohybrid cross 311–16, 441G
Morgan, T.H.
  biography 6B
  contribution to genetics 6–8, 10B
  gene mapping 330–7
  views on genetics 7B
M phase 279–81, 441G
mRNA
  capping 98–100
  chemical modifications 95–6
  definition 53
  intron splicing 101–9
  molecular mass 78T
  organellar 284
  polyadenylation 100
  purification 101B
  RNA editing 109–10
  sedimentation coefficient 78T
  stability 95–6
  transcription in eukaryotes 63B
  turnover 76, 96
Müller, H. 7, 10B
Multigene families
  definition 46, 441G
  evolution 48F, 49
  types 47–9, 48F
Multiple alleles 324, 324F, 441G
Multiple genes 352
Mutagen
  base analogues 201F, 203
  heat 203, 204F
  intercalating agents 203, 204F
  ultraviolet radiation 203–5, 205F
Mutation
  definition 191, 441G
  generation of 191, 192F
  in study of *lac* operon 159B
  mutagens 200–5
  polar 258–9, 258F, 259F
  reversing effects of 198–200
  types 193–8

Nanometre 36B
Neurospora crassa 346B, 347B
Nirenberg, M. 123–5, 124B
NMR, *see nuclear magnetic resonance spectroscopy*
Non-homologous recombination, *see illegitimate recombination*
Non-parental ditype 343, 344F
Non-radioactive labelling 398
Nonsense mutation 195–6, 195F, 442G
Non-viral retroelement 295, 442G
Nuclear magnetic resonance spectroscopy 118B
Nuclease 396B, 407–8, 408F, 442G
Nucleic acid hybridization 395–7, 396F, 442G
Nucleoid 245–8, 246F, 247F, 442G
Nucleolus 98, 442G
Nucleoplasm 98, 442G
Nucleoside 27F, 442G
Nucleosome 269–71, 270F, 272B, 273F, 442G
Nucleotide
  in DNA
    components 26–9
    structure 26, 28F
  in RNA 31–2

Okazaki fragment 182–4, 183F, 184F, 442G
Oligonucleotide 398–9
Oligonucleotide-directed mutagenesis 414B, 442G
Oncogene 240F, 241, 241T, 443G
One-step growth curve 224B
Open promoter complex 66, 66F, 443G
Operator 159, 161F, 163B, 443G
Operon
  definition 46–7, 47F
  lac operon 157–9
  number in *E. coli* 248
  trp operon 163B, 164B
  types 165B
Ordered tetrad 346B, 347B
Overlapping genes 228–9, 228B, 443G

Packaging constraints 388B
Pararetrovirus 236, 443G
Parental ditype 343, 344F
Pauling, L. 36
pBR322 383–5, 383F, 384F, 443G
PCR 49B, 412B, 444G
Penetrance 348–9, 443G
Peptide bond 116, 116F, 443G
Peptide bond formation 143–4, 143F, 144B
Peptidyl site 142, 142F, 443G
Peptidyl transferase 144, 443G
Phage Group 9, 10B
Phage, *see bacteriophage*
Phenotype 193, 193B, 443G
Phosphodiester bond
  structure 30, 30F
  synthesis 60F, 61
Photoreactivation repair 206, 206F, 443G
Plaque 222B, 444G
Plaque hybridization probing 397–8, 397F
Plasmid
  bacterial sex 253–7
  cloning vectors 383–6, 390, 392–3
  copy number 257–8
  description 251–2
  incompatibility 257–8
  sizes 252T, 253
  supercoiling 246F
  types 252–3, 252T
Plasmid amplification 383–4, 383F, 444G
Pleiotropy 348, 444G
Polar mutation 258–9, 258F, 259F, 444G
Polyacrylamide gel electrophoresis 19B
Polyadenylation 100, 100F, 101B, 104B, 444G
Poly(A) polymerase 100, 444G
Polymerase chain reaction 49B, 412B, 444G
Polymorphism 299–300, 299F, 444G
Polynucleotide
  ends 30–1, 30F
  lengths 31
  phosphodiester bond 30, 30F
  RNA 31–2
  structure 29–31, 30F

Polynucleotide kinase 396B
Polynucleotide phosphorylase 123, 444G
Polypeptide, *see protein*
Polyprotein 238, 444G
Polysaccharide, in bacterial capsules 15B
Polysome 144F, 145, 444G
Population group 300–2
Prepriming complex 186B, 445G
Pre-rRNA 83, 84F, 445G
Pre-tRNA 89, 90F, 445G
Pribnowbox 64, 445G
Primary transcript 83, 84F, 89, 90F, 445G
Primase 184, 445G
Primer 183, 183F, 184F, 445G
Primosome 183B, 184, 445G
Prion 241B, 445G
Processed pseudogene 293, 294F, 445G
Proflavin 121–2, 445G
Prokaryote, cell structure 57–9, 58F
Promiscuous DNA 286, 445G
Promoter
  analysis by DNase protection 64B
  E. coli 64–5, 65B
  eukaryotic 71–3, 72F, 73B
  function 63–4
  recognition by σ subunit 65–7
  RNA polymerase I 73B
  RNA polymerase II 71–2, 72F
  RNA polymerase III 73B
Proofreading 184B, 445G
Prophage 223, 225B, 446G
Protease 17
Protein
  amino acid monomers 114–15, 114F, 114T, 115T
  component of chromosomes 12
  labelling with ^{35}S 21–2, 21B, 21F
  link between structure and function 119
  peptide bond 116, 116F
  primary structure 116
  quaternary structure 117–18, 117F
  secondary structure 116–17, 117F
  sedimentation coefficients 78T
  structural predictions 118
  studying function 413–15
  studying structure 118B
  tertiary structure 117, 117F
  types 53–4
Protein synthesis, *see also translation*
  key to gene expression 53–4
Protomer 220, 446G
Protoplast 392B, 446G
Pseudogene 49–50, 293, 446G
Pseudouridine 88, 91F, 92
P-site 142, 142F, 443G
pUC8 385–6, 385F, 386F
Punctuation codon 127, 127T, 446G
Punnett square 315, 446G
Purine 27F, 29, 446G
Pyrimidine 27F, 29, 446G

Quantitative phenotype 352, 446G
Queosine 91F, 92

Radiolabelling 21–2, 21F, 398, 446G
Random spore analysis 340–1, 340F
Reading frame 122, 446G
Recessiveness 314–15, 321–2, 447G
Recipient cell 357, 447G
Recombinant 334, 447G
Recombinant DNA molecule 375–382, 447G
Recombinant DNA technology 23, 375, 447G
Recombinant protein 414–16, 416B, 447G
Recombination
  definition 191, 447G
  in bacteriophage λ infection cycle 223, 224B
  models 208–9
  role in genetics 210–12
  types 207–8
Regulatory mutation 197–8
Regulatory site 164–7, 411–13, 411F
Release factors
  E. coli 145, 145F, 146B
  eukaryotes 147B
Renaturation 119, 119F, 447G
Repetitive DNA
  amounts in different organisms 268
  categories 267–8, 267F
  human 295–6, 297B
Replica-plating 340F, 341, 447G
Replication fork 179, 179F, 181–4, 447G
Replication origin 179, 179F, 447G
Reporter gene 411, 413T, 447G
Repressible operon 163B, 165B
Repressor 159,163B
Resistance plasmid 252, 252T
Restriction endonuclease
  description 49B, 378–9, 378F
  ends generated 379–80, 379T, 380F
  types 379, 379T
Retinoblastoma 348–9, 349F
Retroelement 295, 295B, 296F
Retrovirus
  AIDS virus 238–40, 240F
  cancer 240–2
  genes 234T
  genome 234T
  infection cycle 236–8, 237F
  retroelement 236
  structure 234F
Reverse transcriptase 236, 236F, 396B, 402B, 448G
R-groups of amino acids 114, 114F, 115T
Rho-dependent termination 71, 448G
Ribonuclease
  binding sites for ribosomal proteins 80–1, 81F
  rRNA secondary structure 80, 81F
  testing transforming principle 17
Ribose 31, 32F
Ribosomal proteins 77, 78, 79–80, 79T, 82B
Ribosomal RNA, *see rRNA*
Ribosome

dimensions 77
molecular mass 77, 78T, 79T
proteins 77, 78, 79–80, 79T, 82B
sedimentation coefficients 77, 78T, 79T
sizes 77, 78T
structure 77–83
subunits 77–8
Ribosome binding site 139, 139B, 139F, 140F, 448G
Ribothymidine 90, 91F
Ribozyme
RNase P 89, 108B
rRNA, possible activity 83, 108B
satellite viruses 108B
summary of types 108B
Tetrahymena intron 83, 106F, 107–8, 108B
virusoids 108B
RNA
as genetic material 23, 221
structure 31–2
synthesis in transcription 60–3
RNA, 7SL and 7SK 77B
RNA editing
apolipoprotein-B 109F, 110, 110F
corrects non-standard codons 129B
discovery 109
Trypanosoma brucei 109, 109F
RNA enzyme, *see ribozyme*
RNA polymerase
archaeal 250
E. coli 62, 62F
eukaryotic 62–3, 63B, 71–3, 73B
function of subunits 67B
in recombinant DNA studies 396B
organellar 282–3
RNA polymerase I 62–3, 63B, 73B
RNA polymerase II 62–3, 63B, 71–3, 165–7
RNA polymerase III 62–3, 63B, 71, 73B
role in transcription 65–73
RNase D 89, 90F, 448G
RNase P 89, 90F, 108B, 448G
Rolling circle replication 187F, 188, 448G
Rous sarcoma virus 240, 240F, 241B
rRNA
archaeal 250
5S genes 47, 48F, 85B
gene numbers 85B
lengths 78B, 78T, 79T
organellar 282–3
possible enzymatic activity 83, 108B
processing 83–4, 84F, 104B
ribosome structure 77–83
sedimentation coefficients 78, 78T, 79T
synthesis 83–6
transcription in eukaryotes 63B
transcription unit 83–6, 84F
turnover rate 76

Saccharomyces cerevisiae
cell size 219F
centromere structure 273, 273F
chromosome number 266T, 267
chromosome sequence 406
cloning vectors 390–2
gene mapping 337–47
genetic markers 337T, 338
genome size 266T, 267
genomic library 388T
in molecular genetics 58, 59B
life cycle 338–9, 338F
mitochondrial genome 282T
non-standard mitochondrial codons 128T
number of genes 46B, 85B
recombinant protein 415–16
Salmonella typhimurium 249F, 250, 363, 372B
Sanger, F. 283, 283B, 403, 405
Sanger–Coulson DNA sequencing 403–5
Satellite DNA 295–6, 297B, 448G
Satellite virus 108B, 449G
Scanning 146, 147F, 449G
Schizosaccharomyces pombe 281
Schrödinger, E. 8, 8B, 10B
Second site reversion 199, 199F, 449G
Sedimentation analysis 77
Sedimentation coefficient 77, 449G
Segmented genome 221–2
Segregation 315, 315F, 449G
Selective medium 358, 358F, 449G
Self-fertilization 311, 312B, 449G
Selfish DNA 262, 449G
Semi-conservative replication 175, 175F, 449G
Serotype 14–16, 15B, 449G
Sex cell 198B, 449G
sex combs reduced 170
Sex determination 298B
Sex linkage 331, 449G
Sex pilus 254, 254F, 449G
Shine–Dalgarno sequence 139, 449G
σ subunit 62, 62F, 65–7, 67B
Silent mutation 193, 194F, 194–5, 449G
Simian virus 40, *see SV40*
SINE 295, 449G
Single-strand binding protein 182, 182F, 450G
Site-specific recombination 207, 210–11, 224B, 450G
Small nuclear RNA 103, 450G
Small nuclear RNP 103, 450G
Somatic cell 198B, 450G
Somatostatin 414, 415F
S1 nuclease 102B, 396B, 407–8, 408F, 450G
SOS response 232, 450G
Specialized transduction 368B, 450G
S phase 278–9, 278F, 450G
Spiegelman, S. 95
Spliceosome 104–5, 105F
Splicing
Group I 107–8
Group II 108
GT–AG introns 101–5
Tetrahymena self-splicing intron 107–8
tRNA introns 108–9
Spontaneous mutation rate 200, 450G
Stable RNA 76, 450G

Stahl, F. 176
Start 281, 281F
Stem-loop structure 69, 69F, 450G
Sticky end 380, 380F, 451G
Streptococcus pneumoniae 14–16, 15B
Streptomyces 372, 372B
Streptomycin 144B
Structural gene 159, 451G
Sturtevant, A. 7, 10B, 334–5
Submetacentric 273B
Supercoiling 186, 246, 246B, 246F, 451G
Superwobble 284–5, 285F
Suppression 199–200, 200F, 451G
Sutton, W.S. 7, 10B
S value 77, 451G
Svedberg, T. 19B, 77
SV40 393, 406

TACTAAC box 103
TATA box 73, 165–6, 165F, 168, 451G
Tatum, E.L. 10B, 254
Tautomers 202B
Telomerase 276–7, 277F, 451G
Telomere
  regeneration 276–7, 277F
  roles 274–5
  structure 276, 295
Temin, H. 236, 237B
Temperate infection, *see lysogenic infection cycle*
Temperature-sensitive mutant 197, 197F, 451G
Template strand 45, 45B, 58F, 59, 451G
Terminal deoxynucleotidyl transferase 396B
Termination codon 127, 127T, 451G
Terminator for transcription
  E. coli 69–71
  RNA polymerase I 73B
  RNA polymerase II 73
  RNA polymerase III 73B
Testcross 333B, 451G
Tetrad 338F, 339
Tetrad analysis 341, 341F, 451G
Tetrahymena self-splicing intron 106F, 107–8
Tetratype 343, 344F
Thermophile 250B
θ-form 187F, 188, 451G
Thiostrepton 144B
4-thiouridine 91F, 92
30 nm chromatin fibre 271, 271F, 451G
3′-terminus 30F, 31
Thymine 27F, 29, 452G
Thymine dimer 205, 205F
Ti plasmid 392–3, 452G
Tn3-type transposon 261, 261F, 261T
Trailer segment 68, 68F, 452G
Transcript analysis
  electron microscopy 406–7, 407F
  nuclease analysis 407–8, 408F
Transcription
  definition 52–3
  in *E. coli* 63–71
  in eukaryotes 71–4
  RNA synthesis 60–3
Transcription factor 72F, 73, 168, 452G
Transcription unit 290B
Transduction
  basis 365–6
  discovery 363–5
  mapping 366–8
  specialized transduction 368B
Transduction mapping 363–8, 452G
Transfer RNA, *see tRNA*
Transformation 368–71, 376, 452G
Transformation mapping 368–71, 452G
Transforming principle
  discovery 14–16
  is DNA 17–19, 41B
  is genetic material 16–17
Transient response element 165F, 166F, 167
Transition 193, 453G
Translation
  definition 53, 453G
  E. coli mechanics 138–46
  eukaryotes mechanics 146–7
  role of tRNA 132–8
Translocation 143F, 144, 453G
Transposable element, *see also transposon* 211–12, 212F, 453G
Transposable phage 262, 262F
Transposase 259–60, 453G
Transposition
  definition 191, 453G
  role 211–12, 212F
  transposable elements 211–12, 212F
Transposon
  discovery 258–9
  insertion sequences 259–60
  other types 261–3
  role 263
Transversion 193, 453G
Triplet binding assay 125, 126T, 453G
tRNA
  aminoacylation 132–5
  codon recognition 135–8
  gene numbers 85B
  introns 108–9
  molecular mass 78T
  organellar 283, 284–5
  processing and modification 89–92
  sedimentation coefficient 78T
  specificity 132
  structure 86–9
  suppression 199–200, 200F
  transcription in eukaryotes 63B
  turnover rate 76
  wobble hypothesis 135–8
tRNA deacylase 144, 453G
tRNA nucleotidyl transferase 89, 90F, 453G
Truncated gene fragment 293, 294F
trp operon 163B, 164B, 248F
trp repressor 163B, 167
Trypanosoma brucei 109
Turnover rate 76, 96, 453G
2μm plasmid 390

Ultracentrifugation 18, 19B, 77, 453G
Ultraviolet radiation, mutagenic effect 203–5, 205F
Ultraviolet spectroscopy 18, 19B,

453G
Unit factor 313, 453G
Units of length 36B
Upstream binding site 164–7, 167B
Uracil 31–2, 32F, 453G
U-RNA 63B, 77B, 103–5, 104B, 105F, 453G

Vegetative cell 280, 319, 454G
Velocity gradient centrifugation 19B, 454G
Viral retroelement 236, 454G
Virulence plasmid 252T, 253
Virulent infection, *see lytic infection cycle*
Virus, *see also bacteriophage*
  cloning vectors 393
  description 217–18
  genomes 234, 234T
  replication strategies 236–40
  retroviruses and cancer 240–2
  sizes 219F
  structures 233–4, 234T
Virusoid 108B
Volkin, E. 95
von Tschermak, E. 6, 10B

Watson, J.
  biography 38B
  discovery of double helix 32–3, 35–6, 41B
  discovery of mRNA 95
  views on DNA replication 174–5
What is Life? 8, 8B, 10B
White-eye gene 331–6
Wild-type 193, 454G
Wilkins, M. 35, 36B
Wilson, A. 302, 303B
Wobble hypothesis 135–8, 136F, 137F, 284–5, 285F, 454G
Woese, C. 245
Wollman, E. 359
Wrinkled pea phenotype 321, 322F

X-gal 386, 386F, 387F
X-ray diffraction analysis
  aminoacyl-tRNA recognition studies 135B
  DNA structure analysis 35, 35F
  protein structure analysis 118B
  ribosome structure analysis 83
  tRNA structure analysis 88

Yanofsky, C. 121
Yeast, *see Saccharomyces cerevisiae*
Yeast artificial chromosome 390–2, 391F, 454G
Yeast episomal plasmid 390, 391F, 454G
Yeast integrative plasmid 390, 454G

Z-DNA 38, 38B, 454G
Zinc finger 168, 168F, 170, 455G
Zinder, N. 363
Zygote 338F, 339, 455G